Periodic Table of the Elements

Legend:

23
V
Vanadium
50.942

Atomic Number
Symbol
Name
Atomic Weight

Atomic Weight
Fe: 55.845 — Formal short value, rounded (no uncertainty)
Cm: (247) — Mass number of the most stable isotope

1 IA	2 IIA	3 IIIB	4 IVB	5 VB	6 VIB	7 VIIB	8 VIIIB	9 VIIIB	10 VIIIB	11 IB	12 IIB	13 IIIA	14 IVA	15 VA	16 VIA	17 VIIA	18 VIIIA
1 **H** Hydrogen 1.008																	2 **He** Helium 4.0026
3 **Li** Lithium 6.94	4 **Be** Beryllium 9.0122											5 **B** Boron 10.81	6 **C** Carbon 12.011	7 **N** Nitrogen 14.007	8 **O** Oxygen 15.999	9 **F** Fluoride 18.998	10 **Ne** Neon 20.180
11 **Na** Sodium 22.990	12 **Mg** Magnesium 24.305											13 **Al** Aluminium 26.982	14 **Si** Silicon 28.085	15 **P** Phophorus 30.974	16 **S** Sulfur 32.06	17 **Cl** Chlorine 35.45	18 **Ar** Argon 39.948
19 **K** Potassium 39.098	20 **Ca** Calcium 40.078	21 **Sc** Scandium 44.956	22 **Ti** Titanium 47.867	23 **V** Vanadium 50.942	24 **Cr** Chromium 51.996	25 **Mn** Manganese 54.938	26 **Fe** Iron 55.845	27 **Co** Cobalt 58.933	28 **Ni** Nickel 58.693	29 **Cu** Copper 63.546	30 **Zn** Zinc 65.38	31 **Ga** Gallium 69.723	32 **Ge** Germanium 72.630	33 **As** Arsenic 74.922	34 **Se** Selenium 78.971	35 **Br** Bromine 79.904	36 **Kr** Krypton 83.798
37 **Rb** Rubidium 85.468	38 **Sr** Strontium 87.62	39 **Y** Yttrium 88.906	40 **Zr** Zirconium 91.224	41 **Nb** Niobium 92.906	42 **Mo** Molybdenum 95.95	43 **Tc** Technetium (98)	44 **Ru** Ruthenium 101.07	45 **Rh** Rhodium 102.91	46 **Pd** Palladium 106.42	47 **Ag** Silver 107.87	48 **Cd** Cadmium 112.41	49 **In** Indium 114.82	50 **Sn** Tin 118.71	51 **Sb** Antimony 121.76	52 **Te** Tellurium 127.60	53 **I** Iodine 126.90	54 **Xe** Xenon 131.29
55 **Cs** Cesium 132.91	56 **Ba** Barium 137.33	57-71 Lanthanides	72 **Hf** Hafnium 178.49	73 **Ta** Tantalum 180.95	74 **W** Tungsten 183.84	75 **Re** Rhenium 186.21	76 **Os** Osmium 190.23	77 **Ir** Iridium 192.22	78 **Pt** Platinum 195.08	79 **Au** Gold 196.97	80 **Hg** Mercury 200.59	81 **Tl** Thalium 204.38	82 **Pb** Lead 207.2	83 **Bi** Bismuth 208.98	84 **Po** Polonium (209)	85 **At** Astatine (210)	86 **Rn** Radon (222)
87 **Fr** Francium (223)	88 **Ra** Radium (226)	89-103 Actinides	104 **Rf** Rutherfordium (267)	105 **Db** Dubnium (268)	106 **Sg** Seaborgium (269)	107 **Bh** Bohrium (270)	108 **Hs** Hassium (277)	109 **Mt** Meitnerium (278)	110 **Ds** Darmstadtium (281)	111 **Rg** Roentgenium (282)	112 **Cn** Copernicium (285)	113 **Nh** Nihonium (286)	114 **Fl** Flerovium (289)	115 **Mc** Moscovium (290)	116 **Lv** Livermorium (293)	117 **Ts** Tennessine (294)	118 **Og** Oganesson (294)

Lanthanides (57–71):

57	58	59	60	61	62	63	64	65	66	67	68	69	70	71
La Lanthanum 138.91	**Ce** Cerium 140.12	**Pr** Praseodymium 140.91	**Nd** Neodymium 144.24	**Pm** Promethium (145)	**Sm** Samarium 150.36	**Eu** Europium 151.96	**Gd** Gadolinium 157.25	**Tb** Terbium 158.93	**Dy** Dysprosium 162.50	**Ho** Holmium 164.93	**Er** Erbium 167.26	**Tm** Thulium 168.93	**Yb** Ytterbium 173.05	**Lu** Lutetium 174.97

Actinides (89–103):

89	90	91	92	93	94	95	96	97	98	99	100	101	102	103
Ac Actinium (227)	**Th** Thorium 232.04	**Pa** Protactinium 231.04	**U** Uranium 238.03	**Np** Neptunium (237)	**Pu** Plutonium (244)	**Am** Americium (243)	**Cm** Curium (247)	**Bk** Berkelium (247)	**Cf** Californium (251)	**Es** Einsteinium (252)	**Fm** Fermium (257)	**Md** Mendelevium (258)	**No** Nobelium (259)	**Lr** Lawrencium (266)

Glossary of Terms

SI Prefixes

f	p	n	μ	m	c	k	M	G	T	P	E
femto-	pico-	nano-	micro-	milli-	centi-	kilo-	mega-	giga-	tera-	peta-	exa-
10^{-15}	10^{-12}	10^{-9}	10^{-6}	10^{-3}	10^{-2}	10^3	10^6	10^9	10^{12}	10^{15}	10^{18}

Fundamental Constants

Name	*Symbol*	*Value*
Avogadro's constant	N_A	6.02214×10^{23} mol^{-1}
Bequerel	Bq	1 disintegration s^{-1}
Curie	Ci	3.7×10^{10} disintegrations s^{-1}
Faraday's constant	F	9.64853×10^4 C mol^{-1}
Gas constant	R	8.20574×10^{-2} L atm mol^{-1} K^{-1}
		8.31445 J K^{-1} mol^{-1}
Planck's constant	h	6.62608×10^{-34} J s
Speed of light	c	2.99792×10^8 m s^{-1}
Standard acceleration of free fall	g	9.08665 m s^{-2}
Stephan–Boltzmann constant	k	5.67040×10^{-8} W m^{-2} K^{-4}

Relations between Units

Property

Length	1 in	2.54 cm
	1.000 yard	0.9144 m
	0.6214 mile	1.000 km
Area	1 hectare (ha)	2.47105 acres
		10,000 m^2
Volume	1 quart	0.95 L
Concentration	1 part per million (ppm)	1 mg kg^{-1}
	1 part per billion (ppb)	1 μg kg^{-1}
	1 part per trillion (ppt)	1 ng kg^{-1}
	1 M	1 mol L^{-1}
Energy	1 watt (W)	1 J s^{-1}
	1 kilowatt-hour (kW-h)	3.600×10^3 kJ
	1 J	1 kg m^2 s^{-2}
	1 cal	4.184 J
	1 Food calorie (Cal)	1000 cal
	1 V	1 J C^{-1}
	1 British thermal unit (Btu)	1.055056×10^3 J
	1 quadrillion Btu (quad)	1.06×10^{12} MJ
Mass	1 metric ton	1000 kg
	1 short ton	2000 lbs
	1.000 lb	453.6 g
Pressure	1 atm	1.01325×10^5 Pa
	1 bar	10^5 Pa

Temperature Conversions

Temperature °F = 9/5 (Temperature °C) + 32

Temperature °C = 5/9 [(Temperature °F) – 32]

Temperature K = Temperature °C + 273.15

SOLVE

SOLVE

Problems in Environmental Science

Kathleen L. Purvis-Roberts
Thomas G. Spiro

UNIVERSITY SCIENCE BOOKS
New York

UNIVERSITY SCIENCE BOOKS
An imprint of AIP Publishing
uscibooks.aip.org

Publisher: Jane Ellis
Marketing and Production Manager: Barbara Dickson
Copyedit: Publishers' Design and Production Services, Inc.
Proofreader: Caroline Wofford
Designer, Illustrator, and Compositor: Laurel Muller, Cohographics
Cover Design: Robert Als
Cover Photographer: Paul Souders
Publishing Associate: Felicity Henson
Printer & Binder: Books International

Library of Congress Cataloging-in-Publication Data

Names: Purvis-Roberts, Kathleen, author. | Spiro, Thomas G., 1935- author.
Title: Solve : problems in environmental science / Kathleen Purvis-Roberts,
    Thomas G. Spiro.
Description: New York : University Science Books, [2023] | Includes bibliographical
    references and index. | Summary: "Textbook of quantitative problems about
    environmental issues, covering everything from climate and air pollution
    through water issues to agriculture and toxic materials. Problems are
    accompanied by brief narrative material providing environmental context"
    —Provided by publisher.
Identifiers: LCCN 2022005562 (print) | LCCN 2022005563 (ebook) |
    ISBN 9781940380100 (paperback) | ISBN 9781940380117 (ebook)
Subjects: LCSH: Environmental sciences--Problems, exercises, etc. |
    Environmental chemistry—Problems, exercises, etc.
Classification: LCC GE76 .P87 2023 (print) | LCC GE76 (ebook) |
    DDC 333.7076—dc23/eng20220712
LC record available at https://lccn.loc.gov/2022005562
LC ebook record available at https://lccn.loc.gov/2022005563

Printed in the United States of America
10 9 8 7 6 5 4 3 2 1

Dedicated to our children and grandchildren,
and their generations.

CONTENTS

PART I

INTRODUCTION

PART II

ENERGY AND MATERIALS

PART III

ATMOSPHERE

PART IV

HYDROSPHERE AND LITHOSPHERE

PART V

BIOSPHERE

18. TOXICITY OF CHEMICALS   383

APPENDIXES

LIST OF CONTRIBUTORS

About the Authors

Kathleen L. Purvis-Roberts is a Professor of Chemistry and Environmental Science at the W.M. Keck Science Department of Claremont McKenna, Pitzer, and Scripps Colleges. Her research focuses on atmospheric chemistry, studying the chemical mechanism for particulate matter formation from agricultural sources using analytical chemistry techniques. Recently she was awarded a Jefferson Science Fellowship to spend a year at the U.S. Department of State. While there, she focused on environmental and science policy within the Asia-Pacific Economic Cooperation (APEC). She earned her BA in Chemistry from Westmont College, her PhD in chemistry from Princeton University, and did her postdoctoral research at the National Center for Atmospheric Research.

Thomas G. Spiro is currently at the University of Washington after a long career on the faculty of Princeton University, where he was chair of the Chemistry Department. The role of metal ions in biology has been the principal theme of his research, including pioneering the application of laser resonance Raman spectroscopy, including time-resolved techniques, to the structure and reactivity of metalloproteins, and to the mechanisms of protein folding and allostery. He now studies biomineraliztion in bacteria that oxidize manganese, closing the global manganese cycle. He developed courses in environmental chemistry and co-authored a widely used textbook, *Chemistry of the Environment,* now in its 3rd edition (2012, University Science Books). He helped organize the Malta Conferences, which aim to use science diplomacy as a bridge to peace in the Middle East.

Contributors

We thank the following contributors who wrote chapters, conceived of original problems and solutions, and made helpful suggestions throughout. Without them, this project would not have happened.

Nicole C. Bouvier-Brown (Chapter 6, Nuclear Energy; Chapter 8, Free Radical Chemistry: Nitrogen Oxide, Ozone, and Combustion; Chapter 9, Air Pollution) has a BS in Chemistry/ Biology (with an environmental emphasis) and a PhD in Environmental Science, Policy, and Management from the University of California, Berkeley. She is currently an Associate Professor of Chemistry and Biochemistry at Loyola Marymount University teaching in the Chemistry and Environmental Science programs. Her research interests include developing methodology and measuring air pollution exacerbated by human activity and effectively teaching undergraduate students difficult scientific ideas, such as climate change and sustainability.

Juliane L. Fry (Chapter 2, Energy and Materials; Chapter 5, Fossil Fuels; Chapter 10, Stratospheric Ozone Shield) is an Associate Professor of Air Quality and Atmospheric Chemistry at Wageningen University and the Amsterdam Institute of Advanced Metropolitan Solutions in the Netherlands. Her research focuses on atmospheric and environmental chemistry, specifically on elucidation of interactions between human-produced nitrogen oxides and climate-relevant atmospheric aerosol particles. Julie obtained her PhD in Chemistry from the California Institute of Technology in 2006 and a master's degree in Environmental Law from Lewis and Clark Law School in 2016, and was a Professor of Chemistry and Environmental Studies at Reed College from 2008–2021.

Song Gao (Chapter 12, Water as Solvent: Acids and Bases; Chapter 14, Oxygen and Life; Chapter 18, Toxicity of Chemicals) is a Professor of Environmental Science and Chemistry at Duke Kunshan University and a Professor of the Practice of Global Studies at Duke University. His research interests include atmospheric chemistry (with a focus on secondary aerosol formation mechanisms), environmental analytical chemistry, and developing co-benefits approaches to mitigating plastics pollution, climate change, and stratospheric ozone depletion. He received his BS in Materials Science and Chemistry from the University of Science and Technology of China, his PhD in Chemistry from the University of Washington, and his postdoctoral training on Atmospheric Chemistry from the California Institute of Technology.

Michael T. Mury (Chapter 7, Climate Change; Chapter 11, Water Resources; Chapter 15, Water Pollution and Water Treatment) currently serves as the Science Department Chair and Scheduler at All Saints Academy in Winter Haven, FL. He teaches chemistry, physics, and forensic science to high school students. He earned his BS in Chemistry and BA in Biology from the University of Nebraska at Omaha and his PhD in Computational Chemistry from Clemson University. Michael has previously served on the staff at the American Chemical Society.

FOREWORD

The ability to reason closely is absolutely paramount to dealing with the wide range of environmental problems that threaten our society, which is why this book is important—it's like a workbook to get you thinking in precisely the ways we haven't thought. Many of our deepest problems right now have complex moral, political, and economic dimensions: climate change, paramount among them, is the most unjust and uneconomic thing humans have ever done. But at some level, it's a math problem, grounded in basic physics and chemistry. Solving it is all about pace and rate.

I would caution readers: even if you learn to think very clearly about these questions, that does not mean the debate over these issues will be ultimately resolved by reason alone. I learned many many years ago that winning the scientific argument over global warming was, if not irrelevant, certainly not sufficient: the real fight was, as most fights are, about money and power. Winning those takes a complementary set of skills; the fossil fuel industry bends to power, which is why we've had to build movements. It is irrational that some of us have had to go repeatedly to jail to begin to win this fight—but it is, in some respects, an irrational world.

But that is no reason to dispense with reason. We need to learn to think sharply and cogently, instead of relying on the lazy set of ingrained ideas that a consumer society specializes in passing on to us. Take this seriously; your brain and your heart are both weapons in the most important fights we face, and they need to work in tandem.

Bill McKibben
Middlebury, Vermont

INTRODUCTION

"Here's a short chemistry lesson," wrote climate activist Bill McKibben in a 1995 *New York Times Magazine* feature story. "Grasp it and you will grasp the reason the environmental era has barely begun. . . . Carbon monoxide versus carbon dioxide, one damn oxygen atom, and all the difference in the world." McKibben provides a clear illustration of the power of science to illuminate environmental issues. Carbon monoxide (CO) is a side effect of fuel combustion; it pollutes the air and contributes to smog formation. Better engine performance and catalytic converters have greatly reduced CO emissions from traffic over the years. But carbon dioxide (CO_2) is the product of the combustion reaction. Emission of this most important greenhouse gas can only be stopped by eliminating fossil fuel use.

The two molecules capture two sides of the environmental coin, local versus global effects of human activity. Although much remains to be done, environmental quality has improved in many localities, thanks to environmental controls and new technologies. But the global problems are just beginning to be addressed, and they are much more difficult to solve. Although economics and politics dominate the debates about these issues, science provides the essential foundation for informed discussion.

Environmental science courses have proliferated everywhere, responding to the rising interest among young and old about all things environmental. This book is intended for those interested in a quantitative approach to environmental problem solving. It does not involve higher-level math. Basic algebra and unit conversions are sufficient. A few equations are used from calculus, mainly in considering exponential growth and decay, but they are described as needed. And exponents and logarithms, for which a review is provided, are used as convenient ways of handling orders of magnitude. The main requirement is reasoning ability.

Because molecules are basic to environmental problems, chemical formulas and equations are frequently needed. College-level general chemistry is recommended as a pre-requisite, although a good high school chemistry course would likely be sufficient. However, the problems extend well beyond chemistry, and will be useful broadly in environmental science.

Environmental science courses are diverse in their content and approach. Environmental content has become widely available on the Internet, and many instructors assemble menus of articles or book chapters to supplement their teaching. However, quantitative problems addressing real-world issues in an approachable fashion are hard to find. This book is designed to provide a collection of such problems on the various topics in environmental science. Students can use this book to practice their problem-solving skills. The book is well suited to student-centered learning through guided problem solving. It can also be used for homework assignments in traditional lecture/discussion courses. The worked problems are followed by practice problems, which are starred. The starred answers are provided at the end of the book, while the steps in the solution are available in an accompanying instructor's manual.

SOLVE: Problems in Environmental Science grew out of the textbook *Chemistry of the Environment* in response to repeated requests for more problems. As in that text, the topics range widely—from climate and air pollution through water issues to agriculture and toxic materials. A group of experienced college teachers of environmental science assembled to create the new book. The problems are accompanied by brief narrative material providing the environmental context; some of this material is borrowed from *Chemistry of the Environment,* but much of it is new, and the figures and tables are current. We expect that these brief narratives will be accompanied by topical material of the instructor's choice.

Readers do not need to have read *Chemistry of the Environment* to use *SOLVE,* but the two books can work together. For those wishing to supplement *SOLVE* with the more expansive narrative of *Chemistry of the Environment,* a correspondence of the new and old chapter numbers is provided below.

SOLVE: *Problems in Environmental Science* Chapter Mapping to *Chemistry of the Environment* (CE)

Part I. Introduction
Chapter 1 (CE New). Introduction

Part II. Energy and Materials
Chapter 2 (CE7). Energy Flows and Supplies
Chapter 3 (CE11). Energy Utilization
Chapter 4 (CE10). Renewable Energy
Chapter 5 (CE8.) Fossil Fuels
Chapter 6 (CE9). Nuclear Energy

Part III. Atmosphere
Chapter 7 (CE6). Climate Change
Chapter 8 (CE4). Free Radical Chemistry—Nitrogen Oxide, Ozone, and Combustion
Chapter 9 (CE3). Air Pollution
Chapter 10 (CE5). Stratospheric Ozone Shield

Part IV. Hydrosphere and Lithosphere
Chapter 11 (CE12). Water Resources
Chapter 12 (CE13). Water as Solvent: Acids and Bases
Chapter 13 (CE14). Water and the Lithosphere
Chapter 14 (CE15). Oxygen and Life
Chapter 15 (CE16). Water Pollution and Water Treatment

Part VI. Biosphere
Chapter 16 (CE17). Nitrogen and Food Production
Chapter 17 (CE18). Pest Control
Chapter 18 (CE19). Toxicity of Chemicals

Appendices
Appendix A. Organic Structures
Appendix B. Mathematical Fundamentals
Appendix C. Answers to the Starred Questions

SOLVE

PART I

INTRODUCTION

1

1.1 Sustainability and Chemistry

The idea of sustainability provides an anchor to which all the elements of the environmental story are connected. The most widely cited definition of sustainability comes from the World Commission on Environmental Development: "Sustainable development is development that meets the needs of the present without compromising the ability of future generations to meet their own needs."[1] Concern about sustainability forms many connections among the environmental domains of this book: atmosphere, hydrosphere and lithosphere, biosphere, and energy. Examples include

- **Climate Change:** Chapter 7 presents the evidence for and consequences of global warming, and the facts of anthropogenic greenhouse gas emissions. These emissions are mainly tied to fossil fuel burning (Chapter 5). Limiting the rise in these gases will require difficult choices among many energy options (Chapters 4–6).

- **Smog:** Chapter 8 discusses the mechanism by which the action of sunlight on hydrocarbons and nitrogen oxides in the air leads to the production of tropospheric ozone, whose harmful effects on people and plants are being felt in many regions of the world. Curbing the harm of this photochemistry requires emission controls, and the introduction of low emission vehicles, likely including fuel cells and electric cars (Chapter 3).

- **Acid Rain:** Fossil fuel combustion produces nitrogen and sulfur oxides (Chapter 8), which produce acidic rainfall that can kill fish in lakes and damage plants in the watershed (Chapter 13). Emission controls can reduce acidification, although ironically, climate change may accelerate because the sulfate aerosol from sulfur oxide emissions has somewhat offset the warming effect of greenhouse gases (Chapter 7).

- **Fresh Water:** Localities around the world are running low on freshwater (Chapter 11), and there is increased interest in desalination to tap ocean water. Costs have been coming down, thanks to the development of reverse osmosis membranes (Chapter 15), but substantial energy is required.

- **Food Supply:** Synthetic fertilizer, plant breeding (Chapter 16), and synthetic pesticides (Chapter 17) have enabled the "green revolution" that has dramatically

---

[1] The World Commission on Environment and Development, *Report of the World Commission on Environment and Development: Our Common Future,* 1987, p. 41; available at http://www.un-documents.net/our-common-future.pdf.

increased the food supply and fed an expanding world population in recent decades. However, these methods have also led to increased greenhouse gas emissions (Chapter 7), overfertilization of waterways (Chapter 14), and disruption of ecosystems. They may also have reached a point of diminishing returns in crop yields. Increasing consumption of meat places further stress on the food production system, especially on its water and energy requirements (Chapter 11). The way ahead for agriculture is challenging and has elicited much controversy, especially about the promise and problems of genetically engineered plants (Chapter 17).

- **Avoiding Toxic Chemicals:** The industrial age has spread chemicals around the world, some of which build up in human and animal tissues, and some of which are toxic (Chapter 18). A major aim of "green chemistry" is to devise products and industrial processes that minimize the use and release of toxic chemicals.

1.2 Green Chemistry

Chemists have always striven to make new materials more useful, and to make them more efficiently, but in recent years there has been a growing recognition that bringing industrial civilization into harmony with the natural world will require major efforts by chemists in all sectors of the economy.

1.2.1 Green Chemistry Principles

In *Green Chemistry: Theory and Practice*,[2] the first book on the subject, authors Paul Anastas and John Warner outlined the following 12 principles:

1. **Prevent Waste:** Design chemical syntheses to prevent waste, leaving no waste to treat or clean up.

2. **Design Safer Chemicals and Products:** Design chemical products to be fully effective, yet have little or no toxicity.

3. **Design Less Hazardous Chemical Syntheses:** Design syntheses to use and generate substances with little or no toxicity to humans and the environment.

4. **Use Renewable Feedstocks:** Use raw materials and feedstocks that are renewable rather than depleting. Renewable feedstocks are often made from agricultural products or are the wastes of other processes; depleting feedstocks are made from fossil fuels (petroleum, natural gas, or coal) or are mined.

5. **Use Catalysts, not Stoichiometric Reagents:** Minimize waste by using catalytic reactions. Catalysts are used in small amounts and can carry out a single reaction many times. They are preferable to stoichiometric reagents, which are used in excess and work only once.

6. **Avoid Chemical Derivatives:** Avoid using blocking or protecting groups or any temporary modifications if possible. Derivatives use additional reagents and generate waste.

7. **Maximize Atom Economy:** Design syntheses so that the final product contains the maximum proportion of the starting materials. There should be few, if any, wasted atoms.

---

[2] Oxford University Press, New York, 1998.

8. **Use Safer Solvents and Reaction Conditions:** Avoid using solvents, separation agents, or other auxiliary chemicals. If these chemicals are necessary, use innocuous chemicals.

9. **Increase Energy Efficiency:** Run chemical reactions at ambient temperature and pressure whenever possible.

10. **Design Chemicals and Products to Degrade After Use:** Design chemical products to break down to innocuous substances after use so that they do not accumulate in the environment.

11. **Analyze in Real Time to Prevent Pollution:** Include in-process real-time monitoring and control during syntheses to minimize or eliminate the formation of by-products.

12. **Minimize the Potential for Accidents:** Design chemicals and their forms (solid, liquid, or gas) to minimize the potential for chemical accidents, including explosions, fires, and releases to the environment.

1.2.2 Green Chemistry Examples

Many applications of green chemistry are being actively pursued in industry and academic chemistry departments. We can get an idea of the range of these applications from the annual U.S. Presidential Green Chemistry Challenge Awards, which were initiated in 1996. The winning projects are described on the Web site http://www.epa.gov/greenchemistry. In the examples described briefly below, which of the above principles is involved?

QUESTION 1

A 1996 award went to the Dow Chemical Company for developing a process to substitute CO_2 for ozone-destroying chlorofluorocarbons (CFCs; Chapter 10) as a blowing agent for polystyrene foam, used in all manner of containers and insulating sheets. This substitution eliminated some 3.5 billion lb of CFC-12 and hydrochlorofluorocarbon-22 (HCFC-22), which would eventually have escaped into the atmosphere from the containers. Moreover, the CO_2-blown sheets remain flexible and break less than those blown with CFCs and are easier to recycle. Although CO_2 is a greenhouse gas (Chapter 7), it is obtained as an industrial by-product (from natural gas production and fertilizer manufacture), which would have been vented to the atmosphere.

Q1 ANSWER Principle 2, safer products, is operative here.

---

*QUESTION 2

Carbon dioxide is proving to be a very useful solvent when compressed to the liquid state, or compressed even further to a supercritical fluid, in which the separation of gas and liquid phases disappears. Drying the solvent is accomplished by simply reducing the pressure, whereupon CO_2 is released as a gas and can be recycled. Liquid CO_2 has been introduced as a dry-cleaning solvent, replacing the widely used perchloroethylene (PERC; Cl_2CCCl_2), which frequently leaks and

---

* Answers to starred questions can be found at the end of the book.

contaminates groundwater, and is a suspected human carcinogen (Chapter 18). Also, PERC is volatile and can contribute to smog formation (Chapter 8). Liquid CO_2 is benign, but although it has favorable solvent properties (low viscosity, good wetting ability), it is unable to solubilize the large molecules that comprise dirt and grease. Joseph DeSimone (University of North Carolina) won a 1997 Green Chemistry Award for developing CO_2-soluble surfactants. These surfactants can form micelles around dirt particles, just as detergents do in water, which allows them to float away in the liquid CO_2. Detergents have hydrophilic and hydrophobic regions, and likewise, DeSimone's surfactants contain CO_2-philic and CO_2-phobic regions. The former have chlorine-fluorine (C–F) bonds, which are attracted to the CO_2, while the latter have benzene rings, which are not attracted.

*QUESTION 3

Today's society runs on plastics, found in all manner of products, from grocery bags to cars. They are fabricated from a number of polymers, almost all of which derive from petrochemicals. Although only a small percentage of oil production ends up in plastics, there is much interest in developing alternative polymers based on chemicals from the biosphere (U.S. Energy Information Administration, https://www.eia.gov/tools/faqs/faq.php?id=34&t=6). They would free plastics from their dependence on the dwindling reserves of oil (Chapter 5). In addition, materials made from bio-chemicals are generally biodegradable, whereas those made from petrochemicals are not. Biopolymers can overcome the disposal issues associated with plastics.

NatureWorks LLC, a Cargill subsidiary, won a 2002 Green Chemistry Award for the development of polylactic acid (PLA) from corn. In this process, the starch in corn (Chapter 4) is converted to its constituent sugar molecules, which are then fermented to lactic acid [$(CH_3)(OH)CCOOH$]. The lactic acid is then polymerized with a catalyst to PLA. PLA is biodegradable, and it can also be recycled by hydrolyzing it back to lactic acid and repolymerizing the lactic acid.

PLA can be substituted for petroleum-based polymers in many consumer products and is particularly attractive for food packaging because it can be composted. In 2005, Walmart announced plans to use 114 million PLA containers annually.

*QUESTION 4

Oil-based "alkyd" paints produce durable, high-gloss coatings, but require organic solvents, which then become smog-contributing volatile organic compounds (VOCs; Chapter 8) when they evaporate. Water-borne acrylic latex paints are available, but they have low gloss and are less resistant to corrosion when applied to metals. Procter & Gamble won a 2009 Green Chemistry Award for developing a low-solvent substitute for alkyd paints, based on esterifying sucrose with fatty acids of controlled chain lengths. The resulting "Sefose" sugar esters require little solvent and become cross-linked by oxidation reactions after application, to produce a hard and durable surface.

*QUESTION 5

Chlorine is widely used as a disinfectant and as a bleaching agent (Chapter 15). But it also produces organochlorine molecules as by-products, which have varying toxicities. Tetrachloro dibenzo dioxin

(TCDD, or just "dioxin") and related molecules (Chapter 18) are of particular concern as potent hormone mimics and suspected human carcinogens.

Paper mills have been an important dioxin source because they relied on chlorine to whiten the paper by bleaching the pulp, whose brown color is from phenolic molecules derived from lignin (Chapter 5). Chlorine oxidizes these molecules, destroying the color, but it also chlorinates some of them, which then form dioxin (Chapter 18).

In 2001, the use of chlorine for bleaching paper was banned in the United States. Most paper mills switched to chlorine dioxide (ClO_2), which is a more powerful oxidant, but weaker chlorinating agent than Cl_2. However, some mills are now using the non-chlorinating oxidant hydrogen peroxide (H_2O_2). Hydrogen peroxide is a relatively slow oxidant and requires a catalyst to activate it. In 1999, Terry Collins (Carnegie Mellon University) won a Green Chemistry Award for the development of H_2O_2 catalysts based on an iron ion held by tetra-amido macrocyclic ligands. These ligands adjust the electronic properties of the Fe^{3+}, so that when H_2O_2 binds to it, the O–O bond is readily cleaved, releasing water and creating a powerful FeO-based oxidant. In the environment, the catalyst breaks down into harmless products. This H_2O_2 catalyst system is being used in some paper mills and is also being developed for laundry products and water disinfectants.

*QUESTION 6

Complementing this peroxide-utilizing system is the development of a process for direct preparation of H_2O_2 from O_2 to H_2, for which Headwaters Technology Innovation won a Green Chemistry Award in 2007. This process is based on a Pd/Pt catalyst, made highly active by nanofabrication. It replaces the conventional H_2O_2 synthesis, in which anthraquinone is reacted with H_2 to produce anthrahydroquinone, which is in turn reacted with O_2. The conventional synthesis requires that the H_2O_2 be removed from the solution with an energy-intensive stripping column and concentrated by vacuum distillation. Also, quinone by-products are produced, which require disposal.

*QUESTION 7

Standard chelating agents like the aminocarboxylates nitrilotriacetic acid (NTA) and ethylenediaminetetraacetic acid (EDTA; Chapter 18) degrade slowly in the environment, and are of concern because of their toxicity and their potential for mobilizing toxic metals. In 2001, the Bayer Corporation won a Green Chemistry Award for developing a nontoxic and readily degradable chelating agent, sodium iminodisuccinate (IDS). Its synthesis, from maleic anhydride and ammonia, requires mild conditions in water, while NTA and EDTA require the highly toxic hydrogen cyanide in their synthesis. The IDS is marketed as a builder in detergents, for absorption of nutrient metal ions in agriculture, and for metal ion scavenging in photographic processing and groundwater remediation.

*QUESTION 8

PPG Industries won a Green Chemistry Award in 2001 for replacing lead with yttrium as a corrosion inhibitor in electrodeposited steel coatings. Essentially all primer coats for automobiles are electrodeposited from a bath containing metal ions. Lead has been the metal of choice, because of the corrosion protection offered by Pb_3O_4. Although lead was banned for use in U.S. houses in 1972 (Chapter 18), it won exemption from regulations in automotive coatings, because of the demand for corrosion resistance. Yttrium is found to be twice as effective as lead, and only 1% as toxic. It also eliminates the need for pretreatment of the steel with chromium and nickel, which lead protection requires. Yttrium has now replaced lead in essentially all automotive coatings in the United States and Europe.

*QUESTION 9

Pressure-treated wood is found in 50% of U.S. homes, and in countless decks, fences, piers, bridges, and playground equipment. The treatment, which inhibits decay of the wood for 20–50 yr, involves drawing out moisture, and then injecting the wood under pressure with a preservative solution. The preservative used in the United States has been "chromated copper arsenate," a complex mixture based on these three metals. Although the metals are lodged in the wood, they can leach out over time. Studies of soils under pressure-treated wood decks have found elevated levels of all three metals, and particularly of arsenic (Chapter 18). In 1996, Chemical Specialties, Inc., introduced a replacement wood preservative, based on copper and nontoxic quaternary ammonium salts, for which it won a Green Chemistry Award in 2002. By 2003, wood producers had voluntarily stopped using chromated copper arsenate.

*QUESTION 10

The anti-inflammatory drug ibuprofen (the active ingredient in brands like Advil and Motrin) is one of the most widely used over-the-counter medications and is produced in multimillion-pound quantities annually. A 1997 Green Chemistry Award went to its producer, the BHC Company, for redesigning its synthesis to improve efficiency and reduce waste. The number of separate reactions was reduced from six to three, and the atom economy, defined as the mass fraction of the reagents that end up in the products, increased from 40% to 80%. In addition, the main side product, acetic acid, is used elsewhere. The new synthesis was made possible by the use of anhydrous hydrogen fluoride, and of nickel and palladium catalysts, all of which are recovered.

*QUESTION 11

Recyclable carpet was the innovation that won Shaw Industries a Green Chemistry Award in 2003. As of 2005, only 7% of used carpet was recycled in the United States, with almost all the rest ending in landfills. The scale of the waste is huge; in 2004, 2.3 billion yd^2 of new carpeting was shipped in the United States.

The main barrier to recycling is the poly-vinyl chloride (PVC) backing of most carpets. The chemical properties of PVC are not conducive to recycling. In addition, there are health concerns about the *phthalate* plasticizers used to render PVC pliable, which eventually migrate out of the PVC and have become widespread in the environment. Phthalates may affect hormonal systems (Chapter 18). Also, burning the PVC, either in incinerators or accidental fires, can produce dioxins (Chapter 18).

Shaw's EcoWorx carpeting has polyolefin, instead of PVC, which does not require plasticizers and can be readily recycled. The used carpet is ground up, and a stream of air separates the heavier backing particles from the lighter fiber particles. The backing particles are then re-extruded to form new backing. The fiber particles, made of nylon, are depolymerized to the nylon monomer, *caprolactam*, and then repolymerized into new fibers, which are reattached to the backing. To facilitate recycling, Shaw will take back carpeting at the end of its useful life, having calculated that recycling costs are less than the costs of new carpet materials.

*QUESTION 12

A 2001 award went to EDEN Bioscience Corporation, for the development of *hairpin*, a bacterial protein that elicits a plant's natural defense mechanism against pests. *Hairpin* induces a "hypersensitive response," resulting in cell death at the point of an infestation. The dead cells act as a physical barrier against the pathogen and may also release damaging chemicals. At the same time, *hairpin* acts as a plant growth stimulant, resulting in increased plant biomass and crop yields. *Hairpin* is made by fermentation, using a genetically engineered strain of *Escherichia coli* (*E. coli*) bacteria, and can be applied directly to the stems and leaves of crops. Only small amounts are required, and it has no effect on biota other than plant protection. *Hairpin* is rapidly degraded by UV light and microorganisms.

*QUESTION 13

Professor Sanjoy Banerjee from the City College of New York was given a Green Chemistry award in 2019 for his work to develop a rechargeable alkaline Zn-MnO_2 battery that could be used for applications around grid storage (Chapter 3). These new batteries can be recharged thousands of times without a decrease in the length of a battery's lifetime. Zn and MnO_2 are the primary materials used in the disposable battery market, which are readily available and safe. Chemical dopants, such as copper ions, stabilize the MnO_2 cathodes, which allows them to be recharged thousands of times. Using this type of novel battery for grid-scale energy storage facilitates the use of renewable energy.

*QUESTION 14

In 2020, Vestaron Corporation won a Green Chemistry Award for the development of SPEAR Insecticide, a new type of biopesticide that had efficacy similar to synthetic insecticides. The pesticide is based on a naturally occurring chemical component of spider venom that controls the target pest (Chapter 17), but has no effect on humans, the environment, and non-target wildlife. This biopesticide is a peptide insecticide, developed around a peptide discovered in the venom of the Blue Mountains funnel-web spider. The SPEAR peptide is created from the yeast fermentation of sugar produced from corn, while most other pesticides are made from non-renewable feedstocks. SPEAR also biodegrades into nontoxic amino acids, since it is synthesized from peptides.

1.3 Methods in Environmental Science

In this section, we introduce some concepts and calculational methods that are useful in several areas of environmental science.

1.3.1 *Reservoirs, Flows, and Residence/Replacement Times*

It is often helpful to represent environmental processes in terms of *reservoirs* and *flows*. A reservoir is a region of the environment that holds a significant amount of the material under consideration and is physically or chemically separated from other reservoirs. These amounts can be estimated by multiplying the concentration of the material, obtained by sampling, by the size of the reservoir.

To understand the dynamics of the environment, one needs to understand how much material is transferred from one reservoir to another per unit of time (i.e., the flow). These numbers are generally much harder to determine, and a great deal of ingenuity must go into making flow estimates.

If the flows and reservoirs are known, one can calculate *residence times* and *replacement times* by dividing the reservoir amount by the outflow and inflow rates, respectively:

$$\tau_{res} = \text{reservoir/rate}_{out} \tag{1.1}$$

$$\tau_{repl} = \text{reservoir/rate}_{in} \tag{1.2}$$

The replacement time is the period it takes to replace the material lost to flow out of the reservoir. If these rates are equal, the system is in a steady state (i.e., the amounts do not change with time). Otherwise, the reservoir must be increasing or decreasing with time.

QUESTION 15

As a simple example, consider a new town that was just established near Clear Pond, which holds 90,000,000 L of water. If the town extracts water from the pond at an average rate of 20 L sec^{-1}, what is the residence time of water in Clear Pond, in days?

Q15 ANSWER Residence time (τ) = volume/flux out:

$$\tau = \frac{90,000,000 \text{ L}}{20 \text{ L/S}} = 4,500,000 \text{ s} \left(\frac{1 \text{ min}}{60 \text{ s}}\right)\left(\frac{1 \text{ hr}}{60 \text{ min}}\right)\left(\frac{1 \text{ day}}{24 \text{ hr}}\right) = 52 \text{ days}$$

QUESTION 16

Figure 1.1 shows the cycling of calcium, an essential nutrient for plants. In this case, the amounts are given in mass per unit area (in hectares) of the forest.

From the estimates of the reservoirs and flows, calculate the calcium in

(a) residence time in the trees above ground and

(b) replacement time in the tree below ground, i.e., in the roots.

Q16 ANSWER (a) Residence time above ground:

$$400 \text{ kg} \times \frac{\text{yr}}{40 \text{ kg}} = 10 \text{ years}$$

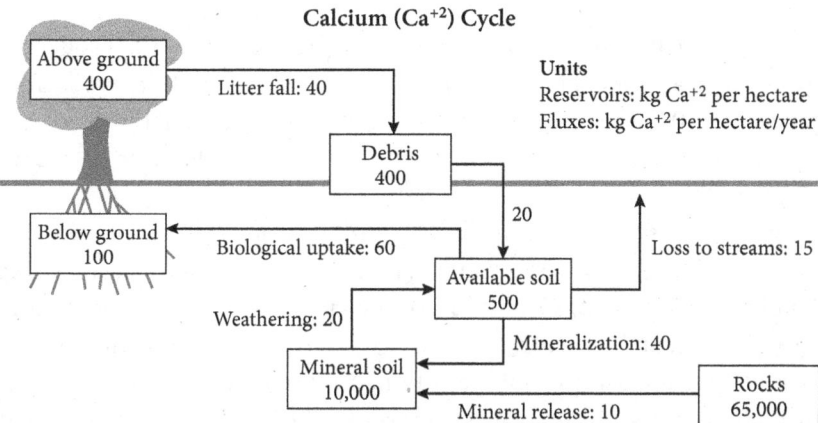

Figure 1.1 Calcium cycle both below and above the earth.
Source: Adapted from Botkin & Keller, *Environmental Science: Earth as a Living Planet, 5th Ed.,* Wiley Publishing, © 2005.

(b) Replacement time in the tree roots:

$$100 \text{ kg} \times \frac{\text{yr}}{60 \text{ kg}} = 1.7 \text{ years}$$

QUESTION 17

Actually, the calcium content of the trees is found not to vary from year to year, either above ground or in the roots. This observation implies that some flows are missing from the diagram. What, and how large, are they?

Q17 ANSWER For one thing, there must be transfer from the roots to the tree, or else the above-ground content would decline. At steady state, this flow must equal that of the leaf fall, 40 kg yr^{-1}. Also, if the root content does not build up with time, there must be an outflow from the roots to the soil (leaching). The leach rate must be the difference between the root uptake rate, and the root to tree rate. For example:

$$60 \frac{\text{kg}}{\text{yr}} - 40 \frac{\text{kg}}{\text{yr}} = 20 \frac{\text{kg}}{\text{yr}}$$

*QUESTION 18

Looking at the flows in and out of "mineral soil," is this reservoir changing over time? In what direction? How long would it take to change the reservoir amount by 1%?

*QUESTION 19

Is the reservoir changing over time for calcium from available soil? In what direction? How long would it take to change the reservoir amount by 1%?

1.3.2 *Exponents*

In order to avoid having to write out many zeros, we can express numbers in *exponential* notation, $n \times b^x$, where b is the base, and x is the exponent (i.e., the number of times b is multiplied by x). Most often we use 10 as the base. If there is a minus sign in front of the exponent, then x is the number of times b is multiplied by x^{-1} (i.e., by $1/x$). For example, 5.1×10^3 means 5,100, while 5.1×10^{-3} means 0.0051.

In addition, we can use prefixes in front of the units in order to modify the amounts. The commonly used prefixes are kilo (k), mega (M), giga (G), tera (T), peta (P), and exa (E). Using base 10, these correspond to $x = 3, 6, 9, 12, 15,$ and 18. Also in common use are centi (c), milli (m), micro (μ), nano (n), pico (p), and femto (f) for $-x = 2, 3, 6, 9, 12,$ and 15 (with 10 as the base). The letters in parentheses are accepted abbreviations for the prefixes. Thus, 5.1×10^3 meters (abbreviated m) is the same as 5.1 kilometers (km), and 5.1×10^{-3} m is the same as 0.51 centimeters (cm) or 5.1 millimeters (mm).

QUESTION 20

Express the number of seconds in a year in exponential form.

Q20 ANSWER

$$\frac{365 \text{ days}}{\text{yr}} \times \frac{24 \text{ hr}}{\text{day}} \times \frac{60 \text{ min}}{\text{hr}} \times \frac{60 \text{ sec}}{\text{min}} = 31{,}536{,}000 \frac{\text{sec}}{\text{yr}} = 3.15 \times 10^7 \frac{\text{sec}}{\text{yr}}$$

*QUESTION 21

How many terajoules (TJ) of electricity is produced in a year by a 1.00 megawatt (MW) power plant? 1 watt (W) is 1 J sec^{-1}.

1.3.3 *Logarithms*

For convenience, exponents are often written as logarithms, or log for short:

$$x = \log_b (b^x) \tag{1.3}$$

The rules for log functions derive from the properties of exponents. Multiplication inside the log is the same as addition outside the log:

$$\log_b (xy) = \log_b (x) + \log_b (y) \tag{1.4}$$

while division inside the log equals subtraction outside of the log:

$$\log_b \left(\frac{x}{y}\right) = \log_b (x) - \log_b (y) \tag{1.5}$$

If there is an exponent inside the log, it can be moved in front of the log as a multiplier, and vice versa:

$$\log_b (x^y) = y\log_b (x) \tag{1.6}$$

For ease of calculations, the number 10 is generally chosen as the base for taking logarithms:

$$x = \log_{10}(10^x) \tag{1.7}$$

Log values can readily be obtained from tables or with a calculator. Handy values to remember are $\log 2 = 0.301 \approx 0.3$ and $\log 5 = 0.699 \approx 0.7$.

QUESTION 22

Recalculate Q20 using logarithms.

Q22 ANSWER Remember from Q20 $\dfrac{365 \text{ days}}{\text{yr}} \times \dfrac{24 \text{ hr}}{\text{day}} \times \dfrac{60 \text{ min}}{\text{hr}} \times \dfrac{60 \text{ sec}}{\text{min}}$

Let $y = \dfrac{\text{number of seconds}}{\text{year}}$

$$\log(y) = \log 365 + \log 24 + 2 \log 60 = 2.56 + 1.38 + 2(1.77) = 7.50$$
$$y = 10^{7.50} = 3.15 \times 10^7$$

A prominent use of logarithms in chemistry is the expression of acidity or basicity as pH or pOH. "p" is the symbol for $-\log_{10}$.

So pH is $-\log[H^+]$, while pOH is $-\log[OH^-]$.

In water, the product $[H^+][OH^-] = K_W$ (the water auto-ionization constant, or ion-product constant) $= 10^{-14}\ M^2$ (M = mol/L, or molar) at 25°C, so

$$pH + pOH = pK_W = 14$$

In Chapter 12, we see that K_W is actually dependent on the temperature, thus pK_W changes slightly when it is above or below 25°C.

QUESTION 23

What are the pH and pOH of a solution containing 0.002 M HCl (a strong acid, that ionizes completely to H^+ and Cl^- in water) at 25°C?

Q23 ANSWER
$$pH = -\log(0.002) = -\log(2 \times 10^{-3}) = 3.0 - \log 2 = 3.0 - 0.3 = 2.7$$

$$pOH = 14.0 - 2.7 = 11.3$$

---

*QUESTION 24

What are the pH and pOH of a solution containing 0.005 M NaOH (a strong base that ionizes completely to Na^+ and OH^- in water) at 25°C?

1.3.4 *Natural Logarithms: Exponential Growth and Population Growth*

Some equations arise naturally that involve exponents of the number e (defined as $\lim_{n\to\infty}(1 + (1/n))^n$ and having the value 2.718281828459045...). An important example is the equation describing exponential growth:

$$Q = Q_0 e^{kt} \tag{1.8}$$

here Q_0 is the initial quantity, and Q is its value at any subsequent time, t. This equation is the integral form of the differential expression for a rate of growth that is proportional to the quantity itself, i.e., constant percentage growth:

$$\frac{dQ}{dt} = kQ \tag{1.9}$$

Equation 1.8 describes curve a in Figure 1.2, which has the quantity Q doubling in constant intervals of time; the doubling time t_2 is the same no matter how large Q gets. As the number of doublings increases, Q gets very large.

To obtain the doubling time, we can take the base e logarithm from both sides of (1.8):

$$kt = \log_e\left(\frac{Q}{Q_0}\right)$$

If $\dfrac{Q}{Q_0} = 2$ to represent the doubling time, then $kt_2 = \log_e 2 = \ln 2 = 0.693$

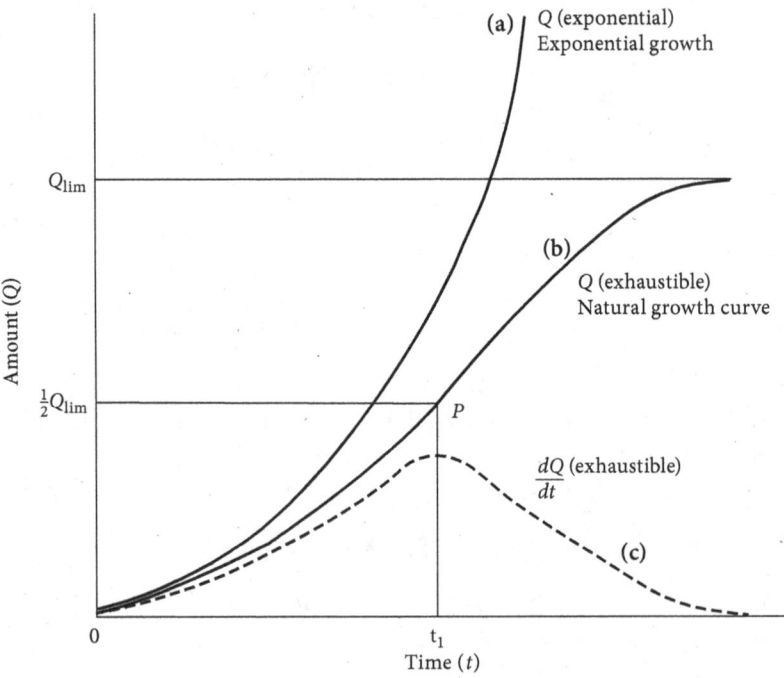

P is an inflection point where the rate of growth shifts from increase to decrease

Figure 1.2 (a) Exponential and (b) natural growth curves. For curve a, $\dfrac{dQ}{dt} = kQ$, while for curve c, $\dfrac{dQ}{dt} = (2\pi)^{-\frac{1}{2}} Q_{\text{lim}} e^{-\frac{(t-t_1)^2}{2}}$. The parameter t_1 is the time when one-half of Q has been used up and the rate of growth shifts from increase to decrease.
Source: Adapted from Spiro et al., *Chemistry of the Environment, 3rd Ed.*, University Science Books, © 2012, all rights reserved.

$\log_{10}$ and $\log_e$ (called common and natural logarithms, respectively) are abbreviated as log and ln, respectively. Values of ln as well as the log can be obtained with calculators or in tables.

A useful value to remember is $\ln 2 = 0.693$, giving the doubling time as:

$$t_2 = 0.693/k \qquad (1.10)$$

(This result is the basis for the commonly used "rule of 70," $t_2 = 70/p$, where p is the constant percentage growth rate constant. Expressed as percent $p = 100k$ and $t_2 = 0.693 \times 100/p = 69.3/p$. This approximate rule is useful since for most purposes 70 is as good as 69.3. Thus, a 7% growth rate gives a doubling time of ~10 yr.)

Another useful number is the conversion factor 2.3 between base e and base 10 logarithms:

$$\ln(x) = 2.3 \log(x) \qquad (1.11)$$

In the real world, things can grow only so far before reaching some kind of limit. Curve in Figure 1.2(b), shows a growth curve with a **limit**, Q_{lim}, arising from exhaustion in the supply of Q (perhaps some natural resource). The rate of growth reaches a peak at a characteristic time, t_1, when half the resource has been used up and then diminishes toward zero. The rate expression is:

$$\frac{dQ}{dt} = (2\pi)^{-\frac{1}{2}} Q_{lim} e^{-\frac{(t-t_1)^2}{2}} \qquad (1.12)$$

Population data for China and India over the last half-century are listed in Table 1.1 and plotted in Fig. 1.3.

TABLE 1.1 Population Data for China and India from 1960 to 2016

Year	China	India
1960	6.67×10^8	4.35×10^8
1970	8.18×10^8	5.48×10^8
1980	9.81×10^8	6.87×10^8
1965	7.15×10^8	4.87×10^8
1975	9.16×10^8	6.13×10^8
1985	1.06×10^9	7.65×10^8
1990	1.14×10^9	8.50×10^8
2010	1.34×10^9	1.21×10^9
2012	1.35×10^9	1.24×10^9
2014	1.36×10^9	1.29×10^9
2016	1.38×10^9	1.32×10^9

Source: World Bank, *Population, Total—India,* available at https://data .worldbank.org/indicator/SP.POP.TOTL?locations=IN-C.

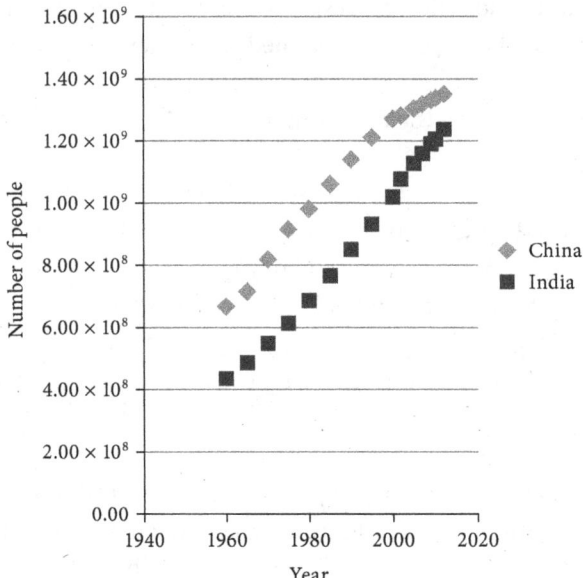

Figure 1.3 Population changes for China and India over the past 50 years.
Source: World Bank Open Data, *Population, Total—India,* available at https://data.worldbank.org/indicator/SP.POP.TOTL?locations=IN-C.

QUESTION 25

Looking at the plots, do you think they indicate exponential growth for these countries?

Q25 ANSWER The upward curvature at the start suggests exponential growth, but the subsequent inflection indicates a fall-off from exponential growth.

QUESTION 26

Use the population size from 1960 to 1985 to estimate the fractional rate of annual growth for each country. Since the data are somewhat scattered, take the five-year fractional differences and average them, then divide by 5. Compare the population size of 2010 with what would have been predicted if growth had continued exponentially at this annual rate.

Q26 ANSWER Reconstruct the table, taking five-year differences and dividing by population at the start of the five-year period, then take the average. Divide the average by 5 to get the average annual growth rate.

$$k = 0.019 \text{ for China and } 0.023 \text{ for India}$$

For exponential growth:

$$\ln\left(\frac{Q}{Q_0}\right) = kt \text{ or } \log\left(\frac{Q}{Q_0}\right) = \frac{kt}{2.3}$$

where Q_0 is the 1960 population, k is the annual fractional growth rate, and $t = 50$ yr for 2010:

$$\text{China: } \log\left(\frac{Q}{Q_0}\right) = \frac{kt}{2.3} = \frac{0.019 \times 50}{2.3} = 0.43$$

$$\text{India: } \log\left(\frac{Q}{Q_0}\right) = \frac{kt}{2.3} = \frac{0.023 \times 50}{2.3} = 0.50$$

Then

$$\text{China: } Q = Q_0 \times 10^{\frac{kt}{2.3}} = 6.67 \times 10^8 \times 10^{0.43} = 1.79 \times 10^9 \text{ people}$$

$$\text{India: } Q = Q_0 \times 10^{\frac{kt}{2.3}} = 4.35 \times 10^8 \times 10^{0.50} = 1.37 \times 10^9 \text{ people}$$

The actual population size of 2010 was 1.34×10^9 people and 1.21×10^9 people, less than the predicted values, illustrating the slowing of growth seen qualitatively in the curves. The discrepancy is greater for China (25%), than India (10%), reflecting the earlier slowdown of the China growth curve.

---

*QUESTION 27

Suppose that, starting in 1960, the annual fractional growth in Table 1.2 had been that seen between 2014 and 2016. What would the population then be in 2010?

TABLE 1.2 Fractional Rate of Population Growth for China and India

Year	China $P(\times 10^8)$	∂ (five-year)	∂/P	India $P(\times 10^8)$	∂ (five-year)	∂/P
1960	6.67			4.35		
1965	7.15	0.48	0.072	4.87	0.52	0.120
1970	8.18	1.03	0.144	5.48	0.61	0.125
1975	9.16	0.98	0.120	6.13	0.65	0.119
1980	9.81	0.65	0.071	6.87	0.64	0.104
1985	10.60	0.79	0.080	7.65	0.78	0.102
1990	11.40	0.80	0.076	8.50	0.85	0.111
Total			0.526			0.681
T/6			0.0937			0.113

1.3.5 *Exponential Decay and Population Decline*

Something decays exponentially if it disappears in constant proportion to the amount of the substance still left. The equation is the same as for exponential growth, but with a minus sign:

$$\frac{dQ}{dt} = -kQ \tag{1.13}$$

and the integral form is

$$Q = Q_0 e^{-kt} = -\ln\left(\frac{Q}{Q_0}\right) \tag{1.14}$$

Instead of a constant doubling time, there is a constant half-life, $t_{\frac{1}{2}}$, reached when $Q = \frac{Q_0}{2}$. Again the rule of 70 applies, this time to the half-life $t_{\frac{1}{2}} = \frac{0.693}{kt}$.

QUESTION 28

According to the United Nations, the population of Russia peaked at 148,689,000 in 1991, just before the breakup of the Soviet Union. Afterward, it began to fall, presumably due to economic and social dislocation, stabilizing again in the mid-2000s. The population was 147,915,000 in 1997 and 142,487,000 in 2006. Calculate the annual rate of decrease during that interval, assuming an exponential decay of population.

Q28 ANSWER Assuming that the rate was proportional to the population (i.e., constant fractional rate, the decrease would have been exponential:

$$kt = -\ln\left(\frac{Q}{Q_0}\right) = -\ln\left(\frac{142,487,000 \text{ people}}{147,915,000 \text{ people}}\right) = 0.0374$$

with $t = 9$ yr.

Then $k = \dfrac{0.0374}{9} = 0.00415$ per year or 0.415% per yr.

Note: The answer is nearly the same if we assumed a linear decline, that is

$$\frac{(147,915,000 - 142,487,000) \text{ people}}{9 \text{ yr}} = 603,000 \frac{\text{people}}{\text{yr}}$$

corresponding to a fractional rate of:

$$\frac{603,000}{147,915,000} \times 100\% = 0.408\%$$

The discrepancy gets larger as the rate increases.

---

*QUESTION 29

In 2005, the United Nations warned that the population could fall by one-third by 2050, if trends did not improve. Does this date accord with your decay rate calculated in Q28?

1.3.6 *First-Order Reactions*

A first-order reaction involves a single reactant, and its rate is proportional to the reactant concentration

$$\text{Reactant} \rightarrow \text{Products}$$

The reaction rate is the rate of disappearance of the reactant.

$$\text{Rate} = -\frac{d[\text{reactant}]}{dt} = k[\text{reactant}] \tag{1.15}$$

(The minus sign is because the reactant is disappearing.)

This is the differential form of exponential decay, and the integral form is:

$$kt = -\ln\frac{[\text{reactant}]}{[\text{reactant}]_0} \tag{1.16}$$

As an example, consider the splitting of ozone molecules by UV radiation, a key reaction in the stratosphere (see Chapter 10)

$$O_3 \rightarrow O + O_2 \text{ (at radiation wavelengths 200–320 nm)}$$

$$-\frac{d[O_2]}{dt} = k[O_2]$$

---

*QUESTION 30

At 30 km altitude, the middle of the stratosphere, the radiation flux is such that $k = 10^{-3}$ sec^{-1}. How long would it take to lose 10% of the O_3 (ignoring the processes that build up ozone)?

1.4 Conclusions

Sustainability is the concept around which measures for environmental protection are now being organized. We want to meet people's needs now, without compromising those of future generations. Chemists are adopting green chemistry principles to develop processes and products that have the least possible environmental impact.

With the widely applicable tools of the reservoir and flow analysis, exponents and logarithms, and exponential growth and decay in hand, we are ready to tackle the environmental issues raised in the following chapters.

PART II

ENERGY AND MATERIALS

ENERGY FLOWS AND SUPPLIES

2.1 Human Energy Consumption

Humans have harnessed diverse sources of energy to fuel our development. By examining global energy fluxes, we can propose human energy use in the context of the total energy supplied by the sun to earth. Table 2.1 shows these global energy fluxes, demonstrating that human usage of energy is a very minor fraction of the total energy absorbed by the earth's surface. However, energy use is rapidly rising, requiring the development of ever more energy resources. Furthermore, to address concerns about global climate change, humans must work to reduce the amount of the energy that comes from fossil sources, which emit CO_2.

TABLE 2.1 Global Energy Fluxes

Sources	Rates (10^{20} kJ yr^{-1})
Solar energy incident on earth	54.4
Solar energy affecting earth's climate and biosphere	38.1
Energy taken up in global evaporation of water	12.5
Energy in wind	0.109
Solar energy taken up in photosynthesis	0.0850
Energy conducted from the earth's interior to its surface	0.0100
Total primary energy consumed by humans, 2016	0.00612
Total energy produced in the United States, 2018	0.00129
Total energy consumed in the United States, 2018	0.00107
Fossil energy produced in the United States, 2018	0.00079

Sources: Energy Information Administration, *International Energy Outlook 2007,* available at http://www.eia.gov/oiaf/ieo/; Energy Information Administration, *Annual Energy Review 2007,* available at http://www.eia.gov/emeu/aer/contents.html; and Energy Information Agency, International Energy Statistics, *International,* available at https://www.eia.gov/beta/international/data/browser/#/?c=41000000020000600000000000000g0002000000000000000001&vs=INTL.44-1-AFRC-QBTU.A&vo=0&v=H&end=2016 and *Total Energy,* available at https://www.eia.gov/totalenergy/data/browser/index.php?tbl=T01.01#/?f=A&start=1949&end=2018&charted=4-6-7-14 [given in units of quad BTU per year; 1 quad BTU – 1.05×10^{15} kJ].

QUESTION 1

To put these energy flows in context, calculate what fraction of incoming solar radiation is taken up in plant photosynthesis.

Q1 ANSWER The annual amount of energy taken up by photosynthesis is 0.0850 kJ, and the total solar energy incident on earth is 54.4×10^{20} kJ. Taking the ratio of these two numbers:

$$\frac{0.0850 \times 10^{20} \text{ kJ}}{54.4 \times 10^{20} \text{ kJ}} \times 100\% = 0.00156 \times 100\% = 0.156\%$$

*QUESTION 2

What fraction of total primary energy on earth is consumed by humans?

*QUESTION 3

What fraction of energy consumed in the United States is produced from fossil energy?

QUESTION 4

In what direction would you expect the fraction of fossil fuel use to energy consumed in the United States to be trending?

Q4 ANSWER Coal has been decreasing in the United States and renewables have been increasing. But oil and gas have been increasing even faster, as shown in Figure 2.1. If these trends continue, the fossil energy fraction will increase further. Reversing this trend would require stronger regulations and/or a price on carbon release.

Our present civilization is dependent on the efficient extraction and distribution of energy supply among end-users. One convenient depiction of this energy system (for the case of the United States) is shown in Figure 2.2, showing estimated 2018 energy use, split out by both sources and end-uses. The figure shows, among other things, that U.S. energy supply substantially exceeds consumption, showing that we are a net energy exporter.

QUESTION 5

What fraction of domestic energy production in the United States is nuclear?

Q5 ANSWER Using the numbers on the left-hand side of Figure 2.2 which describes the sources of energy:

$$\frac{8.44 \text{ quad BTUs}}{95.70 \text{ quad BTUs}} \times 100\% = 0.0882 \times 100\% = 8.82\%$$

---

* Answers to starred questions can be found at the end of the book.

History of energy consumption in the United States (1776–2012)

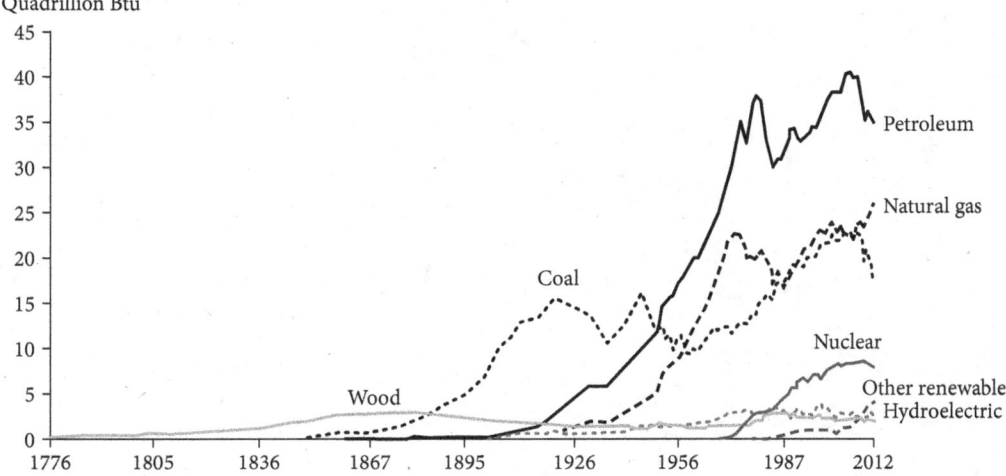

Energy production (reference case)

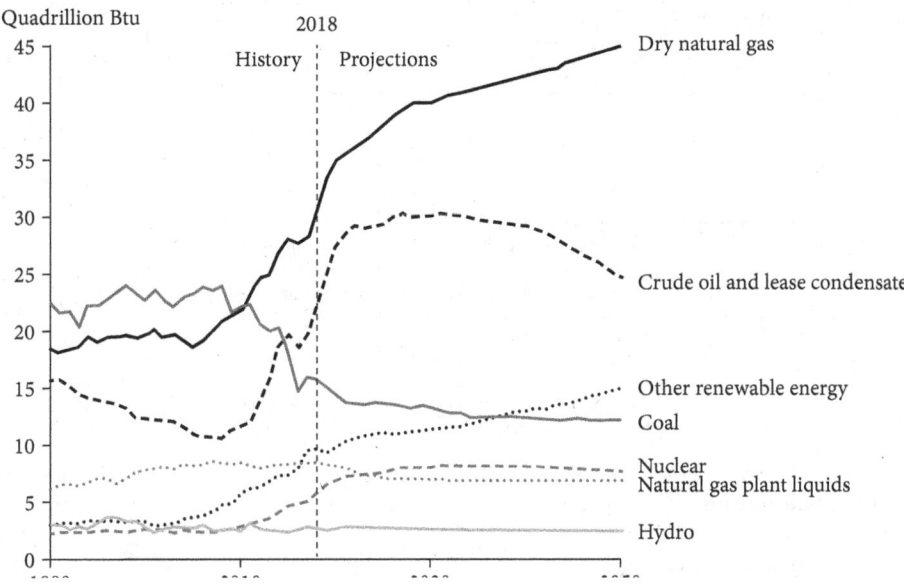

Figure 2.1 The fuel mix of energy consumption in the United States has changed dramatically over the past centuries, including a recent dramatic increase in renewable feedstocks.
Sources: The data in the plots are from the U.S. Energy Information Agency, *Energy Sources Have Changed Throughout the History of the United States,* 2013, available at https://www.eia.gov/todayinenergy/detail .php?id=11951 and *Annual Energy Outlook 2019,* 2019, available at https://www.eia.gov/pressroom /presentations/capuano_01242019.pdf

U.S. energy flow, 2018

quadrillion Btu

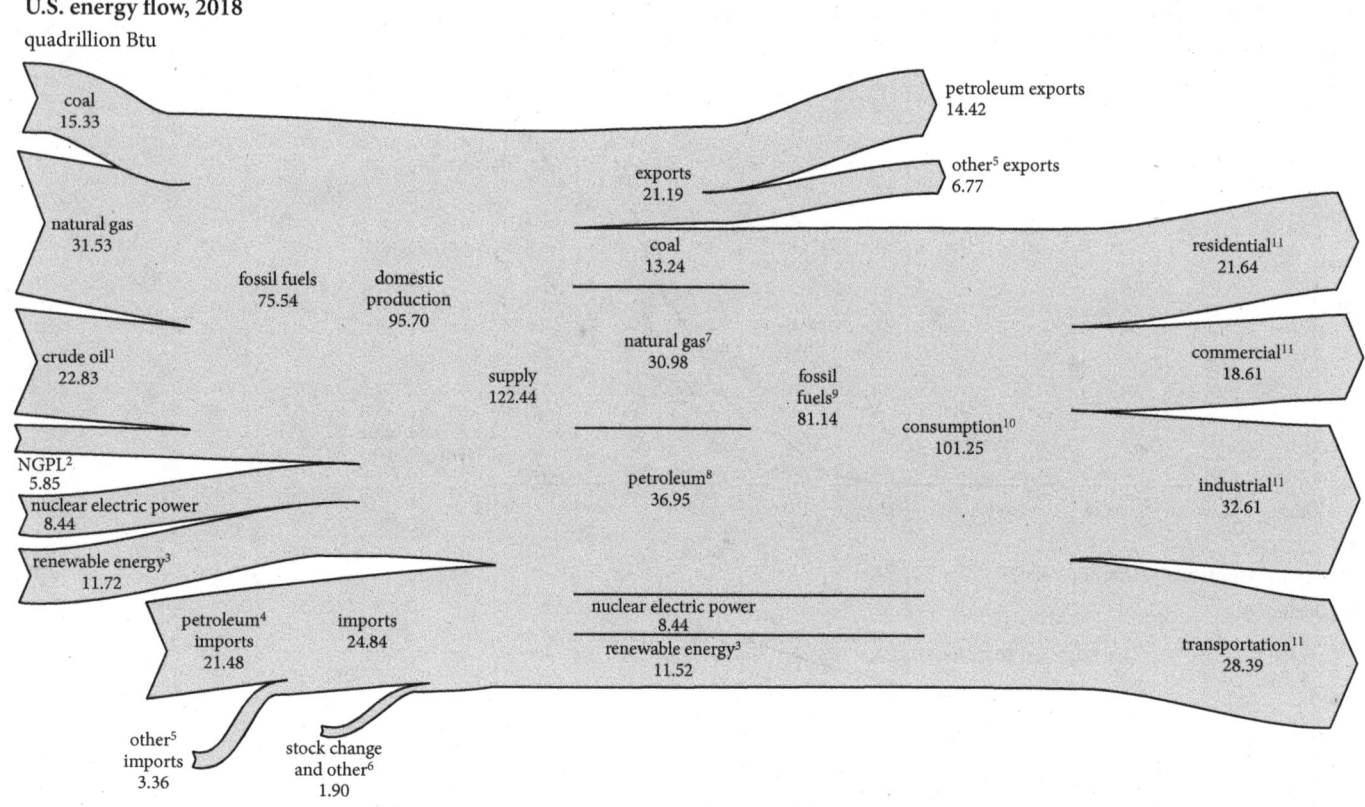

---

[1] Includes lease condensate.
[2] Natural gas plant liquids.
[3] Conventional hydroelectric power, biomass, geothermal, solar, and wind.
[4] Crude oil and petroleum products. Includes imports into the Strategic Petroleum Reserve.
[5] Natural gas, coal, coal coke, biomass, and electricity.
[6] Adjustments, losses, and unaccounted for.
[7] Natural gas only; excludes supplemental gaseous fuels.
[8] Petroleum products supplied.
[9] Includes −0.03 quadrillion Btu of coal coke net imports.

[10] Includes 0.15 quadrillion Btu of electricity net imports.
[11] Total energy consumption, which is the sum of primary energy consumption, electricity retail sales, and electrical system energy losses. Losses are allocated to the end-use sectors in proportion to each sector's share of total electricity retail sales. See Note 1, "Electrical System Energy Losses," at the end of U.S. Energy Information Administration (EIA), *Monthly Energy Review* (April 2019), Section 2.

Notes: • Data are preliminary. • Values are derived from source data prior to rounding for publication. • Totals may not equal sum of components due to independent rounding.

Sources: EIA, *Monthly Energy Review* (April 2019), Tables 1.1, 1.2, 1.3, 1.4a, 1.4b, and 2.1.

Figure 2.2 The U.S. Energy Information Association (EIA) produces this Sankey diagram annually to show the relative contributions of various energy sources and sectors of energy consumption. The units here (quadrillion BTU) are the preferred energy unit used by the U.S. EIA; see conversion factor in the notes to Table 2.1.
Source: U.S. Energy Information Association, *Total Energy,* available at https://www.eia.gov/totalenergy/data/.

This snapshot in time obscures a key element of the energy challenge: with exponentially growing population, global energy demands are increasing apace. Figure 2.1 shows U.S. energy consumption over time.

---

*QUESTION 6

What fraction of energy consumed in the United States goes to transportation?

QUESTION 7

Which source of energy was dominant in 1900? Which is dominant today? Which has grown the fastest in the last 20 yr?

Q7 ANSWER Looking at Figure 2.1, we can see that coal was dominant in 1900, but petroleum is dominant today. Natural gas has recently been the fastest-growing energy source in the United States.

*QUESTION 8

What decade saw the fastest growth in consumption of energy from coal? From petroleum? Natural gas? Nuclear?

2.2 Exponential Growth

In Chapter 1, you encountered the concept of exponential growth as it pertains to human populations and consequently to human activities such as energy consumption. Here we will grapple with exponential growth quantitatively. Exponential growth arises from a quantity that grows as a constant *percentage* rate, so that a quantity Q grows in proportion to how much is present at a given time, shown by differential equation (2.1), or integrated equation (2.2), where k is the fractional growth rate:

$$\frac{dQ}{dt} = kQ \tag{2.1}$$

$$Q = Q_0 e^{kt} \tag{2.2}$$

For exponential growth, the doubling time (time elapsed for $Q/Q_0 = 2$) is constant and equal to $0.693\ k^{-1}$.

While energy use is increasing everywhere, the annual rate of increase (and thus the doubling time) varies by country. This is shown in Figure 2.3, which outlines the projected energy growth in the fastest-growing countries of the Organization for Economic Co-operation and Development (OECD), contrasted with the faster-growing non-OECD countries.

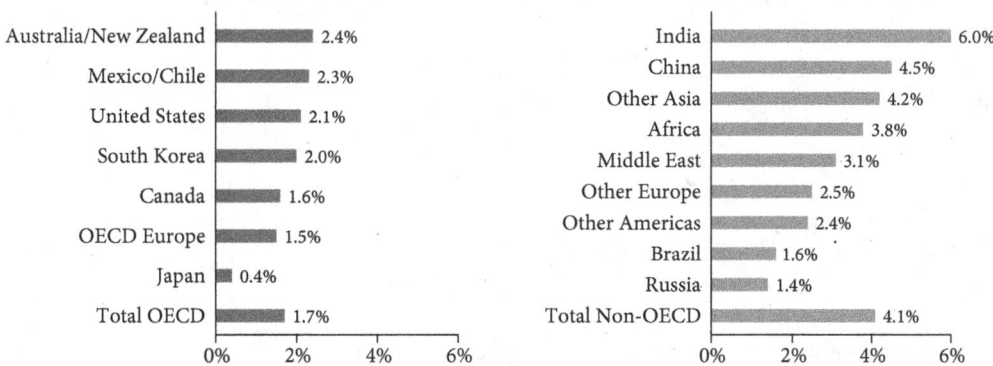

Figure 2.3 Projected energy trends (2015–2040) in selected OECD (left) and non-OECD (right) countries. *Source:* The Energy Information Administration, *International Energy Outlook 2018,* available at https://www.iea.org/reports/world-energy-outlook-2018.

QUESTION 9

In 2017, global energy demand grew 2.3%, according to the International Energy Agency. For this growth rate, what would be the doubling time of energy demand?

Q9 ANSWER Using the relationship where doubling time equals $0.693\ k^{-1}$ we find:

$$t_{double} = \frac{0.693}{0.023\ \text{yr}^{-1}} = 30\ \text{yr}$$

*QUESTION 10

If the world energy consumption were to increase at 2.8% per year, how long would it take to double?

QUESTION 11

If this rate of growth (2.8%) continued for the next 100 yr, how would human energy consumption compare with all the solar energy incident on earth?

Q11 ANSWER According to Table 2.1, human consumption of energy is currently 0.00612×10^{20} kJ yr^{-1}. We calculated the doubling time for a rate of growth of 2.8% to be 25 yr in Q10. If this rate of growth continues for 100 yr, this will have doubled four times, and thus would be 16 times larger:

$$16 \times 0.00612 \times 10^{20}\ \text{kJ yr}^{-1} = 0.0979 \times 10^{20}\ \text{kJ yr}^{-1}$$

Comparing this to total solar energy incident on earth (54.4×10^{20} kJ yr^{-1}):

$$\frac{0.0979 \times 10^{20}\ \text{kJ yr}^{-1}}{54.4 \times 10^{20}\ \text{kJ yr}^{-1}} \times 100\% = 0.00180 \times 100\% = 0.180\%$$

This is still a very small fraction!

QUESTION 12

If this rate of growth (2.8%) continued longer, starting at today's value, how long would it take to equal the solar energy incident on earth?

Q12 ANSWER For this question, we need to use the equation for exponential growth:

$$Q = Q_0 e^{kt}$$

Taking the natural logarithm, we get:

$$\ln\left(\frac{Q}{Q_0}\right) = kt$$

where $k = 0.028$ yr^{-1}, $Q = 54.4 \times 10^{20}$ kJ yr^{-1}, and $Q_0 = 0.00612 \times 10^{20}$ kJ yr^{-1}.
Solving for t:

$$t = \frac{\ln\left(\frac{Q}{Q_0}\right)}{k} = \frac{\ln\left(\frac{54.4 \times 10^{20}\ \text{kJ yr}^{-1}}{0.00612 \times 10^{20}\ \text{kJ yr}^{-1}}\right)}{0.028\ \text{yr}^{-1}} = 325\ \text{yr}$$

This takes a while!

TABLE 2.2 Selected Country Energy Use and Projected Increase

Country	Population (million)	Energy Use (kWh per capita)	Projected Annual Increase (%)
India	1,366	805	6.0
China	1,433	3,927	4.5
Australia	25.2	10,071	2.4
United States	329	12,994	2.1
South Korea	51.2	10,417	2.0
Brazil	211	2,620	1.6
Canada	37.4	15,588	1.6
Russia	146	6,603	1.4

Source: Energy Information Administration, *International Energy Outlook 2018*, available at https://www.iea.org/reports/world-energy-outlook-2018

QUESTION 13

Table 2.2 lists 2019 per capita energy use for selected countries and the current growth rate. What are the top three countries in total energy use?

Q13 ANSWER Multiplying the per capita energy use by the population shows that China was the top energy user in 2019 with 5.63×10^{12} kWh. The United States (1/4 of China's population) was second, with 4.28×10^{12} kWh, while India (almost as populous as China) was third, with 1.10×10^{12} kWh. However, India's energy use is growing fastest, followed by China.

India: $\qquad 1366 \times 10^6 \text{ people} \times \dfrac{805 \text{ kWh}}{\text{person}} = 1.10 \times 10^{12} \text{ kWh}$

China: $\qquad 1433 \times 10^6 \text{ people} \times \dfrac{3{,}927 \text{ kWh}}{\text{person}} = 5.63 \times 10^{12} \text{ kWh}$

Australia: $\qquad 25.2 \times 10^6 \text{ people} \times \dfrac{10{,}071 \text{ kWh}}{\text{person}} = 0.254 \times 10^{12} \text{ kWh}$

United States: $\qquad 329 \times 10^6 \text{ people} \times \dfrac{12{,}994 \text{ kWh}}{\text{person}} = 4.28 \times 10^{12} \text{ kWh}$

South Korea: $\qquad 51.2 \times 10^6 \text{ people} \times \dfrac{10{,}417 \text{ kWh}}{\text{person}} = 0.533 \times 10^{12} \text{ kWh}$

Brazil: $\qquad 211 \times 10^6 \text{ people} \times \dfrac{2{,}620 \text{ kWh}}{\text{person}} = 0.553 \times 10^{12} \text{ kWh}$

Canada: $\qquad 37.4 \times 10^6 \text{ people} \times \dfrac{15{,}558 \text{ kWh}}{\text{person}} = 0.582 \times 10^{12} \text{ kWh}$

Russia: $\qquad 146 \times 10^6 \text{ people} \times \dfrac{6{,}603 \text{ kWh}}{\text{person}} = 0.964 \times 10^{12} \text{ kWh}$

QUESTION 14

If the current growth rates were to continue, how long would it take for China's per capita energy use to equal that of the United States?

Q14 ANSWER Continued constant percentage rate leads to exponential growth:

$$Q = Q_0 e^{kt}$$

If Q_1 and Q_2 are per capita energy use for the United States and China, respectively, then when they are equal:

$$Q_1 e^{k_1 t} = Q_2 e^{k_2 t}$$

$$\frac{Q_1}{Q_2} = \frac{e^{k_2 t}}{e^{k_1 t}} = e^{(k_2 - k_1)t}$$

$$Q_1 - \ln Q_2 = (k_2 - k_1)t$$

$$t = \frac{\ln Q_1 - \ln Q_2}{k_2 - k_1} = \frac{\ln 12994 - \ln 3927}{0.045 - 0.021} = 49.9 \text{ yr}$$

China would catch up to the United States in per capita energy use in 2069.

*QUESTION 15

How long would it take India to catch up to the United States?

QUESTION 16

The growth in energy consumption has diminished in the west, partly due to the shift of manufacturing overseas, where it contributes to other countries' energy consumption. If you were working for the Energy Information Agency and tasked with producing a ranking of energy use by country, but attributing the energy consumption to the end-user of all goods, what additional data would you seek to use? How would you define the ranking?

Q16 ANSWER Each country's *effective gross* energy consumption would be adjusted downward according to the energy embedded in exported goods, and upward according to the energy embedded in imported goods. Estimating these "embedded energy" values would be difficult, but both could possibly be estimated by determining the average energy intensity of traded goods and using this as a multiplicative factor on the trade volumes in dollars.

Exponential growth applies to areas other than energy. For example, during an early age of rapid agricultural expansion (1810–1850), the farmed area of the planet increased from 1.43 billion hectares (ha) to 1.78 billion ha, or on average, 0.6% per year.

Agricultural area over the long term

Total areal land use for agriculture, measured as the combination of land for arable farming (cropland) and grazing in hectares

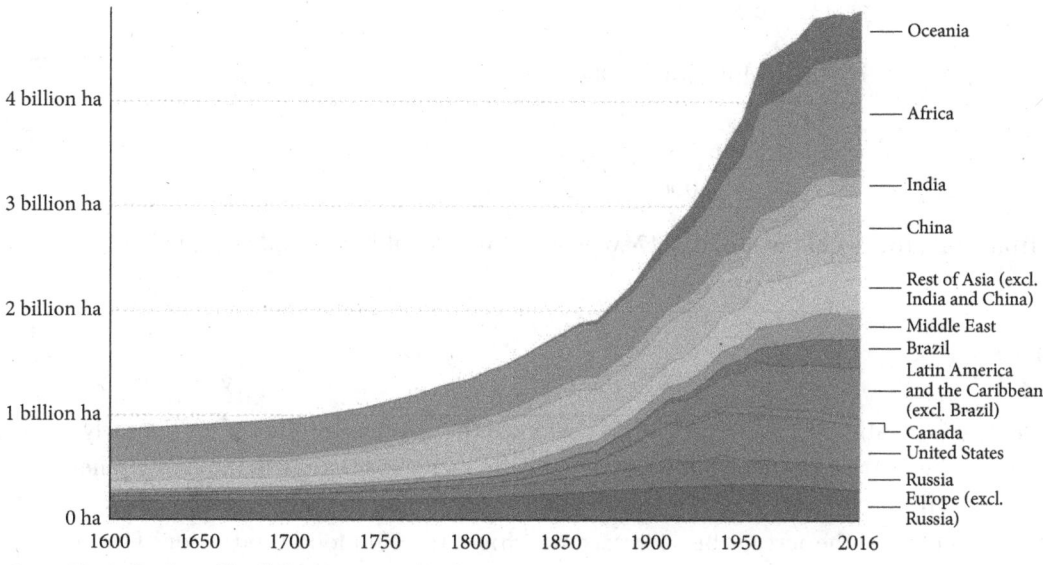

Source: History Database of the Global Environment (2017)

Figure 2.4 The farmed area of the planet increased rapidly beginning in about 1800.
Source: Adapted from *Our World in Data: Crop Yields,* available at https://ourworldindata.org/yields-and-land-use-in-agriculture.

*QUESTION 17

At this expansion rate, how long would it take to double from the area farmed in 1810, 1.43 billion ha? Is your answer consistent with Figure 2.4 in the years after 1850?

QUESTION 18

If the rate of agricultural expansion referenced in Q17 were to continue to exponentially grow, how long would it take for all of the habitable land areas of the earth to be covered by crops? The earth's total land area is 150 million km^2, of which about 70% is designated as habitable; 1 km^2 is equal to 100 ha.

Q18 ANSWER Use $Q = Q_0 e^{kt}$ to solve for t.
70% of 150 million km^2 gives 105 million km^2 habitable land:

$$0.70 \times 150 \times 10^6 \ km^2 \times \frac{100 \ ha}{km^2} = 1.05 \times 10^{10} \ ha$$

$$Q = Q_0 e^{kt}$$

$$1.05 \times 10^{10} \ ha = 1.43 \times 10^9 \ ha \ e^{(0.006)t}$$

$$\frac{1.05 \times 10^{10} \text{ ha}}{1.43 \times 10^9 \text{ ha}} = 7.34 = e^{(0.006)t}$$

$$\ln(7.34) = 0.006t$$

$$t = \frac{\ln(7.34)}{0.006} = 332 \text{ yr}$$

Starting from 1810, this would mean in 2142 we would have all habitable land occupied by agriculture.

QUESTION 19

Crop productivity has greatly increased since 1850, thanks to the "Green Revolution" of the early twentieth century, when agricultural output was dramatically increased thanks to the development of crop breeding and the use of fertilizer. If the graph in Figure 2.5 were representative of crops worldwide, how much could the acreage be decreased to achieve the 2014 food production? Is there any evidence of this effect in Figure 2.4?

Q19 ANSWER Crop productivity was 3–4 times higher in 2014 than in 1850, according to Figure 2.5, suggesting that the 2014 acreage requirement was 3–4 times lower than it would otherwise have been. In fact, the acreage growth curve does bend over (Fig. 2.4) starting around 1980, and acreage was substantially lower in 2016 than would be predicted from the earlier exponential growth.

Long-term cereal yields in the United Kingdom

Average agricultural yields in key crops in the United Kingdom from 1270–2014, measured in tons per hectare.

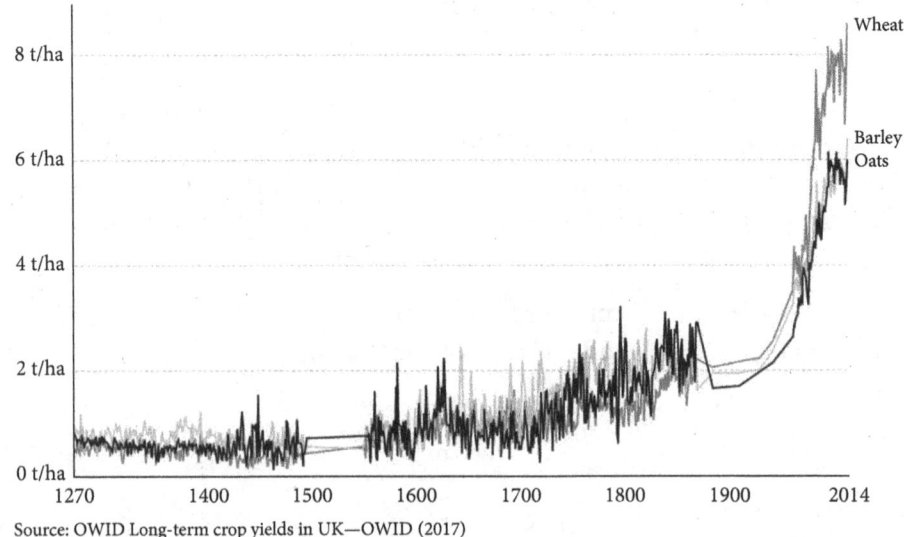

Source: OWID Long-term crop yields in UK—OWID (2017)

Figure 2.5 These data for the United Kingdom are representative of crops worldwide, where yields dramatically increased with the advent of fertilized usage in the "Green Revolution" of the early twentieth century.
Source: Adapted from *Our World in Data,* available at https://ourworldindata.org/grapher/long-term-cereal-yields-in-the-united-kingdom.

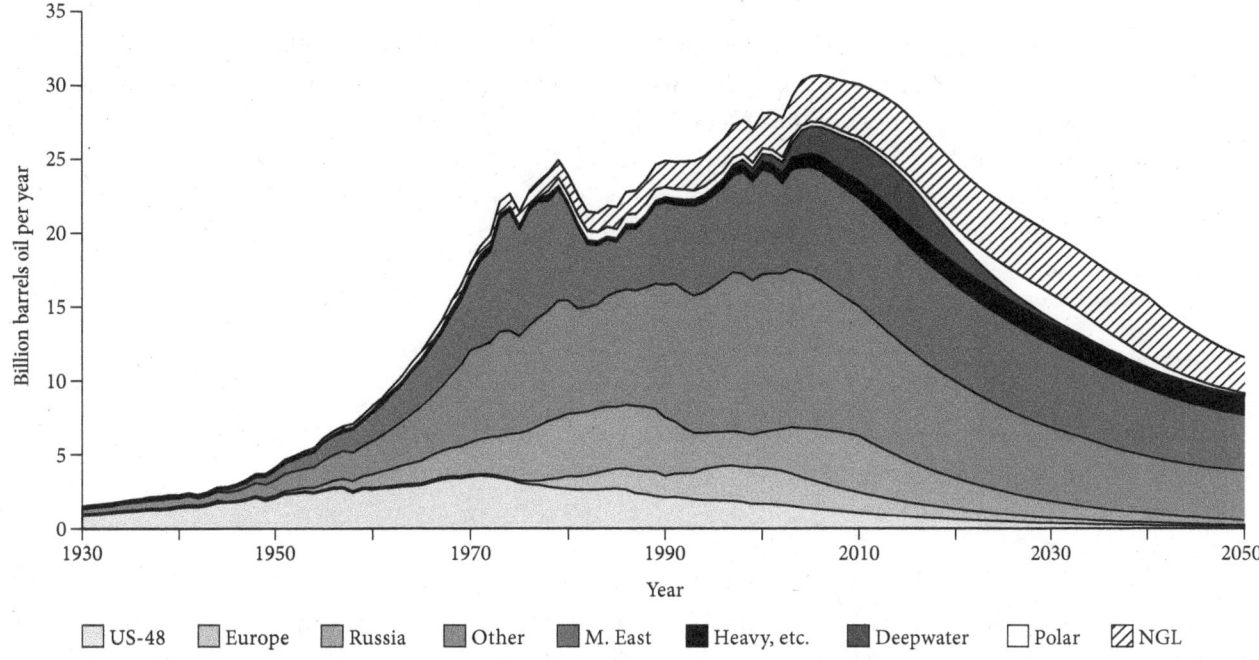

Figure 2.6 A prediction of global peak oil production in billion barrels of oil per year. "Other" denotes oil shales and coal sources; "heavy" includes bitumen and heavy oil; "deepwater" denotes oil in the water of a depth of more than 500m; "NGL" represents liquids from natural gas plants and gas fields.
Source: Adapted from Association for the Study of Peak Oil (ASPO) Newsletter, 2004, and from Spiro et al., *Chemistry of the Environment, 3rd Ed.,* University Science Books, © 2012, all rights reserved.

The exponential rise cannot go on forever; it is constrained by some sort of limit (e.g., arable land for agricultural acreage). This is just as true for fossil energy resources as for agriculture. As we will see in Chapter 5, several fossil sources of energy have been shown to "peak," meaning their extraction rate decreased after reaching a maximum, as described by the differential equation below:

$$\frac{dQ}{dt} = (2\pi)^{-\frac{1}{2}} Q_{\text{lim}} e^{-\frac{(t-t_1)^2}{2}} \tag{2.3}$$

where Q_{lim} is the limiting (maximum) rate of use and t_1 is the time it takes to reach the maximum extraction rate for an exhaustible resource. After t_1, the rate declines, forming a bell-shaped curve that has been observed in resource extraction rates.

For example, oil extraction in the contiguous 48 states of the United States peaked in about 1970 at about 4 billion barrels (bbl) of oil per year (see Fig. 2.6).

QUESTION 20

U.S. oil extraction began in earnest in ~1920. Use Figure 2.6 to estimate t_1 for US-48 (recall, this will be the *length* of time since extraction began), and then calculate what Q_{lim} this would correspond to. Can you find U.S. oil reserve data to check your estimate against?

Q20 ANSWER Estimating t_1 from the graph, ~50 yr (from the start of oil exploration in 1920 until the US-48 peak in 1970). At the peak of the curve, where $t = t_1$, dQ/dt is the amount of oil produced in US-48. Reading off the peak of the US-48 shaded area, $dQ/dt = 4$ bbl yr^{-1}, and because at this point in time (where $t = t_1$) the exponential term is $e^0 = 1$, we have:

$$\frac{dQ}{dt} = (2\pi)^{-\frac{1}{2}} Q_{\text{lim}} = 4 \text{ bbl yr}^{-1}$$

Which we can solve for Q_{lim}; $Q_{\text{lim}} = 10$ bbl of oil per year.

*QUESTION 21

Conduct a similar analysis for Russia, where oil exploration began ~1930. (It looks more double-peaked, but use the first peak for this calculation.)

2.3 Human Energy Sources

As you calculated in question Q3, energy consumption is presently heavily dependent on fossil fuels (oil, gas, and coal). Not only do these fossil sources produce CO_2 emissions that cause anthropogenic climate change, but they are non-renewable sources that will eventually run out. By examining data on our best estimates of resource bases and usage rates, we can estimate how long each of these finite fossil energy sources will last.

The data from Figure 2.6 were taken in 2004 before hydraulic fracking technology was developed. After drilling into the earth, fracking uses high-pressure chemicals, water, and sand aimed at rock to release the natural gas inside. This technology allows access to reservoirs of natural gas that were not available previously and demonstrates how new technologies can change predictions in available resources. After 2004, natural gas was more abundant and a less expensive source of energy.

QUESTION 22

According to the EIA, as of January 2018, the demonstrated U.S. reserve base of coal was 475 billion short tons, larger than natural gas and oil resources in terms of British Thermal Units (BTUs) production potential. Annual U.S. coal production in 2017 was 775 million short tons.[1] Based on this information, will the United States have run out of coal during your lifetime?

Q22 ANSWER No, unless you are very ambitious about the length of your life! There will be no peak coal for quite some time. If we divide the reserve base by the annual production, we can estimate the time until the resource runs out at the current rate:

$$\frac{475 \times 10^9 \text{ short tons}}{775 \times 10^6 \text{ short tons per year}} = 613 \text{ yr}$$

However, the coal production rate is dropping (Fig. 2.2). Coal will likely continue to be replaced in power plants by gas and by renewables. Some of the reserves may remain in the ground.

---

[1] U.S. Energy Information Administration, *Annual Coal Report 2018,* available at https://www.eia.gov /coal/annual/.

*QUESTION 23

Energy growth rates. The EIA regularly predicts future energy consumption. One recent projection estimated that world petroleum consumption will increase from 197 quadrillion BTUs in 2018 to 229 quadrillion BTUs in 2040. Assuming exponential growth this period 2005–2030, what average growth rate does this imply for total energy use?

QUESTION 24

According to the EIA, there were about 2,459 trillion cubic feet (Tcf) of technically recoverable resources (TRRs) of dry natural gas in the United States in 2018. Using the 2018 annual rate of U.S. natural gas production of 30.4 Tcf, how long with the U.S. natural gas TRR last?

Q24 ANSWER Dividing the recoverable resource by the annual production, the TRR will last:

$$\frac{2{,}459 \text{ Tcf}}{30.4 \text{ Tcf yr}^{-1}} = 81.7 \text{ yr}$$

The United States has enough dry natural gas to last about 80 yr at the current production rate. Renewables are starting to compete with gas, as the costs of wind and solar energy continue to decrease, and as restrictions on carbon emissions are strengthened.

*QUESTION 25

If the TRR were to increase to 3,000 Tcf, and the annual rate of production were to be reduced to 20 Tcf yr^{-1}, in response to the increasing competitiveness of alternate renewable energy sources, how long would the natural gas reserve last?

2.4 Conclusions

The amount of energy that humans *consume* is a small fraction of total energy that is *supplied* to the planet in the form of solar radiation, but human energy use has constantly increased over human history. The energy resource mix used to generate electricity and fuel transportation is constantly changing, but the sum total increases with the exponential growth of the human population. Economies of scale mean that larger countries are typically more energy efficient. Fossil energy resources are by their nature finite and nonrenewable, and the production of these feedstocks will peak.

3

ENERGY UTILIZATION

3.1 Energy Utilization

How we use energy is even more important than how we produce it. Because of inefficiencies in energy production, 1 kJ of energy saved is worth more than 1 kJ produced. Energy can be saved in many ways, from its production to its end uses. In this chapter, we consider limits on the efficiency of energy production, and some opportunities for conservation.

3.1.1 Energy Conversions and Efficiency

The many forms of energy can be interconverted to produce useful work. However, some of the energy is inevitably converted to waste heat, which is lost to the environment. The efficiency of the conversion is the fraction of the input energy that results in useful work.

We first consider energy efficiency in the context of gasoline and electric cars. A gasoline car is powered by burning the fuel; its power is limited by the inherent inefficiency of heat engines (see Section 3.2) and by additional losses in the mechanical system. Typically, the fraction of fuel energy delivered to the drive train, which transfers power from the transmission to the wheels (taking account of the energy needed to keep the engine running), is only about 20%. Then the drive train has to deliver the energy to the wheels to keep the car moving forward through a gearing system, losing another 10% of the energy in the process.

QUESTION 1

What is the overall fuel efficiency of the gasoline car?

Q1 ANSWER The individual conversion efficiencies multiply. If the first and second stages have efficiencies of e_1 and e_2, then the overall efficiency, $e_T = e_1 \times e_2$

Since the drive train loses 10% of the energy, 90% is left, so the overall efficiency is:

$$(0.20 \times 0.90) \times 100\% = 18\%$$

On the other hand, an electric motor is highly efficient, delivering about 90% of the energy directly to the wheels. However, the battery loses about 10% of its energy through leakage, and another 10% is lost during charging and discharging the battery.

---

*QUESTION 2

What is the overall electrical efficiency of the electric car?

---

* Answers to starred questions can be found at the end of the book.

The electricity used to power the electric car has to be generated somewhere, usually from coal, gas, or nuclear energy. Fuel to electricity energy efficiencies of power plants are about 40%. About 10% of the energy is lost in transmitting it to the car charger.

---

*QUESTION 3

What is the overall fuel efficiency of the electric car with electricity from fossil fuels?

Next, we consider how best to convert the energy of fuels into electricity, an issue central to modern society.

3.2 Heat into Energy: Entropy

The highest efficiency attainable in energy conversion processes is set by the *second law of thermodynamics*. (The *first law* is simply a statement of energy conservation: energy is neither destroyed nor created, provided we remember that heat is a form of energy.) The second law of thermodynamics states that the *entropy* always increases in a spontaneous process. Entropy is a measure of disorder among molecules, and there is a close connection between entropy and heat. Heat arises from the motion of atoms and molecules; its quantity, Q, is their total kinetic energy, while the temperature, T, is their average kinetic energy. Entropy is defined as:

$$S = \frac{Q}{T} \tag{3.1}$$

T is the absolute temperature, in Kelvin: K = °C + 273. A given quantity of heat produces greater disorder of the molecules at low rather than at high temperatures.

QUESTION 4

Recalling that it takes 1 calorie (cal) [4.18 joules (J)] to raise the temperature of 1 milliliter (ml) of water by 1°C (or K), how much heat is required to heat a 100. mL cup of tea from room temperature (25°C) to boiling (100.°C)? How much entropy is gained when the tea cools back down to room temperature? (The volume of the tea, 100. mL, and the boiling temperature, 100.°C, includes a period after the number to show that all three digits are significant. This allows us to have three places in our final answer below.)

Q4 ANSWER $Q = 100.\ \mathrm{mL} \times ((100. - 25)°C + 273) \times 1\dfrac{\mathrm{cal}}{\mathrm{mL\ K}} = 3.48 \times 10^4\ \mathrm{cal}$:

$$\Delta S = S_2 - S_1 = \frac{Q}{T_2} - \frac{Q}{T_1} = \frac{3.48 \times 10^4\ \mathrm{cal}}{298\ \mathrm{K}} - \frac{3.48 \times 10^4\ \mathrm{cal}}{373\ \mathrm{K}} = 23.5\frac{\mathrm{cal}}{\mathrm{K}}$$

The entropy is gained by the tea and its surroundings.

Heat can be converted to useful energy, as in an internal combustion engine, or a steam or gas turbine, but the fraction of the heat that can be converted is limited by the second law, since some entropy is always gained. The total heat is divided into work, W, and the heat lost to the surroundings:

$$Q_{\text{total}} = W + Q_{\text{loss}} \qquad (3.2)$$

and since entropy must increase, it follows that:

$$\frac{Q_{\text{loss}}}{T_s} > \frac{Q_{\text{total}}}{T_h} \qquad (3.3)$$

where T_s and T_h are the temperatures of the surroundings and the heat source, respectively.

QUESTION 5

Show that combining these two equations (3.2) and (3.3) and rearranging terms leads to this formula for the ratio of the work extracted to the heat input:

$$\frac{W}{Q_{\text{total}}} < 1 - \frac{T_s}{T_h} \qquad (3.4)$$

Q5 ANSWER We start with the relationship between work and heat from (3.2):

$$W = Q_{\text{total}} - Q_{\text{loss}}$$

and rearrange the relationship by dividing through by the Q_{total}:

$$\frac{W}{Q_{\text{total}}} = 1 - \frac{Q_{\text{loss}}}{Q_{\text{total}}} \qquad (3.5)$$

Using (3.3), $\dfrac{Q_{\text{loss}}}{T_s} > \dfrac{Q_{\text{total}}}{T_h}$, we multiply both sides by $\dfrac{T_s}{Q_{\text{total}}}$ to bring the temperatures to the same side. This gives:

$$\frac{Q_{\text{loss}}}{Q_{\text{total}}} > \frac{T_s}{T_h}$$

When we substitute this relationship into (3.5), we yield (3.4):

$$\frac{W}{Q_{\text{total}}} < 1 - \frac{T_s}{T_h}$$

This formula tells us that the maximum theoretical efficiency of a heat engine is $\dfrac{W_{\text{max}}}{Q_{\text{total}}} = 1 - \dfrac{T_s}{T_h}$

The colder the surroundings and the hotter the heat source, the greater is the efficiency.

Consider a coal-fired steam power plant, in which the burning coal heats a boiler whose steam drives an electricity generator, and is then condensed with cooling water (Fig. 3.1).

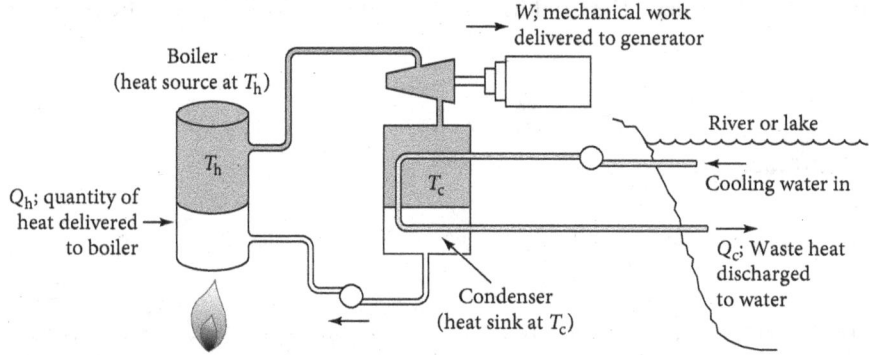

Figure 3.1 Maximum work and waste heat from a steam engine power plant.
Source: Adapted from Spiro et al., *Chemistry of the Environment, 3rd Ed.*, University Science Books, © 2012, all rights reserved.

QUESTION 6

If the temperatures of the boiler-produced steam (at high pressure) and the cooling water are 550°C and 27°C, respectively, what is the maximum efficiency of the power plant?

Q6 ANSWER To convert T from °C to K add 273:

$$T_s = 27°C + 273 = 300.K$$

$$T_h = 550°C + 273 = 823K$$

Then

$$\frac{W_{mass}}{Q_{total}} = 1 - \frac{300.K}{823K} = 0.635 \text{ or } 63.5\%$$

Of course, there are other heat losses than to the cooling water, from the boiler and from the various machinery parts. Heat recovery engineering can reduce but not eliminate these. The actual efficiency is always less than the theoretical efficiency. In a modern coal-steam plant it is about 40%.

In principle, a gas generator, in which the hot gases from a flame directly propel a turbine (as in a jet engine), can be more efficient, since the hot gas can achieve a higher temperature than pressurized steam. In current turbines, a temperature of 1260°C can be reached. But the temperature of the exhaust gas is also high, typically 500°C, so the maximum efficiency, 1 − [(500°C + 273)/(1,260°C + 273)] = 0.496, 49.6%, and actually ~33% in practice. However, the exhaust gas can be used to heat a steam boiler in a "combined cycle" power plant (Fig. 3.2), for higher overall fuel efficiency. Many coal-fired power plants have been retrofitted with a natural gas-fueled "topping cycle."

---

*QUESTION 7

Show that a combined-cycle plant can have a theoretical efficiency of 80%.

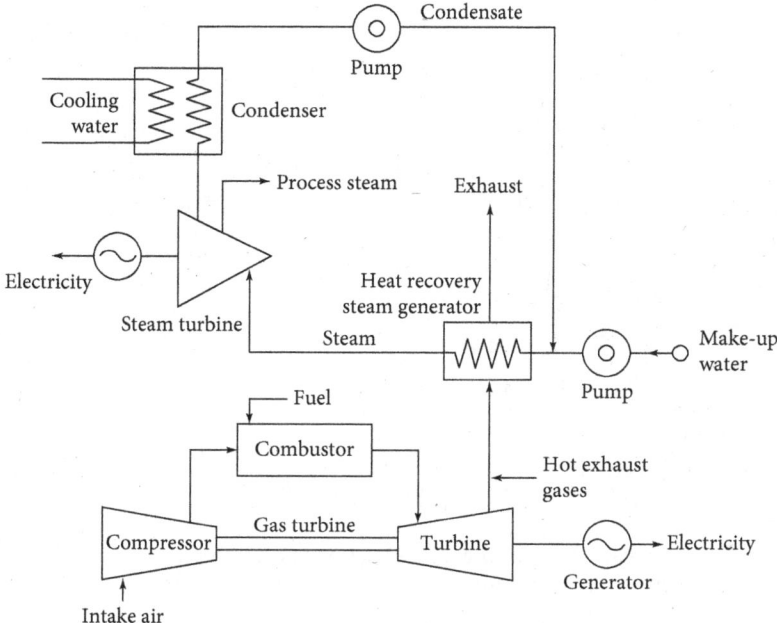

Figure 3.2 Schematic of gas/steam turbine combined-cycle power plant.
Sources: Williams & Larson, "Advanced Gasification-based Biomass Power Generation," in *Renewable Energy, Sources for Fuels and Electricity*, Johansson (Ed.), Island Press, 1993, pp. 729–785. Adapted from Spiro et al., *Chemistry of the Environment, 3rd Ed.*, University Science Books, © 2012, all rights reserved.

The second law of thermodynamics also offers guidance on how best to use electricity to provide heat, when that is required.

3.3 Energy to Heat: Heat Pumps

Converting electricity to heat through resistive heating is wasteful, since so much of the fuel energy has been lost to the environment at the power plant. A more efficient use of the electricity is to run a heat engine in reverse, using heat from the environment to evaporate a working liquid, and extracting the heat by using mechanical work to condense it (Fig. 3.3). The second law still applies, and the maximum conversion is given by the inverse of (3.6).

$$\frac{Q_{max}}{W} = \frac{T_h}{(T_h - T_c)} \tag{3.6}$$

where T_c and T_h are the temperatures of the environment and the temperature of the delivered heat.

QUESTION 8

If the indoor and outdoor air temperature is 27°C and 7°C, what is the theoretical efficiency of the required heat pump?

Q8 ANSWER $\dfrac{Q_{max}}{W} = \dfrac{(27°C + 273)}{[(27°C + 273) - (7°C + 273)]} = 15.0 \text{ or } 1{,}500\%$

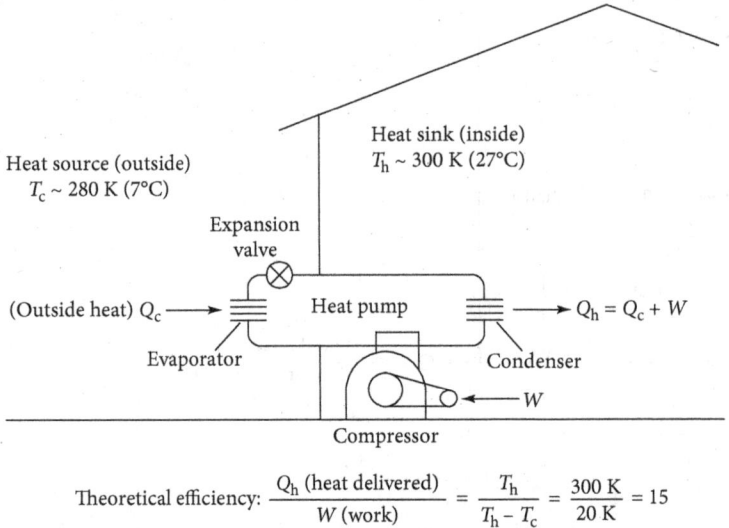

Theoretical efficiency: $\dfrac{Q_h \text{ (heat delivered)}}{W \text{ (work)}} = \dfrac{T_h}{T_h - T_c} = \dfrac{300 \text{ K}}{20 \text{ K}} = 15$

Figure 3.3 The heat pump: an efficient means of residential space heating and cooling.
Source: Adapted from Spiro et al., *Chemistry of the Environment, 3rd Ed.*, University Science Books, © 2012, all rights reserved.

In principle, 15 cal of (low grade) heat could be extracted from the cold outside air by 1 cal of (high grade) work. Of course, such a high efficiency cannot be practically achieved, but ratios of 3–5 are attainable, much better than the ratio of 1 from a resistive heater.

The efficiency degrades when the weather is cold.

---

*QUESTION 9

What is the theoretical efficiency of an ambient air heat pump if the indoor air is kept at 27°C, but the outdoor air cools to −10°C?

Higher efficiency can be obtained by circulating the working liquid through a reservoir in the ground, which stores heat from the summer sun. Below 6m, the earth's crust maintains a constant temperature, corresponding to the average yearly temperature year-round, typically ~10°C–15°C in temperate climates.

---

*QUESTION 10

What is the theoretical efficiency if the ambient air heat pump in Q9 is replaced by a ground effect (sometimes called "geothermal") heat pump operating at 10°C?

An air conditioner is just a heat pump operating in reverse. Ground effect heat pumps can be reversed in summer, again with improved efficiency because of the constant ground temperature.

It would be better not to use electricity for heating at all, but to use the heat that is otherwise wasted in the initial generation of the electricity.

3.4 Cogeneration

Energy can be saved if the waste heat from a power plant is used to provide heat for industry, institutions, or residential housing, instead of dumping it directly into the environment. Cogeneration plants, producing both electricity and steam heat, are in operation at hospitals, universities, and industrial installations. They also provide a number of communities with "district heating," with steam pipes carrying heat to a cluster of buildings. Of course, it is harder to transmit heat than electricity over long distances. Also installing steam pipes is expensive, and the upfront capital costs can be daunting. Nevertheless, district heating is being widely employed, especially in Northern Europe. Half or more of houses are heated this way in Denmark, Sweden, Finland, Estonia, Poland, and the Czech Republic (the figure is 98% in Iceland, which has exploited its abundant hydrothermal sources for the purpose).[1]

QUESTION 11

According to the U.S. Department of Energy, the average U.S. house uses 100 million British thermal units (BTUs are still used by the heating industry) annually. How many such houses could be heated by the steam coolant of a 1 megawatt (MW) thermal electric generation plant, operating at 40% fuel conversion efficiency?

Q11 ANSWER 1 BTU is equivalent to 1.055 kJ, so 100 million BTU:

$$1.00 \times 10^8 \text{ BTU} \times \frac{1{,}055 \text{ kJ}}{1 \text{ BTU}} \times \frac{1{,}000 \text{ J}}{\text{kJ}} = 1.055 \times 10^{11} \text{ J}$$

At 40% efficiency, a 1-MW electric plant requires:

$$\frac{1 \text{ MW}}{0.40} = 2.5 \text{ MW of fuel energy}$$

If we assume that the energy loss of 60% is entirely transferred to the steam coolant, then heat production is 0.6×2.5 MW = 1.5 MW.

Since 1 W = 1 J sec^{-1} and a year lasts:

$$\frac{60 \text{ sec}}{\text{min}} \times \frac{60 \text{ min}}{\text{hr}} \times \frac{24 \text{ hr}}{\text{day}} \times \frac{365 \text{ days}}{\text{yr}} = 3.15 \times 10^7 \text{ sec}$$

the annual heat provided is:

$$1.5 \text{ MW} \times \frac{(1 \times 10^6 \text{ W})}{\text{MW}} \times (3.15 \times 10^7 \text{ sec}) \times \frac{1\frac{\text{J}}{\text{sec}}}{1 \text{ W}} = 4.73 \times 10^{13} \text{ J}$$

The ratio of the available heat to the heat required per house is:

$$\frac{4.73 \times 10^{13} \text{ J}}{1.055 \times 10^{11} \text{ J}} = 448 \text{ houses}$$

---

[1] Calcea, *Europe Makes Progress Integrating Renewables in District Heating,* 2020, available at https://www.energymonitor.ai/sectors/heating-cooling/europe-makes-progress-integrating-renewables-in-district-heating.

Of course, the number will be reduced by heat losses from the steam pipes (~10% depending on design and distances covered), and on the times the plant operates at less than full capacity (the "capacity factor" is typically 75% for coal plants but 98% for nuclear plants, which also tend to be much larger).

Next, we turn to a different approach to harnessing the energy in fuels, through electrochemistry.

3.5 Electricity from Chemical Energy: Fuel Cells

Instead of combustion, the energy in fuels can, in principle, be converted directly to electricity, thereby avoiding the limitations of heat engines. Electrons from the fuel can be passed to oxygen through an electric circuit, using a fuel cell. An example is shown in Figure 3.4, in which the fuel is hydrogen (the type of cell currently being developed for transportation). The overall reaction:

$$2\,H_2 + O_2 \rightarrow 2\,H_2O \tag{3.7}$$

is carried out as two half-reactions:

$$2\,H_2 \rightarrow 4\,H^+ + 4\,e^- \tag{3.8}$$

$$4\,e^- + O_2 + 4\,H^+ \rightarrow 2\,H_2O \tag{3.9}$$

Electrons are transferred from H_2 at one electrode (the cathode) and, after traveling through the circuit, are transferred to O_2 at the other electrode (the anode). Protons produced in the first half-reaction diffuse from the cathode through a membrane that facilitates proton

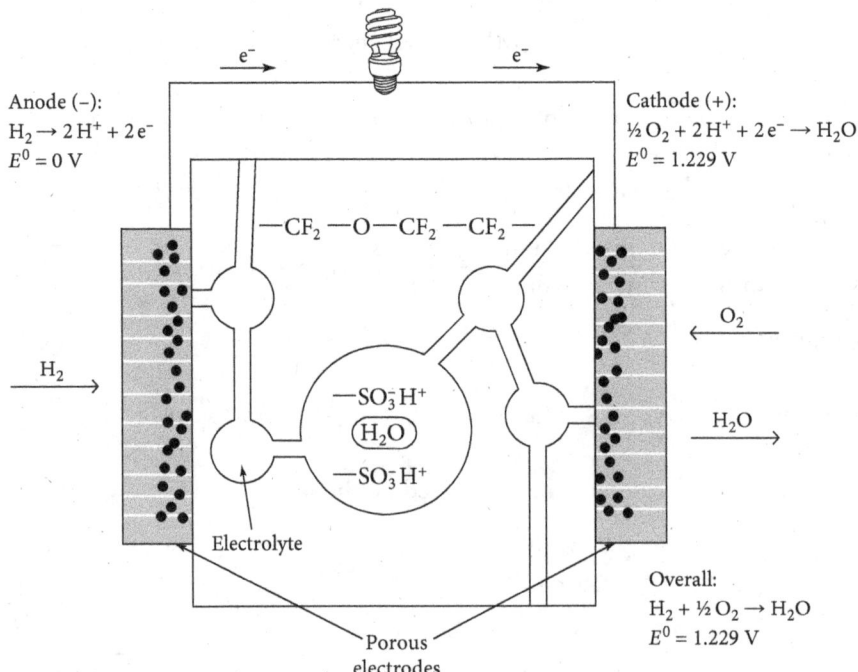

Figure 3.4 Schematic of a proton exchange membrane (PEM) fuel cell using hydrogen as a fuel. *Source:* Adapted from Spiro et al., *Chemistry of the Environment, 3rd Ed.,* University Science Books, © 2012, all rights reserved.

transfer (in this case a fluorocarbon polymer having pores lined with anionic sulfate groups) to the anode, where they are taken up in the second half-reaction.

The energy associated with the overall reaction is -241.8 kJ mol^{-1} of H_2O produced, if the water is in the gas phase, and -285.7 kJ/mol if the water is liquid, the extra energy being released upon condensation of the water. (By convention, the sign is negative when energy is released because the reactant energy is subtracted from the product energy.) All of this energy is converted to heat in combustion, but only a fraction is available to the electrons in the fuel cell because the reaction decreases entropy. Order increases because three molecules are converted to only two, and more order is created when gaseous molecules are confined to the liquid phase. The energy equivalent to the gain in entropy ($T\Delta S$) is subtracted from the total energy change (ΔH), leaving the remainder, the free energy change (ΔG), available to the electrons:

$$\Delta G = \Delta H - T\Delta S \tag{3.10}$$

Thus, the maximum theoretical efficiency of the fuel cell is:

$$\frac{\Delta G}{\Delta H} = 1 - \frac{T\Delta S}{\Delta H} \tag{3.11}$$

QUESTION 12

The free energy change for reaction (3.8), ΔG, has been determined at 25°C (298K) to be -228.6 and -237.2 kJ mol^{-1} when the water is in the gas and liquid states. Calculate the theoretical efficiency of a fuel cell operating at 25°C.

Q12 ANSWER At 25°C water is liquid, so:

$$T\Delta S = \Delta H - \Delta G = \left(-285.7\,\frac{kJ}{mol}\right) - \left(-237.2\,\frac{kJ}{mol}\right) = -48.5\,\frac{kJ}{mol}$$

$$1 - \frac{T\Delta S}{\Delta H} = 1 - \left(\frac{-48.5\,\frac{kJ}{mol}}{-285.7\,\frac{kJ}{mol}}\right) = 0.830 \text{ or } 83\%$$

---

*QUESTION 13

From the data in Q12, calculate the theoretical efficiency of a fuel cell operating at 100°C.

The free energy change determines the cell potential, in volts (V), through the equation:

$$\Delta G° = -nFE° \tag{3.12}$$

(The superscript ° signifies standard conditions: 25°C, and reactants and products in their standard states; in this case, 1 atm pressure for the gases H_2 and O_2),

F (the Faraday) = the amount of charge in a mole of electrons [96,500 coulombs (C), or 96.5 with the free energy expressed in kJ, since 1 V = 1 J/C]

n = the number of electrons transferred in the reaction

$E°$ = standard potential of the cell in volts

QUESTION 14

Compute the standard potential of the hydrogen fuel cell.

Q14 ANSWER To determine n, we note that both half-reactions as written above [Eqs. (3.8) and (3.9)] involve four electrons, but two molecules of H_2O are produced, so moles of electrons per mole of H_2O = 2. Since the standard potential is for 25°C, we choose the liquid water free energy, $\Delta G° = -237.2$ kJ mol^{-1} for liquid water:

$$E° = \frac{-\Delta G°}{nF} = -\left(\frac{-237.2 \frac{kJ}{mol}}{2\left(95.6 \frac{kJ}{mol\,V}\right)} \right) = 1.24V$$

$E°$ can also be obtained from tabulations of standard *reduction* potentials, like Table 3.1.

QUESTION 15

Calculate the hydrogen fuel cell standard potential from the data entries in Table 3.1 and using (3.8) and (3.9).

Q15 ANSWER The half-reactions and potentials we need are:

$2H^+ + 2e^- = H_2$ $E° = 0.00V$ (the H^+ reduction potential is set to zero, in order to set a scale for the other reduction potentials.)

$$O_2 + 4H^+ + 4e^- = 2H_2O \qquad E° = 1.23V$$

In the fuel cell, hydrogen gas is being oxidized, so we multiply the first reaction by 2, to equalize the number of electrons ($4e^-$). The equation is also reversed because it is being oxidized (Table 3.1 provides the potentials for reduction equations), thus altering the sign of $E°$ (when the half-reaction is flipped, the sign of the potential is the opposite as well). The two chemical equations are added together as the reduction potentials:

$$
\begin{array}{lll}
& 2H_2 = 4H^+ + 4e^- & E° = -0.00V \\
+ & O_2 + 4H^+ + 4e^- = 2H_2O & E° = 1.23V \\
\hline
& 2H_2 + O_2 \rightarrow 2H_2O & E° = 1.23V
\end{array}
$$

***QUESTION 16**

Devise an electrochemical cell based on the $Fe^{3+/2+}$ and MnO_4^-/Mn^{2+} half-reactions in Table 3.1. What is the cell potential, and what are the ions in the anode and cathode cell compartments?

The fuel cell's standard potential is a maximum theoretical voltage. There are inevitable losses in practice, as shown in Figure 3.5. The "entropy" region of the diagram is the inevitable thermodynamic loss, which we have already treated, above. The "mixed potential" loss results from leakage of the gaseous fuels, hydrogen and oxygen, across the cell membrane, reacting with one another directly, without producing electrons. This leakage lowers the cell potential

TABLE 3.1 Standard Electrode Potentials in Aqueous Solution at 25°C

Cathode (Reduction) Half-Reaction	Standard Potential E° (Volts)
$Li^+(aq) + e^- \rightarrow Li(s)$	−3.04
$K^+(aq) + e^- \rightarrow K(s)$	−2.92
$Ca^{2+}(aq) + 2e^- \rightarrow Ca(s)$	−2.76
$Na^+(aq) + e^- \rightarrow Na(s)$	−2.71
$Mg^{2+}(aq) + 2e^- \rightarrow Mg(s)$	−2.38
$Al^{3+}(aq) + 3e^- \rightarrow Al(s)$	−1.66
$Zn^{2+}(aq) + 2e^- \rightarrow Zn(s)$	−0.76
$Cr^{3+}(aq) + 3e^- \rightarrow Cr(s)$	−0.74
$Fe^{2+}(aq) + 2e^- \rightarrow Fe(s)$	−0.41
$Cd^{2+}(aq) + 2e^- \rightarrow Cd(s)$	−0.40
$PbSO_4(s) + 2e^- \rightarrow Pb(s) + SO_4^{2-}(aq)$	−0.36
$V^{3+}(aq) + e^- \rightarrow V^{2+}(aq)$	−0.26
$Ni^{2+}(aq) + 2e^- \rightarrow Ni(s)$	−0.23
$Sn^{2+}(aq) + 2e^- \rightarrow Sn(s)$	−0.14
$Pb^{2+}(aq) + 2e^- \rightarrow Pb(s)$	−0.13
$Fe^{3+}(aq) + 3e^- \rightarrow Fe(s)$	−0.04
$2H^+(aq) + 2e^- \rightarrow H_2(g)$	0.00
$Sn^{4+}(aq) + 2e^- \rightarrow Sn^{2+}(aq)$	0.15
$Cu^{2+}(aq) + e^- \rightarrow Cu^+(aq)$	0.16
$ClO_4^-(aq) + H_2O(l) + 2e^- \rightarrow ClO_3^-(aq) + 2OH^-(aq)$	0.17
$AgCl(s) + e^- \rightarrow Ag(s) + Cl^-(aq)$	0.22
$Cu^{2+}(aq) + 2e^- \rightarrow Cu(s)$	0.34
$ClO_3^-(aq) + H_2O(l) + 2e^- \rightarrow ClO_2^-(aq) + 2OH^-(aq)$	0.35
$IO^-(aq) + H_2O(l) + 2e^- \rightarrow I^-(aq) + 2OH^-(aq)$	0.49
$Cu^+(aq) + e^- \rightarrow Cu(s)$	0.52
$I_2(s) + 2e^- \rightarrow 2I^-(aq)$	0.54
$ClO_2^-(aq) + H_2O(l) + 2e^- \rightarrow ClO^-(aq) + 2OH^-(aq)$	0.59
$Fe^{3+}(aq) + e^- \rightarrow Fe^{2+}(aq)$	0.77
$Ag^+(aq) + e^- \rightarrow Ag(s)$	0.80
$Hg^{2+}(aq) + 2e^- \rightarrow Hg(l)$	0.85
$ClO^-(aq) + H_2O(l) + 2e^- \rightarrow Cl^-(aq) + 2OH^-(aq)$	0.90
$NO_3^-(aq) + 4H^+(aq) + 3e^- \rightarrow NO(g) + 2H_2O(l)$	0.96
$VO_2^{2+}(aq) + 2H^+(aq) + e^- \rightarrow VO^{2+}(aq) + H_2O(l)$	1.00
$Br_2(l) + 2e^- \rightarrow 2Br^-(aq)$	1.07
$O_2(g) + 4H^+(aq) + 4e^- \rightarrow 2H_2O(l)$	1.23
$MnO_2(g) + 4H^+(aq) + 2e^- \rightarrow Mn^{2+} + 2H_2O(l)$	1.23
$Cr_2O_7^{2-}(aq) + 14H^+(aq) + 6e^- \rightarrow 2Cr^{3+}(aq) + 7H_2O(l)$	1.33
$Cl_2(g) + 2e^- \rightarrow 2Cl^-(aq)$	1.36
$Ce^{4+}(aq) + e^- \rightarrow Ce^{3+}(aq)$	1.44
$MnO_4^-(aq) + 8H^+(aq) + 5e^- \rightarrow Mn^{2+}(aq) + 4H_2O(l)$	1.49
$PbO_2(s) + SO_4^{2-}(aq) + 4H^+(aq) + 2e^- \rightarrow PbSO_4(s) + 2H_2O(l)$	1.69
$H_2O_2(aq) + 2H^+(aq) + 2e^- \rightarrow 2H_2O(l)$	1.78

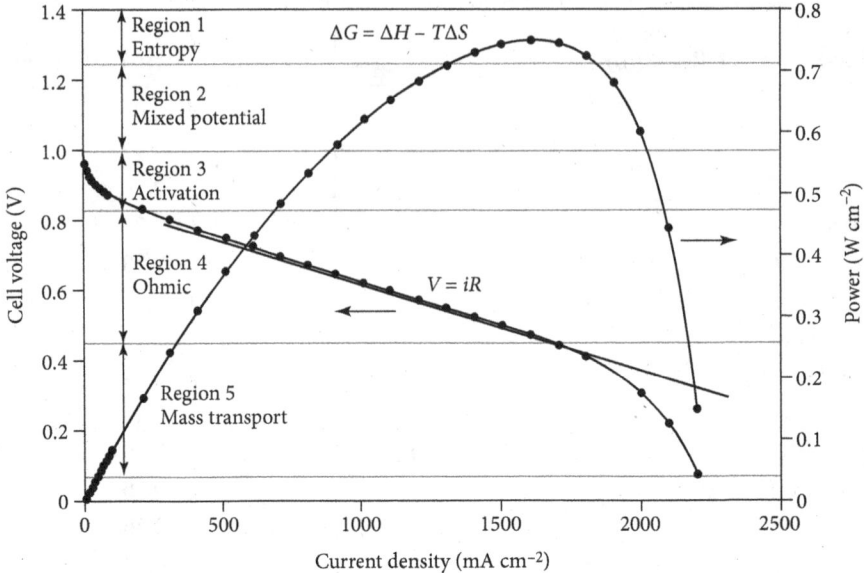

Figure 3.5 Graph of cell voltage versus current density for a hydrogen/air polymer electrolyte membrane fuel cell.
Source: Thomas & Zalbowitz, *Fuel Cells: Green Power*, Los Alamos National Laboratory, 1999; adapted from Spiro et al., *Chemistry of the Environment, 3rd Ed.*, University Science Books, © 2012, all rights reserved.

from the theoretical maximum of 1.23V to about 1V. Further losses occur when current is drawn from the cell. The onset of current leads to a drop in potential ("activation" region) because the electrode reactions are slow. Hydrogen and oxygen have strong bonds, which need to be activated in the reaction, even though energy is eventually produced. Catalysts are added to the electrodes in order to reduce the activation energy. Increasing the current further reduces the potential proportionally because of the cell's electrical resistance, R. For a given current, i, the potential drop is $V = iR$, reflecting the limited rate at which the ions can cross the membrane. Finally, at the highest currents, the supply of fuel cannot keep up, and the potential is limited by the rate of "mass transport." Except for the entropy limitation, these losses can be minimized, but not eliminated through improved chemistry and engineering. In applications of the fuel cell, the important measure of performance is power, the product of current and voltage, which is the rate at which energy is delivered (in watts, W, with units of J sec^{-1}). Figure 3.5 also shows the power curve for the cell.

QUESTION 17

Examining the current/voltage curve in Figure 3.5, estimate the resistance of the cell. How might this be improved?

Q17 ANSWER By reading the graph, we can see that in the ohmic (linear) region, the voltage falls from ~0.7V to ~0.5V between 500 and 1,500 mA/cm^2. Since $V = iR$, the resistance is the slope:

$$\frac{0.2\text{V}}{1{,}000\,\frac{\text{mA}}{\text{cm}^2}} \times \frac{1\,\Omega}{1\,\frac{\text{V}}{\text{A}}} \times \frac{1{,}000\,\text{mA}}{\text{A}} = \sim 0.2\ \Omega\ \text{cm}^2$$

The current increases with the area of the electrodes, so it has been normalized in this graph to mA/cm^2. Since the cell resistance is mainly determined by the migration rate of the ions through the cell membrane, it would be decreased by improving this rate, e.g., by increasing the density of ion pores in the membrane.

---

*QUESTION 18

Estimate the loss in voltage associated with the activation energy. How might this be improved?

*QUESTION 19

Estimate the current at which mass transport of the fuels becomes important. How might this limitation be improved?

*QUESTION 20

Estimate the practical efficiency (ratio of actual to theoretical potential) of the fuel cell at the point of maximum power.

Unfortunately, H_2 is not itself a fuel resource. It is currently derived from natural gas, whose principal component, methane, CH_4, can provide two molecules of H_2. However, this conversion itself consumes some of the energy in natural gas and also produces a molecule of CO_2. Fuel efficiency would be improved if hydrocarbons could themselves be employed in fuel cells, but unfortunately their oxidation at electrodes is very sluggish, and good catalysts have not yet been found. Research continues in this area, and there are some promising devices, but none are currently practical except in specialized applications.

3.6 Energy Storage: Batteries

Another important topic is energy storage, since electricity is evanescent and disappears when its source is unavailable. The fuel itself is an energy storage medium, but renewable sources, wind and solar, require storage to maintain the electricity supply when the wind dies or the sun sets. A number of storage methods are available all with advantages and disadvantages. Batteries are among the most prominent.

Batteries are electrochemical energy storage devices. They operate as a fuel cell, passing electrons from an anode to a cathode through an external circuit, but the electrodes are chosen to be easily reversible (low activation energies for the electrode reactions) so that the cell reaction can be reversed by an external electricity source. The most familiar example is the lead-acid battery used in cars (Fig. 3.6). At the anode, Pb is oxidized to Pb^{2+}, while at the cathode, PbO_2 is reduced to Pb^{2+}. The electrolyte is sulfuric acid, and the sulfate precipitates Pb^{2+} as $PbSO_4$.

QUESTION 21

What is the overall cell reaction for the lead-acid battery? Calculate its standard potential, using Table 3.1.

Q21 ANSWER The needed half-reactions and potentials are:

$$PbSO_4 + 2e^- = Pb + SO_4^{2-} \qquad\qquad E° = -0.36V \qquad\qquad (3.13)$$

$$PbO_2 + SO_4^{2-} + 4H^+ + 2e^- = PbSO_4 + 2H_2O \qquad E° = 1.69V \qquad\qquad (3.14)$$

Since Pb is being oxidized, we reverse the first reaction:

$$Pb + SO_4^{2-} = PbSO_4 + 2e^- \qquad\qquad E° = +0.36V \qquad\qquad (3.15)$$

and add it to the second to give the overall cell reaction:

$$PbO_2 + Pb + SO_4^{2-} + 4H^+ = 2\,PbSO_4 + H_2O$$

The overall standard potential can be calculated by adding the potentials of each of the two half reactions:

$$E° = 1.69V + 0.36V = 2.05V$$

*QUESTION 22

You can make a battery out of a copper electrode in contact with a solution of $CuSO_4$ and a zinc electrode in contact with a solution of $ZnSO_4$. Look up the reduction potentials in Table 3.1 and determine which is the anode and cathode. What is the cell reaction and what is its standard potential.

Although there are losses in practice, as with the fuel cell, they are much smaller for the lead-acid battery. Because there is no gaseous fuel to migrate and short-circuit the cell, the electrode reactions are much faster, and the resistance is lower. The cells operate near 2V. To obtain higher voltages, several cells are connected in series (a stack) in batteries.

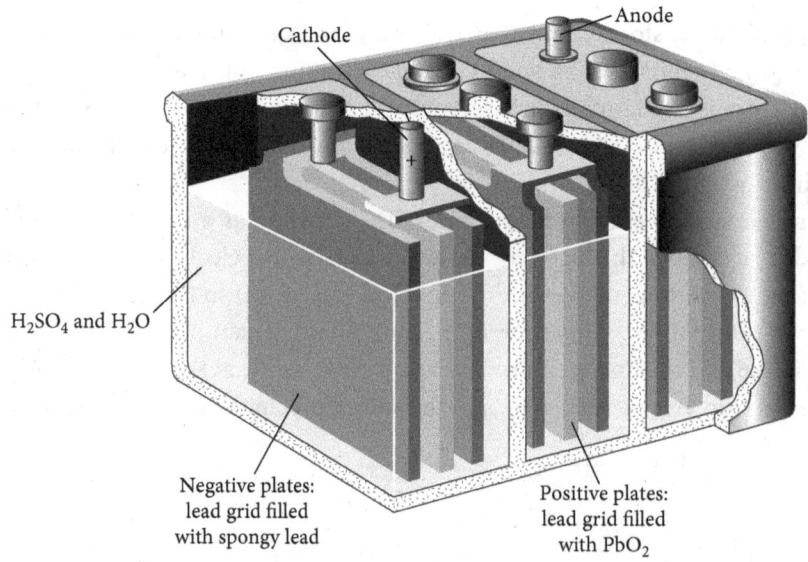

Figure 3.6 Cross-section of lead-acid battery.
Source: Buell & Gerard, *Chemistry in Environmental Perspective*, Prentice-Hall, 1994; adapted from Spiro et al., *Chemistry of the Environment, 3rd Ed.*, University Science Books, © 2012, all rights reserved.

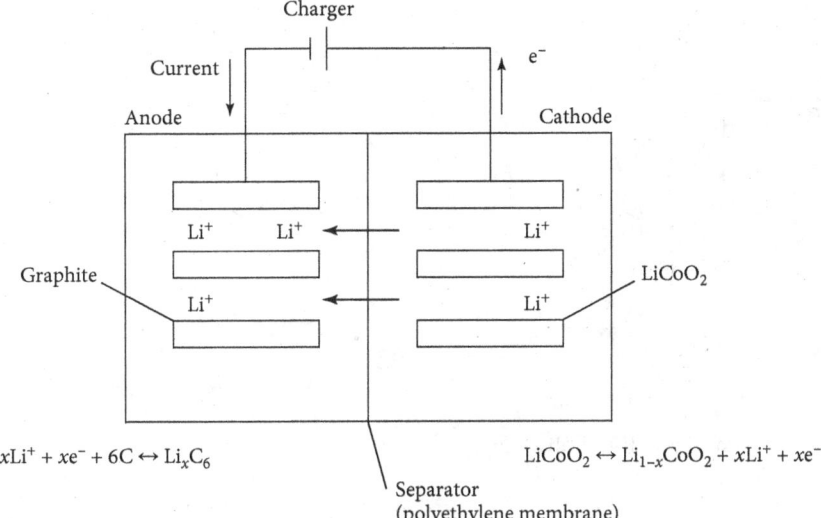

$$xLi^+ + xe^- + 6C \leftrightarrow Li_xC_6 \qquad\qquad LiCoO_2 \leftrightarrow Li_{1-x}CoO_2 + xLi^+ + xe^-$$

Figure 3.7 Figure of a rechargeable lithium-ion battery while charging.
Source: Adapted from Spiro et al., *Chemistry of the Environment, 3rd Ed.,* University Science Books, © 2012, all rights reserved.

The disadvantage of lead batteries is their weight. They are useful in providing auxiliary electricity to vehicles (lights, ignition, and so on), but are too heavy to power electric vehicles. Much lighter batteries are being developed for this purpose, especially the lithium-ion battery. In this battery, the anode is a mixed oxide, such as $LiCoO_2$, which liberates Li^+, and passes electrons to a graphite cathode, where the Li^+ ions intercalate between the negatively charged carbon sheets (Fig. 3.7). Lithium-ion batteries are used in cell phones, laptops, and many consumer products. It is also emerging as the battery of choice in very large battery stacks designed to back up electric generating stations, in order to even out fluctuations both in electricity demand (e.g., air conditioning on hot days) and in supply (from renewable sources).

However, there are other options for "grid-storage." One is the "flow battery" in which solutions of redox-active ions flow through the anode and cathode compartments from reservoirs, which can hold large quantities of the reactants, thereby storing a large quantity of energy. A commercially available example is the vanadium redox flow battery (Fig. 3.8). Vanadium can exist in several oxidation states in an aqueous solution, allowing it to transfer electrons over a large range of potentials. When the flow battery is discharged electrons are released from V^{2+}, which is left as V^{3+}, at the right-hand electrode, and reduce VO_2^+ to VO^+ at the left-hand electrode. Charging the battery with current from the grid reverses these half-reactions.

---

*QUESTION 23

What are the oxidation states of vanadium in VO_2^+ to VO^{2+}?

*QUESTION 24

From Table 3.1, what are the standard potentials for the two electrodes, and for vanadium flow battery? Write the cell reaction when the battery is discharging.

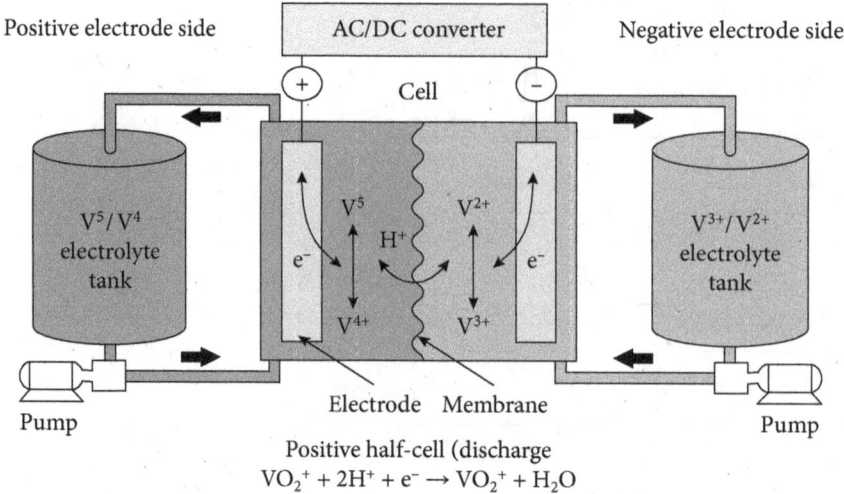

Figure 3.8 A schematic of a vanadium redox flow battery.
Source: Adapted from Sumitomo Electric Industries, Ltd. (SEI), copyright © 2001.

The fuel cell is a lightweight alternative to storage batteries, but its adoption has been slow, because its losses pose difficult challenges, as discussed above, and because of the difficulty of transporting and storing H_2 gas (high-pressure tanks, or metal hydrides, both of them being heavy). Nevertheless, some bus fleets now run on electricity from H_2 fuel cells. If these can be overcome, then one can envision a "hydrogen economy," in which electricity is used to split H_2O into H_2 and O_2 (a water electrolyzer is just the fuel cell run in reverse), and H_2 is stored and transported to where electricity is needed. H_2 then becomes the storage medium. An alternative to the electrolyzer is "artificial photosynthesis," in which sunlight is captured by assemblies of molecules capable of using the energy to split water directly into H_2 and O_2. Although many devices have been demonstrated in laboratories, it is uncertain whether they can compete with the combination of solar electricity and electrolysis.

3.7 Materials and Energy

Our use of materials is an important aspect of energy utilization because of the energy required for their preparation, their assembly into products, and their eventual disposal. There are many opportunities for saving energy at each step, and the choice of material can itself have important consequences for energy use.

3.7.1 Recycling

Recycling offers energy savings because it generally requires more energy to extract materials from the environment than from discarded products (Fig. 3.9). A good example is aluminum, the energy required to reprocess aluminum cans being only 5% of that required to extract the metal from ore. The corresponding ratio for steel (refined iron) is much higher, 48%. One reason for the disparity is that the metal–oxygen bonds are much stronger for Al than for Fe in the oxide ores, Al_2O_3 and Fe_2O_3, from which the metals are refined (because Al^{3+} is much smaller, and, therefore, more strongly attracted to O^{2-} than Fe^{3+}). This can be seen from the oxide formation energies, −1,670 and −822 kJ mol^{-1}, for Al^{3+} and Fe^{3+} for the reactions $2M + (3/2)O_2 \rightarrow M_2O_3$ (M = Al or Fe).

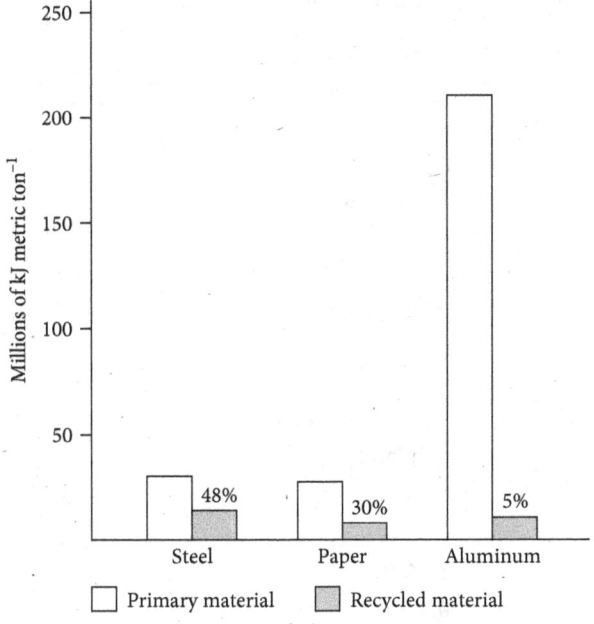

Figure 3.9 Comparison of energy requirements for the production of steel, paper, and aluminum from primary and recycled materials.
Source: Adapted from Spiro et al., *Chemistry of the Environment, 3rd Ed.,* University Science Books, © 2012, all rights reserved.

QUESTION 25

If these formation energies were the whole story, how would the ratios in Figure 3.9 compare? What other factors might be important?

Q25 ANSWER If the only factor is the energy required to convert the oxides to the metals, then twice as much energy would be saved by not having to refine Al than Fe, and the reprocessing energy/refining energy ratio would be expected to double for Fe, from 5% to 10%, instead of the 48% actual value. Additional terms that go into the ratio would be the energy required to mine, ship, and process the ore from the rock containing it (the denominator of the ratio), and the energy required to collect, sort, and reprocess the material to be recycled (the numerator).

*QUESTION 26

Figure 3.9 shows the actual energy requirements for refining and recycling Al and Fe, as well as the ratios. From the length of the bars, what can you deduce about the relative magnitude of the factors other than the oxide formation energies?

3.7.2 *Materials Properties: Paper Versus Plastic*

The choice of material for a given application can also affect the energy required to produce a product. A contentious example is the choice of paper versus plastic in bags and cups. Paper is usually the popular choice, but plastic makes more sense from an energy use standpoint. Consider a study of what goes into making a paper or a styrofoam cup, summarized in Table 3.2.

TABLE 3.2 Raw Materials, Utility, and Environmental Summary for Hot Drink Containers

Item	Paper Cup[*]	Polyfoam Cup[†]
Per cup		
Raw materials		
Wood and bark	25–27g	0g
Petroleum fractions	1.5–2.9g	3.4g
Other chemicals	1.1–1.7g	0.07–0.12g
Finished weight	10.1g	1.5g
Per metric ton of material		
Utilities		
Steam	9,000–12,000 kg	5,500–7,000 kg
Power	980 kWh	260–300 kWh
Cooling water	50 m^3	130–140 m^3
Water effluent		
Volume	50–190 m^3	1–4 m^3
Suspended solids	4–16 kg	0.4–0.6 kg
Biological oxygen demand (BOD)	2–20 kg	0.2 kg
Organochlorines	2–4 kg	0 kg
Metal salts	40–80 kg	10–20 kg
Air emissions		
Chlorine	0.2 kg	0 kg
Chlorine dioxide	0.2 kg	0 kg
Reduced sulfides	1–2 kg	0 kg
Particulates	2–3 kg	0.3–0.5 kg
Chlorofluorocarbons (CFCs)	0 kg	0 kg[‡]
Pentane	0 kg	35–50 kg
Sulfur dioxide	~10 kg	3–4 kg
Recycle potential		
To primary user	Possible. Washing can destroy.	Easy. Negligible water uptake.
After use	Possible. Hot melt adhesive or coating difficulties.	Good. Resin reuse in other applications (though this has yet to be implemented).
Ultimate disposal		
Proper incineration	Clean	Clean
Heat recovery	20 MJ kg^{-1}	40 MJ kg^{-1}
Mass to landfill	10.1 g/cup	1.5 g/cup
Biodegradable	Yes. BOD to leachate, methane to air.	No. Essentially inert.

[*]Uncoated fully bleached kraft paper cup.

[†]Molded polystyrene foam bead (seamless) cup.

[‡]Many producers of foamable beads have never used CFCs.

Source: Updated and adapted by Hocking, from original article in Hocking, "Paper Versus Polystyrene: A Complex Choice," *Science*, Vol. 251, 1991, pp. 504–505. Adapted from Spiro et al., *Chemistry of the Environment, 3rd Edition*, University Science Books, © 2012, all rights reserved.

QUESTION 27

Table 3.2 lists the power requirements in kWh per metric ton. What is the ratio of power requirement for a paper versus a Styrofoam cup?

Q27 ANSWER The power per ton is 980 kWh for paper and ~300 kWh for Styrofoam, but the cups weigh 10.1g and 1.5g, respectively. So the per cup ratio is:

$$980\frac{\text{kWh}}{\text{ton}} \times \frac{10.1\text{g}}{300\frac{\text{kWh}}{\text{ton}}} \times 1.5\text{g} = 22$$

Much more energy is expended in producing paper versus styrofoam cups.

The main reason a styrofoam cup requires much less energy to produce is that it holds hot liquids better than paper and, therefore, requires much less material. The cellulose fibers of paper weaken and separate when exposed to water.

*QUESTION 28

Table 3.2 also lists the amount of heat energy recoverable from incinerating the cups. If the incinerator is used to run a power plant, operating at 30% efficiency, compare the electrical energy recovered with each material with the energy required to produce it.

However, energy is not the whole story. Plastic is notorious for littering public spaces and also for ending up in waterways and the ocean. Indeed, the oceans harbor an enormous amount of plastic, by some estimates more by weight than all of the fish! Furthermore, the plastic eventually breaks up into very small pieces, which can be ingested by fish, invertebrates, and birds, sometimes carrying adsorbed pollutants with them. Of course, plastic cups, bags, and bottles can be recycled, but most of them are not. Also, plastic manufacturers are developing new compositions that biodegrade in the environment, but these new plastics have not penetrated the market very far. In the meantime, many communities have banned or taxed plastic bags and containers, and encouraged people to use reusable bags and refillable mugs.

3.8 Conclusions

Overall, there are many ways to think about energy usage in the world. With more efficient technologies and gaining more work out of energy sources already in use, we can gain more energy from sources already in place. Batteries and electrodes are both being used as energy sources and storage for renewable energy too. Finally, thinking about the materials that we use and the energy needed to create and recycle those products can contribute to saving energy resources.

4

RENEWABLE ENERGY

4.1 Forms of Renewable Energy

The alternative to fossil and nuclear fuel use is to harness renewable energy resources. With the exception of geothermal energy, renewable energy is derived directly or indirectly from sunlight. The annual energy from sunlight striking the earth's surface is >10,000 times our global annual energy consumption from all energy sources. The annual sunlight striking each square meter of the earth's surface is equal to the energy content of 190 kg of high-grade bituminous coal. The sun already provides us, in fact, with our most basic energy needs, including heat, freshwater, and plant life. There is plenty left over for our other energy needs if we could learn how to use it effectively.

The difficulty is that sunlight is diffuse and intermittent, in other words, very spread out and not constant. The technologies required to store solar energy are currently expensive. However, this is rapidly changing. Technological and material revolutions are making renewable energy generation and energy storage systems competitive with fossil fuel systems. Global renewable energy production is increasing dramatically. In 2019, for the first time, renewable energy represented the largest percent of all new electricity generation systems built.[1] Also, for the first time ever, electricity generated from renewable energy (excluding large-scale hydropower systems) exceeded that generated by nuclear power.[2] Depending on site-specific conditions, energy prices from new renewable energy generation plants can be cheaper than new coal- and gas-fired power plants. The increasing environmental burdens associated with fossil and nuclear fuel use are driving renewable energy technologies. There is reason to believe that with further technological improvements, renewable energy will provide most of our energy needs.

Global renewable energy statistics are shown in Table 4.1. The first column shows the estimated amount of energy that could be feasibly produced from each renewable resource. The second column shows the total installed electricity generating capacity for each renewable resource in 2019. The third column shows the percent of total global energy use (including both electric and non-electric uses).

---

[1] International Energy Agency, *Report: Extract Renewables,* 2019, available at https://www.iea.org/reports/global-energy-review-2019/renewables

[2] U.S. Energy Information Administration, *Renewables Became the Second-Most Prevalent U.S. Electricity Source in 2020,* 2021, available at https://www.eia.gov/todayinenergy/detail.php?id=48896

TABLE 4.1 Global Renewable Energy Resources and Electricity Generation

Energy Source	Estimated Available Renewable Energy Resource (TWh yr^{-1})	2019 Electricity Generation (TWh yr^{-1})	2019 Share of Total Electricity Generation (%)
Wind	170,000	1,265	5%
Solar	460,000	575	2%
Hydropower	17,000	4,156	16%
Geothermal	1,400,000	83	0.3%
Biomass	69,000	638	3%
Marine	1,900	26	0.1%
Total renewables	*2,100,000*	*6,743*	*26.4%*
Total		**25,398**	**100%**

Sources: Estimated available energy resource from Sims et al., "In Climate Change 2007: Mitigation. Contribution of Working Group III to the Fourth Assessment Report of the Intergovernmental Panel on Climate Change," *2007: Energy Supply, Table 4.2.,* Cambridge University Press, 2007. Electricity generation data from U.S. Energy Information Agency, *Electricity Generation—International Energy Statistics,* available at https://www.eia.gov/international/data/world /electricity/electricity-generation.

Renewable energy currently accounts for 26.4% of the global electricity generation (Table 4.1). Renewable energy generation is attractive for developing, as well as developed, countries. The dispersed nature of renewable sources can be an advantage for applications that are far from the existing electrical grid or fuel transportation systems, or where the existing energy infrastructure has insufficient capacity or is not well developed. The economics of smaller, distributed renewable generation systems powered by sunlight, wind, or biomass-derived fuels are often financially attractive compared to the extending power lines over long distances. Distributed power generation, where electricity is generated near where it is used, is gaining popularity in both developed and developing countries, and provides improved resilience as the consequences of power failures and blackouts become painfully clear in an age of increased climatic instability.

QUESTION 1

Compare the annual solar energy incident upon the earth's surface with the total annual global energy consumption in 2019 (1.62×10^8 GWh yr^{-1}). Assume that the solar energy flux hitting the earth, S_0, is 1,370 watts (W) m^{-2}, and the earth's albedo (the proportion of light reflected by a surface) is 0.3 (i.e., 30% of the incident flux is reflected and 70% is absorbed). Earth's radius is 6.37×10^6 m.

Q1 ANSWER The energy incident on the earth from the sun per year can be calculated using the average solar energy flux striking the earth, the area of the earth, the number of hours in a year, and the fraction of energy absorbed by the earth. As the earth orbits the sun and rotates about its axis, it is constantly intercepting a solar flux of 1,370 W m^{-2}. The amount of flux intercepted is equal to the area of a flat disk, or circle, with the same diameter of the earth – i.e., the earth's disk.

The area of the earth's disk can be calculated:

$$A_{\text{earth disk}} = \pi r^2 = \pi (6.37 \times 10^6 \text{ m})^2 = 1.275 \times 10^{14} \text{ m}^2$$

The amount of solar radiation intercepted is the solar flux (in W m^{-2}) multiplied by the area of the earth's disk:

$$S_{\text{intercepted}} = S_0 \times A_{\text{earth disk}} = 1{,}370 \, \frac{\text{W}}{\text{m}^2} \times 1.275 \times 10^{14} \text{ m}^2 = 1.746 \times 10^{17} \text{ W}$$

Of the intercepted flux, 30% is reflected (the earth's albedo, or reflectivity). Multiply by the one minus the albedo to get the absorbed solar flux.

$$S_{\text{absorbed}} = S_{\text{intercepted}} \times (1 - \text{Albedo}) = (1 - 0.3) \times 1.746 \times 10^{17} = 1.22 \times 10^{17} \text{ W}$$

Multiply by the hours (hr) in a year to get the annual solar energy and convert from watts to gigawatts (GW):

$$365 \text{ days} \times \frac{24 \text{ hr}}{\text{day}} = 8{,}760 \, \frac{\text{hr}}{\text{yr}}$$

$$1.22 \times 10^{17} \text{ W} \times \frac{8{,}760 \text{ hr}}{\text{yr}} \times \frac{1 \text{ GW}}{1 \times 10^9 \text{ W}} = 1.07 \times 10^{12} \text{ GWh yr}^{-1}$$

Compared to the annual global energy consumption:

$$\frac{1.62 \times 10^8 \text{ GHh}}{1.07 \times 10^{12} \text{ GHh}} \times 100\% = 0.0151\%$$

We could utilize a lot more of the sun's energy for our global energy consumption.

*QUESTION 2

How does the solar flux in Arizona compare with total yearly U.S. energy consumption (approximately 4.13×10^6 GWh in 2019)? Assume the average solar flux in Arizona is the same as Phoenix, approximately 7.35 kilowatt-hours (kWh) m^{-2} day^{-1}. Arizona has an approximate area of 2.95×10^{11} m^2.

4.2 Solar Heating

One of the oldest applications of solar energy is heating. For millennia, humans have selected building sites and designed buildings that take advantage of sunlight to heat buildings in the winter and to shade unwanted solar energy during the summer. Currently, space heating and hot water heating constitute ~62% of U.S. household energy needs.[3] There is significant potential to increase our use of solar energy to meet these heating needs.

Solar heating can be accomplished by either passive or active designs. Passive design involves no moving parts. For example, a building and its windows can be oriented so that solar radiation is admitted during the winter when the sun is low in the sky and shaded

---

* Answers to starred questions can be found at the end of the book.

[3] U.S. Energy Information Administration, *Use of Energy Explained*, available at https://www.eia.gov /energyexplained/use-of-energy/homes.php

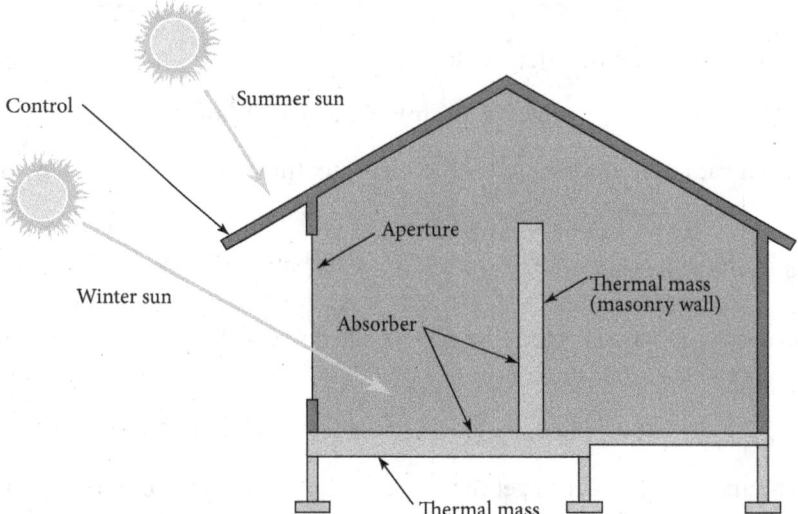

Figure 4.1 House with passive solar heating.

during the summer by overhangs and other design features when the sun is high in the sky. Thermally dense materials (e.g., concrete, stone) can also be used to provide thermal storage. Sunlight passing through windows can be absorbed by dark interior surfaces and stored in thermally dense walls and floors (Fig. 4.1). The thermal mass helps interior spaces remain comfortable throughout the day. Building colors can be selected to either reflect or absorb incident solar energy. Specially designed "cool roofs" are increasingly used in hot climate zones to reflect unwanted solar radiation. Specially designed window coatings are used to adjust, or "tune," the amount of sunlight, visible light, and thermal radiation (heat) that are absorbed, transmitted, and emitted from windows. This helps maximize the use of sunlight for desirable heating and daylighting and minimize unwanted heat gains or losses.

In contrast to passive solar energy systems, active solar energy systems utilize some type of active control, movement, or other "action." Examples of active solar system include solar trackers that keep solar panels oriented directly toward the sun; solar water heating systems that use a system of temperature sensors, valves, and pumps to circulate water through solar collectors that are controlled by some type of electronic controller (there are passive solar water heaters that involve no pumps, valves, or controls); operating louvers on windows that open or close as needed; optical coatings on windows that can lighten or darken to control solar heat gains; and many other systems. Figure 4.2 illustrates an active solar heating system that provides both water and space heating. Water in a storage tank is circulated through solar collectors when the sun is shining. The circulation pump is turned on when the temperature of the solar collectors is greater than the water temperature in the solar storage tank. Potable water flows through a heat exchanger in this tank to create hot water, and then through a backup heater to provide supplemental heating. Hot water in the solar storage tank can also be circulated through another space heater to heat up air in the ventilation system.

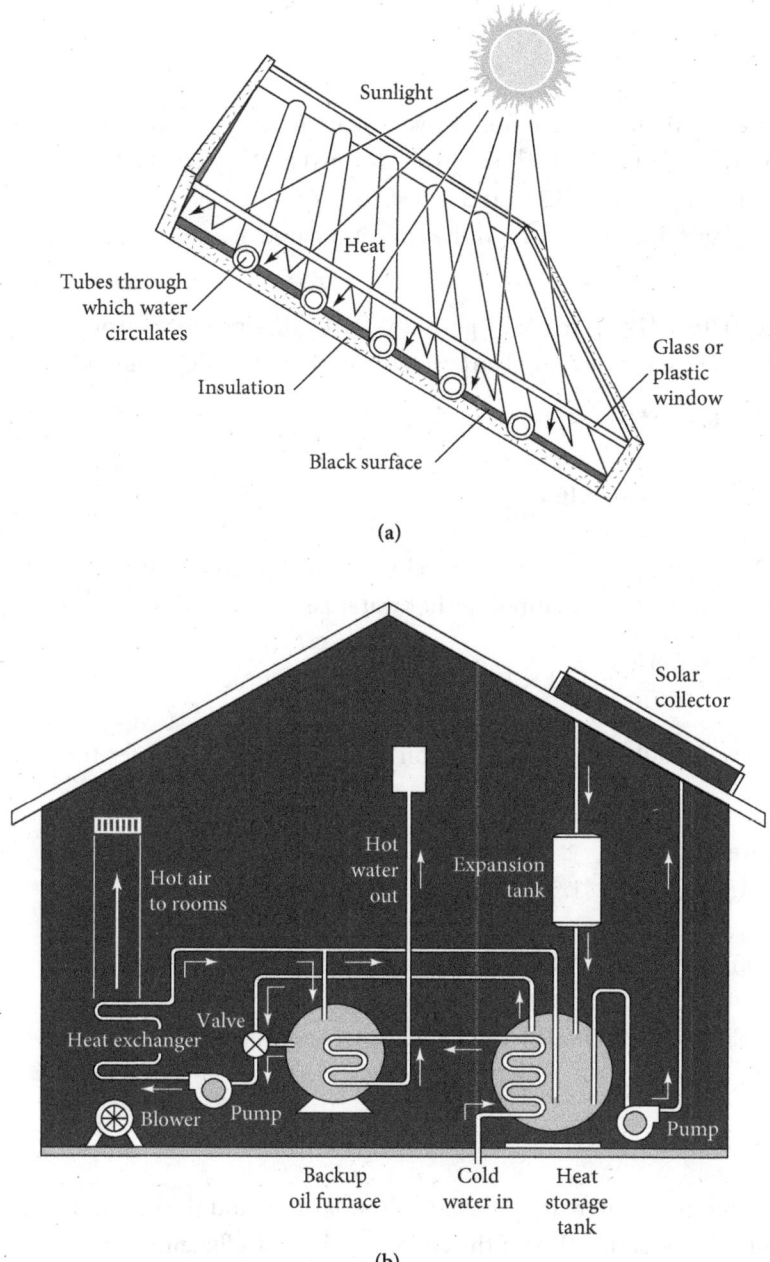

(a)

(b)

Figure 4.2 (a) Solar collector on the roof of an active solar-heated house. (b) Pumping system in an active solar-heated house for provision of space heating and hot water.
Sources: Adapted from (a) Nebel, *Environmental Science: The Way the World Works,* Prentice-Hall, 1993; and (b) Anderson & Riordan, *The Solar Home Book: Heating, Cooling and Designing with the Sun,* Brick House Publishing, 1976. Adapted from Spiro et al., *Chemistry of the Environment, 3rd Ed.,* University Science Books, © 2012, all rights reserved.

QUESTION 3

Should you install a solar hot water heater? Will it save you money over the long run? The average home has a 200 liter (L) (50–60 gal) hot water tank, which is effectively drained and replenished three times a day. Assume that the entering tap water is 15°C and is heated to 55°C. Given average energy from the sunlight of 1.53 kJ cm^{-2} day^{-1}, how large would the collection area of a solar water heater need to be if its efficiency is 30.0%?

Q3 ANSWER Since it takes 1 calorie (cal) (0.00419 kJ) to heat 1 gram (g) [or 1 mL, since the density of water is 1 g per milliliter (g mL^{-1})] of water by 1°C, heating 1 mL water from 15°C to 55°C requires:

$$\Delta T = (55°C - 15°C) = 40°C$$

$$\frac{0.00419 \text{ kJ}}{°C \text{ mL}} \times 40. \ °C = 0.168 \frac{kJ}{mL}$$

600L (equivalent to 600,000 mL or 6.00×10^5 mL) of water per day is heated, since a 200-L tank is drained and replenished three times per day, the heat required for hot water is:

$$6.00 \times 10^5 \text{ mL} \times \frac{0.168 \text{ kJ}}{mL} = 1.01 \times 10^5 \text{ kJ}$$

Given a collection efficiency of 30.0% (0.300), the solar collection area required to transfer this much heat is calculated:

$$\text{collection area} = \frac{\text{heat required}}{\text{solar heat per unit area} \times 0.300} = \frac{1.01 \times 10^5 \text{ kJ}}{1.53 \text{ kJ cm}^{-2} \times 0.300} = 219,080 \text{ cm}^2$$

$$219,080 \text{ cm}^2 \frac{1 \text{ m}^2}{(100 \text{ cm})^2} = 21.9 \text{ m}^2$$

The size of the water heater is a reasonable size to put on the roof of a house. See Questions 5 through 7 to think about cost.

*QUESTION 4

Say that the efficiency of the solar hot water heater increased to 35.0%, how large would the collection area have to be then? What percent smaller is this than if the collector is 30.0% efficient?

QUESTION 5

Assume that the price of a solar collector is $650 m^{-2}. How much would it cost to install the hot-water system in Q3?

Q5 ANSWER If the cost of the solar collector is $650 m^{-2}:

$$\text{installation cost} = \frac{\$650}{m^2} \times 21.9 \text{ m}^2 = \$14,235$$

*QUESTION 6

How much money would you save if the efficiency of the collector was 35.0% instead of 30.0%?

QUESTION 7

If the price of oil remains at \$0.55 L^{-1} for 20 yr, how much would the solar collector save? Assume that the heating content of oil is 2.51×10^4 kJ L^{-1} and it can be burned with 90% efficiency. Use the energy needed to heat water per day calculated in Q3.

Q7 ANSWER The energy required to heat water over a period of 20 yr would be:

$$\frac{1.01 \times 10^5 \text{ kJ}}{\text{day}} \times \frac{365 \text{ days}}{\text{yr}} \times 20 \text{ yr} = 7.37 \times 10^8 \text{ kJ}$$

The oil required to supply this much energy at 90% (0.90) efficiency is:

$$\text{liters of oil} = 7.37 \times 10^8 \text{ kJ} \times \frac{L}{2.51 \times 10^8 \text{ kJ} \times 0.90} = 3.26 \times 10^4 \text{ L}$$

Thus, over the 20-year period:

$$\text{cost} = \frac{\$0.55}{L} \times (3.26 \times 10^4 \text{ L}) = \$1.8 \times 10^4 = \%18,000$$

for a net saving of: \$18,000 − \$14,235 = \$3,765.

Depending on the climate, a well-designed and properly sized solar water heater can provide up to two-thirds of a household's hot water needs. It can save 50%–85% of the hot water portion of monthly electricity utility bills if the backup element is kept at 50°C (122°F).

4.3 Solar Thermal Electricity

One approach to create electricity from the sun is to use concentrated solar radiation to heat water and run a steam generator. This is similar to traditional fossil fuel thermal power plants (Chapter 5), except that the sun is used to create steam rather than burning coal or natural gas. One solar thermal electric design currently in operation is the "power tower." This technology uses mirrors to concentrate the sun's rays by up to 5,000 times onto a central receiving tower. A fluid is circulated through the receiver where it is heated up. This hot fluid is then used to create steam, which drives a steam turbine to generate electricity. The Ivanpah Solar Power Facility outside of Las Vegas, NV, is an example of a commercially operating power tower technology (Fig. 4.3). The power station uses a field of 173,000 heliostats, each with two mirrors, to track and reflect the sun's rays onto three power tower receivers located on top of 140m tall towers for a capacity of 392 MW. Water is circulated through the receivers when the sun is shining, creating steam which drives steam turbines to produce electricity. Other designs, such as the currently non-operational Crescent Dunes project north of Las Vegas, circulate molten salt through the solar receiver. The molten salts can be heated to much higher temperatures and can be used to store significant amounts of thermal energy, allowing the plant to generate steam to create electricity even when the sun is not shining. There are currently 10 power towers in operation around the world, with others in various stages of planning and construction.

Another type of solar thermal electric power technology is the "parabolic trough" solar thermal power plant. This design uses a series of parabolic trough mirrors that concentrate sun-

Figure 4.3 Ivanpah Solar Power Facility outside Las Vegas, NV.
Source: By Aioannides. Taken from the side of the road near I15. Previously published: None.
CC BY-SA 3.0, https://commons.wikimedia.org/w/index.php?curid=25841974.

light on a pipe running through the light focus. Water or another heat transfer fluid is circulated through this pipe, heating up and creating steam. The steam is used to run a steam-powered turbine spinning electric generators. This system operates at much lower temperatures than the power tower design. There are approximately 45 parabolic trough power plants in operation worldwide. One of the oldest is the Solar Energy Generating System (SEGS) outside Boron, CA (Fig. 4.4). This system was first constructed in 1984 and currently has a capacity of 310 MW.

Figure 4.4 A 354 MW SEGS Parabolic Trough Solar Thermal Power Plant, CA.
Source: U.S. Bureau of Land Management, retrieved from http://www.ca.blm.gov/cdd/alternative_energy.html

QUESTION 8

What is the theoretical efficiency (Chapter 3) of a power tower solar plant if the boiler temperature reaches 535°C (995°F) and the cooling water is at 25°C (77°F)?

Q8 ANSWER Apply equation from Energy Utilization (Chapter 3) after converting to absolute temperatures:

$$535°C + 273 = 808K \quad 25°C + 273 = 298K$$

$$\frac{W_{max}}{Q_h} < 1 - \frac{T_c}{T_h} = 1 - \frac{298K}{808K} = 0.63$$

The maximum efficiency is 63%.

*QUESTION 9

Calculate the theoretical efficiency of a power tower solar plant where the molten salt reaches 649°C (1,200°F) and the cooling temperature is 18°C (64°F).

4.4 Photovoltaic Electricity

Sunlight can be transformed directly into electricity via the photovoltaic (PV) effect. When light is absorbed in a PV material, positive and negative charges are created, which can be collected by electrodes on either side and passed to an external electric circuit. The most highly developed PV material, and the one currently used in large-scale applications, is silicon (Si), the same material on which the electronics industry is based.

PV materials are semiconductors, solids with electrical properties between metals and insulators. In solids, the electronic levels spread into energy bands, because of the mutual interactions of the electronic levels on all the atoms in the solid lattice. The filled levels together produce the valence band, while the empty levels produce the conduction band, as illustrated in Figure 4.5. If an electron is injected into the conduction band of a solid, it can move freely throughout the lattice, since all the orbitals in the band are empty.

The energy difference between the top of the valence band and the bottom of the conduction band is called the bandgap. Semiconductors have bandgaps of intermediate energies, energies that are well matched to solar photons. When light is absorbed by a semiconductor, an electron is promoted from the valence to the conduction band. This is the basis of the PV effect (Fig. 4.5). To be absorbed, however, the light must have energy equal to or greater than the bandgap. Light from the sun has a distribution of wavelengths, which peaks near 500

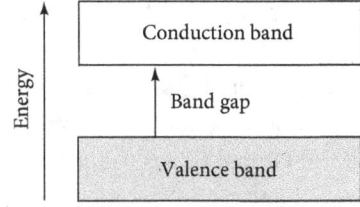

Figure 4.5 Valence and conduction bands in semiconductors. *Source:* Adapted from Spiro et al., *Chemistry of the Environment, 3rd Ed.,* University Science Books, © 2012, all rights reserved.

nanometers (nm) in the green–blue region of the spectrum. The distribution has a broad tail, extending far into the infrared (IR) region, but two-thirds of the photons have wavelengths <1,140 nm, which corresponds to the energy of the Si bandgap, 1.09 electron volts (eV). Consequently, silicon can capture most of the solar photons.

QUESTION 10

The bandgap for a Si semiconductor is 1.09 eV. If sunlight peaks at 483 nm, does this provide enough energy to excite an electron from the valance band to the conduction band? Remember that $1 J = 6.24 \times 10^{18}$ eV.

Q10 ANSWER The energy for a photo of light can be calculated using Planck's equation:

$$E = h\nu \quad h = 6.626 \times 10^{-34} \, J \, sec$$

$$\nu = \frac{c}{\lambda} = \frac{3.00 \times 10^8 \, m \, sec^{-1}}{483 \, nm} \times \frac{1 \times 10^9 \, nm}{1 \, m} = 6.21 \times 10^{14} \, sec^{-1}$$

$$E = 6.626 \times 10^{-34} \, J \, sec \times 6.21 \times 10^{14} \, sec^{-1} = 4.11 \times 10^{-19} \, J$$

$$1 \, J = 6.24 \, 10^{18} \, eV$$

$$E = 4.11 \times 10^{-19} \, J \times \frac{6.24 \times 10^{18} \, eV}{1 \, J} = 2.56 \, eV$$

Therefore, a wavelength of 483 nm has enough energy to excite an electron into the conduction band for a Si semiconductor.

---

*QUESTION 11

Part of the solar spectrum includes IR wavelengths. Does IR radiation at 1,725 nm provide enough energy to excite an electron from the valance band to the conduction band in a PV cell? Remember that the bandgap for a Si semiconductor is 1.09 eV.

It is not enough, however, simply for sunlight to shine on a piece of Si. The electrons, which are promoted to the conduction band, have no reason to travel in any particular direction. After a short time, they fall back to the valence band, refilling the holes that had been created by the light. In order to produce a flow of electrons in an external circuit, a PV material must have a built-in electrical potential, which is created by an intrinsic charge asymmetry. This charge asymmetry is produced by doping, or adding small amounts of foreign atoms to the semiconductor that have either more or fewer valence electrons than the native atoms.

Significant progress has been made in commercializing PV technology, and PV systems are now the largest source solar electricity (see Table 4.1). Utility-scale PV systems generate electricity commercially at prices that directly compete with conventional fossil fuel power systems. PV panels are also being widely installed on commercial and residential rooftops throughout the world. In 2020, California was the first state in the United States to mandate that all new homes install solar PV systems. California also has "solar-ready" requirements in its building code, which require other new constructions to be designed to readily add PV and/or solar water heating systems at a later date. This anticipates the continued drop in solar prices and future installation of solar systems. PV systems are also widely used for remote and distributed power systems.

A limitation of PV is that it only generates power when the sun is out. However, a related advantage of PV is that it that daylight hours coincide with high electricity demand for air-conditioning, so PV is well matched to the peak loads of many electrical utilities.

QUESTION 12

How much power can a typical 290 W solar panel produce in Las Vegas on a typical day with an average of 7.5 hr of sunlight?

Q12 ANSWER The amount of power that solar panels produce is dependent on the amount of direct sunlight striking the panel (which depends on collector mounting orientation, whether it has a tracker, time of day, weather), inverter, and other system efficiencies, and to a lesser extent the panel temperature (output decreases at high temperatures). There are many detailed calculators that account for this complexity, but as a first-order approximation, we can estimate daily power generation by multiplying the panel's rated power by the average daily hours of sunlight:

$$7.5 \text{ hr} \times 290\text{W} = 2{,}175 \text{ watt-hours (Wh)} = 2.2 \text{ kWh of energy}$$

---

*QUESTION 13

The average amount of direct sunlight in Malaysia is 6.5 hr. How much less energy is created in Malaysia than Nevada using the same solar panels in Q12?

QUESTION 14

How much land would it take to power the United States with solar power only if the United States uses about 4.1×10^6 GWh of electricity and approximately 2.8 acres of solar panels can create a GWh of electricity?

Q14 ANSWER If the United States uses about 4.1×10^6 GWh of electricity and approximately 2.8 acres of solar panels can create a GWh of electricity:

$$4.1 \times 10^6 \text{ GWh} \times \frac{2.8 \text{ acres}}{\text{GWh}} = 11{,}500{,}000 \text{ acres}$$

QUESTION 15

What percent of the continental U.S. does this amount to?

Q15 ANSWER The continental U.S. is approximately 1.9 billion acres, so the percentage of land necessary to create all of the U.S. power needs from solar is:

$$\frac{11{,}500{,}000 \text{ acres}}{1.9 \times 10^9 \text{ acres in U.S.}} \times 100\% = 0.60\%$$

*QUESTION 16

The Mojave Desert receives a lot of direct sunlight every year, so it is a good location for a solar farm. What percentage of the Mojave Desert (approximately 47,877 miles^2 (mi^2) would need to be used to cover the electricity needs of the United States with solar electricity?

4.5 Photosynthesis

Hundreds of millions of years before the invention of the PV solar cell, nature solved the problem of harnessing photoinduced electron excitation to produce useful energy. In photosynthesis, as in the Si solar cell, electrons are excited by the absorption of solar photons. Nature has evolved a photoreaction center, at which charge is separated across a biological membrane. The key components of the photoreaction center are a collection of chlorophyll molecules capable of absorbing sunlight and generating electrons in their photoexcited states. The electrons hop from one molecule to another along an energy gradient of empty energy levels. Instead of an electric current, the charge separation leads to the production of energy-storing chemicals. The electron is taken up in a series of biochemical steps that lead to the reduction of CO_2 to carbohydrates, while other processes involve a series of steps that oxidize H_2O to O_2.

Because of its extended system of conjugated double bonds, chlorophyll absorbs light strongly in the 400–700 nm region of the solar spectrum. This region contains ~50% of the total energy in solar photons. A green plant absorbs ~80% of the incident photons in this range (the rest are lost to reflection, transmission, and absorption by other molecules), leaving ~40% of the total solar energy available for photosynthesis. Of this, 28% actually ends up in carbohydrates; the rest is lost at the various electron transfers and chemical steps. For example, a plant uses ~40% of the conversion from sunlight to carbohydrates for metabolic needs.

QUESTION 17

Calculate the conversion efficiency from sunlight to carbohydrates. Chapter 3 covers additive conversion efficiencies.

Q17 ANSWER Taking into account the conversion efficiencies mentioned above:
   The 400–700-nm region contains ~50% of total energy in solar photons.
   Green plants absorb ~80% of photons in the same solar range:

$$0.50 \times 0.80 = 0.40$$

Therefore, ~40% of the total solar energy is available for photosynthesis.
   Of all the energy that makes it into the plant to do photosynthesis, only 28% actually forms carbohydrates:

$$0.40 \times 0.28 = 0.11$$

Thus, the conversion efficiency from sunlight to carbohydrates is approximately 11%.

However, the plant uses ~40% of the energy for its own metabolic needs, so only ~60% of the energy is used for energy storage. This leaves only:

$$0.11 \times 0.6 = 0.067 \ (6.7\%)$$

as the fraction of sunlight stored as photosynthetic energy.

This maximum conversion efficiency applies to the so-called C4 plants, in which the first product of photosynthesis is a C4 sugar; these include corn, sorghum, and sugarcane, and some other plants that grow best in hot climates. The C3 plants, in which the first photosynthetic product is a C3 sugar, include wheat, rice, soybeans, trees, and other plants that dominate temperate climates and account for 95% of global plant biomass. These plants are about one-half as efficient in photosynthesis as the C4 plants. To these inefficiencies must be added the limitations on plant growth imposed by temperatures that are too low or too high, insufficient water, and insufficient nutrients. These limitations and inefficiencies explain why only ~0.3% of the global insolation reaching the earth's surface is used by green plants and algae in photosynthesis.

4.6 Hydroelectricity

Hydroelectric power plants utilize part of the energy of the solar-driven hydrological cycle. The continents of the world receive more rain than the water they lose by evapotranspiration, and the excess runs off in rivers to the ocean. The running water can be used to turn a turbine and generate electricity. Although there are many small hydroelectric facilities that utilize river flow directly, the larger installations, which account for most of the available hydropower, rely on dams to increase the hydraulic head (water pressure) and to even out the flow, thereby allowing the continuous production of electricity. Dams serve other purposes as well, including the provision of water for residential, industrial, and agricultural purposes; facilitating flood control and/or navigation; and providing recreational facilities. In fact, most dams do not generate electricity, although many could be retrofitted to do so. Hydroelectricity is produced by the force of falling water (Fig. 4.6). The power produced depends on the volume of water that turns the turbine generator and on the height from which the water falls (water head).

$$P = d \times g \times h \times f \times C_p \tag{4.1}$$

d = water density (1,000 kg m^{-3})
g = acceleration due to gravity (9.8 m sec^{-2})
h = water head (in m)
f = flow rate (volume of water flowing per second, m^3 sec^{-1})
C_p = hydroelectric plant operation efficiency

One joule (J) of energy is equal to 1 kg m^2 sec^{-2} and 1 J sec^{-1} is equal to a watt (W) of energy. Multiplying the units of water flow together gives:

$$\frac{\text{kg}}{\text{m}^3} \times \frac{\text{m}}{\text{sec}^2} \times \text{m} \times \frac{\text{m}^3}{\text{sec}} \times \frac{\text{kg m}^2}{\text{sec}^3} \times \frac{\text{J}}{\text{sec}} = \text{W}$$

About 70%–80% of the energy in the falling water can be captured by the turbines, and further losses in converting the mechanical energy into electrical energy reduce the system's efficiency (C_p) to ~50%–60%.

Hydroelectric Power Generation

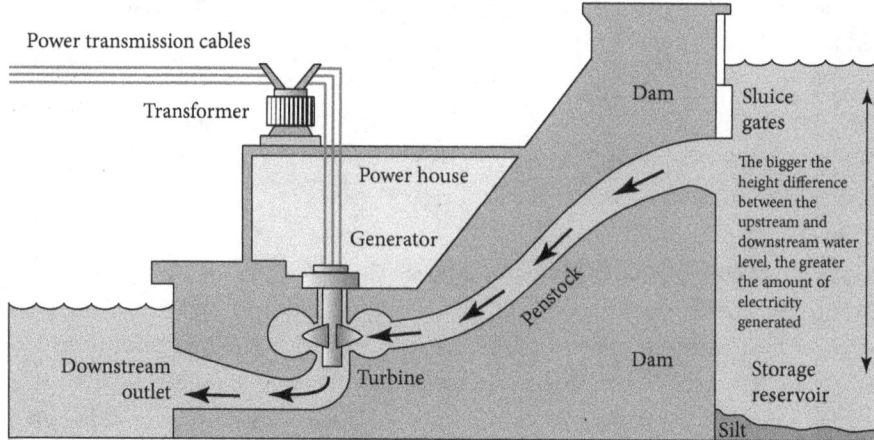

Figure 4.6 Hydroelectric power generation plant.
Source: Adapted from Spiro et al., *Chemistry of the Environment, 3rd Ed.,* University Science Books, © 2012, all rights reserved.

The share of the world energy production by hydropower is expected to be ~16% by 2025. This is a small fraction of the available hydrological potential, but probably represents one-third to one-fourth of the potential sites that could be developed economically.[4] Like other forms of solar energy, hydroelectricity adds no CO_2 or other emissions to the atmosphere, but it is not without environmental costs. For example, the concrete used to build the dam is a large source of CO_2 emissions. Other costs include flooding of towns and villages, artifacts, the disruption of fish migration, and removal of nutrients for downriver vegetation and farms.

Smaller pumped storage hydropower systems can be used to mitigate fluctuations in renewable energy production, such as wind and solar power. When the wind is blowing or the sun is shining and excess electricity is being produced, water can be pumped uphill to a storage reservoir. When wind speeds are low or at night, the water can flow downhill through a hydroelectric turbine to create electricity.

QUESTION 18

Water flows over 50.m high in Niagara Falls at an average rate of 1,800 m^3 sec^{-1}. If all the energy of the falling water could be harnessed by a hydroelectric power plant having 60.% efficiency, what would be the power output from that plant?

Q18 ANSWER Use equation (4.1) to obtain the answer:

$$P = d \times g \times h \times f \times C_p$$
$$= 1,000.\ kg\ m^{-3} \times 9.8\ m\ sec^{-2} \times 50.\ m \times 1,800\ m^3\ sec^{-1} \times 0.60$$
$$= 5.3 \times 10^8\ W\ (or\ 0.53\ GW)$$

This is enough energy to power about 100,000 homes.

---

[4] International Energy Agency, *Hydropower, Bioenergy, CSP and Geothermal,* 2020, available at https://www.iea.org/reports/renewables-2020/hydropower-bioenergy-csp-and-geothermal

*QUESTION 19

Victoria Falls in southern Africa has one of the longest curtains of water in the world. If the height of the waterfall is 108m and an average flow of 1,088 m^3 sec^{-1} occurs over the falls, what is the potential energy that could be harnessed with hydroelectric power, if the process is 55% efficient?

QUESTION 20

What is the percent difference in power input from the plant at Niagara Falls if it has an efficiency of 63% versus 60.%?

Q20 ANSWER The power for a 60.% efficient plant at Niagara Falls was already calculated to be 5.3 × 10^8 W (or 0.53 GW).

If the plant is 63% efficient, then:

$$P = d \times g \times h \times f \times C_p$$

$$= 1{,}000 \text{ kg m}^{-3} \times 9.8 \text{ m sec}^{-2} \times 50. \text{ m} \times 1{,}800 \text{ m}^3 \text{ sec}^{-1} \times 0.63$$

$$= 6.3 \times 10^8 \text{ W (or 0.63 GW)}$$

$$\text{percent difference} = \frac{|0.63 - 0.53|}{0.53} \times 100\% = 19\%$$

This shows that a small increase (or decrease) in the efficiency of the plant can make a difference in energy output.

*QUESTION 21

What is the percent difference in power output for a hydroelectric plant that is 59% efficient, in comparison to 55% efficient, at Victoria Falls?

4.7 Wind Power

The wind represents another form of solar power because wind results from air-temperature differences associated with different rates of solar heating. A global circulation of air, the Hadley circulation, is created by moist hot air rising at the equator and being replaced by drier air flowing in from the region of 30° north and south latitude. At higher latitudes, the air flows toward the poles and is deflected westward by the earth's rotation, creating a wavelike pattern known as the Rossby circulation. Regional variations in atmospheric temperature superimpose smaller circulation systems on the global pattern. Locally strong winds are created by sharp temperature differences between, for example, the land and the sea, and they can be channeled by mountains and valleys. Many regions have steady prevailing winds as a result of these conditions.

Wind power system costs have fallen steadily, and now wind electricity is the lowest cost alternative to electricity from fossil fuel and nuclear plants and is growing rapidly. Installed capacity by the end of 2019 was >622,408 MW worldwide.[5]

---

[5] ResearchandMarkets.com, *The Worldwide Wind Turbine Industry Is Expected to Grow at a 5.34% Between 2019 and 2025,* 2020, available at https://www.globenewswire.com/news-release/2020/11/25/2133532/0/en /The-Worldwide-Wind-Turbine-Industry-is-Expected-to-Grow-at-a-5-34-Between-2019-and-2025.html

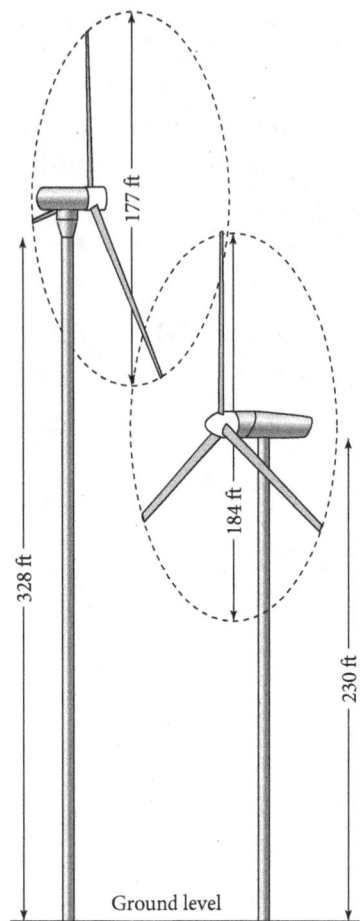

177 ft

328 ft

184 ft

230 ft

Ground level

Figure 4.7 State-of-the-art wind turbines with high hub heights and large rotor diameters.
Sources: Adapted from U.S. Department of Energy, Wind, and Hydropower Technologies Program, 2010, originally available at http://www1.eere.energy.gov/windandhydro/wind_how.html, and Spiro et al., *Chemistry of the Environment, 3rd Ed.,* University Science Books, © 2012, all rights reserved.

Current advanced turbines are of the horizontal axis design (Fig. 4.7). They capture the wind's energy with two or three propeller-like blades, which are mounted on a rotor to generate electricity. The turbines sit atop tall towers, taking advantage of increased wind speed and decreased turbulence with altitude. Strong winds are critical because power generation increases with the cube of wind speed. It increases linearly with the area swept by the propellers, which is proportional to the square of their length. Consequently, there is a premium on using long propellers mounted on tall towers. The limiting factor on size is the possibility of structural failure in high winds. Innovative tower designs using stronger lightweight materials now allow taller towers to be built at a reduced cost. A wind turbine installed in the United States in the 1980s had a hub height of 20m with a 20m rotor length. As a comparison, the average hub height of installations in the United States in 2019 was 88m with an average rotor length of 116m. Wind turbine sizes are determined by the site where they are located to maximize energy generation.

Wind turbines operate by converting the kinetic energy (KE) of the wind into mechanical energy of the rotor. The equation for power produced (4.2) is analogous to (4.1) for hydropower and is obtained from the KE per unit time (Δt) of a parcel of air moving through the rotor.

The air mass is calculated by the relationship of air density, the area swept by the rotor, the wind speed, and change in time:

$$m_{air} = d \times A \times v \times \Delta t \tag{4.2}$$

d = air density, in kilograms per cubic meter (kg m^{-3})(1.204 kg m^{-3} for sea-level dry air
   at 20°C)

A = area swept by the rotor, in square meters (m^2)($\pi \times r^2$, where r is rotor radius)

v = wind speed, in meters per second (m sec^{-1})

Δt = change in time

The KE is calculated with the following relationship:

$$\text{KE} = \frac{1}{2} \times m \times v^2 \qquad (4.3)$$

m = mass of air moving through the turbine (kg)

v = wind speed, in meters per second (m sec^{-1})

m_{air} (4.2) can be used to calculate the KE of the air mass by substituting the relationship
for m_{air} into the KE relationship (4.3):

$$\text{KE} = \frac{1}{2} \times m \times v^2 = \frac{1}{2} \times d \times A \times v \times \Delta t \times v^2 \qquad (4.4)$$

Dividing by Δt and multiplying by the efficiency factor (C_p) yields (4.5), the power generated by a wind turbine:

$$P = \frac{\text{KE}}{\Delta t} \times C_p = \frac{1}{2} \times d \times A \times v^3 \times C_p \qquad (4.5)$$

Again, like a hydroelectric plant, multiplication of the units yields J sec^{-1} = W.

QUESTION 22

How much power can be generated from a single turbine with a rotor length of 109m, a wind speed
of 4.0 m sec^{-1}, and average efficiency of 42%? Assume that the air is for sea-level dry air at 20°C.

Q22 ANSWER   First, the area of the turbine needs to be calculated:

$$A = \pi \times r^2 = \pi \times (109 \text{ m})^2 = 3.73 \times 10^4 \text{ m}^2$$

$$P = \frac{1}{2} \times d \times A \times v^3 \times C_p = \frac{1}{2} \times 1.204\frac{\text{kg}}{\text{m}^3} \times (3.73 \times 10^4 \text{ m}^2) \times \left(\frac{4.0 \text{ m}}{\text{s}}\right)^3 \times 0.42$$

$$= 6.0 \times 10^5 \frac{\text{kg m}^2}{\text{s}^3} = 6.0 \times 10^5 \frac{\text{J}}{\text{s}} = 6.0 \times 10^5 \text{ W}$$

Not all of the wind's energy is available because the wind velocity at the turbine falls as
the energy is extracted. The theoretical limit for the extractable fraction is 59%. In practice,
current wind turbines reach 50% efficiency.

Average wind speeds vary greatly from place to place and determine the practicality of
wind power. Figure 4.8 is a map of the United States showing regions having class 3 or higher
winds. (Class 3 winds have average power densities of 150–250 W m^{-2} at 10m height and
300–400 W m^{-2} at 50m.) Although wind power was first developed in California, the potential is much higher in the Great Plains, where a swath of class 3 or higher winds extends from
the Dakotas to Texas.

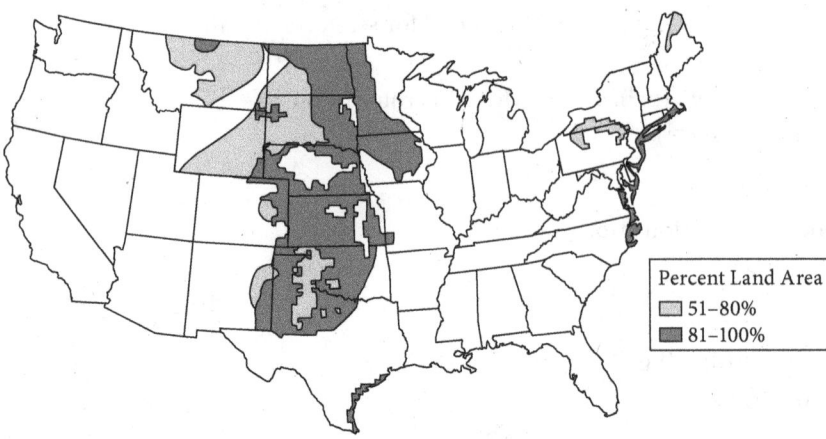

Figure 4.8 Percentage of the land area estimated to have class 3 or higher wind power in the contiguous United States. *Sources:* National Renewable Energy Laboratory, U.S. Department of Energy, *Wind Energy Resource Atlas of the United States,* 1986. Adapted from Spiro et al., *Chemistry of the Environment, 3rd Ed.,* University Science Books, © 2012, all rights reserved.

---

*QUESTION 23

Question 22 was for a turbine in California experiencing lower wind speeds. Calculate the percent difference in power generated by the same turbine in the Texas panhandle, with an average wind speed of 7.5 m sec^{-1}. Again, assume that the air is for sea-level dry air at 20°C.

Because the power generated by the wind is dependent on the velocity cubed, a small change in velocity of 3.5 m sec^{-1} on average between California and Texas can greatly increase the power input.

Here are a few more calculations one can do with wind energy.

QUESTION 24

Assume a wind turbine with a hub 50. m above the ground, a rotor diameter of 50. m, and a wind conversion efficiency of 25%. The turbine operates in an area with an annual average wind-power density of 500. W m^{-2}. How much electrical energy (in kWh) can the turbine generate per year? Compare this with the average household use of ~20 kWh day^{-1}.

Q22 ANSWER With a power density of 500. W m^{-2}:

$$\text{energy generated} = \frac{0.500 \text{ kW}}{\text{m}^2} \times \frac{24 \text{ hr}}{\text{day}} \times \frac{365 \text{ days}}{\text{yr}} = 4{,}380 \frac{\text{kWh}}{\text{m}^2 \text{ yr}}$$

The area covered by the wind turbine = πr^2, where r = 50. m; thus:

$$\text{area} = \pi \times (50. \text{ m})^2 = 7.9 \times 10^3 \text{ m}^2$$

Incorporating 25% efficiency:

$$\text{electrical energy per turbine} = 7.9 \times 10^3 \text{ m}^2 \times 4{,}380 \frac{\text{kWh}}{\text{m}^2 \text{ yr}} \times 0.25 = 8.6 \times 10^6 \frac{\text{kWh}}{\text{yr}}$$

Per day, a turbine output is:

$$8.6 \times 10^6 \frac{\text{kWh}}{\text{yr}} \times \frac{\text{yr}}{365 \text{ days}} = 2.4 \times 10^4 \frac{\text{kWh}}{\text{day}}$$

This is enough to power:

$$2.4 \times 10^4 \frac{\text{kWh}}{\text{day}} \times \frac{\text{day}}{24 \text{ kWh}} = 1.2 \times 10^3 \text{ households}$$

QUESTION 25

Wind densities ≥ 500 W m^{-2} at an altitude of 50. m are exploitable with today's technologies; ~1.2% of the land area of the contiguous United States (the total land area is 7,827,989 km^2; 1 TWh = 10^9 kWh; 1 kWh = 3.6×10^3 kJ) possesses such wind densities. If, on average, wind farms contain eight turbines per square kilometer, what is the U.S. potential for electrical energy production from wind power? (Assume a uniform power density of 500. W m^{-2}, and the same specifications as for the turbine in Q24.) Compare this potential with the 4,000 TWh of electricity produced in the United States in 2019.

Q25 ANSWER Land area covered by wind farms = (7,827,989 km^2) × 0.012 = 9.4×10^4 km^2. Based on the above calculation of the energy created per turbine:

$$\frac{\text{power generated}}{\text{km}^2} = \frac{8 \text{ turbines}}{\text{km}^2} \times \frac{8.6 \times 10^6 \text{ kWh}}{\text{yr}} = 6.9 \times 10^7 \frac{\text{kWh}}{\text{km}^2 \text{ yr}}$$

The total wind energy potential in the United States:

$$\frac{6.9 \times 10^7 \text{ kWh}}{\text{km}^2 \text{ yr}} \times 9.4 \times 10^4 \text{ km}^2 = 6.5 \times 10^2 \frac{\text{kWh}}{\text{yr}}$$

Given that U.S. electricity production in 2019 = 4,000 TWh = 4.00×10^{12} kWh, the potential percent electricity that could derive from the wind is:

$$\frac{6.5 \times 10^{12} \text{ kWh}}{4.00 \times 10^{12} \text{ kWh}} \times 100\% = 160\%$$

*QUESTION 26

Continuing technical advances will allow wind-power generation on lands where the wind power density is 300. W m^{-2} at 50. m (assume air density is 1.22 kg m^{-3}). In this case, wind power could be harvested on 21% of U.S. land. Assuming that one-third of the land was covered by wind farms (with a density of eight turbines per square kilometer), how much electricity would be generated? What percentage of the U.S. demand could be met? (The total land area of the contiguous United States is 7,827,989 km^2; 1 TWh = 10^9 kWh; 1 kWh = 3.6×10^3 kJ.) Assume a uniform power density of 300 W m^{-2}, and wind turbines with the same design as in Q24.

4.8 Conclusions

The use of renewable energy sources such as solar thermal, solar PV, and wind are growing as a percentage of energy generation around the world. Large-scale hydroelectric power is already extensively used in many countries, but still has potential in many developing countries. Currently, we use a small percentage of the energy resources available with regard to renewable energy. As the technological challenges of energy storage are solved, we will be able to utilize energy from the sun when it is not shining and electricity from the wind when it is not blowing.

5

FOSSIL FUELS

5.1 Closing the Carbon Budget:
Fossil Fuel Burning and Increasing Atmospheric CO_2 Concentration

Fossil fuels have provided the majority of the energy to enable global industrialization. Because their combustion results in a net transfer of carbon from underground reserves to the atmosphere (see Fig. 5.1), fossil energy use trends and fuel source choices are of great relevance to understanding current and future climate change. It is, thus, highly relevant to understand the relative energy content of different fuels, and the technological prospects for removing and storing atmospheric carbon, effectively returning it below ground. The following problems address some aspects of the fossil fuel energy problem.

The representative carbon cycle diagram shows the large fluxes (transfer) of carbon between the atmosphere and ocean and the atmosphere and vegetation. In both cases, the (smaller) difference between up and down fluxes indicates a net transfer of carbon to the biosphere and ocean (e.g., for the biosphere cycling, $119.6 - 120 - 2.6 = -3$ Gt C yr^{-1} means a net downward flux from the atmosphere to the vegetation of 3 Gt C yr^{-1}).

The transfer of carbon from fossil reservoirs to the atmosphere, which can be seen in comparing fluxes in the global carbon cycle (Fig. 5.1), has resulted in an increasing atmospheric concentration of CO_2 (Fig. 5.2).

In these problems, we will assess whether the flux of carbon from fossil fuel reservoirs into the atmosphere accords with the observed annual increases in atmospheric CO_2.

QUESTION 1

Look at the global carbon cycle depicted in Figure 5.1 to find an estimate of the annual fossil fuel emissions to the atmosphere and then translate this to an expected annual increase in ppm CO_2, assuming all other fluxes balance one another. Then, compare your number to what you see in Figure 5.2, showing historic increases in atmospheric carbon. Do they agree? *Useful information:* The atmosphere contains 1.80×10^{20} mol of gas.

Q1 ANSWER In Figure 5.1, we see a flux of carbon from fossil fuel reservoirs into the atmosphere of 6.4 Gt C yr^{-1}. To determine how much this input of carbon dioxide would increase the atmospheric mixing ratio (concentration, ppm) of CO_2, we first convert the flux to moles CO_2:

$$\frac{6.4\ \text{Gt C}}{\text{yr}} \times \frac{1 \times 10^{15}\ \text{g}}{1\ \text{Gt}} \times \frac{1\ \text{mol C}}{12\ \text{g C}} \times \frac{1\ \text{mol CO}_2}{1\ \text{mol C}} = 5.3 \times 10^{14}\ \text{mol CO}_2 \text{ increase per year}$$

Now divide this increase by the total moles of gas in the atmosphere and multiply by 1 million to determine the ppm increase:

$$\frac{5.3 \times 10^{14}\ \text{mol CO}_2}{1.8 \times 10^{20}\ \text{mol atmos}} \times 1 \times 10^6 \text{ parts per million (ppm)} = 3.0 \text{ ppm increase per year}$$

Inspecting Figure 5.2, the increase appears to be ~20 ppm over the most recent ~10 yr, or about 2 ppm yr^{-1}, which is only two-thirds of the calculated increase.

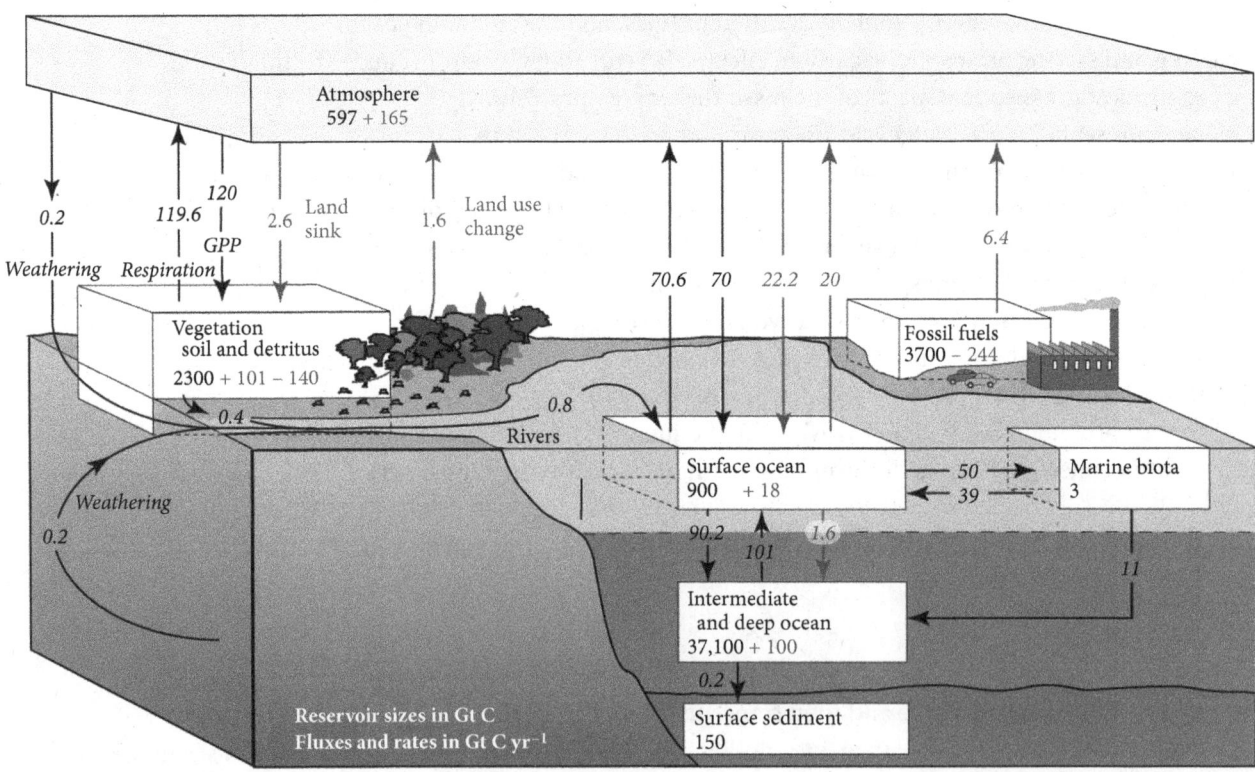

Figure 5.1 This diagram of the global carbon cycle shows the carbon reservoirs (Gt C) and major annual fluxes [Gt C per year (Gt C yr^{-1})]. Preindustrial estimates of values are in black, with estimated anthropogenic contributions in gray.
Sources: Adapted from a report of Working Group I of the Intergovernmental Panel on Climate Change (IPCC), 2007; and Spiro et al., *Chemistry of the Environment, 3rd Ed.*, University Science Books, © 2012, all rights reserved.

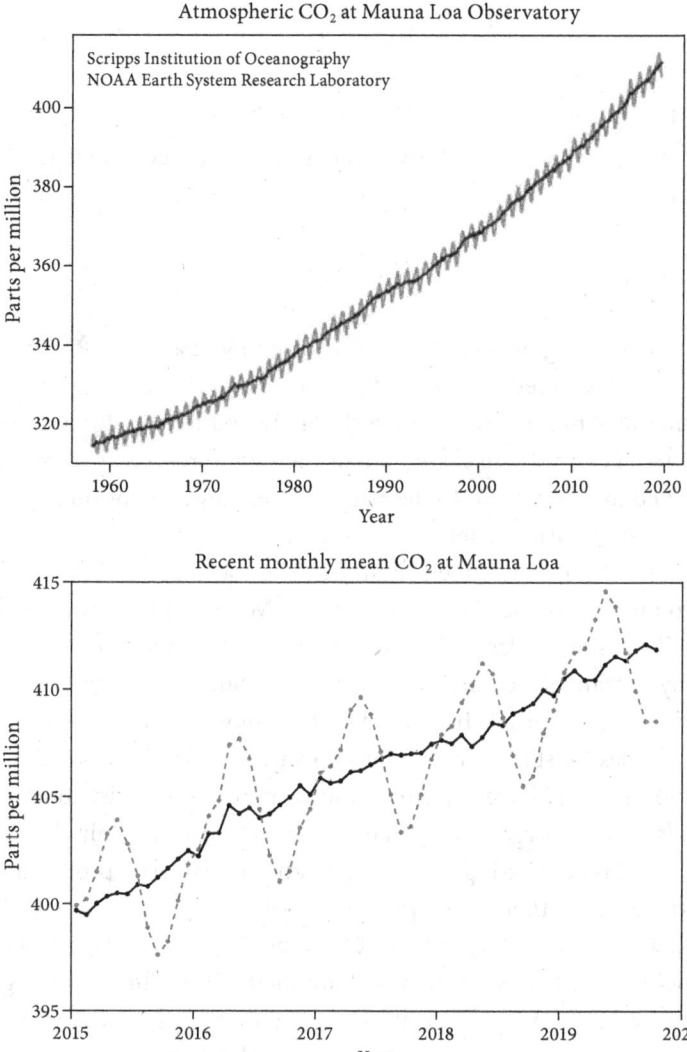

Figure 5.2 Historic increases in the atmospheric concentration of CO_2.
Source: Adapted from The Global Monitoring Laboratory, National Oceanic & Atmospheric Administration, *Trends in Atmospheric Carbon Dioxide,* available at https://www.esrl.noaa.gov/gmd/ccgg/trends/.

QUESTION 2

How would you explain this discrepancy between the calculated and observed concentration increase?

Q2 ANSWER The assumption that other fluxes balance one another must be in error. In fact, we saw that the fluxes between the atmosphere and vegetation and the oceans do not balance. A net downward flux of 3 Gt C yr^{-1} occurs. If we subtract that from the 6.4 Gt C yr^{-1} of fossil fuel emissions, the net increase is 3.4 Gt C yr^{-1}. The predicted increase is then 3 ppm × 3.4/6.4 = 1.6 ppm, which is close to the observed ~2 ppm.

---

*QUESTION 3

Land-use change also contributes to carbon in the atmosphere. Look at the global carbon cycle depicted in Figure 5.1 to find an estimate of land-use emissions to the atmosphere and then translate this to an annual increase in ppm CO_2.

5.2 Origins of Fossil Fuels

Fossil fuels can be seen as a "detour" in the global carbon cycle. About 1%–2% of dead organic matter does not decay, but rather is preserved and buried underground. Here it undergoes anaerobic biochemical and geochemical processing to convert organic matter into fossil fuels. Because these fossil fuels derive from once-living plants, fossil fuels are a reservoir of stored solar energy. When these are burned to release CO_2 to the atmosphere, they add another route for the movement of carbon from land to atmosphere (see Fig. 5.3).

Petroleum and natural gas deposits are of marine origin. In the oceans, photosynthesis is estimated to produce 25–50 billion tons of reduced carbon annually. Most of this is recycled to the atmosphere as CO_2, but a small fraction settles to the bottom, where there is no access to O_2. This biological debris is covered by clay and sand particles and forms a compacted organic layer in a matrix of porous clay or sandstone. Anaerobic bacteria digest the biological matter, releasing CH_4, NH_3, and H_2O. The molecules most resistant to digestion are hydrocarbon-based lipids. The saturated hydrocarbons found in oil have structural and carbon number distributions similar to those found in the lipids of living organisms. All petroleum deposits contain derivatives of the hydrocarbon hopane ($C_{30}H_{52}$), attesting to the importance of bacterial processing, since bacteria contain hopane derivatives in their membranes (Fig. 5.4).

In contrast to oil and gas, coal is of terrestrial origin. Coal deposits are the remains of plant matter from the huge, thickly wooded swamps that flourished 250 million years ago during a period of mild and moist climate. Woody plants are made up mainly of lignin and cellulose. While aerobic bacteria rapidly oxidize cellulose (a carbohydrate) to CO_2 and H_2O when the plant dies, lignin is much more resistant to bacterial action. Lignin is a complex, three-dimensional (3D) polymer based on benzene rings (Fig. 5.5). The building units are coniferyl and sinapyl alcohol for lignins from coniferous and deciduous plants, respectively.

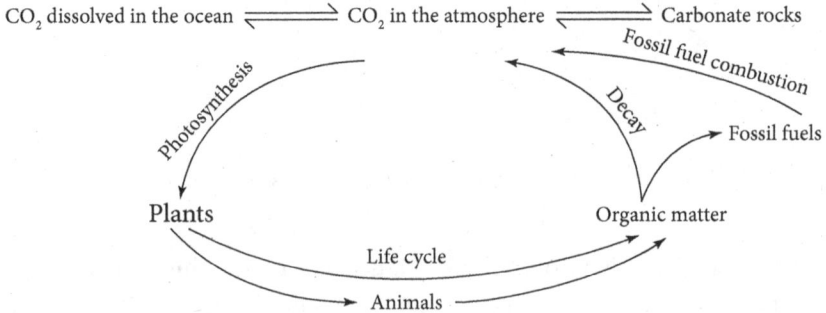

Figure 5.3 Carbon cycle with the fossil fuel "detour" shown.
Source: Adapted from H. Schobert, *Chemistry of Fossil Fuels and Biofuels,* Cambridge University Press, 2013, Fig. 1.5.

---

* Answers to starred questions can be found at the end of the book.

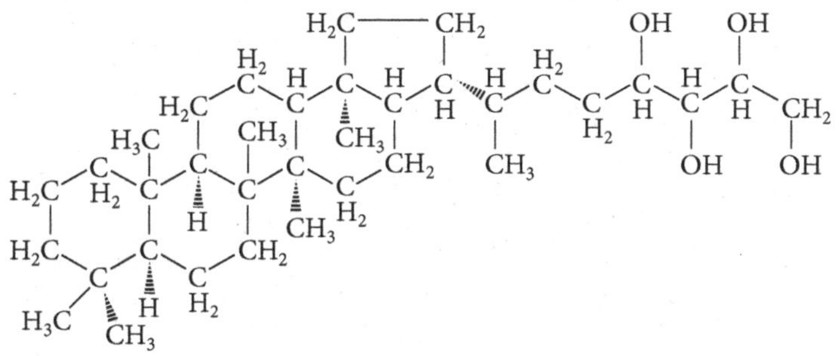

Figure 5.4 The structure of bacteriohopanetetrol.
Source: Adapted from Spiro et al., *Chemistry of the Environment, 3rd Ed.*, University Science Books, ©
2012, all rights reserved.

5.3 Fuel Energy

Different fuel types combust to release greater or less energy per mass of fuel burned. This
can be rationalized in terms of the energy contained in various types of chemical bonds and
chemical structures of the components of the fuels. Chemical energy is released by combustion based on the net energy change of combustion reactions:

$$\text{Fuel} + O_2 \rightarrow H_2O + CO_2 \tag{5.1}$$

We know that the large exothermic energy release arises because the formation of strong
new C=O and O–H bonds in CO_2 and H_2O releases substantially more energy than is
required to break C–C, C=C, and C–H bonds in hydrocarbon fuels (see Table 5.1). However,
C–O and O–H bonds are stronger than C–C bonds (although not as strong as the aromatic
1.5 bond order carbon–carbon bonds), so their presence in fuel materials reduces the net
energy release. In addition, since oxygen atoms are 30% more massive than carbon atoms, the
fuel energy density (kJ g^{-1}) penalty of adding oxygen is larger than simply the bond enthalpy
differences.

Table 5.1 shows the bond dissociation energies for different types of bonds, and Figure 5.5
shows some of the structural subunits of coal. In general, multiple bonds are stronger than

TABLE 5.1 Bond Dissociation Enthalpies for Selected Bonds

Bond	Enthalpy (kJ mol^{-1})	Bond	Enthalpy (kJ mol^{-1})
H–H	432	C≡O	1071
O=O	494	C–C	347
O–H	460	C=O	611
C–H	410	C⋯C^a	519
C–O	360	N=O	623
C=O	799	N≡N	941

^{a}Aromatic, 1.5 bond order

Source: Adapted from Spiro et al., *Chemistry of the Environment, 3rd Ed.*, University
Science Books, © 2012, all rights reserved.

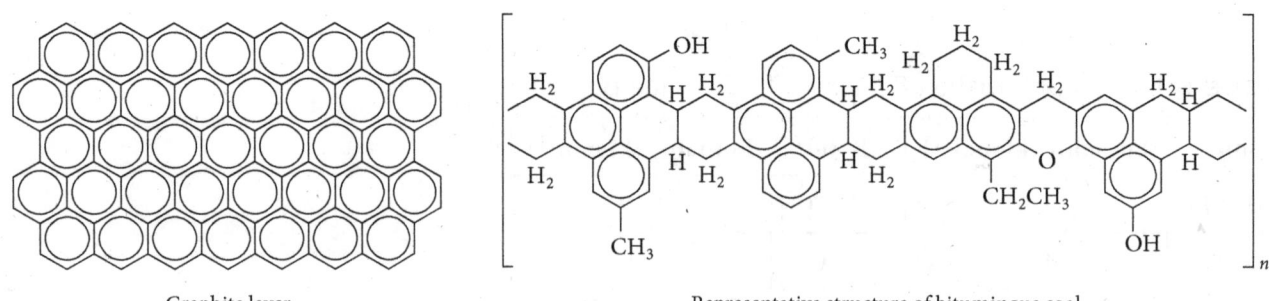

Figure 5.5 Structural units of coal.
Source: Adapted from Spiro et al., *Chemistry of the Environment, 3rd Ed.,* University Science Books, © 2012, all rights reserved.

single bonds (because a large number of electrons are shared), and more polar bonds (e.g., bonds including O) are stronger than the carbon–carbon bond. Because all combustion reactions produce CO_2 and H_2O, the differences in energy content of fuels are determined by the strength of the bonds that must be broken in the fuels and the number of CO_2 and H_2O molecules produced (reaction stoichiometry).

5.3.1 Comparing the Components of Fossil Fuel Energy Contents

Table 5.2 shows the relative energy content and CO_2 emissions from different fossil fuels, demonstrating that choices among different fossil fuel mixes can affect global CO_2 emissions. We see that methane (CH_4) has the lowest CO_2 emission rate per unit of energy among the hydrocarbons listed in Table 5.2. Because burning CH_4 emits less CO_2 per energy produced than petroleum, CH_4 is a "cleaner" fuel source from a greenhouse gas perspective.

TABLE 5.2 Combustion Energetics Estimated from Bond Energies

	Energy content (kJ)				
	Reaction enthalpy	Per mol O_2	Per mol fuel	Per g fuel	Mol CO_2/ 100kJ
Hydrogen $2H_2 + O_2 \rightarrow 2H_2O$	482	482	241	120	0
Methane $CH_4 + 2O_2 \rightarrow CO_2 + 2H_2O$	810	405	810	51.6	1.2
Petroleum $2(-CH_2-) + 3O_2 \rightarrow 2CO_2 + 2H_2O$	1220	407	610	43.6	1.6
Coal $4(-CH-) + 5O_2 \rightarrow 4CO_2 + 2H_2O$	2046	409	512	39.3	2.0
Ethanol $C_2H_5OH + 3O_2 \rightarrow 2CO_2 + 3H_2O$	1257	419	1257	27.3	1.6
Carbohydrate $4(-CHOH-) + O_2 \rightarrow CO_2 + H_2O$	447	447	447	14.9	2.2

Source: Adapted from Spiro et al., *Chemistry of the Environment, 3rd Ed.,* University Science Books, © 2012, all rights reserved.

QUESTION 4

Check the energy values for methane combustion in Table 5.2, using the bond energies in Table 5.1.

Q4 ANSWER The equation for methane combustion:

$$CH_4 + 2O_2 \rightarrow CO_2 + 2H_2O$$

has one molecule of methane and two molecules of oxygen as reactants and one molecule of carbon dioxide and two molecules of water as products.

The product bonds are:

$$\text{Two C=O bonds, } 2 \text{ mol} \times 799 \frac{kJ}{mol} = 1{,}598 \text{ kJ}$$

$$\text{Four O–H bonds, } 4 \text{ mol} \times 460 \frac{kJ}{mol} = 1{,}840 \text{ kJ}$$

$$\text{Total bonds formed} = 1{,}598 \text{ kJ} + 1{,}840 \text{ kJ} = 3{,}438 \text{ kJ}$$

The reactant bonds are:

$$\text{Four C–H bonds } 4 \text{ mol} \times 410 \frac{kJ}{mol} = 1{,}640 \text{ kJ}$$

$$\text{Two O=O bonds } 2 \text{ mol} \times 494 \frac{kJ}{mol} = 988 \text{ kJ}$$

$$\text{Total bonds broken} = 1{,}640 \text{ kJ} + 988 \text{ kJ} = 2{,}628 \text{ kJ}$$

Subtracting reactant energies from product energies:

$$2{,}628 \text{ kJ} - 3{,}438 \text{ kJ} = -810 \text{ kJ}$$

This is also the energy per mol of methane, but per mol of oxygen, the energy is halved to −405 kJ. To obtain the energy per g of fuel:

$$\frac{810 \frac{kJ}{mol}}{16 \frac{g}{mol}} = 51.6 \text{ kJ of energy given off per g of methane}$$

One mole of CO_2 is released per mol of methane, but to obtain 1,000 kJ of energy requires:

$$\frac{1{,}000 \text{ kJ}}{810 \text{ kJ}} = 1.2$$

an additional 1.2 mol CO_2.

*QUESTION 5

The major component of natural gas used by utility companies is methane (CH_4); the gas used in gas barbecues, and so on, is propane (C_3H_8). Assuming that each of these gases burns completely, compare the amount of energy released by each

(a) in terms of kJ mol^{-1} of CO_2 produced; and

(b) in terms of kJ g^{-1} of fuel.

Do the calculations based on the chemical bonds broken and formed in the combustion reactions. Use Table 5.1 to obtain bond energies.

*QUESTION 6

Check the energy values for ethanol combustion in Table 5.2, using the bond energies in Table 5.1. A car driven on ethanol gets lower mileage [miles per gallon (miles gal^{-1})] than an equivalent car driven on gasoline. The heat content of pure gasoline is 47.0 kJ g^{-1}; assume a molecular formula of C_8H_{18}. (Note: This number is higher than the aggregate petroleum energy content reported in Table 5.2 because these are the refined products.) Estimate how much less (assume for the purpose of this question that the densities of ethanol and gasoline are the same).

QUESTION 7

Combustion reactions have a large driving force because the O=O bond energy is less than the C=O bond energy in CO_2, and less than twice the O–H bond energy in H_2O (Table 5.1). What electronic factors are responsible for these inequalities?

Q7 ANSWER The O atom is more electronegative than H (in the O–H bond) and C in the C=O bond, which gives the bond's ionic characteristic. A negative charge accumulates on the O, whereas a positive charge accumulates on the H or C. This attraction of positive and negative charges leads to an increase in the bond energy. In general, ionic bonds are stronger than nonionic ones.

QUESTION 8

Based on the above discussion of bond enthalpies and fuel energies, rank the three structural units of coal shown on the left of Figure 5.5 by fuel energy density (kJ g^{-1}). Explain your reasoning by calculating the energy density for each using the bond energy calculations.

Q8 ANSWER Figure 5.5 shows the chemical structures of coniferyl alcohol, sinapyl alcohol, and graphite. Based on the discussion of chemical energy release by combustion, we know that the formation of strong new C=O and O–H bonds in CO_2 and H_2O releases more energy than is required to break C–C, C=C, and C–H bonds in hydrocarbon fuels (Table 5.1). However, C–O and O–H bonds are stronger than C–C bonds (although not as strong as the aromatic 1.5 bond order bonds). The presence of C–O and O–H bonds in fuels reduces the net energy released. Oxygen atoms are 30% larger in size than carbon atoms, so the fuel energy density (kJ g^{-1}) penalty of adding oxygen is larger than simply the bond enthalpy differences. By this qualitative argument, we might expect the trend of fuel energy density to be opposite the trend in the O:C ratio of the structural components:

Coniferyl alcohol	Sinapyl alcohol	Graphite
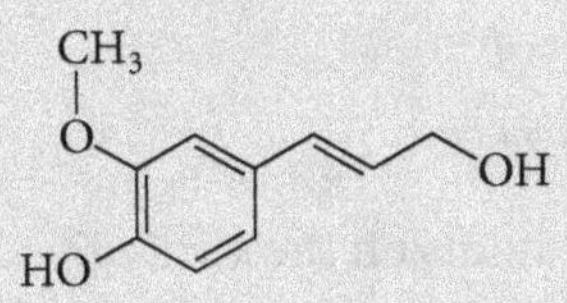	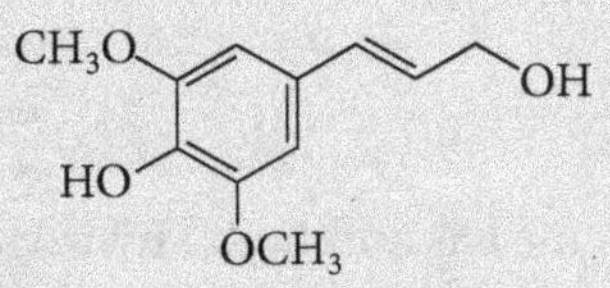	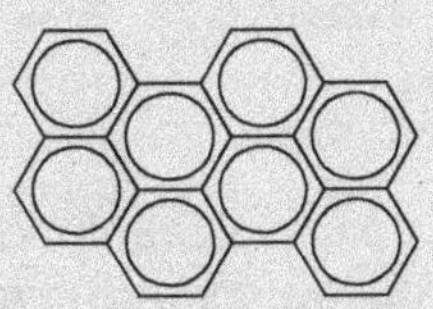
$C_{10}H_{12}O_3$; O:C = 0.30	$C_{11}H_{14}O_4$; O:C = 0.36	O:C = 0
Middle	Lowest energy density	Highest energy density

Let us check this predicted trend using the actual bond energy calculations:

(a) Coniferyl alcohol:

First, write a balanced chemical equation for the combustion reaction:

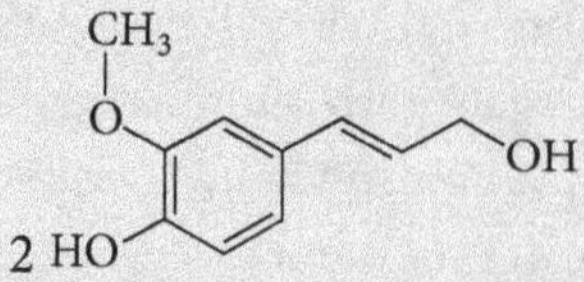

$+ 23\,O_2 \rightarrow 20\,CO_2 + 12\,H_2O$

Remember that the coniferyl alcohol molecule has additional hydrogen atoms on the structure. Check out Appendix A section 6 for a discussion of carbon framework representations.

Now calculate the energy of all bonds broken and formed:

Bonds broken:

$$2\,\text{mol}\,[4\,(\text{C–O bond}) + 2\,(\text{O–H bond}) + 10\,(\text{C–H bond}) + 5\,(\text{C–C bond}) + 4\,(\text{C=C bond})]$$
$$+ 23\,\text{mol}\,(\text{O=O bond})$$
$$= 2\,\text{mol}\,[4\,(360\,\text{kJ mol}^{-1}) + 2(460\,\text{kJ mol}^{-1}) + 10(410\,\text{kJ mol}^{-1}) + 5(347\,\text{kJ mol}^{-1})$$
$$+ 4(611\,\text{kJ mol}^{-1})] + 23\,\text{mol}\,(494\,\text{kJ mol}^{-1}) = 32{,}640\,\text{kJ per 2 mol coniferyl combusted}$$

Bonds formed:

$$(20\,\text{mol}\,[2\,(\text{C=O bond})] + 12\,\text{mol}\,[2(\text{O–H bond})])$$
$$= (20\,\text{mol}\,[2(799\,\text{kJ mol}^{-1})] + 12\,\text{mol}\,[2(460\,\text{kJ mol}^{-1})])$$
$$= 43{,}000\,\text{kJ per 2 mol coniferyl combusted}$$

Subtracting reactant energies from product energies:

$$32{,}640\,\text{kJ} + (-43{,}000\,\text{kJ}) = -10{,}360\,\text{kJ released per 2 mol coniferyl combusted}$$

$$2\,\text{mol} \times \frac{180.20\,\text{g}}{\text{mol}} = 360.4\,\text{g coniferyl}$$

Therefore, the energy density $= \left| \dfrac{-10{,}360\,\text{kJ}}{360.4\,\text{g}} \right| = 28.7\,\text{kJ g}^{-1}$

(b) Sinapyl alcohol:

First, balance the chemical reaction:

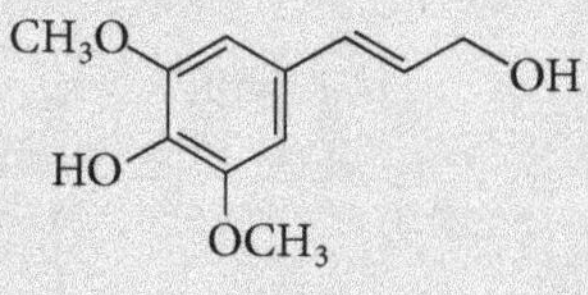

$+ 25\,O_2 \rightarrow 22\,CO_2 + 14\,H_2O$

Remember that the sinapyl alcohol molecule has additional hydrogen atoms on the structure. Check out Appendix A section 6 for a discussion of carbon framework representations.

Now calculate the energy of all bonds broken and formed:

Bonds broken:

$$2 \text{ mol } [6 \text{ (C–O bond)} + 2 \text{ (O–H bond)} + 12 \text{ (C–H bond)} + 5 \text{ (C–C bond)} + 4 \text{ (C=C bond)}]$$
$$+ 25 \text{ mol (O=O bond)}$$
$$= 2 \text{ mol } [6 \text{ (360 kJ mol}^{-1}) + 2 \text{ (460 kJ mol}^{-1}) + 12 \text{ (410 kJ mol}^{-1}) + 5 \text{ (347 kJ mol}^{-1})$$
$$+ 4 \text{ (611 kJ mol}^{-1})] + 25 \text{ mol (494 kJ mol}^{-1}) = 36{,}708 \text{ kJ per 2 mol sinapyl combusted}$$

Bonds formed:

$$(22 \text{ mol } [2 \text{ (C=O bond)}] + 14 \text{ mol } [2 \text{ (O–H bond)}])$$
$$= (22 \text{ mol } [2 \text{ (799 kJ mol}^{-1})] + 14 \text{ mol } [2 \text{ (460 kJ mol}^{-1})])$$
$$= 48{,}036 \text{ kJ per 2 mol sinapyl combusted}$$

Subtracting reactant energies from product energies:

$$32{,}708 \text{ kJ} - 48{,}036 \text{ kJ} = -11{,}328 \text{ kJ released per 2 mol sinapyl combusted}$$

$$2 \text{ mol} \times \frac{210.23 \text{ g}}{\text{mol}} = 420.46 \text{ g sinapyl}$$

Therefore, the energy density $= \left| \dfrac{-11{,}328 \text{ kJ}}{420.46 \text{ g}} \right| = 26.9 \text{ kJ g}^{-1}$

(c) Graphite:
First, balance the chemical reaction; for simplicity, just consider one 6-C unit:

$$+ 6 \, O_2 \rightarrow 6 \, CO_2$$

Remember that the graphite molecule has carbon atoms in the structure. Check out Appendix A section 6 for a discussion of carbon framework representations.

Now calculate the energy of all bonds broken and formed:

Bonds broken:

$$6 \text{ mol (aromatic 1.5 C–C bond)} + 6 \text{ mol (O=O bond)} =$$
$$6 \text{ mol (519 kJ mol}^{-1}) + 6 \text{ mol (494 kJ mol}^{-1}) = 6{,}078 \text{ kJ per ring combusted}$$

Bonds formed:

$$6 \text{ mol } [2 \text{ (C=O bond)}] = 6 \text{ mol } [2 \text{ (799 kJ mol}^{-1})] = 9{,}588 \text{ kJ per benzene combusted}$$

Subtracting reactant energies from product energies:

$$6{,}078 \text{ kJ} - 9{,}588 \text{ kJ} = -3{,}510 \text{ kJ released per benzene combusted}$$

$$1 \text{ ring} \times \frac{72.00 \text{ g}}{\text{mol}} = 72.00 \text{ g graphite}$$

Therefore, the energy density $= \left| \dfrac{-3510 \text{ kJ}}{72.00 \text{ g}} \right| = 48.8 \text{ kJ g}^{-1}$

Summary:

Coniferyl alcohol	Sinapyl alcohol	Graphite
$C_{10}H_{12}O_3$; O:C = 0.30	$C_{11}H_{14}O_4$; O:C = 0.36	O:C = 0
28.7 kJ g^{-1}	26.9 kJ g^{-1}	48.8 kJ g^{-1}

As predicted, energy density is inversely related to the O:C ratio.

*QUESTION 9

Rank the three structural units of lubrication oil, kerosene, and gasoline by fuel energy density (kJ g^{-1}) (qualitatively, since we do not know the exact molecular formulae, gasoline has the shortest average hydrocarbon chain length, followed by kerosene and then lubrication oil). See Figure 5.6 for approximate sizes of these petroleum fractions. Start by using bond energies to calculate the energy released per gram of fuel for pentane and decane to provide a general trend for hydrocarbon chain length versus energy released. Explain your reasoning.

5.3.2 Comparing Fossil Transport Fuels to an Alternative: H$_2$

In the discussion of alternative fuels for transportation, hydrogen (H$_2$) is the poster child for "green" fuels. While it is true that the combustion of hydrogen does not produce any CO$_2$, the impact of how hydrogen is produced is often not accounted for in these calculations.

QUESTION 10

The most common commercial route to hydrogen gas is steam reforming of natural gas, whereby methane reacts with high-temperature water vapor in the presence of a catalyst to form carbon monoxide and H$_2$ gas. Write a balanced chemical equation for steam reforming.

Q10 ANSWER The balanced chemical reaction for H$_2$ production described above would be:

$$CH_4 + H_2O \rightarrow CO + 3H_2 \quad \text{Steam reforming}$$

*QUESTION 11

Additional hydrogen can be formed by reacting carbon monoxide with water vapor in the water–gas shift reaction, which also oxidizes carbon monoxide to carbon dioxide. Write a balanced chemical equation for the water–gas shift reaction. Finally, write a balanced *net* chemical reaction for the formation of hydrogen from methane considering both steam reforming and the water–gas shift reactions.

QUESTION 12

Hydrogen has the highest energy density of any combustible fuel (heat content of $H_2 = 142$ kJ g^{-1}). Calculate the mass of CO_2 produced per kJ of energy for combustion of hydrogen from the synthesis of H_2 from methane. *Hint:* Consider burning 1.00 kg of H_2 and the CO_2 costs of producing 1.00 kg H_2.

Q12 ANSWER The amount of energy produced (kJ) from 1.00 kg of H_2 combustion is:

$$1.00 \text{ kg } H_2 \times \frac{1,000 \text{ g}}{1 \text{ kg}} \times \frac{142 \text{ kJ}}{1 \text{ g}} = 1.42 \times 10^5 \text{ kJ energy released}$$

Then, using the net chemical reaction for the formation of H_2 from methane determined in Q11:

$$CH_4 + 2 H_2O \rightarrow CO_2 + 4 H_2$$

Calculate the amount of CO_2 produced from 1.00 kg of H_2:

$$1.00 \text{ kg } H_2 \times \frac{1,000 \text{ g}}{1 \text{ kg}} \times \frac{1 \text{ mol } H_2}{2.016 \text{ g}} \times \frac{1 \text{ mol } CO_2}{4 \text{ mol } H_2} \times \frac{44.01 \text{ g } CO_2}{1 \text{ mol } CO_2} = 5,485 \text{ g } CO_2$$

Then, we can use the mass of CO_2 and the energy given off to calculate the ratio:

$$\frac{5,485 \text{ g } CO_2}{1.42 \times 10^5 \text{ kJ}} = 0.0384 \frac{\text{g } CO_2}{\text{kJ}}$$

QUESTION 13

Compare the mass of CO_2 produced per kJ of energy for combustion of methane-derived H_2 and gasoline. The heat content of pure gasoline is 47.0 kJ g^{-1}; assume a molecular formula of C_8H_{18}. (Note: This number is higher than the aggregate petroleum energy content reported in Table 5.2 because these are the refined products.) How do these fuels compare in terms of their relative impact on carbon emissions to the atmosphere?

Q13 ANSWER Gasoline: $C_8H_{18} + {}^{25}\!/_2 O_2 \rightarrow 8 CO_2 + 9 H_2O$

$$1.00 \text{ kg } C_8H_{18} \times \frac{1,000 \text{ g}}{1 \text{ kg}} \times \frac{47.0 \text{ kJ}}{1 \text{ g}} = 4.70 \times 10^4 \text{ kJ energy released}$$

Then, calculate the mass of CO_2 formed in the combustion of 1.00 kg of C_8H_{18}:

$$1.00 \text{ kg } C_8H_{18} \times \frac{1,000 \text{ g}}{1 \text{ kg}} \times \frac{1 \text{ mol } C_8H_{18}}{114.22 \text{ g}} \times \frac{8 \text{ mol } CO_2}{1 \text{ mol } C_8H_{18}} \times \frac{44.01 \text{ g } CO_2}{1 \text{ mol } CO_2} = 3,082 \text{ g } CO_2$$

Then, we can use the mass of CO_2 and the energy given off to calculate the ratio:

$$\frac{3,082 \text{ g } CO_2}{4.70 \times 10^4 \text{ kJ}} = 0.0656 \frac{\text{g } CO_2}{\text{kJ}}$$

The greenest fuel from a carbon emissions standpoint has the *lowest* mass of CO_2 released per kJ of energy created. Therefore, H_2 is more green than gasoline, according to this "green factor." If H_2 is derived from methane, then there are other environmental factors to think about, such as the energy to create H_2 and the high greenhouse potential for CH_4. We will discuss greenhouse potential in detail in Chapter 7.

*QUESTION 14

Compare the mass of CO_2 produced per kJ of energy for combustion of methane-derived H_2 and diesel fuel. The heat content of diesel fuel is 46 kJ g^{-1}; assume a molecular formula of $C_{12}H_{23}$. (Note: this number is higher than the aggregate petroleum energy content reported in Table 5.2 because these are the refined products.) How do these fuels compare in terms of their relative impact on carbon emissions to the atmosphere?

5.4 Petroleum

Oil is a complex mixture of hydrocarbons, molecules that contain mostly carbon and hydrogen. There are also small quantities of sulfur (up to 10%), oxygen (up to 5%), and nitrogen (up to 1%), bound in complex organic molecules. Several metallic elements (V, Ni, Fe, Al, Na, Ca, Cu, and U) are present in trace amounts. Most of the hydrocarbon molecules are saturated (no multiple bonds), but an appreciable fraction, ~30%, is aromatic (at least one benzene ring). The molecules range widely in size and are separated in refineries based on their boiling points. Figure 5.6 shows the distillation process for dividing petroleum into its various fractions and indicates the uses to which these fractions are placed.

A particularly important chemical transformation is "cracking," whereby a larger hydrocarbon, in the kerosene or gas–oil range, is broken down into two smaller hydrocarbons in the gasoline range. Cracking is accomplished at high temperatures (400–600°C) with the aid of a catalyst, an aluminosilicate material impregnated with potassium:

$$C_{(m+n)}H_{2(m+n)+2} \rightarrow C_mH_{2m} + C_nH_{2n+2}$$

$$\text{Alkane} \qquad\qquad \text{Alkene} \qquad \text{Alkane}$$

$$\text{(kerosene or gas-oil size)} \qquad \text{(gasoline size)} \tag{5.2}$$

Another way to enhance the gasoline fraction is to build up a midsize molecule from two smaller ones, in the process of "alkylation." The R groups on these molecules represent a chain of carbon atoms. In this reaction, the R group is added to one carbon of a double bond, and the H atom is added to the other carbon of the double bond:

$$\underset{\text{H}}{\overset{\text{R}}{>}}C=C\underset{\text{R}'}{\overset{\text{H}}{<}} + \text{R}''\text{H} \rightarrow H\underset{\text{H}}{\overset{\text{R}}{>}}C-C\underset{\text{R}'}{\overset{\text{H}}{<}}\text{R}''$$

$$\text{Alkene} \qquad \text{Alkane} \qquad\qquad \text{Alkane}$$

$$\text{(3, 4, or 5 C atoms)} \qquad\qquad \text{(gasoline size)} \tag{5.3}$$

This process is catalyzed by strong acids. The cracking and alkylation reactions increase the gasoline fraction of crude oil, typically 20% by volume, to 40%–45%.

QUESTION 15

Following the framework of generic equations [Eq. (5.2)] showing the chemical mechanisms to enhance the gasoline fraction of petroleum, write a balanced reaction for the cracking of *n*-dodecane.

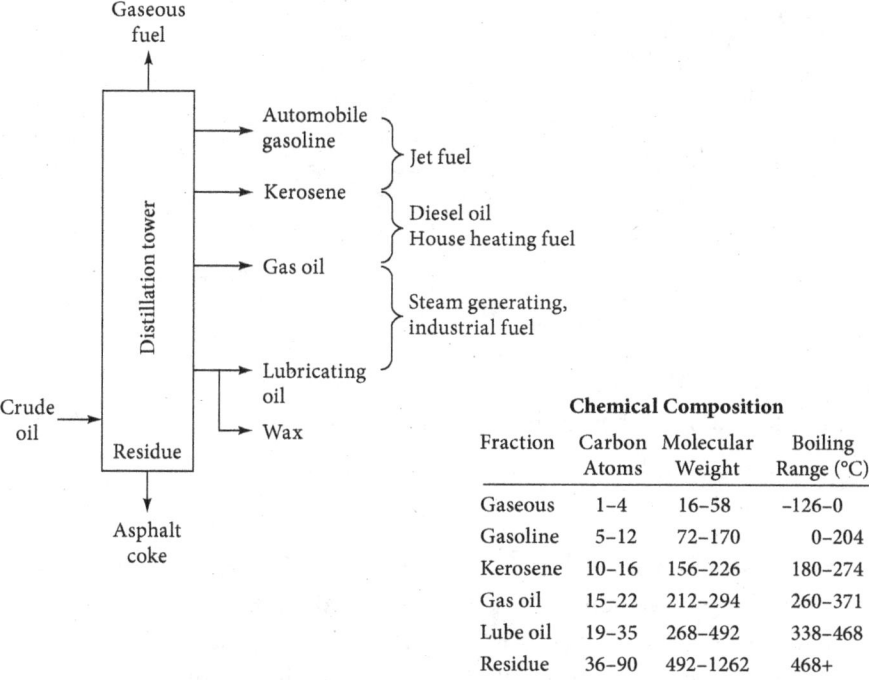

Chemical Composition

Fraction	Carbon Atoms	Molecular Weight	Boiling Range (°C)
Gaseous	1–4	16–58	–126–0
Gasoline	5–12	72–170	0–204
Kerosene	10–16	156–226	180–274
Gas oil	15–22	212–294	260–371
Lube oil	19–35	268–492	338–468
Residue	36–90	492–1262	468+

Figure 5.6 Distillation process for crude oil refining.
Source: Adapted from Spiro et al., *Chemistry of the Environment, 3rd Edition,* University Science Books, © 2012, all rights reserved.

Q15 ANSWER 400–600°C
Catalyst:

$$CH_3(CH_2)_9CH_3 \rightarrow CH_3CH_2CHCHCH_2CH_2CH_3 + CH_3CH_2CH_2CH_2CH_3$$

or

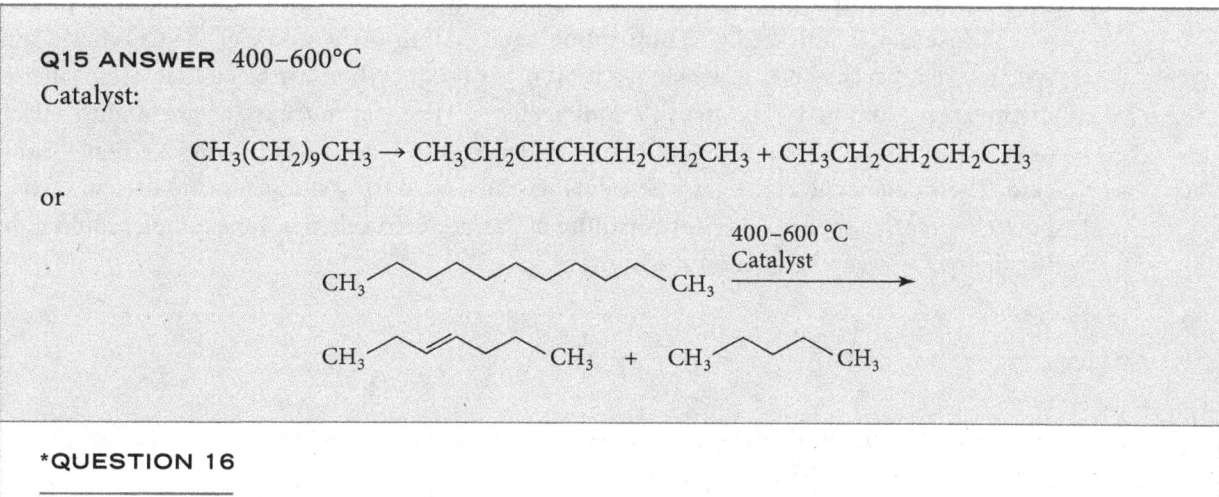

---

*QUESTION 16

Write a balanced reaction for the cracking of *n*-octane.

QUESTION 17

Following the framework of generic equations above [Eq. (5.3)] showing the chemical mechanisms to enhance the gasoline fraction of petroleum, write a balanced reaction for the alkylation of 3-octene by *n*-hexane.

Q17 ANSWER

*QUESTION 18
_____

Write a balanced reaction for the alkylation of 2-hexene by *n*-pentane.

5.5 Natural Gas

Although natural gas represents as large an energy resource as petroleum, it has historically been considered a by-product of oil exploration and production. Indeed, the magnitude of the potential gas supply has only recently become appreciated, thanks to better information about gas-bearing formations and to improved recovery techniques. Along with oil, large quantities of gas are locked in impermeable shale, from which it is being released by new technologies of hydraulic fracturing ("fracking") and horizontal drilling. Natural gas provides a substantial fraction of the U.S. energy consumption (>32%). Most of it is produced domestically, but some is shipped by pipeline from Canada. A small fraction is imported by tankers, as liquefied natural gas (LNG).

Although natural gas produces less CO_2 than other fossil fuels, CH_4 is itself a potent greenhouse gas. Its infrared (IR) absorption bands fall in the window of the CO_2 and H_2O spectra, and, because it is less reactive than other hydrocarbons, it has a long atmospheric lifetime. An additional CH_4 molecule contributes ~20 times as much to the greenhouse effect as an additional CO_2 molecule. Consequently, leaks of CH_4 are a serious environmental concern. These can occur at the gas wells, during transfers, during storage before use, and from power sources. Escape of CH_4 during idling of gas-powered vehicles, for example, could nullify their CO_2-related greenhouse advantage.

QUESTION 19
_____

How much CH_4 leakage would nullify the greenhouse advantage of switching from petroleum to natural gas for vehicle fuel? Use the values in Table 5.2 to make your argument.

Q19 ANSWER In order to answer this, we want to estimate at what fractional leak rate of CH_4 the CO_2-related greenhouse advantages of a natural gas engine versus petroleum engine are nullified.

Given the cited 20-times higher global warming potential of methane versus CO_2, we can answer this question by dividing the CO_2 advantage of methane over petroleum by 20. That amount would be the methane emissions at which the CO_2 advantage would be offset. We have these values for both fuel types in mole CO_2 emitted per 1,000 kJ energy generated in Table 5.2:

Methane: 1.2 mol CO_2 per 1,000 kJ
Petroleum: 1.6 mol CO_2 per 1,000 kJ

The difference between the two gases:

$$1.6 \text{ mol CO}_2 - 1.2 \text{ mol CO}_2 = 0.4 \text{ mol CO}_2$$

This would be equivalent to:

$$\frac{0.4}{20} = 0.02 \text{ mol CH}_4$$

Now, in order to express this as the fractional leak rate, we need to know how many moles of CH_4 were burned to produce 1,000 kJ of energy.

Also from Table 5.2, we learn that methane has an energy density of 51.6 kJ g^{-1} fuel. Thus, to produce 1,000 kJ of energy, we would need to combust:

$$1,000 \text{ kJ} \times \frac{1 \text{ g CH}_4}{51.6 \text{ kJ}} \times \frac{1 \text{ mol CH}_4}{16.04 \text{ g CH}_4} = 1.21 \text{ mol CH}_4$$

Then, the above number of moles CH_4 at which benefit of natural gas would be offset divided by the CH_4 combusted gives us a fractional leak rate. The fractional leak rate is then:

$$\frac{0.02 \text{ mol loss}}{1.21 \text{ mol combusted}} \times 100\% = 1.7\% \text{ leak rate}$$

Not much tolerance for leaks!

5.6 Coal

Figure 5.7 shows the variety of coal deposits around the United States, based on the regional variation in the biomass that was the original source of the coal seam. Lignite is the softest coal; its name recognizes the close similarity to the parent wood component, lignin. Over one-third of the lignite mass is moisture, while the remaining carbonaceous material is almost evenly divided between "volatile matter," hydrocarbons that are released upon heating, and "fixed carbon," the nonvolatile carbon fraction. Sub-bituminous coal is harder than lignite,

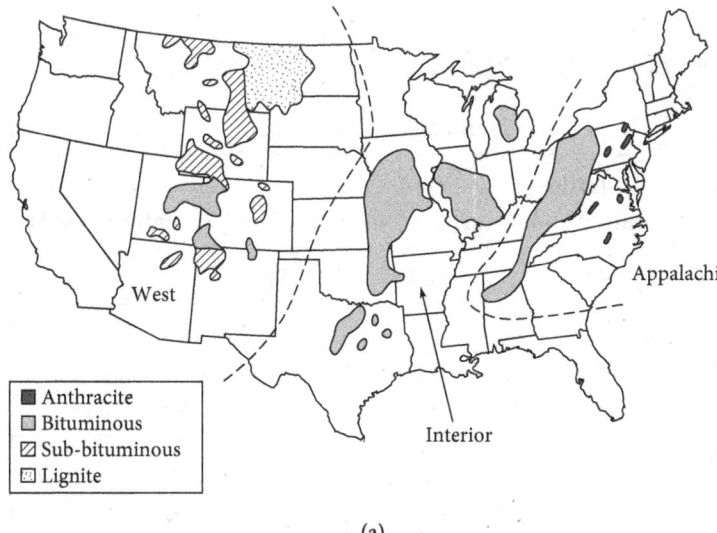

Average Sulfur Content of Coal by Rank

Rank	% of Total Coal Reserves	% with S Content > 1%
Anthracite	0.9	2.9
Bituminous	46.0	70.2
Sub-bituminous	24.7	0.4
Lignite	28.4	9.3
Total, all ranks	100.0	35.0

(a)　　　　　　　　　　　　　　　(b)

Figure 5.7 (a) Coal distribution in the United States. (b) Table of average sulfur content of coal.
Sources: Adapted from Averitt, "U.S. Geological Survey," *Bulletin 1136*, U.S. Department of Interior, 1960; and Spiro et al., *Chemistry of the Environment, 3rd Ed.,* University Science Books, © 2012, all rights reserved.

containing ~20% moisture and 40% fixed carbon, but is softer than bituminous coal, which contains very little moisture. The hardest coal is anthracite, which is ~80% fixed carbon. The heating value of the coal varies with the fraction of reduced carbon and hydrogen and is much lower for soft than hard coals because of their high moisture content.

Different coals have variable amounts of ash, the mineral residue left after complete combustion, reflecting different amounts of minerals (such as pyrite and FeS_2) incorporated during the metamorphic processes. Some sulfur is also bound in the complex organic molecules of coal. When coal is burned, both inorganic and organically bound S is oxidized to SO_2, a significant air pollutant (see Section 3 in Chapter 9 for more details). Figure 5.7 shows the distribution of the coals in the United States and their sulfur content.

QUESTION 20

Coal transport: Look at Figure 5.7 and Table 5.3. Despite its location 6 mi from the Deer Run coal mine, since the 1990s, Illinois' 915-MW Coffeen Power Station burns coal transported 900 mi by rail from Wyoming. Why do think this might be the case? What are the advantages and disadvantages of this fuel switching?

Q20 ANSWER The map in Figure 5.7 shows that Illinois has its own reserves of coal, but they are bituminous, while the coal seams in Wyoming are sub-bituminous. While the local bituminous coal has a higher heating value (30.7–32.7 kJ g^{-1}, compared to Wyoming's sub-bituminous coals, 22.3 kJ g^{-1}), this advantage is outweighed by the also much higher sulfur content (>70% of bituminous coal has sulfur content greater than 1% compared to only 0.4% of sub-bituminous coal). This sulfur is oxidized during combustion to SO_2, a harmful air pollutant that contributes to respiratory disease (see Chapter 9). Since the 1990 Clean Air Amendments made the air pollution regulation more stringent, including for existing coal plants, the intra-United States import of coal from lower sulfur seams became an economic incentive.

TABLE 5.3 Composition and Heat Content of Common Coals Found in the United States

Rank	Location by state	Moisture (%)	Volatile matter (%)	Fixed carbon (%)	Ash (%)	Heating value (kJ g^{-1})
			Chemical analysis			
Anthracite	Pennsylvania	4.4	4.8	81.8	9.0	30.5
Bituminous						
Low volatile	Maryland	2.3	19.6	65.8	12.3	30.7
High volatile	Kentucky	3.2	36.8	56.4	3.6	32.7
Sub-bituminous	Wyoming	22.2	32.2	40.3	4.3	22.3
Lignite	North Dakota	36.8	27.8	30.2	5.2	16.2

Sources: Data from U.S. Bureau of Mines, *Information Circular No. 769*, 1954, U.S. Department of the Interior; adapted from Spiro et al., *Chemistry of the Environment, 3rd Ed.*, University Science Books, © 2012, all rights reserved.

---

*QUESTION 21

The Elmer Smith Power Plant in Owensboro, KY, uses coal mined within the state of Kentucky. Why do you think this might be the case? What air quality issues might this introduce (see Section 4 in Chapter 9 for details about sulfur dioxide air pollution)? Suggest ways that the emissions from the power plant might be mitigated.

5.7 Environmental Repercussions of Fossil Fuel Use

A major environmental downside of using any type of fossil fuel for energy production is the net addition of CO_2 into the atmosphere from geologic reservoirs, resulting in long-term climate warming impacts. This is a disadvantage of all fossil fuel use, albeit with varying amounts of CO_2 emitted per kJ of energy produced for each energy source, with coal being the worst (highest CO_2 emitted per kJ) and natural gas being the best.

Petroleum fossil fuel use has the additional disadvantage of contamination from oil spills. Coastal waterways are particularly vulnerable because they are fragile ecosystems that are often in harm's way. Tanker ships transport oil through near-coastal shipping lanes, load and unload cargo in ports, and occasionally (although decreasing in frequency since 1970, thanks to tighter regulations) have accidents. Oil spills from offshore drilling also contribute to coastal oil spills, killing seabirds and contaminating fish and mollusks. Oil spills are contained using skimmers and booms, or chemical dispersants, but the efficacy of these countermeasures varies, and they are never complete. Petroleum-derived gasoline and diesel fuel are also major contributors to air pollution, especially ozone and small particles formed from atmospheric reactions of transportation exhaust gases.

Natural gas, in addition to producing CO_2 when combusted, is a potent greenhouse in its unburned state. Because it is gaseous, it is difficult to contain and can leak from wellheads during extraction. As discussed in Question 19, even a modest leak rate can nullify the energy density-related CO_2 emissions advantage of natural gas. Recently, the advent of new hydraulic fracturing techniques has raised the risk of contamination of aquifers by fracking fluids, which include lubricants (such as polyacrylamide and diesel oil), acids to prevent scaling (e.g., sulfuric and citric acid), corrosion inhibitors (e.g., ethylene glycol, dimethylformamide, and ammonium bisulfite), and antimicrobials (e.g., glutaraldehyde and pesticides). Because the injections of these fluids increase pressure in rock formations and intentionally cause small earthquakes to enhance reservoir permeability, there have been examples of larger earthquakes inadvertently spurred by fracking.

Although coal is perhaps the most convenient form of fossil fuel to transport, traditional coal mines are hazardous workplaces, where miners are subject to black-lung disease from coal dust. Mine drainage also is typically highly acidic and can contaminate surface waters. Furthermore, coal contains substantial sulfur and heavy metal (including mercury) impurities which can be emitted into the exhaust stream or must be scrubbed and disposed of as toxic waste.

5.8 Carbon Capture and Storage

As described above, one major downside of fossil energy production is the emission of CO_2 into the atmosphere and climate impacts. A potential solution to the CO_2 emission problem is to capture the CO_2 and store it out of harm's way. Although conceptually attractive, carbon capture and storage (CCS) faces major hurdles. The amounts of material are huge, since each gram of carbon produces (44 g mol^{-1} CO_2/12 g mol^{-1} C=) 3.5 g CO_2. The cost of capture and handling will be large, adding to the cost of producing energy. Another hurdle is ensuring that storage is safe and that CO_2 does not escape back into the atmosphere.

In a conventional fuel-fired plant, the CO_2 would have to be separated from the exhaust gases after combustion. Because the air is mostly N_2, the CO_2 is diluted, typically 15% of the gas, and would have to be concentrated, thereby incurring an energy penalty. Organic amines are the most commonly used absorbents for CO_2 sequestration, following this reaction to produce a carbamate:

$$R_3N + CO_2 \rightarrow R_3NCO_2 \qquad (5.4)$$

QUESTION 22

Estimate the number of railcars worth of carbamate waste that would be produced annually if all of the Coffeen Power Station's CO_2 emissions (see Table 5.2) were sequestered using this chemistry with monoethanolamine (C_2H_7NO) as the absorbent. A railcar can hold about 125.5 tons. Assume Coffeen operates year-round at 915 MW.

Q22 ANSWER Using the estimate of coal CO_2 emissions per kJ in Table 5.2 and the potential carbon capture and sequestration reaction forming carbamates shown above, Table 5.2 shows an average of 2.0 mol CO_2 emitted per 1,000 kJ coal energy production. For a full year (365 days), operating at 915 MW, Coffeen would thus produce:

$$915 \text{ MW} = \frac{915 \times 10^6 \text{ J}}{\text{sec}} \times 365 \text{ days} \times \frac{24 \text{ hr}}{1 \text{ day}} \times \frac{3,600 \text{ sec}}{1 \text{ hr}} = 2.9 \times 10^{16} \text{ J} = 2.9 \times 10^{13} \text{ kJ}$$

and the resulting CO_2 emissions would be:

$$2.9 \times 10^{13} \text{ kJ} \times \frac{2 \text{ mol } CO_2}{\text{kJ}} = 5.8 \times 10^{13} \text{ mol } CO_2$$

Using the sequestration chemistry:

$$(HOCH_2CH_2)H_2N + CO_2 \rightarrow (HOCH_2CH_2)H_2NCO_2$$

Because the molecular weight of the carbamate product is 105 g mol^{-1}, this would produce:

$$5.8 \times 10^{13} \text{ mol } CO_2 \times \frac{1 \text{ mol carbamate}}{1 \text{ mol } CO_2} \times \frac{105 \text{ g}}{1 \text{ mol carbamate}} = 6.1 \times 10^{15} \text{ g carbamate}$$

$$6.1 \times 10^{15} \text{ g carbamate} \times \frac{\text{ton}}{1 \times 10^6 \text{ g}} = 6.1 \times 10^9 \text{ tons carbamate}$$

A typical railcar has the capacity to hold 125.5 tons of material, so:

$$6.1 \times 10^9 \text{ tons carbamate} \times \frac{\text{railcar}}{125.5 \text{ tons}} = 4.9 \times 10^7 \text{ railcars of carbamate!}$$

For reference, a 1,000 MW coal plant requires approximately 30,000 railcars of coal annually.

In practice, the carbamate is heated to regenerate the amine and capture the CO_2 in pure form, so the challenge is to pressurize and transport (by pipeline or rail) 5.8×10^{13} mol $CO_2 \times 44$ g mol^{-1} $CO_2 \times$ ton $\times 10^{-6}$ g^{-1} = 2.6 billion tons of CO_2 and inject it at a suitable underground storage site. The heating, pressurization, transportation, and injection would all add to the cost of the coal energy.

*QUESTION 23

Estimate the number of railcars worth of carbamate waste that would be produced annually if all of the Elmer Smith Power Station's CO_2 emissions were sequestered using this chemistry with methyl-diethanolamine as the absorbent. A railcar can hold about 125.5 tons. Assume Elmer Smith operates year-round at 400 MW.

In addition to these engineered solutions, carbon can also be naturally sequestered by absorption by trees (afforestation) and sequestration in healthy soils.

5.9 Conclusions

Fossil fuels, such as petroleum, coal, and natural gas, have provided inexpensive energy to aid in the world's development. They are easy to transport and we have a distribution network throughout the world for accessing the energy. However, fossil fuels do have environmental consequences, from mining and extraction, leaks and spills, air pollution, and most importantly emission of CO_2 (and CH_4) into the atmosphere. Even though it is possible to extract the CO_2 from the waste streams of power plants, this can cost energy and create massive amounts of materials that need to be disposed of. Greener types of energy, such as hydrogen gas, can be used as we transition to new energy sources.

NUCLEAR ENERGY

6.1 Isotopes and Nuclear Equations

Apart from geothermal energy, the one significant form of energy on earth that is not related to the sun, either directly or indirectly, is the energy that resides in the nuclei of atoms.

All nuclei (except for $_1^1H$) have at least one proton and one neutron in their nuclei. The nucleus of an element can have a variable number of neutrons, and this is why the atomic masses are not necessarily close to being integral numbers. For example, chlorine (Cl) has an atomic mass of 35.453 amu, while all Cl atoms have 17 protons, roughly one-half of them have 18 neutrons, and most of the rest have 19 neutrons. Nuclei of an element having a different number of neutrons are called isotopes. The two major isotopes of chlorine are Cl-35 (17 protons + 18 neutrons) and Cl-36 (17 protons + 19 neutrons). All elements have several isotopes, but most of them are present in low abundance.

QUESTION 1

Potassium has two stable isotopes, K-39 and K-41. Using the periodic table, show which isotope is more abundant in nature?

Q1 ANSWER The atomic mass of potassium on the periodic table is 39.098 amu; thus, we know that K-39 is much more abundant.

---

*QUESTION 2

Carbon has two stable isotopes, C-12 and C-13 (and C-14 which is unstable). Using the periodic table, show which isotope is most abundant in nature?

QUESTION 3

Rubidium is used in "atomic clocks" and other precise electronic equipment. There are two major Rb isotopes, Rb-84 (84.12 amu) and Rb-87 (86.909 amu). The isotopic abundance of Rb-84 and Rb-87 is 51.67% and 48.33%, respectively. Calculate the average atomic mass of rubidium.

---

* Answers to starred questions can be found at the end of the book.

Q3 ANSWER (0.517)(84.12 amu) + (0.4833)(86.909 amu) = 85.47 amu

*QUESTION 4

Oxygen has three major isotopes, O-16 (15.995 amu), O-17 (16.995 amu), and O-18 (17.999 amu). The isotopic abundance of O-16, O-17, and O-18 are 99.759%, 0.037%, and 0.204%, respectively. Calculate the average atomic mass of oxygen.

6.1.1 *Stable Isotopes*

Stable isotopes have been key to understanding biogeochemical cycles and paleoclimate data. Carbon has two stable isotopes and each carbon reservoir in the carbon biogeochemical cycle has its own isotopic signature. For example, terrestrial plants prefer carbon dioxide with C-12, while the ocean exchanges atmospheric carbon dioxide with both C-12 and C-13 equally.

Stable isotope abundances are reported as delta (δ) values in parts per thousand, or per mil (‰). The isotopic ratio is normalized (divided) by a standard to calculate the delta value. For example:

$$\delta^{13}C = \left[\frac{\left(^{13}C/^{12}C\right)_{sample}}{\left(^{13}C/^{12}C\right)_{standard}} \right] \times 1{,}000$$

As you can see, the greater the delta value, the more the sample is enriched in C-13. This standardized value allows researchers to compare isotopic data among different laboratories and over time.

QUESTION 5

If the ocean C-13/C-12 ratio is 0.011142 and the standard ratio is 0.011237, what is the $\delta^{13}C$ value? (Data from NOAA's Global Monitoring Laboratory, Earth System Research Laboratories.[1])

Q3 ANSWER

$$\delta^{13}C = \left[\frac{0.011142}{0.011237} - 1 \right] \times 1{,}000 = -8.4542‰$$

This means that the ocean is slightly C-13 depleted; there is more C-12 than C-13 by only eight parts per thousand, relative to the standard sample.

*QUESTION 6

If the terrestrial biosphere has a C-13/C-12 ratio of 0.010945, then show that the biosphere prefers C-12 over C-13. (Assume the standard is the same as in Q5.)

Carbon-13/carbon-12 ratios can even distinguish between plants that use C_3 and C_4 modes of photosynthesis; thus, it can be used to track plant migration or animal dietary changes.

---

[1] https://www.esrl.noaa.gov/gmd/ccgg/isotopes/deltavalues.html

Stable isotopes of nitrogen have shed light on the nitrogen biogeochemical cycle. For example, nitrogen isotopic ratios (N-15/N-14) allow ecologists to track nitrogen sources in an ecosystem, including how much N_2 from the atmosphere is fixed biologically.

Stable oxygen isotopes have played a critical role in understanding earth's paleoclimate. Oxygen isotopes in calcium carbonate from fossilized shells that have been deposited in marine sediments over time provide information about past climate changes. During glacial periods, O-16 was stored in the ice sheets and the oceans were enriched in O-18 (and thus more O-18 accumulated into the fossils and sediment).[2]

6.1.2 Radioactive Isotopes

Some nuclei are unstable, and referred to as "radioactive," meaning that they spontaneously break down by the emission of particles and/or radiation. We can write chemical equations to summarize the reactants and products of nuclear reactions. Because some radioactive decay processes interconvert protons and neutrons, we must specify all protons, neutrons, and electrons in addition to writing the symbols for the various elements involved in the reaction. Each element or particle will include a superscript denoting the mass number (total number of neutrons + protons) and a subscript indicating the atomic number (the number of protons).

The $_{-1}^{0}\beta$ particle has the same charge and mass as an electron (see Table 6.1) but is produced or consumed in a nuclear reaction. The α particle is effectively a helium nucleus; it is a good way for a nucleus to shed mass without changing the p/n ratio.

TABLE 6.1 The Symbols for Elementary Particles

Symbol	Particle	Mass (amu)
$_1^1H$; $_1^1p$	Proton	1.007825
$_0^1n$	Neutron	1.008665
$_{-1}^0e$; $_{-1}^0\beta$	Electron (beta particle)	0.00055
$_{+1}^0e$; $_{+1}^0\beta$	Positron	0.00055
$_2^4He$; $_2^4\alpha$	Alpha particle	4.00260

To balance a nuclear equation, conserve both the mass number and the atomic number. There should be the same number of protons plus neutrons (mass) on each side of the equation. There should be the same number of charges in the products and in the reactants.

QUESTION 7

Oxygen-20 naturally decays to yield fluorine-20. Write a balanced reaction to represent this reaction.

Q7 ANSWER Because we know the identities of each element, we also know their atomic numbers from the periodic table. Oxygen must have eight protons and fluorine will always have nine protons:

$$_8^{20}O \rightarrow\ _9^{20}F + ?$$

---

[2] For more information, visit NASA's Earth Observatory webpage on Palteoclimate and the Oxygen Balance at https://earthobservatory.nasa.gov/features/Paleoclimatology_OxygenBalance.

The difference between the reactants and products in this equation is one proton, but no corresponding mass; thus it must be a β particle (or electron):

$$^{20}_{8}O \rightarrow ^{20}_{9}F + ^{0}_{-1}\beta$$

The starting isotope, called the "parent," yields the "daughter" isotope. In this case, the oxygen-20 parent yields fluorine-20 daughter via β emission.

Isotopes can also be bombarded with particles to create other isotopes.

QUESTION 8

Identify the isotope (X) produced when magnesium-26 is bombarded with a proton via the following equation:

$$^{26}_{12}Mg + ^{1}_{1}p \rightarrow ^{4}_{2}\alpha + X$$

Q8 ANSWER If we add up the mass (protons + neutrons) on each side, we get 27 on the reactant side and 4 on the product side; thus, we know X has a mass of 23 (27 − 4). Adding up the atomic numbers on each side, we get 13 on the reactant side and 2 on the product side; thus, we know X has an atomic number of 11 (13 − 2). The atomic number tells us that X is sodium:

$$^{26}_{12}Mg + ^{1}_{1}p \rightarrow ^{4}_{2}\alpha + ^{23}_{11}Na$$

*QUESTION 9

These reactions occur during an explosion of a uranium atomic bomb. Identify X in each nuclear equation.

(i) $^{235}_{92}U + ^{1}_{0}n \rightarrow ^{140}_{56}Ba + 3X + ^{93}_{36}Kr$

(ii) $^{235}_{92}U + ^{1}_{0}n \rightarrow X + ^{90}_{37}Rb + 2^{1}_{0}n$

6.2 Nuclear Stability

Most of the atom's mass is concentrated in the nucleus, but its volume is small; thus, nuclei are extremely dense ($\sim 2 \times 10^{14}$ g cm^{-3}). What is holding it together? Protons so close together repel each other electrostatically. However, there are also short-range attractions between proton and proton, proton and neutron, and neutron and neutron. The stability of the nucleus is determined by the difference between the electrostatic repulsion and short-range "strong" attractive forces. If the repulsion outweighs attraction, the nucleus disintegrates the emitting particles and/or radiation.

QUESTION 10

Calculate the density of a uranium-235 nucleus (in g cm^{-3}). Assume the nucleus is a sphere. Radius $R = R_0 A^{1/3}$, where $R_0 = 1.2 \times 10^{-15}$ m and A is the number of nucleons (protons and neutrons in the nucleus).

Q10 ANSWER Density is mass divided by volume. For the mass, we know that the uranium atom is 235 amu, but we need this in grams:

$$235 \text{ amu} \left(\frac{1 \text{ g}}{6.022 \times 10^{23} \text{ amu}} \right) = 3.90 \times 10^{-22} \text{ g}$$

For the volume, let us recall the volume of a sphere as $V = (4/3) \pi R^3$.

As mentioned in the question, radius $R = R_0 A^{1/3}$, where $R_0 = 1.2 \times 10^{-15}$ m and A is the number of nucleons (protons and neutrons in the nucleus). This is based on the approximate size of protons and neutrons:

$$R = R_0 A^{1/3} = (1.2 \times 10^{-15} \text{ m})(235)^{1/3} = 7.4 \times 10^{-15} \text{ m}$$

$$V = \frac{4}{3}\pi \left[(7.4 \times 10^{-15} \text{ m}) \left(\frac{100 \text{ cm}}{1 \text{ m}} \right) \right]^3 = 1.7 \times 10^{-36} \text{ cm}^3$$

Thus, density is:

$$\frac{3.90 \times 10^{-22} \text{ g}}{1.7 \times 10^{-36} \text{ cm}^3} = 2.3 \times 10^{14} \frac{\text{g}}{\text{cm}^3}$$

*QUESTION 11

Calculate the density of a carbon-13 nucleus (in g cm^{-3}) if the radius is about 2.8×10^{-15} m. Assume the nucleus is a sphere.

We can look at the proton-to-neutron ratio (p/n) for a clue to nuclear stability. Stable atoms with a low atomic number (<~20) have p/n close to 1 (Fig. 6.1). This trend deviates at higher atomic numbers because a larger number of neutrons, with their strong attractive force to other subatomic particles, are needed to counteract the proton–proton electrostatic repulsion and stabilize the nucleus.

Below the area of stability means that a nucleus has a very low p/n ratio; to increase this ratio (and move toward stability), the nucleus will undergo β-particle emission. This emission leads to an increase in the number of protons in the nucleus and a simultaneous decrease in the number of neutrons:

$$^1_0 n \rightarrow {}^1_1 p + {}^0_{-1}\beta$$

For example:

$$^{40}_{19}\text{K} \rightarrow {}^{40}_{20}\text{Ca} + {}^0_{-1}\beta$$

Above the area of stability means that a nucleus has a higher p/n ratio; to lower this ratio, the nuclei either emit a positron:

$$^1_1 p \rightarrow {}^1_0 n + {}^0_{+1}\beta$$

For example:

$$^{38}_{19}\text{K} \rightarrow {}^{38}_{18}\text{Ar} + {}^0_{+1}\beta$$

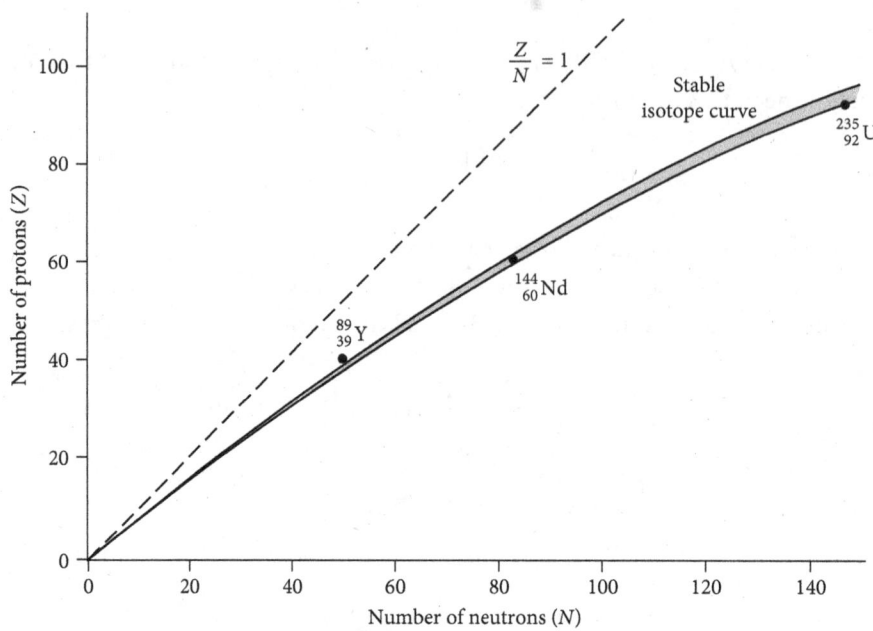

Figure 6.1 The proton–neutron stability curve.
Source: Adapted from Spiro et al., *Chemistry of the Environment, 3rd Ed.,* University Science Books, © 2012, all rights reserved.

Or it can undergo electron capture, a capture of an inner shell electron by the nucleus. The electron combines with a proton to form a neutron.

For example:

$$^{37}_{18}\text{Ar} + ^{0}_{-1}e \rightarrow ^{37}_{17}\text{Cl}$$

A nucleus can also simply have too much energy or mass, even if it has a stable number of protons and neutrons. This situation is often encountered in the products of β or α decay.

For example:

$$^{238}_{92}\text{U} \rightarrow ^{4}_{2}\text{He} + ^{234}_{90}\text{Th}$$

Even though energy is carried off by the β or α particle, the transformed nucleus may be in an excited state; decaying to its ground state will release a gamma (γ) ray. A γ ray is a form of electromagnetic radiation with very high energy.

For example:

$$^{238}_{92}\text{U} \rightarrow ^{4}_{2}\text{He} + ^{234}_{90}\text{Th} + 2^{0}_{0}\gamma$$

QUESTION 12

Determine whether the following isotopes would undergo β particle or positron emission to reach stability. Then, write a balanced nuclear reaction to show the decay.

(i) $^{15}_{8}\text{O}$

(ii) $^{40}_{19}\text{K}$

Q12 ANSWER Given that these isotopes have low atomic numbers ($< \sim 20$), we know that the isotope will decay toward a p/n ratio of 1.

(i) Oxygen has eight protons, and with an atomic mass of 15, this isotope has seven neutrons. Thus, the p/n ratio is 1.14. Since it is above 1, this isotope will emit a positron.

$$^{15}_{8}O \rightarrow {}^{15}_{7}N + {}^{0}_{1}\beta$$

(ii) Potassium has 19 protons, and with a mass of 40, this isotope has 21 neutrons. Thus, the p/n ratio is 0.90. Since it is below 1, this isotope will emit a β particle.

$$^{40}_{19}K \rightarrow {}^{40}_{20}Ca + {}^{0}_{-1}\beta$$

*QUESTION 13

Determine whether the following isotopes would undergo β particle emission or electron capture to reach stability. Then, write a balanced nuclear reaction to show the decay.

(i) $^{24}_{11}Na$

(ii) $^{7}_{4}Be$

6.2.1 Nuclear Binding Energy

A quantitative measure of nuclear stability is the nuclear binding energy, which is the energy that would be required to split the nucleus into its nucleons (the constituent protons and neutrons). Comparing the stability of any two nuclei must account for the fact that they have different numbers of nucleons. Figure 6.2 plots the binding energy per nucleon as the mass of the nucleus increases. At first, the binding energy increases strongly, but it reaches a maximum at elemental iron (Fe), and then slowly declines. The slow decline shows that the electrostatic repulsion among the positively charged protons within large elements gradually overwhelm the nuclear force. Eventually, the nuclei become unstable with respect to the loss of a He nucleus (also known as an α particle). The special stability of the He nucleus (Fig. 6.2) explains why the favored decay route for heavy nuclei is the ejection of an α particle. All elements heavier than bismuth (Bi), with 83 protons, are unstable, and thus radioactive.

Another decay route for heavy nuclei is to fission into two daughter nuclei, with a very large release of energy. There is only one fissionable isotope that is found naturally in the crust of the earth: the uranium isotope with a mass of 235. The decay of Ur-235 is the basis of nuclear energy.

The concept of nuclear binding energy evolved from studies that showed the masses of nuclei are always less than the sum of the masses of the nucleons (its protons and neutrons). This is because when a nucleus is formed, there is a "loss" of mass due to the binding energy. Einstein's relativity theory tells us the equivalence of this mass defect.

$$E = mc^2 \tag{6.1}$$

We can calculate the amount of energy released using:

$$\Delta E = \Delta mc^2 \tag{6.2}$$

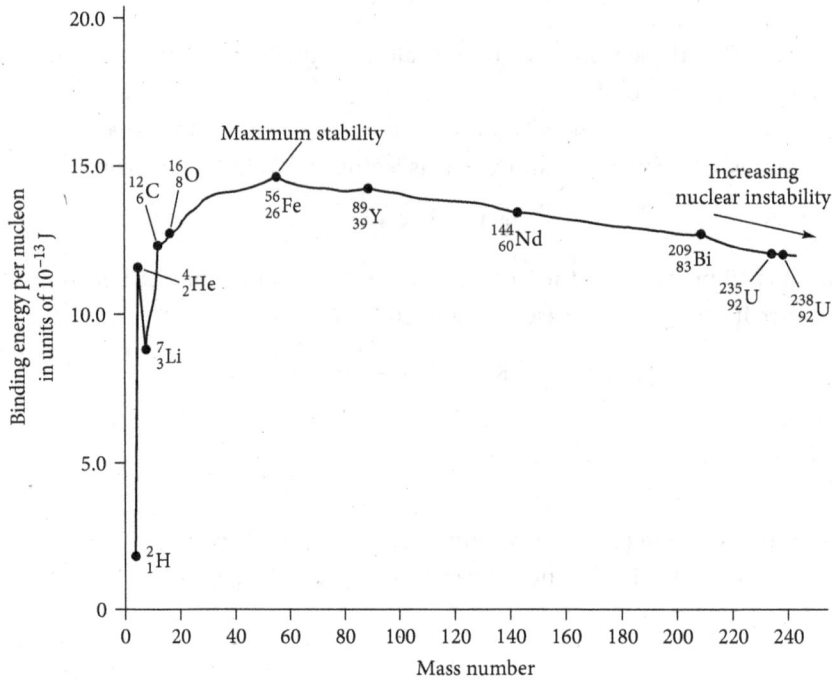

Figure 6.2 The nuclear binding energy curve.
Source: Adapted from Spiro et al., *Chemistry of the Environment, 3rd Ed.,* University Science Books, © 2012, all rights reserved.

where the change in energy equals the change in mass times the speed of light (2.998×10^8 m sec^{-1}) squared.

    ΔE is defined as (energy of products – energy of reactants)

    Δm is (mass of the products – mass of the reactants)

    This energy (heat) is given off to the surroundings; thus, the formation of nuclei is an exothermic process. The nuclear binding energy is calculated as the energy of the bound nuclei minus the summed energies of its constituents.

QUESTION 14

Calculate the nuclear binding energy (in J) and the nuclear binding energy per nucleon for $^{209}_{83}$Bi if its atomic mass is 208.9804 amu.

Q14 ANSWER Treat this problem as if the individual nucleons were the reactants going to form the isotope as the product. From the notation, we know that bismuth has 83 protons and 126 neutrons (209 – 83). Next, we need the total mass of the nucleons using the mass values provided in Table 6.1:

$$83 \times 1.007825 \text{ amu} = 83.649475 \text{ amu}$$

$$126 \times 1.008665 \text{ amu} = 127.09179 \text{ amu}$$

Thus, the total atomic mass of a $^{209}_{83}$Bi atom calculated from its constituent nucleons is:

$$83.649475 \text{ amu} + 127.09179 \text{ amu} = 210.741265 \text{ amu}$$

which is larger than the measured mass of 208.9804 amu found on the periodic table. Thus, when creating this atom, there was a mass defect of:

$$\Delta m = (208.9804 \text{ amu}) - (210.741265 \text{ amu}) = -1.7609 \text{ amu}$$

We relate this mass to energy:

$$\Delta E = (-1.7609 \text{ amu})\left(2.998 \times 10^8 \frac{\text{m}}{\text{sec}}\right)^2 = -1.58 \times 10^{17} \frac{\text{amu m}^2}{\text{sec}^2}$$

Now adjust the units to joules. Remember that 1 joule (J) is 1 kg m^2 sec^{-2}:

$$-1.58 \times 10^{17} \frac{\text{amu m}^2}{\text{sec}^2}\left(\frac{1.00 \text{ g}}{6.022 \times 10^{23} \text{ amu}}\right)\left(\frac{1 \text{ kg}}{1,000 \text{ g}}\right)\left(\frac{1 \text{ J}}{1\frac{\text{kg m}^2}{\text{sec}^2}}\right) = -2.62 \times 10^{-10} \text{ J}$$

This is the amount of energy released when one bismuth-209 nucleus is formed from its constituent protons and neutrons. Thus, the nuclear binding energy is 2.62×10^{-10} J because it is always a positive quantity. This is the amount of energy needed to decompose the nucleus into separate nucleons, or the nuclear binding energy.

Since this bismuth ion has 209 nucleons:

$$\frac{2.62 \times 10^{-10} \text{ J}}{209 \text{ nucleons}} = 1.25 \times 10^{-12} \frac{\text{J}}{\text{nucleon}}$$

This value matches that shown in Figure 6.2.

---

*QUESTION 15

_____

Calculate the nuclear binding energy (in J) and the nuclear binding energy per nucleon for Fe-56 if its atomic mass is 55.9349 amu.

6.3 Radioactive Decay

Nuclei outside the area of stability, as well as any nuclei with more than 83 protons, tend to be unstable. The spontaneous emission of particles, radiation, or both, is known as radioactivity. It can take many steps to get from a radioactive isotope to a stable isotope; thus, the nucleus undergoes a radioactive decay series. Each step begins with a parent isotope that forms a daughter isotope.

All radioactive decays obey first-order kinetics. The proportion of parent atoms that decay during each unit of time is always the same. The number of decaying parent atoms continuously decreases, while the number of daughter atoms continuously increases. The rate of decay at time t is given by:

$$\text{rate} = kQ \tag{6.3}$$

where k is the rate constant and Q is the number of decaying nuclei at time t. But this is a decay, so the rate is negative:

$$\text{rate} = -kQ \tag{6.4}$$

Since the rate is a change in nuclei per change in time, we can write (6.4) as:

$$\frac{dQ}{dt} = -kQ_0 \tag{6.5}$$

Rearranging the equation and integrating from time zero to time t gives:

$$\frac{dQ}{Q_0} = -kdt \tag{6.6}$$

$$\int_{Q_0}^{Q_t} \frac{dQ}{Q_0} = -k\int_0^t dt \tag{6.7}$$

$$\ln Q - \ln Q_0 = -k(t - 0) \tag{6.8}$$

$$\ln Q_t = -kt + \ln Q_0 \tag{6.9}$$

$$\ln\left(\frac{Q_t}{Q_0}\right) = -kt \tag{6.10}$$

$$Q_t = Q_0 e^{-kt} \tag{6.11}$$

One important characteristic of a radioactive decay is its half-life ($t_{1/2}$), or the time required for the parent isotope to decrease by one-half of its initial concentration. Thus, by definition, when $t = t_{1/2}$, $Q_t = Q_0/2$.

Rearranging (6.10) and solving for time:

$$t = \frac{1}{k}\ln\left(\frac{Q_t}{Q_0}\right) \tag{6.12}$$

If $t = t_{1/2}$, then:

$$t_{\frac{1}{2}} = \frac{1}{k}\ln\left(\frac{Q_0}{\frac{Q_0}{2}}\right) = \frac{\ln 2}{k} \tag{6.13}$$

And the rate constant would be:

$$k = \frac{\ln 2}{t_{\frac{1}{2}}} \tag{6.14}$$

QUESTION 16

The radioactive decay of Tl-206 to Pb-206 has a half-life of 4.20 min. Starting with 1 mol of Tl-206, how many parent atoms are left after 1.00 hour (hr).

Q16 ANSWER Using the half-life, we can calculate the rate constant (6.14):

$$k = \frac{\ln 2}{4.20 \text{ min}} = \frac{0.165}{\text{min}}$$

Now solve for Q_t using (6.11). Watch to make sure all of the units cancel in the exponent:

$$Q_t = (6.022 \times 10^{23} \text{ atoms})e^{-\left[\left(\frac{0.165}{\text{min}}\right)(1.00 \text{ hr})\left(\frac{60 \text{ min}}{1 \text{ hr}}\right)\right]} = 3.02 \times 10^{19} \text{ atoms}$$

---

*QUESTION 17

Strontium-90 is one of the products of Ur-235 fission and it has a half-life of 28.1 yr. How many years will it take for 1.00g of the isotope to become 0.750g by decay?

6.3.1 *Dating Using Radioactive Decay*

The rate of radioactive decay is unique for each isotope. Radioactive decay rates are constant and not affected by the chemical or physical environment (thus, it is the same all over and within the earth). Thus, we can use radioactive decay to determine the actual age of an object. The best-known example is radiocarbon dating.

The carbon isotope with mass number 14 $\left(^{14}_{6}C\right)$ has six protons and eight neutrons. It decays by β emission, producing $^{14}_{7}N$, which is a stable isotope:

$$^{14}_{6}C \rightarrow {}^{14}_{7}N + {}^{0}_{-1}\beta$$

$^{14}_{6}C$ is produced in the atmosphere by cosmic rays, which are ultrahigh-energy particles that rain down continuously from outer space. A $^{14}_{7}N$ atom hit by a cosmic ray will form $^{14}_{6}C$:

$$^{14}_{7}N + {}^{1}_{0}n \rightarrow {}^{14}_{6}C + {}^{1}_{1}H$$

$^{14}_{6}C$ produced this way immediately reacts with O_2 molecules forming $^{14}_{6}CO_2$. Thus, $^{14}_{6}C$ enters the carbon cycle and is incorporated into all living things. When life stops, so does its exchange of carbon with the atmosphere, and $^{14}_{6}C$ of preserved organic matter gradually decreases according to the decay rate. The half-life of $^{14}_{6}C$ is 5,730 yr, so a 6,000-yr-old sample of organic matter will have about half of $^{14}_{6}C$ left.

QUESTION 18

As mountaintop glaciers retreat, material once buried underneath the ice is exposed. A hiker stumbles on the remains of a dog that was trapped in the ice many years ago. Archeological evidence shows that there was a human settlement present in this area approximately 4,500 yr ago. If that dog was a part of the village community (and thus died about 4,500 yr ago), what percentage of the original $^{14}_{6}C$ would still be remaining in the dog's body?

Q18 ANSWER From above, we know the half-life of $^{14}_{6}C$ is 5,730 yr. Using the half-life, we can calculate the rate constant [Eq. (6.14)]:

$$k = \frac{\ln 2}{5,730 \text{ yr}} = \frac{1.2097 \times 10^{-4}}{\text{yr}}$$

Now solve for the ratio of nuclei using (6.10):

$$\ln\left(\frac{Q_t}{Q_0}\right) = -\left(\frac{1.2097 \times 10^{-4}}{yr}\right)(4,500 \text{ yr})$$

$$\ln\left(\frac{Q_t}{Q_0}\right) = -0.5444$$

$$e^{\ln\left(\frac{Q_t}{Q_0}\right)} = e^{-0.5444}$$

$$\left(\frac{Q_t}{Q_0}\right) = 0.5802$$

58% of the C-14 would be remaining.

*QUESTION 19

At Stonehenge, a charcoal sample was dug up, presumably the remains of a fire; its $^{14}_{6}C$ activity was measured at 9.6 disintegrations per minute per gram (disintegrations min^{-1} g^{-1}) of C (see Section 6.4 for a more detailed discussion of disintegrations). Living tissue has a $^{14}_{6}C$ activity of 15.3 disintegrations min^{-1} g^{-1}. When did the Stonehenge fire burn?

QUESTION 20

Would it be possible to use $^{14}_{6}C$ to estimate the age of a rock containing carbon from the time of the dinosaurs? Explain. (Dinosaurs became extinct ~65 million years ago.)

Q20 ANSWER No, it would not make sense to use $^{14}_{6}C$ to estimate the age of a sample that is more than 65 million years old. The lifetime of $^{14}_{6}C$ is 5,730 yr, so 65 million years (65 × 10^6 yr) is 11,000 half-lives. After that amount of time, there would essentially be no carbon-14 remaining in the sample. Radiocarbon dating is limited to objects less than 50,000 yr old.

In fact, even fossil fuels have very little $^{14}_{6}C$ left because they were formed millions of years ago. Thus, when fossil fuels are combusted, the carbon emitted into the atmosphere as carbon dioxide is depleted of $^{14}_{6}C$. Researchers can use the decreasing proportion of $^{14}_{6}C$ in atmospheric carbon dioxide to determine how much new carbon dioxide is coming from fossil fuel combustion.

*QUESTION 21

A bottle of wine is sold for a high price because it is supposedly 50 yr old, but you suspect it is much younger. In the lab, your instrumentation can detect four different isotopes, sulfur-35, carbon-14, nitrogen-13, and hydrogen-3. The half-lives of these isotopes are 88 days, 5,370 yr, 598 sec, and 12.5 yr, respectively. Assuming that the activities of the isotopes were known (or could be found) at the time the bottle was sealed, which isotope will help you determine the age of this wine?

Most unstable isotopes do not have a continuous source (like $^{14}_{6}C$ does), so many have long since disappeared from earth since its formation 4.5 billion years ago. However, a few have slow-enough decay rates that are still present in significant abundance.

6.4 Biological Effects of Radioactivity

When unstable nuclei decay, α, β, or γ particles or rays are released, in the range of millions of electron volts (MeV) where 1 eV $= 1.602 \times 10^{-19}$ J. Table 6.2 summarizes the penetration depth for the different radiation types and their relative effectiveness in producing damage.

Alpha particles, being larger particles, produce intense damage over a short distance. These are the least dangerous type of radiation from a source outside of the body because they cannot penetrate the skin. However, if an α emitter is ingested or inhaled, tissue damage can be severe; α particles are more effective in forming ions in the surrounding tissues.

A β particle, being an energetic electron, is much lighter and is only singly charged. It can penetrate the outer layers of skin and clothing to produce severe burns. Inside the body, β particles are less disruptive to individual cells than α particles because they transfer their energy over a wider area of tissues.

A γ ray is a high-energy photon and has a different mode of interaction with matter than charged particles. Because of their great penetrating power, γ rays are more dangerous than α or β particles outside the body. The probability of a γ ray hitting an atom in its path is quite low, but when it collides, it transfers a large amount of energy, and the electron that has been ionized carries away enough energy to ionize many other atoms.

Neutrons interact with matter by penetrating the electron shells and reacting directly with the nuclei, causing ionization or producing radioisotopes. Bare neutrons decay spontaneously ($t_{1/2} = 12$ min) into protons and electrons, and are therefore of concern only in the immediate vicinity of the nuclear reactors where they are produced.

Radiation can remove electrons from atoms and molecules in its path, leading to the formation of ions and radicals. Radicals are short-lived, highly reactive molecules that have an unpaired electron. In biological tissues, the main damage mechanisms are associated with the production of hydroxyl radicals ($OH\cdot$) and superoxide ions ($O_2^-\cdot$) from the ionization of water.

TABLE 6.2 Paths of Energetic Particles in Biological Tissue

Type of Radiation	Range in Biological Tissue[a] (cm)	Relative Biological Effectiveness[b]
alpha	0.005	10–20
beta	3	1
gamma	~20	1

Some Hazardous Radioactive Isotopes			
Element	Type of Radiation	Half-Life	Site of Concentration
$^{239}_{94}$Pu	alpha	24,360 years	Bone, lung
$^{90}_{38}$Sr	beta	28.8 years	Bone, teeth
$^{131}_{53}$I	beta, gamma	8 days	Thyroid
$^{137}_{55}$Cs	beta, gamma	30 years	Whole body

[a] For a 6-MeV particle.
[b] Accounts for the fact that cell damage increases as the density of the damage site increases.

Source: Adapted from Spiro et al., *Chemistry of the Environment, 3rd Ed.*, University Science Books, © 2012, all rights reserved.

$$H_2O \xrightarrow{\text{radiation}} H_2O^+ + e^-$$

$$H_2O^+ + H_2O \rightarrow H_3O^+ + OH\cdot$$

$$e^- + O_2 \rightarrow O_2^-\cdot$$

These radicals attack cell membranes and a host of organic molecules, such as enzymes and DNA molecules. The nucleic acids, which make up DNA, are particularly vulnerable to attack by the hydroxyl radical. This attack can alter the genetic code and thus transform normal cells into cancerous ones; there are well-established correlations between radiation exposure and the incidence of cancer. The potential for damage is dependent on the location of the isotope emitting radiation, relative to the biological tissue. If the isotope is external to the body, then the issue is what kind of shielding is needed to absorb the rays. If the isotope is ingested, then the issue becomes one of transport and elimination, as well as the decay rate.

Radioactivity was discovered at the beginning of the twentieth century, but its harmful effects were not yet recognized. Marie Curie famously worked for many years with radioactive materials; she discovered radium and polonium (named after her homeland of Poland). She was the first woman to win a Nobel Prize and the first person (and only woman) to win the Nobel Prize twice. Due to her work, she suffered from anemia and died of leukemia. Other early sufferers of radiation poisoning were women working in factories painting radium on watch dials to make them glow. The workers often licked their paintbrushes to obtain a fine point; thus, they developed lip cancers and bone cancers or leukemia from ingesting the material.

Aside from the location of the radiation source and the type of radiation, exposure depends on the concentration of a given radioactive isotope and its half-life. The shorter the half-life, the greater the disintegration rate, and the more intense the exposure. On the other hand, if the half-life is very short, then the exposure is also brief.

Radioactive disintegrations are measured in curies (after Marie Curie, Ci) or becquerels (Bq).

1 Ci = 3.70×10^{10} nuclear disintegrations sec^{-1} (the decay rate of 1g of radium)

1 Bq = 1 nuclear disintegration sec^{-1}

QUESTION 22

Given that $^{226}_{88}$Ra has a half-life of 1.6×10^3 yr, derive the curie (Ci) unit. Remember that 1 Ci is the decay rate of 1.00 g of radium-226.

Q22 ANSWER   Let us solve for the rate of nuclear disintegration in the radium nucleus. First, we will find the rate constant for the decay using the half-life and (6.14):

$$k = \frac{\ln 2}{1.6 \times 10^3 \text{ yr}} = \frac{4.33 \times 10^{-4}}{\text{yr}}$$

Now convert the units since the time component of the Ci is in seconds:

$$\frac{4.33 \times 10^{-4}}{\text{yr}}\left(\frac{1 \text{ yr}}{365 \text{ days}}\right)\left(\frac{1 \text{ day}}{24 \text{ hr}}\right)\left(\frac{1 \text{ hr}}{60 \text{ min}}\right)\left(\frac{1 \text{ min}}{60 \text{ sec}}\right) = 1.37 \times 10^{-11} \frac{1}{\text{sec}}$$

Since Ci is the decay rate of 1g of radium, how many atoms is that?

$$1.00 \text{ g}\left(\frac{1 \text{ mol}}{226 \text{ g}}\right)\left(\frac{6.022 \times 10^{23} \text{ atoms}}{1 \text{ mol}}\right) = 2.6645 \times 10^{21} \text{ atoms}$$

Each atom will experience 1 disintegration. Using (6.3), we can solve for the rate of radium's disintegration:

$$\text{rate} = kQ = \left(\frac{1.37 \times 10^{-11}}{\sec}\right)(2.6645 \times 10^{21} \text{ atoms}) = 3.70 \times 10^{10} \frac{\text{atoms}}{\sec} = 3.70 \times 10^{10} \frac{\text{disintegrations}}{\sec}$$

Indeed, 1 Ci is 3.70×10^{10} nuclear disintegrations $\sec^{-1}$.

*QUESTION 23

Compare the activity (in mCi) of a 0.500-g sample of these thorium isotopes

(i)   Th-234, half-life = 24.1 days.

(ii) Th-232, half-life = 1.4×10^{10} yr

QUESTION 24

If a 1.00-g sample of cobalt-60 (59.92 g mol^{-1}) has an activity of 1.3×10^3 Ci, determine the half-life of this radioisotope in years.

Q24 ANSWER We want to use the rate equation to determine the rate constant because it is directly related to the half-life of a first-order reaction, such as nuclear decay. First, let us figure out how many atoms are in the sample:

$$1.00 \text{ g}\left(\frac{1 \text{ mol}}{59.92 \text{ g}}\right)\left(\frac{6.022 \times 10^{23} \text{ atoms}}{1 \text{ mol}}\right) = 1.005 \times 10^{22} \text{ atoms}$$

Each atom will experience 1 disintegration:

$$1.005 \times 10^{22} \text{ atoms} = 1.005 \times 10^{22} \text{ disintegrations}$$

Now we need the rate in a unit that will complement the equation:

$$1.3 \times 10^3 \text{ Ci}\left(\frac{3.70 \times 10^{10} \frac{\text{disintegrations}}{\sec}}{1 \text{ Ci}}\right) = 4.81 \times 10^{13} \frac{\text{disintegrations}}{\sec}$$

Using (6.3), we can solve for the rate constant:

$$\text{rate} = kQ$$

$$4.81 \times 10^{13} \frac{\text{disintegrations}}{\sec} = k(1.005 \times 10^{22} \text{ disintegrations})$$

$$k = 4.79 \times 10^{-9} \frac{1}{\sec}$$

Now convert this rate into a half-life using (6.13).

$$t_{\frac{1}{2}} = \frac{\ln 2}{4.79 \times 10^{-9} \frac{1}{\text{sec}}} = 1.45 \times 10^8 \text{ sec}$$

Now convert these units to years:

$$1.45 \times 10^8 \text{ sec} \left(\frac{1 \text{ min}}{60 \text{ sec}}\right)\left(\frac{1 \text{ hr}}{60 \text{ min}}\right)\left(\frac{1 \text{ day}}{24 \text{ hr}}\right)\left(\frac{1 \text{ yr}}{365 \text{ days}}\right) = 4.6 \text{ yr}$$

*QUESTION 25

A 0.0100-g sample of a radioactive isotope with a half-life of 102.1 yr decays at a rate of 0.1676 Ci. Calculate the molar mass and identify the isotope.

Exposure to radiation is measured in several different units. A roentgen (R) is the amount of radiation that, on passing through 1 cm^3 of air (at 0°C and 1 atm), would create one electrostatic unit. A rad (radiation absorbed dose) is the amount of radiation that results in the absorption of 0.01 J kg^{-1} of irradiated material:

$$1R = 0.877 \text{ rad}$$

Finally, rem (roentgen equivalent man) quantifies the effect that the radiation has on different tissues:

$$\text{rem} = \text{rad} \times \text{RBE factor}$$

where the relative biological effectiveness (RBE) factor takes into account the type of radioactive emission and the tissue being exposed. For β and γ rays, the factor is 1. For α particles striking most tissues, the factor is 10, but it can be higher.

(There are also international units, the gray (Gy), equivalent to 100 rad, and the Sievert (Sv), equivalent to 100 rem.)

QUESTION 26

If a person absorbs 50 mrad of α-particle radiation in her eye, what is the dose in mrem? The biological factor of α radiation in the eye is 30.

Q26 ANSWER Since (rem = rad × RBE factor), we simply plug into the equation. Since the rad value is in millirad, then the rem would also be in millirem:

$$\text{mrem} = 50 \text{ mrad} \times 30 = 1,500 \text{ mrem}$$

*QUESTION 27

Two technicians were accidentally exposed to radiation. If Mary was exposed to 15 mGy and Joe was exposed to 3 rad, which person received more radiation?

QUESTION 28

Plutonium-239 (239.0522 g mol^{-1}, half-life = 2.41 × 10^4 yr) decays by α emission and emits 6.02 million electron volts (MeV) in each disintegration. If you inhale 1.00 mg of this isotope and it affects 1.00 kg of your lungs, what is your daily dose in rem, assuming only 50% of the energy is absorbed? Assume an RBE factor of 10, the lowest for α particles.

Q28 ANSWER First, let us figure out the rate of Pu-239's decay. Using the half-life, we can calculate the rate constant [Eq. (6.14)]:

$$k = \frac{\ln 2}{2.41 \times 10^4 \text{ yr}} = \frac{2.876 \times 10^{-5}}{\text{yr}}$$

Next, figure out how many atoms are in the inhaled sample:

$$1.00 \text{ mg}\left(\frac{1 \text{ g}}{1{,}000 \text{ mg}}\right)\left(\frac{1 \text{ mol}}{239.0522 \text{ g}}\right)\left(\frac{6.022 \times 10^{23} \text{ atoms}}{1 \text{ mol}}\right) = 2.52 \times 10^{18} \text{ atoms}$$

Each atom will experience one disintegration. Using (6.3), we can solve for the rate of plutonium's disintegration:

$$\text{rate} = kQ = \left(\frac{2.876 \times 10^{-5}}{\text{yr}}\right)(2.52 \times 10^{18} \text{ atoms}) = 7.247 \times 10^{13} \frac{\text{atoms}}{\text{yr}} = 7.247 \times 10^{13} \frac{\text{disintegrations}}{\text{yr}}$$

Each disintegration emits energy, so let us find how much radiation comes from this material:

$$7.247 \times 10^{13} \frac{\text{disintegrations}}{\text{yr}}\left(\frac{6.02 \times 10^6 \text{ eV}}{1 \text{ disintegration}}\right)\left(\frac{1.602 \times 10^{-19} \text{ J}}{1 \text{ eV}}\right) = 69.89 \frac{\text{J}}{\text{yr}}$$

The problem states that only 50% of that energy is absorbed by the 1.00 kg lung tissue:

$$\frac{0.50\left(69.89 \frac{\text{J}}{\text{yr}}\right)}{1.00 \text{ kg}} = 34.945 \frac{\text{J}}{\text{kg yr}}$$

Converting to rad:

$$34.945 \frac{\text{J}}{\text{kg yr}}\left(\frac{1 \text{ rad}}{\frac{0.01 \text{ J}}{\text{kg}}}\right) = 3494.5 \frac{\text{rad}}{\text{yr}}$$

Convert this to rem, using the RBE factor of 10 since this is the smallest value for α-particle emission:

$$\text{rem} = \left(3494.5 \frac{\text{rad}}{\text{yr}}\right)(10) = 34945 \frac{\text{rem}}{\text{yr}}$$

Lastly, calculate the daily dose:

$$\left(34945 \frac{\text{rem}}{\text{yr}}\right)\left(\frac{1 \text{ yr}}{365 \text{ days}}\right) = 95.7 \frac{\text{rem}}{\text{day}}$$

6.5 Everyday Exposure to Radiation

It is impossible to avoid exposure to every source of radiation in daily life. Low-level radiation is all around us. Table 6.3 lists the average annual doses for people in the United States at sea level and how that dosage has changed since the early 1980s. In the 1980s, the vast majority of radiation came from natural sources, such as radon gas, but now there is a large amount of exposure from advanced medical techniques, such as computed tomography (CT) scans. A CT scan emits a powerful dose of radiation, in some cases equivalent to about 200 chest X-rays, or the amount most people would be exposed to from natural sources over 7 yr.

Exposure is based on an individual's behavior, occupation, location, and medical treatments. For example, the occupational exposure for airline personnel could be ~200 mrem yr^{-1}.

TABLE 6.3 Average Annual Exposure to Radiation for People in the United States[a]

Source of Radiation	Early 1980s		2006	
	Dose (mrem)	%Total Dose	Dose (mrem)	%Total Dose
Natural				
Radon gas	198	55	229	37
Cosmic rays	29	8	31	5
Terrestrial (other than radon)	29	8	19	3
Internal (radioisotopes in food and water)	40	11	31	5
Total natural	**295**	**82**	**310**	**50**
Medical				
Computed tomography (CT scans)			149	24
Nuclear medicine	14	4	74	12
Interventional fluoroscopy			43	7
Conventional radioscopy/fluoroscopy (X-rays)	40	11	31	5
Total medical	**54**	**15**	**298**	**48**
Consumer activities				
Cigarette smoke			4	1
Building materials			3	1
Commercial air travel			3	1
Other			2	<1
Total consumer products	**11**	**3**	**12**	**2**
Other				
Occupational (miners, medical workers, aviation personnel, nuclear plant workers)	<1	<0.1	<1	<0.1
Nuclear power generation	<1	<0.1	<1	<0.1
Research activities	<1	<0.1	<1	<0.1
Miscellaneous	<1	<0.1	<1	<0.1
Total radiation exposure (natural and human generated)	**360**	**100**	**620**	**100**

[a]Comparison between the early 1980s and 2006.

Sources: National Council on Radiation Protection and Measurements, *Ionizing Radiation Exposure of the Population of the United States* (Report No. 93), Author, 1987; National Council on Radiation Protection and Measurements, *Ionizing Radiation Exposure of the Population of the United States* (Report No. 160), Author, 2009. Adapted from Spiro et al., *Chemistry of the Environment, 3rd Ed.,* University Science Books, © 2012, all rights reserved.

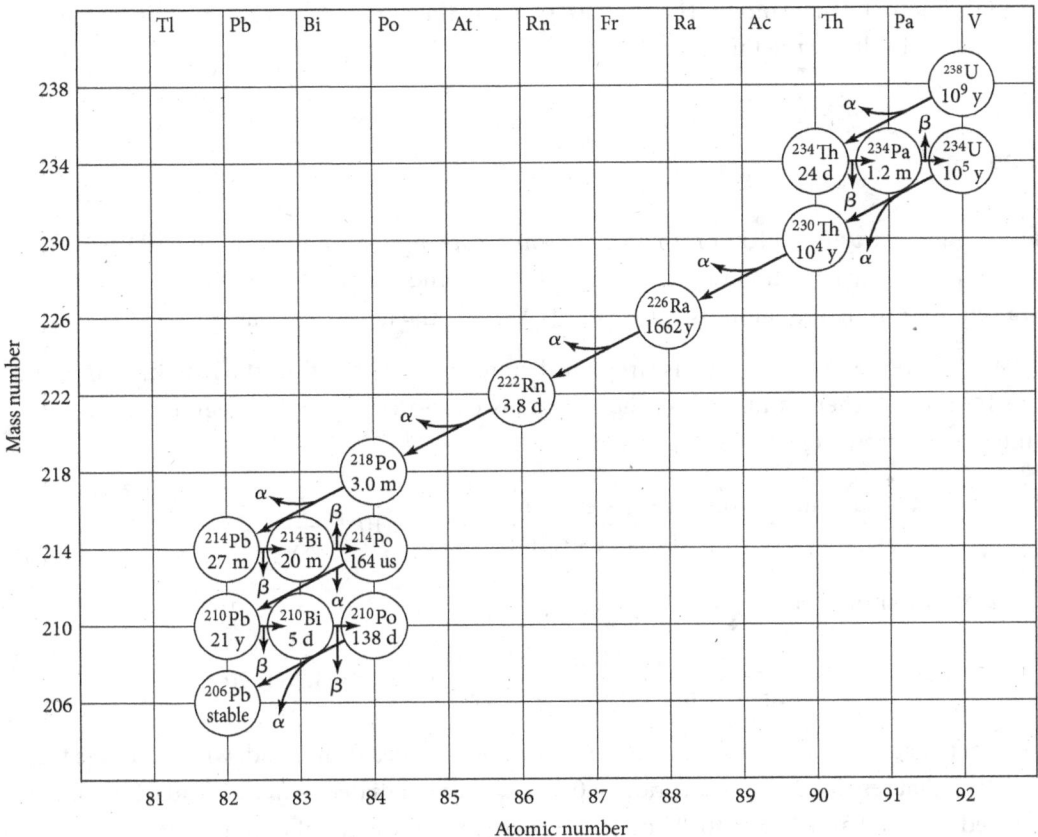

Figure 6.3 Uranium-238 decay chain through Pb-206. The half-lives of isotopes are given in the circles. *Source:* Adapted from Spiro et al., *Chemistry of the Environment, 3rd Ed.*, University Science Books, © 2012, all rights reserved.

As already mentioned, the large variability of radon exposure depends on the type of bedrock, which varies by location. In addition, cosmic-ray exposure increases for people living at higher altitudes. Lastly, people who live in locations contaminated by anthropogenic radioisotopes can be exposed to higher doses of radiation. These areas include sites of previous nuclear accidents, nuclear test sites, and uranium mining or processing sites. Even years after the nuclear activity, there can be small doses of strontium-90 and cesium-137 in the food sources (produce and animal products) as well as in the water.

The decay of heavy elements generally proceeds in a sequential cascade, as shown for ^{238}U in Figure 6.3. One isotope in this decay chain is of special environmental significance, ^{222}Rn. Radon is a noble gas, thus it does not form chemical bonds, and it is free to escape from the site where it was formed. Its half-life is 3.82 days, so ^{222}Rn can enter buildings by seeping through foundations near uranium-containing rocks. It can also infiltrate wells and enter the nearby water supply. Fortunately, inexpensive tests for radon are readily available, and remediation is generally straightforward to vent the soil emissions away from the buildings.

Uranium miners have been at a high risk of developing lung cancer, mostly attributed to Rn and its daughter isotopes. Radon itself is expelled from the lungs as fast as it is inhaled, but the daughter isotopes are incorporated into dust particles that the miners breathe. These particles stick to the lining of the lungs and have time to damage the tissue with their radioactive decay. This same magnified effect occurs for those who live in houses near uranium-containing rocks that smoke. The smoke can act as the particulate transport for the daughter isotopes to infiltrate the lungs.

QUESTION 29

Calculate the radiation exposure to Rn α rays for a person breathing air for a year at the EPA working limit of 0.15 Bq L^{-1} if the person's lungs contains 1.0L of air and weighs 3.0 kg. The energy carried by an α ray emitted from a Rn atom is 9.0×10^{-13} J. Assume the lowest RBE factor.

Q29 ANSWER First, remember that 1 Bq is simply 1 disintegration sec^{-1}; thus the EPA limit is 0.15 disintegrations L^{-1} sec^{-1}. If there is 1.0L of air, then the α ray exposure is 0.15 disintegrations sec^{-1}. Each disintegration produces 9.0×10^{-13} J, thus:

$$\frac{0.15 \text{ disintegrations}}{\text{sec}}\left(\frac{9.0 \times 10^{-13} \text{ J}}{1 \text{ disintegration}}\right) = 1.35 \times 10^{-13} \frac{\text{J}}{\text{sec}}$$

Our time frame is 1 yr of exposure, so:

$$1.35 \times 10^{-13} \frac{\text{J}}{\text{sec}}\left(\frac{60 \text{ sec}}{1 \text{ min}}\right)\left(\frac{60 \text{ min}}{1 \text{ hr}}\right)\left(\frac{24 \text{ hr}}{1 \text{ day}}\right)\left(\frac{365 \text{ days}}{1 \text{ yr}}\right) \times 1 \text{ yr} = 4.26 \times 10^{-6} \text{ J}$$

Next, remember that the unit for exposure is rem and a rem comes from a rad, where a rad is the amount of radiation that results in the absorption of 0.01 J kg^{-1} of irradiated material. Thus, if we have 3.0 kg of irradiated material and 4.26×10^{-6} J of energy, we can find the number of rads:

$$4.26 \times 10^{-6} \text{ J}\left(\frac{1 \text{ rad}}{\frac{0.01 \text{ J}}{\text{kg}}}\right)\left(\frac{1}{3.0 \text{ kg}}\right) = 1.4 \times 10^{-4} \text{ rad}$$

Convert this to rem, using the RBE factor of 10 since this is the smallest value for α-particle emission:

$$\text{rem} = (1.4 \times 10^{-4} \text{ rad})(10) = 0.0014 \text{ rem} = 1.4 \text{ mrem}$$

---

*QUESTION 30

If a full-body CT scan can give a dose of 1,000 mrem and a person is 81 kg, how much energy is this person exposed to from this one medical procedure? Note that the RBE factor for X-rays is 1.

6.6 Power from Fission

If a heavy nucleus (>200 amu) splits to form smaller nuclei, then there is a release of energy because of the greater stability of the lighter nuclei (Fig. 6.2). The difference between the mass of all products and the initial atom's mass is converted to energy and released from the following fission reaction (6.2). One way to represent this change in energy during fission is to compare the binding energies of the products with that of the reactant. Although many heavy nuclei can be made to undergo fission, the fission of naturally occurring U-235 and the anthropogenic isotope plutonium-239 have the most practical importance for nuclear power generation.

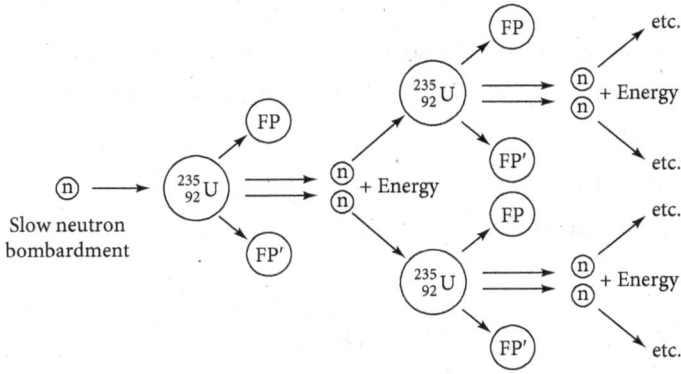

Figure 6.4 Uranium-235 chain reaction induced with neutrons. FP and FP′ represent various fission products.
Source: Adapted from Spiro et al., *Chemistry of the Environment, 3rd Ed.,* University Science Books, © 2012, all rights reserved.

The fission of uranium-235 is induced when the nucleus absorbs a neutron. Not only does the nucleus then split into two lighter nuclei, but two or three neutrons are also released, initiating a chain reaction. The lighter nuclei, called fission products, can have a range of masses, which is the reason sometimes two and other times three neutrons are released (Fig. 6.4). More than 30 different elements have been found among the fission products, but a representative reaction is:

$$^{235}_{92}U + {}^{1}_{0}n \longrightarrow {}^{90}_{38}Sr + {}^{143}_{54}Xe + 3\,{}^{1}_{0}n$$

Note that the above balanced nuclear equation, using whole numbers, shows that the total mass remains the same. This is not true when we consider the masses out to six or seven decimal places.

TABLE 6.4 Nuclear Binding Energies of Uranium-235 and Some of Its Fission Products

Isotope	Nuclear Binding Energy (J)	Mass (amu)
$^{235}_{92}U$	2.82×10^{-10}	235.0439299
$^{90}_{38}Sr$	1.23×10^{-10}	89.9077379
$^{143}_{54}Xe$	1.92×10^{-10}	142.93511

Source: Adapted from Spiro et al., *Chemistry of the Environment, 3rd Ed.,* University Science Books, © 2012, all rights reserved.

QUESTION 31

Estimate the amount of energy released from the representative reaction given the nuclear binding energies in Table 6.4:

$$^{235}_{92}U + {}^{1}_{0}n \longrightarrow {}^{90}_{38}Sr + {}^{143}_{54}Xe + 3\,{}^{1}_{0}n$$

Q31 ANSWER Note that the nuclear binding energies given in Table 6.4 are per nucleus (or per atom), not per nucleon like that shown in Figure 6.2. We want the difference in energies between the products and the reactants:

$$\Delta E = (1.23 \times 10^{-10} \text{ J} + 1.92 \times 10^{-10} \text{ J}) - 2.82 \times 10^{-10} \text{ J} = 3.30 \times 10^{-11} \text{ J}$$

Thus, there is 3.30×10^{-11} J released per uranium nucleus.

To put this in perspective, 1g of Ur-235 would release:

$$\left(3.30 \times 10^{-11} \frac{\text{J}}{\text{atom}}\right)\left(\frac{6.022 \times 10^{23} \text{ atoms}}{1 \text{ mol}}\right)\left(\frac{1 \text{ mol}}{235 \text{ g}}\right) = 8.46 \times 10^{10} \text{ J}$$

This can be compared to the combustion of 1 ton of coal which releases 5×10^7 J.

---

*QUESTION 32

Estimate the amount of energy released from the representative reaction given the masses in Table 6.4 and the mass of a neutron in Table 6.1:

$$^{235}_{92}\text{U} + ^{1}_{0}n \rightarrow ^{90}_{38}\text{Sr} + ^{143}_{54}\text{Xe} + 3\,^{1}_{0}n$$

The fission products are highly radioactive because they have nearly the same neutron/proton ratio as U-235. Thus, they have too many neutrons and move toward stability through β emission.

Because more neutrons are produced than are originally captured in the initiation step, U-235 undergoes a nuclear chain reaction, a self-sustaining sequence of fission reactions (Fig. 6.4). For the chain reaction to occur, enough uranium-235 must be present in the sample to capture the neutrons. A critical mass is the amount of uranium-235 for which the probability that a neutron produced in a fission reaction induces another fission reaction is 1. For pure U-235, the critical mass is 15 kg, while it is only 4.4 kg for pure plutonium-239.

6.6.1 Reactors

Research into atomic bombs also led the way for the development of reactors to produce electric power. The first application was on naval submarines. Today, about 20% of the electric power generated in the United States comes from nuclear reactors[3]. Most reactors in the United States are light water reactors ("light" distinguishes this from "heavy" water reactors that use D_2O, see below).

Figure 6.5 shows a schematic of a nuclear reactor. To generate electricity, one must find a way to turn a turbine. In a traditional fossil fuel power plant, a combustion reaction creates heat and generates steam that turns the turbine. The nuclear reactor is simply using a nuclear reaction, instead of a combustion reaction, to generate heat and create steam that turns the turbine. As seen below, nuclear fission is more efficient at generating energy than combustion.

Instead of coal or oil, a nuclear reactor contains fuel rods with pellets of uranium or ura-

---

[3] U.S. Energy Information Administration, *Nuclear Explained: U.S. Nuclear Industry*, 2021, available at https://www.eia.gov/energyexplained/nuclear/us-nuclear-industry.php.

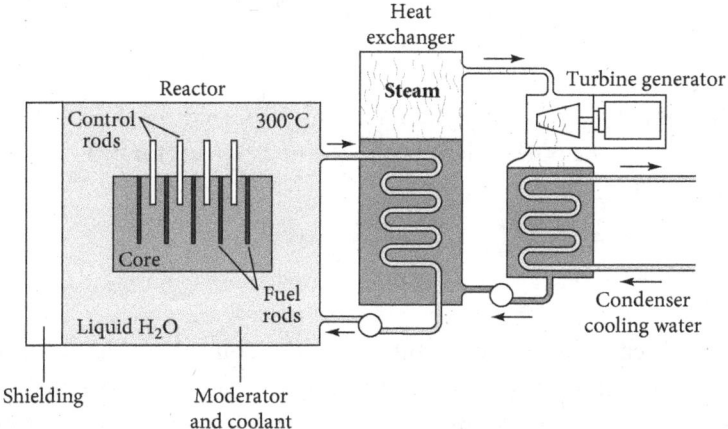

Figure 6.5 Schematic of the pressurized light water reactor.
Source: Adapted from Spiro et al., *Chemistry of the Environment, 3rd Ed.,* University Science Books, © 2012, all rights reserved.

nium oxide. Naturally occurring uranium contains about 0.7% of the U-235 isotope, which is too low to sustain a chain reaction. The reactor requires enrichment of U-235 to a concentration of 3%–4.5% (see Section 6.2.1 for more information).

In order to ensure that the chain reaction is always under control, operators must regulate the number and speed of the neutrons. To limit the number of neutrons present, control rods containing cadmium (Cd) or boron (B), which absorb neutrons effectively, are lowered among the fuel rods. Without the control rods, the reactor core would melt from the heat generated. To regulate the speed of the neutrons, a moderator surrounds the fuel and controls rods. Slow neutrons split U-235 more efficiently, but because the fission reaction is so highly exothermic, the neutrons it produces travel very quickly. The moderator increases the efficiency of the fission reaction. Water is commonly used as the moderator; thus, it can also be used as a coolant.

An alternative to the "light" water reactor is a "heavy" water reactor. It uses D_2O instead of H_2O as the coolant and moderator. Heavy water is less effective as a moderator because of the larger deuterium (D or ^2H) mass, but D absorbs neutrons much less than H. As a result, it is possible to maintain a chain reaction even with unenriched uranium (0.7% U-235).

QUESTION 33

A nuclear power plant, like coal and natural gas-burning power plants, heats water to form steam to turn a turbine and create electricity. If the temperature of the nuclear reactor-produced steam (at high pressure) and of the cooling water are 250°C and 75°C, respectively, what is the maximum efficiency of the power plant? Refer to Section 2 in Chapter 3 as a reminder of how to calculate efficiencies.

Q33 ANSWER To convert T from °C to K add 273:

$$T_s = 75°C + 273 = 348 \text{ K}$$

$$T_h = 250°C + 273 = 523 \text{ K}$$

Then $\dfrac{W_{mass}}{Q_{total}} = 1 - \dfrac{348 \text{ K}}{523 \text{ K}} = 0.335$ or 33.5%.

---

*QUESTION 34

Nuclear power plants that use liquid metals as a heat transfer liquid have higher efficiencies than those that use only water. Calculate the efficiency of a power plant where the nuclear reactor produced steam is 575°C and the cooling water is 20°C.

6.6.1 *Breeder Reactors*

It is possible to extend the nuclear fuel supply by converting the dominant uranium isotope (U-238) to another fissionable isotope, plutonium-239. Figure 6.6 shows the production pathway. Pu-239 can be used in a power reactor just as U-235 can. The efficient production of Pu-239 requires a "breeder" reactor, operating with fast neutrons. The fast breeder differs from an ordinary fission reactor in having the water coolant replaced by liquid sodium (Na). Being much heavier than water, the Na atoms slow the neutrons to a much smaller extent. At the same time, it is still a liquid, so it can still act as a heat sink.

Breeder reactors are more expensive to build than conventional reactors. To date, the United States does not have any breeder reactors, while only a few have been built in other countries.

6.6.2 *Hazards and Sustainability*

6.6.2.1 **Mining and Enrichment**

There are several points in the entire nuclear energy cycle (Fig. 6.7) where the dispersal of radioactive materials is an actual or potential problem. Uranium mining is itself a hazardous occupation. Miners have a high risk of developing lung cancer because they inhale dust, radon, and radon decay products. Not only does the mining process itself contaminate the environment by exposing radioactive elements, but the very large volume of waste material that remains after the uranium is chemically extracted from the ore is itself radioactive. As in other mining operations, the liquid tailings are held in ponds until the solids separate, but pollution of local groundwater can occur if these ponds overflow or leak. In the United States, current mining operations occur in Utah, Nebraska, and Wyoming.

Uranium occurs naturally as uranium dioxide (UO_2) ore. Major suppliers and processors of uranium ore are Canada, Australia, Russia, Niger, Ukraine, and Kazakhstan. The ore contains

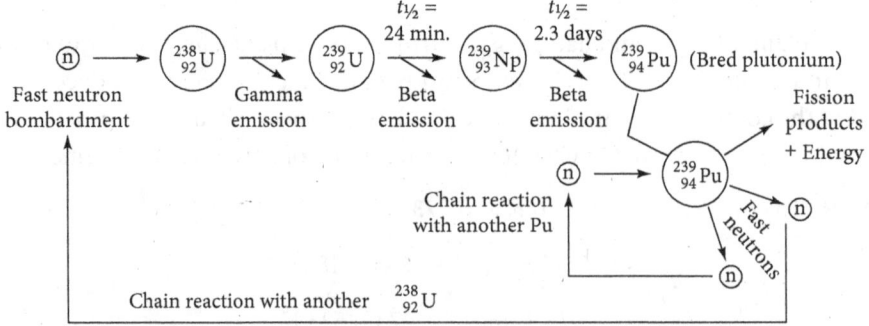

Figure 6.6 Production of plutonium-239 from uranium-238 bombardment.
Source: Adapted from Spiro et al., *Chemistry of the Environment, 3rd Ed.*, University Science Books, © 2012, all rights reserved.

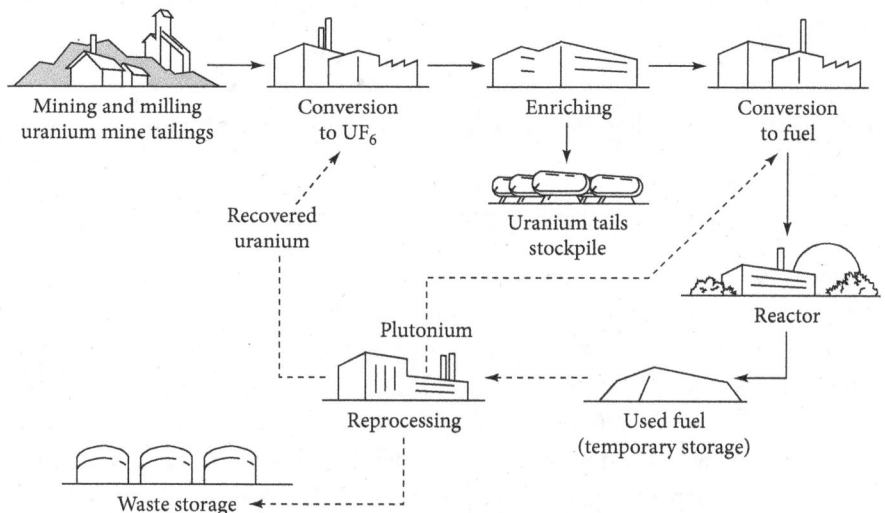

Figure 6.7 Fuel cycle for the light-water nuclear reactor; dashed lines are not yet part of the U.S. cycle. *Source:* Adapted from Spiro et al., *Chemistry of the Environment, 3rd Ed.*, University Science Books, © 2012, all rights reserved.

other actinide metals and various impurities, so a multistage extraction process is required to isolate the uranium. In refining, uranium is first oxidized to UO_2^{+2} and $U_2O_7^{-2}$, then partially reduced by heating to yellowcake (U_3O_8–a 1:2 combination of UO_2 and UO_3). To enrich the uranium (see below), the yellowcake is first oxidized completely to UO_3 and then fluorinated to UF_6. Finally, it is reduced back into UO_2 and fabricated into fuel pellets for the reactor. The pellets are stacked and sealed in long metal tubes to form the fuel rods. The fuel rods are bundled together to make up a fuel assembly. Depending on the reactor type, each fuel assembly has about 179–264 fuel rods. Each reactor core can hold 121–193 fuel assemblies.

Elements can exist in multiple oxidation states, depending on the number of electrons removed or added to the valence shell of the atoms. Since redox reactions frequently involve electron transfer between complex molecules, it is important to keep track of the atomic oxidation state, in order to balance the reaction equations. See Chapter 14 for a more detailed discussion.

QUESTION 35

Using the oxidation numbers of uranium, show that there is a reduction from gaseous UF_6 to the UO_2 pellet form.

Q35 ANSWER Recall from general chemistry that reduction is a gain of electrons; thus, the oxidation number/oxidation state will become more negative.

Fluorine always has an oxidation state of –1; thus to balance, uranium must have an oxidation state of +6 in UF_6.

Oxygen usually has an oxidation number of –2; thus, the two O atoms would provide a total of –4. To balance the neutral molecule, uranium must have an oxidation state of +4 in UO_2.

Thus, the oxidation state changes from +6 to +4, and thus, uranium gains electrons.

*QUESTION 36

Using the oxidation numbers of uranium, show that there is oxidation in the first uranium refining step from UO_2^{+2} and $U_2O_7^{-2}$.

The U-235 isotope is somewhat scarce because 6.4 of its half-lives have elapsed since the formation of Earth. Thus, it only accounts for 0.7% of the uranium found naturally (Fig. 6.8 shows this is 1.75/250 tons). Thus, most nuclear power plants (and all weapons-grade uranium) require a concentration (or enrichment) of the uranium-235 isotope. Physical separations can be operated in successive states to produce gradual enrichment. For example, gaseous UF_6 is passed through a succession of porous diffusion barriers. At each of these, the lighter $^{235}_{92}UF_6$ is enriched by a factor equal to the square root of the mass ratio.

QUESTION 37

How many barriers would be needed to increase the enrichment by four times (0.7%–2.8%)?

Q37 ANSWER   Using Graham's law of diffusion, we can calculate an enrichment factor for each barrier:

$$\frac{r_{235}}{r_{238}} = \sqrt{\frac{\text{mass of }^{238}_{92}UF_6}{\text{mass of }^{235}_{92}UF_6}} = \sqrt{\frac{[238 + 6(19.00)] \text{ amu}}{[235 + 6(19.00)] \text{ amu}}} = 1.004$$

Now we want a total enrichment of 4, so:

$$(1.004)^n = 4$$
$$n \log(1.004) = \log (4)$$
$$n = 347.3$$

In other words, it would take 348 diffusion stages to increase the enrichment by four times (0.7%–2.8%).

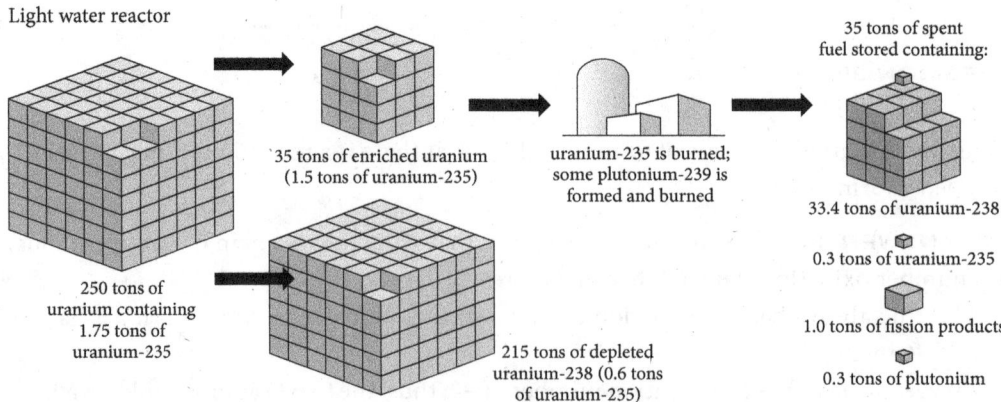

Figure 6.8 Amount of uranium-235 present throughout the fuel cycle.
Source: Adapted from Hargraves & Moir, "Liquid Fluoride Thorium Reactors: An Old Idea in Nuclear Power Gets Reexamined," *American Scientist*, Vol. 98, No. 4, 2010, pp. 304–313, https://www.jstor.org/stable/27859537.

*QUESTION 38

How many barriers would be needed to enrich U-235 to 90% (weapons-grade)?

It should also be noted that gas diffusion has been replaced largely by gas centrifuge. Centrifugal force offers a more effective route to isotope separation. This force is proportional to mass, not the square-root of mass, as in Graham's law above.

6.6.2.2 Greenhouse Gas Emissions and Water Use

Although the nuclear power generation process does not emit any greenhouse gases (as combustion does), the other steps in the fuel cycle (Fig. 6.7) do. In addition, there are greenhouse gas emission costs in the construction and decommission of the reactor itself. Over the lifetime of a nuclear power facility, Sovacool (2008)[4] estimates there is 66g of CO_2 equivalent emitted per kilowatt-hour produced. This is compared to 443–1,050g of CO_2e/kWh for traditional fossil fuels.

Also, not only does a typical nuclear reactor use water as a moderator, it uses large quantities of water to cool and condense the steam for reuse. Macknick and team[5] estimate that between 270 and 670 gallons (gal) of water is consumed per megawatt hour produced. In this way, a nuclear power plant is similar to a conventional power plant that burns fossil fuel. Most power plants are thus built near a river, lake, or ocean; unfortunately, this method of cooling causes thermal pollution. Furthermore, a substantial amount of water is used for the ore mining and extraction process.

6.6.2.3 Heavy Versus Light Water Compromise

The relative merits of light and heavy water reactors depend in large measure on the relative costs of separating D_2O from H_2O versus separating U-235 from U-238. The costs of the latter have been somewhat hidden by the preexistence of large U-235 separation plants constructed for the production of nuclear weapons, which have been used to supply the fuel for commercial reactor programs.

Using heavy water means that uranium does not need to be enriched, but energy will have to be supplied to create and concentrate D_2O. D_2O is prepared by either fractional distillation or electrolysis of ordinary water. In countries where hydroelectric power is abundant, the production cost can be reasonable. Canada is currently one of the few nations successfully using heavy water nuclear reactors.

One additional disadvantage of heavy water reactors is the creation of radioactive tritium. Deuterium can absorb a proton and become tritium (^3H), which has a 12.3-yr half-life. This radioisotope can be dangerous if inhaled or ingested.

6.6.2.4 Nuclear Waste

The uranium-235 in a reactor cannot be completely used up, because of the buildup of fission products, which themselves absorb the neutrons and eventually slow down the chain

---

[4] Sovacool, "Valuing the Greenhouse Gas Emissions from Nuclear Power: A Critical Survey," *Energy Policy*, Vol. 36, No. 8, 2008, pp. 2940–2953.

[5] Macknick et al., "Operational Water Consumption and Withdrawal Factors for Electricity Generating Technologies: A Review of Existing Literature," *Environmental Research Letters*, Vol. 7, No. 4, 2012, pp. 045802, doi:10.1088/1748-9326/7/4/045802.

reaction. After ~1–3 yr, the fuel rods must be replaced with new ones. Spent fuel can be repro-cessed by chemical extraction of the fission products, separation of the accumulated Pu, and re-concentration of U. The chemistry is straightforward, but the technology is complicated by the need for remote handling of intensely radioactive material. There is also a danger that the recovered plutonium could be stolen from a reprocessing plant and used to make nuclear weapons. Reprocessing produces high-level waste (HLW) that is highly radioactive and of moderately high temperature. Since no method has been approved for disposing of this waste permanently, it has been accumulating at reprocessing sites for almost 50 yr. The intention is to immobilize the HLW in an insoluble matrix, seal it inside a corrosion-resistant container, and bury it deep underground in a geologically stable rock structure. The waste must also be isolated from the human environment for exceedingly long periods of time.

TABLE 6.5 Half-Lives of Example Fission Products and Reactor-Generated Isotopes from Uranium-235 Decay

Isotope	Half-Life (year)
$^{143}_{54}Xe$	1.62×10^{-8}
$^{90}_{38}Sr$	28.1
$^{239}_{94}Pu$	24,110

QUESTION 39

The length of time for safety can be assessed by the rule of thumb that radiation sinks to negligible levels after 10 half-lives. Is it reasonable to find a burial site where disturbance by earthquakes and groundwater infiltration can be excluded for this period of time? Consider that the waste will be made up of many radioactive isotopes; use fissionable products, such as Xe-143 and Sr-90, as well as the reactor-generated product Pu-239 for this example.

Q39 ANSWER Using Table 6.5, multiply each half-life by 10:

$$1.62 \times 10^{-8} \text{ yr} \times 10 = 1.62 \times 10^{-7} \text{ yr}$$

$$1.62 \times 10^{-7} \text{ yr}\left(\frac{365 \text{ days}}{1 \text{ yr}}\right)\left(\frac{24 \text{ hr}}{1 \text{ day}}\right)\left(\frac{60 \text{ min}}{1 \text{ hr}}\right)\left(\frac{60 \text{ sec}}{1 \text{ min}}\right) = 5.11 \text{ sec}$$

$$28.1 \text{ yr} \times 10 = 281 \text{ yr}$$

$$24,110 \text{ yr} \times 10 = 241,100 \text{ yr}$$

These examples range from 5 sec to 241,000 yr. Considering that plutonium would require almost a quarter of a million years of stability, this is not a very reasonable task.

For the time being, spent fuel rods are being stored temporarily at reactor sites, in pools of water to dissipate the heat and allow the most intense radioactivity to decay away.

*QUESTION 40

How much heat is generated from the decay of radioactive waste? As an example, let us consider the β decay of Sr-90 (89.907738 amu) which has a half-life of 28.1 yr:

$$^{90}_{38}Sr \rightarrow \,^{90}_{39}Y + \,^{0}_{-1}\beta$$

The yttrium (89.907152 amu) further decays with a half-life of 2.66 days to a stable isotope (Zr-90, 89.9047044 amu):

$$^{90}_{39}Y \rightarrow \,^{90}_{40}Zr + \,^{0}_{-1}\beta$$

(i) Starting with 1 mol of Sr-90, how many atoms will decay in 1 yr?

(ii) How much energy is generated from the decay of 1 mol of Sr-90, considering both decay reactions? Use Table 6.1 for the mass of the β particle.

6.6.2.5 Reactor Safety

Under normal operating conditions, the radiation released by nuclear reactors is very low. The danger is not that a nuclear explosion could be set off; the fissionable material is too dilute. But a great deal of heat is generated by an operating fuel rod assembly, even after the control rods have been lowered. This heat is carried away by the water circulating through the reactor, but if the water is somehow not replenished, the temperature can rise to disastrous levels, allowing the radioactive materials to be released. Many fission products, such as Pu-239 and Sr-90, are dangerous radioactive isotopes with long half-lives.

The most famous nuclear accident occurred on April 26, 1986, in the Ukrainian town of Chernobyl. Reactors of the Chernobyl class, which are still in use throughout the former Soviet Union, are of a different design than the reactor described in Section 6.1. They employ graphite, instead of water, as the moderator. Graphite contains only C atoms and it slows neutrons quite effectively; it can sustain a chain reaction when only 1.8% of the fuel is fissionable U-235. The heat is carried away by flowing water around the individual fuel rods. Because the rods are cooled individually, they can be replaced one at a time, without shutting down the reactor. Consequently, this type of reactor has one of the highest rates of productive time online in the nuclear industry. Unfortunately, graphite, unlike water, is flammable.

When power levels dip very low, the reactor becomes unstable and can race out of control, which is what happened on the day of the accident during a test of a reactor's response to a simulated power failure. The loss of power, in combination with control rod malfunctions and operator misjudgments, allowed the reactor to surge out of control, vaporizing the water, blowing off the roof of the reactor, and igniting the graphite. Due to cost limitations at the time, there was not a secondary containment building around the reactor as was used in other countries. Thus, a great cloud of radioactive debris rained out radioisotopes over much of Europe and parts of Asia. The main fission products found were ^{131}Xe, ^{85}Kr, ^{131}I, ^{134}Cs, and ^{137}Cs isotopes. The reactor continued to burn for 10 days, eventually releasing tens or hundreds of millions of curies of radioactivity into the environment.

The initial explosion and response by the fire fighters resulted in 31 deaths from acute radiation poisoning. A major chronic health consequence has been an increase in thyroid cancer among children in the area. It is presumed that the I-131 radioisotope is concentrated in the thyroid and then decays via β emission with a half-life of eight days. Thankfully, thyroid cancer has a high cure rate. Most of the iodine in the children's systems came from drinking milk from cows that had grazed on contaminated plants and from eating vegetables that had radioactive deposits.

QUESTION 41

After the accident at Chernobyl, residents were given tablets of potassium iodide to help flush the radioactive iodine from their bodies. If they started taking the tablet five days after the accident, what percentage of ^{131}I released by the explosion would be remaining in their thyroid? The half-life of I-131 is eight days.

Q41 ANSWER Using the half-life of eight days, we can calculate the rate constant (6.14):

$$k = \frac{\ln 2}{8 \text{ days}} = \frac{0.0866}{\text{days}}$$

Now solve for the ratio of nuclei using (6.10):

$$\ln\left(\frac{Q_t}{Q_0}\right) = -\left(\frac{0.0866}{\text{days}}\right)(5 \text{ days})$$

$$\ln\left(\frac{Q_t}{Q_0}\right) = -0.4430$$

$$e^{\ln\left(\frac{Q_t}{Q_0}\right)} = e^{-0.4430}$$

$$\left(\frac{Q_t}{Q_0}\right) = 0.6486$$

65% of I-131 would remain after five days.

Radiation experts predict between 4,000 and 9,000 premature deaths among people who were living in the heavily contaminated regions.[6] Cancers other than thyroid cancer have much longer development times. The World Health Organization (WHO) also notes that another long-term effect of this disaster is the mental stability of the residents of Chernobyl. There were substantial traumas from perceived health risks and relocation. Eventually, more than 300,000 people were told to leave their homes for what they thought was a temporary relocation. Instead of returning home, the area around the nuclear plant was declared uninhabitable.

A more recent nuclear accident occurred in March 2011. A magnitude 9.0 earthquake shook Japan on March 11th and caused a core meltdown at the Fukushima Daiichi nuclear

---

[6] Blakemore, "The Chernobyl Disaster: What Happened, and the Long-Term Impacts," *National Geographic*, 2019, available at https://www.nationalgeographic.com/culture/topics/reference/chernobyl-disaster/.

power plant. The six reactors at the site were mainly undamaged, but the power grid in northern Japan failed. Diesel generators were started to supply emergency cooling systems for
the core, but the ocean-side power plant took a direct hit from a massive tsunami with a
wave height of about 6m, so ocean water flooded the generators. This caused a failure of the
emergency power supply to most of the cooling systems. Three of the reactors had no way to
remove the heat. The residual heat arising from the radioactive decay of the fission products
raised the temperature of the fuel rods in the core to thousands of degrees, melting them and
allowing the escape of radioactive products from the rods. The temperatures also converted
the remaining cooling water to high-pressure steam.

As a result, the bare zirconium (Zr) metal of the fuel rods reacted with the high-temperature (>1,000°C) steam to produce hydrogen gas:

$$Zr(s) + 2\,H_2O(g) \rightarrow ZrO_2(s) + 2\,H_2(g)$$

Gases, including the hydrogen, were released and explosions were observed several times
in the weeks following the earthquake. Unlike the Chernobyl power plant design, these reactor cores do have primary and secondary containment, so there was no concentrated cloud of
radioactive debris.

There was still a need to provide cooling water on the reactor cores so that a complete
meltdown did not occur. About 3 million gal of highly radioactive water waste was released
into the Pacific Ocean to make room at the storage site for even more contaminated water. In
addition, radioactive water was seeping into the Pacific Ocean from leaks in the reactor and
spent fuel ponds.

Radioactive contaminants escaped the plant, contaminating drinking water and agricultural products near the plant with I-131 and Cs-137. The WHO estimates that the average
radiation dose among the 530,000 recovery operation workers was 12 rem, while the evacuees
experienced 3 rem[7] (over 1,000 times more powerful than a typical X-ray). Low concentrations of radioactive materials spread throughout the world in the oceans and atmosphere.

---

*QUESTION 42

The people of Fukushima were moved out of their houses two days after the accident, and after 10
yr, some of them have started to return. What percentage of ^{137}Cs released by the explosion would
remain in the environment? Is it safe for the residents to return to that area to live? The half-life of
C-137 is 30.2 yr.

6.6.2.6 Weapons Proliferation

The first application of nuclear fission was in the development of the atomic bomb. A
fission bomb can be made from either $^{235}_{92}U$ or $^{239}_{94}Pu$. The bomb is made by separating two
masses of fissionable material, each incapable of sustaining a chain reaction separately. The

---

[7] World Health Organization, *Radiation: The Chernobyl Accident*, 2011, available at https://www.who.int
/news-room/questions-and-answers/item/radiation-the-chernobyl-accident.

two masses are brought together using another explosive, like TNT, to form a critical mass at the moment of detonation.

If uranium is used, it must be enriched (Section 6.6.2.1) to >93%. As seen above, isotope enrichment is a technically demanding process that involves a major commitment of resources. However, the same isotope enrichment plants that produce fuel-grade enrichment can be operated to produce weapons-grade enrichment. That is why there is always a concern when foreign adversaries start developing isotope enrichment facilities, even if they say it is only for nuclear power generation. Uranium-235 was the fissionable material in the bomb dropped on Hiroshima, Japan, on August 6, 1945.

Plutonium is a different matter. Since Pu-239 is produced in the uranium reactor, no isotope enrichment is needed to produce weapons-grade fuel. The Pu extracted in a reprocessing plant can be used in nuclear weapons. Plutonium-239 was used in the bomb that exploded over Nagasaki on August 9, 1945.

6.6.3 *Future of Nuclear Power*

According to the World Nuclear Association, in 2019 there were about 445 nuclear reactors around the world providing about 10% of the world's electricity.[8] The countries generating the most nuclear power are, in order, the United States, France, China, Russia, and South Korea. Although France does not produce as much energy as the United States from nuclear reactions, the amount that it does produce accounts for more than 70% of its national electricity production.

According to the U.S. Energy Information Administration (EIA), at the end of December 2020, the United States had 94 nuclear reactors operating at 56 nuclear power plants in 28 states. The United States has the largest nuclear electricity generation capacity and generated more nuclear electricity than any other country in 2020. The average age of the U.S. nuclear reactors is 39 yr.[9] Since 1990, nuclear power has made up about 20% of the total electricity generated by the United States. Although the nuclear energy capacity has been steady since 1990, the electricity generation was able to continue to climb until 1998, where it has held steady at ~800,000 megawatt-hours (MWh) (Fig. 6.9).

2016 saw the first new U.S. nuclear energy reactor to come online since 1996. The new facility is in Tennessee and two more reactors are now under construction in Georgia that are likely to come online by 2023. Increasing demands for electricity, as well as rapidly rising demands in countries like China and India, have increased the demand for nuclear power (Fig. 6.10).

6.7 Power from Fusion

Nuclear energy can potentially be obtained by fusing nuclei. Small nuclei are less stable than heavier ones, up to iron (Fig. 6.2). This behavior suggests that if two light nuclei combine through fusion and form a larger, more stable nucleus, the gain in stability will release an appreciable amount of energy.

Nuclear fusion powers our sun. When two H nuclei fuse and form a He nucleus, an enormous amount of energy is released. However, fusing the nuclei requires extreme reaction conditions in order to overcome the huge energy barrier caused by the repulsion between

---

[8] World Nuclear Association, *Nuclear Power in the World Today*, available at https://www.world-nuclear.org/information-library/current-and-future-generation/nuclear-power-in-the-world-today.aspx.

[9] U.S. Energy Information Administration, *Nuclear Explained: U.S. Nuclear Industry*, available at https://www.eia.gov/energyexplained/nuclear/us-nuclear-industry.php.

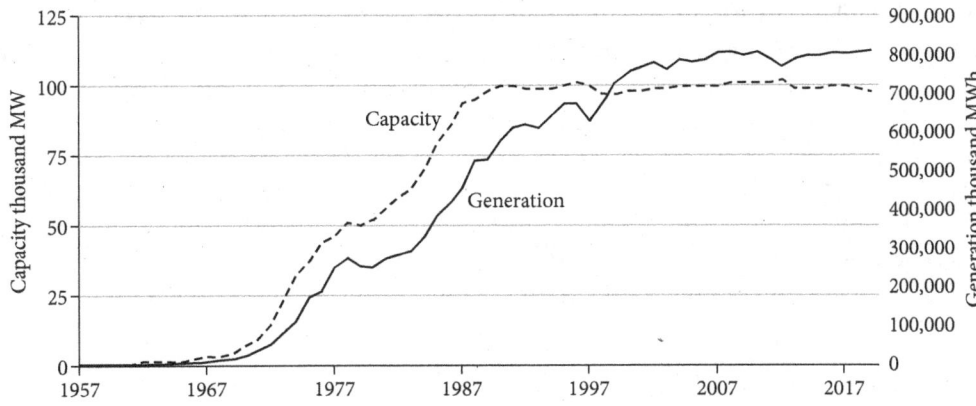

Figure 6.9 U.S. nuclear energy capacity and generation from 1957 to 2019. Capacity is net summer, MW is megawatts, and MWh is megawatt-hours.
Source: U.S. Energy Information Administration, *Monthly Energy Review*, Table 8.1, 2019, available at https://www.eia.gov/energyexplained/index.php?page=nuclear_use.

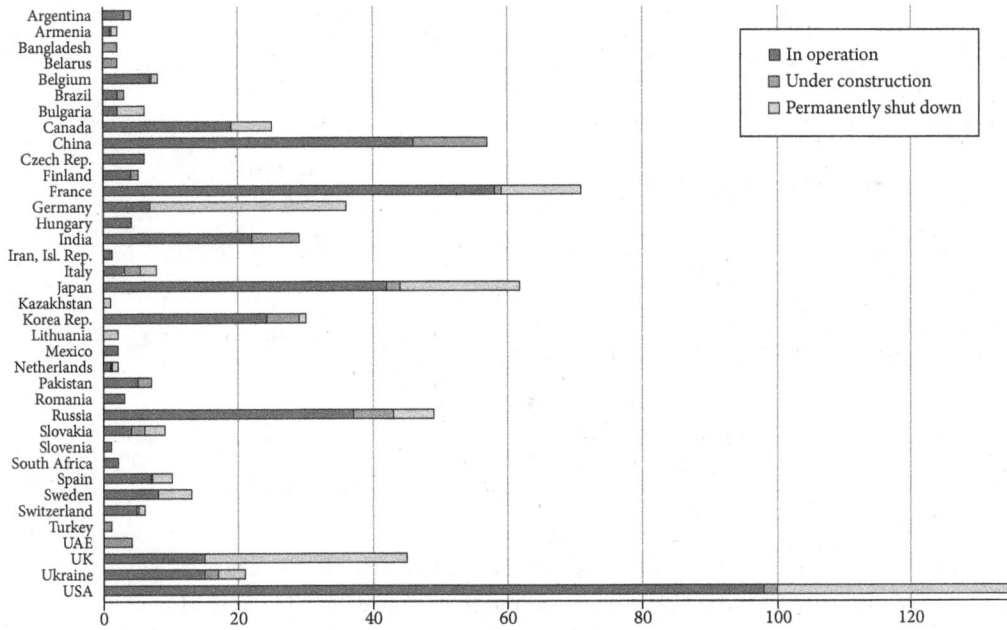

Figure 6.10 The number of nuclear power reactors by country and status.
Source: International Atomic Energy Agency Power Reactor Information System, *Country Statistics*, available at https://pris.iaea.org/PRIS/CountryStatistics/CountryStatisticsLandingPage.aspx

the two H protons. Such extreme conditions (100–1000 million °C) are found in the center of stars. Similar conditions are also obtainable in hydrogen bombs.

Fusion cannot be carried out on ordinary H atoms, because two protons cannot fuse without neutrons being available to stabilize the product nucleus. This means that heavy isotopes of H must be used (e.g., deuterium ^{2}H or tritium ^{3}H). For example, we could fuse two deuterium isotopes together:

$$^2_1H + {}^2_1H \rightarrow {}^3_1H + {}^1_1H$$

This is particularly an attractive option because deuterium makes up 0.015% of all hydrogen on Earth and enriching it would pale in comparison to how much energy is generated from the reaction. Unfortunately, the ignition temperature for this reaction is $5 \times 10^8\,°C$.

QUESTION 43

How much energy is generated from the fusion of two deuterium isotopes? A deuterium isotope has a mass of 2.0141018 amu, tritium has a mass of 3.0160493 amu, and hydrogen has a mass of 1.0078250 amu:

$$\ce{^2_1H + ^2_1H \rightarrow ^3_1H + ^1_1H}$$

Q43 ANSWER We can calculate the amount of energy released using (6.2):

$$\Delta E = \Delta mc^2$$

We need the change in mass or (mass of products) − (mass of reactants)

$$\text{mass of products} = (3.0160493\ \text{amu} + 1.0078250\ \text{amu}) = 4.0238743\ \text{amu}$$

$$\text{mass of reactants} = 2(2.0141018\ \text{amu}) = 4.0282036\ \text{amu}$$

$$\Delta m = 4.0238743\ \text{amu} - 4.0282036\ \text{amu} = -0.0043293\ \text{amu}$$

Relate this mass to energy:

$$\Delta E = (-0.0043293\ \text{amu})\left(2.998 \times 10^8\,\frac{\text{m}}{\text{sec}}\right)^2 = -3.891 \times 10^{14}\frac{\text{amu m}^2}{\text{sec}^2}$$

Now adjust the units to joules. Remember that 1J is 1 kg m² sec⁻²:

$$-3.891 \times 10^{14}\frac{\text{amu m}^2}{\text{sec}^2}\left(\frac{1.00\ \text{g}}{6.022 \times 10^{23}\ \text{amu}}\right)\left(\frac{1\ \text{kg}}{1{,}000\ \text{g}}\right)\left(\frac{1\ \text{J}}{1\frac{\text{kg m}^2}{\text{sec}^2}}\right) = -6.46 \times 10^{-13}\ \text{J}$$

The fusion of two H-2 atoms would release 6.46×10^{-13} J.

In contrast to fission, nuclear fusion is attractive because the fuels are cheap and almost inexhaustible and if the process produces radioactive waste, it is short-lived. If a fusion generator was turned off, the reaction will stop right away without the danger of meltdown. The problems with fusion involve finding a way to hold the nuclei together long enough, and at extremely high temperatures, for the reaction to occur.

---

*QUESTION 44

The reaction between deuterium and tritium isotopes has a much lower ignition temperature (1–2 $\times 10^8\,°C$). How much energy is generated from this fusion reaction? Compare this to the fission of two deuterium isotopes seen in Q43. A deuterium isotope has a mass of 2.0141018 amu, tritium has a mass of 3.0160493 amu, and the masses of helium and the neutron are found in Table 6.1.

$$\ce{^2_1H + ^3_1H \rightarrow ^4_2H + ^1_0n}$$

6.8 Other Uses of Isotopes and Related Particles

Radioactive isotopes have many uses in science and medicine [partially why our exposure to radioactivity has increased over time (Table 6.3)]. In addition to using radioactive decay to date objects (see Section 6.3), scientists can use radioisotopes as tracers to determine reaction mechanisms or to diagnose diseases. Lastly, the newest elements on the periodic table made using particle accelerators are all radioactive.

6.8.1 Tracers

6.8.1.1 Determining Structure and Mechanisms

How do we know the exact structure of a molecule? Scientists have to use chemical analysis tools to elucidate mass fragments or vibrational modes of functional groups. But they can also use radioactive isotopes to "label" a specific element and trace its behavior. Radioactive isotopes are easy to trace because their decay is easy to detect. Changing an element from one isotope to another will not affect its chemical behavior in the molecule because isotopes only differ in the number of neutrons.

Tracers can be used to investigate the actual steps involved in a reaction mechanism or to predict behavior at a molecular level. This is the fundamental basis for the field of organic chemistry. Radioactive isotopes, such as ^{14}C, can be used to "label" a specific reactant. Then the reactant can be traced throughout the mechanism. Since the isotope will have a different mass, it may even be possible to isolate intermediate products.

In the same way, radioisotopes can also be used to understand behavior and transport in the environment. They can be used to search for leaks, measure flows, and follow pollution propagation.

6.8.1.2 Diagnosing Disease

The benefit of using radioisotopes to diagnose disease is that they are easy to detect and can be used in very small amounts. A specific radioisotope chosen for its ability to follow blood flow or concentrate in a certain organ can be given to a patient. Then, the decay signature will be detected using body scans. The first type of detection was a γ camera that detected single photons; the camera built an image from the location of the emission. A more recent detection method is positron emission tomography (PET) scans. A positron-emitting radioisotope is ingested by the patient and as it decays, it emits a positron that quickly combines with a nearby electron, resulting in simultaneous emission of two identifiable γ rays in opposite directions. These rays are detected by a PET camera to give a precise three-dimensional location. PET's most important clinical role is using fluorine-18 to detect most cancers. Even better diagnosis comes with combining PET and CT scans.

QUESTION 45

Write a nuclear reaction for the fluorine-18 decay via positron emission.

Q45 ANSWER A positron is $_{+1}^{0}\beta$, thus the resulting equation would be:

$$^{18}_{9}F \rightarrow \, ^{0}_{+1}\beta + \, ^{18}_{8}O$$

The decay product is oxygen-18.

*QUESTION 46

What are the medical implications for using fluorine-18 if its half-life is only 1.8 hr?

Ideally, the isotope will be localized to address the issue, produce radiation that can be detected externally, but expose the body to a minimal dose of radiation. Gamma emitters are thus ideal radiopharmaceuticals because γ rays penetrate the body tissues. In addition, the radioisotope must have a half-life long enough to follow the biological process in question, but short enough to avoid unnecessary exposure.

The most common radioisotope used in diagnosis is technetium-99m (Tc-99m), where m denotes that the Tc-99 isotope is produced in its excited nuclear state. Tc-99m accounts for about 80% of all nuclear medicinal procedures worldwide. Its half-life is 6 hr and its decay involves the emission of γ rays and low energy electrons to yield Tc-99 in its nuclear ground state. The γ rays are 140 keV, about the same wavelength as conventional X-ray emissions, so the response can be readily detected.

QUESTION 47

A typical bone scan using Tc-99m will subject the patient to 19–30 mCi of radiation. If a person only needs 19 mCi of radiation and the half-life of Tc-99m is 6 hr, how much of the isotope must the individual ingest (in nanograms)? Assume the molar mass is 98.9 g mol^{-1}.

Q47 ANSWER First, recall the definition of a curie: 1 Ci = 3.70×10^{10} nuclear disintegrations sec^{-1}. Thus, convert the amount of radiation into nuclear disintegrations per second:

$$19 \text{ mCi} \left(\frac{1 \text{ Ci}}{1{,}000 \text{ mCi}} \right) \left(\frac{3.70 \times 10^{10} \text{ disintegrations/sec}}{1 \text{ Ci}} \right) = 7.03 \times 10^8 \frac{\text{disintegrations}}{\text{sec}}$$

This value is the rate of decay. Since we have the half-life, we can also calculate the rate constant using (6.14):

$$k = \frac{\ln 2}{6 \text{ hr}} = \frac{0.116}{\text{hr}}$$

Because we know that rate = kQ [Eq. (6.3)], and there is 1 disintegration atom^{-1}, we can thus calculate the amount of Tc-99m present. We must also resolve the time units:

$$\frac{\text{rate}}{k} = Q$$

$$\frac{\left(7.03 \times 10^8 \frac{\text{disintegrations}}{\text{sec}}\right)\left(\frac{60 \text{ sec}}{1 \text{ min}}\right)\left(\frac{60 \text{ min}}{1 \text{ hr}}\right)}{0.116 \frac{1}{\text{hr}}} = 2.18 \times 10^{13} \text{ disintegrations}$$

Thus, we would need 2.18×10^{13} atoms:

$$2.18 \times 10^{13} \text{ atoms} \left(\frac{1 \text{ mol}}{6.022 \times 10^{23} \text{ atoms}} \right) \left(\frac{98.9 \text{ g}}{\text{mol}} \right) \left(\frac{1 \times 10^9 \text{ ng}}{1 \text{ g}} \right) = 3.60 \text{ ng}$$

*QUESTION 48

How much Tc-99m is left one day after the isotope is ingested?

*QUESTION 49

Cobalt-60 is an isotope used in cancer treatment. It decays via γ emission at 2.4×10^{-13} J photon^{-1}. Calculate the wavelength (in nm) of this γ emission.

6.8.2 Transuranium Isotopes

The so-called "transuranium" elements are those that are larger than uranium (at atomic number 92). These elements are synthesized using particle accelerators and all of their isotopes are radioactive. Most of the transuranium elements also have very short half-lives, so their use is limited. Studying this process allows scientists to understand nuclear properties.

QUESTION 50

Element 113 (nihonium, Nh) has a half-life of 1 min. It is formed by α decay of element 115 (moscovium-288, ^{288}Mc). ^{288}Mc is created by colliding ^{243}Am with ^{48}Ca and its half-life is 1 sec. Write nuclear equations to represent these two syntheses.

Q50 ANSWER Start with the synthesis of Mc. Using the identities of the two reactant elements, we know how many protons there are in this reaction. Those protons are all accounted for in the product element, but the mass does not add up. Because the additional particle must have mass, but no additional protons, it must be a neutron.

$$^{243}_{95}\text{Am} + ^{48}_{20}\text{Ca} \rightarrow ^{288}_{115}\text{Mc} + 3\,^{1}_{0}n$$

Now that we know the mass number of Mc, we can find that of Nh because there is an emission of an α particle:

$$^{288}_{115}\text{Mc} \rightarrow ^{284}_{113}\text{Nh} + ^{4}_{2}\alpha$$

Thus, nuclear chemistry provides a way to continuously expand the periodic table.

6.9 Conclusions

Nuclear chemistry can be used for a wide range of purposes, from creating energy to diagnosing cancer. A large part of the world's electricity derives from nuclear power. Although the actual chemical process of creating nuclear energy does not give off carbon dioxide, the construction of the power plants and mining the uranium necessary to run them do have environmental costs, including CO_2 emissions. We have to think about the half-life of radioactive materials that are created as waste and how the different particles and energy given off in nuclear reactions affect living organisms. Nuclear fusion could be a good option for future energy sources with little toxic waste, if technical difficulties could be overcome. The use of nuclear medicine to diagnose and treat illnesses is an important tool in improving human health.

PART III

ATMOSPHERE

7

CLIMATE CHANGE

No environmental issue is more momentous than the extent of human influence on the earth's climate. Temperatures at the earth's surface have been rising for the last century and the rate of increase seems to be accelerating. Glaciers and sea ice are melting, continental interiors are drying, storms are increasing in intensity, and ecosystems are altering dramatically. How much of this change is dependent on human actions and how much can be controlled in the future?

Climate change is often mentioned on the news when changes in temperature and weather patterns are discussed. Overall, weather in particular areas has been changing over time and that change has been more drastic in the last few years. Warmer or colder temperatures, more violent storms, and more droughts year after year are some of the noticeable changes currently. Energy flow on Earth, the greenhouse effect, water's effect on climate and energy flow, the carbon cycle, temperature predictions, and limiting human impact are explored in this chapter.

7.1 Energy Flows

Weather and climate result from the effect of energy flows on the earth's surface and in its atmosphere (Fig. 7.1). Nearly all of this energy comes from the sun, with small increments from the earth's molten core and tidal energy. About 30% of the sun's rays are reflected or scattered back into space, either from the earth's surface or from its atmosphere. The rest of the light is absorbed and converted to heat before being reradiated back into space. This heat flow drives the earth's weather system via wind, ocean currents, rain, and snow.

QUESTION 1

Figure 7.1 shows that 15.1×10^8 TWh of energy flows from the sun to the earth. Describe where all this energy is dispersed.

Q1 ANSWER The energy from the sun disperses into many places around the earth:

Reflected by the atmosphere → 3.9×10^8 TWh

Photosynthesis → 0.024×10^8 TWh

Heating of the land → 2.1×10^8 TWh

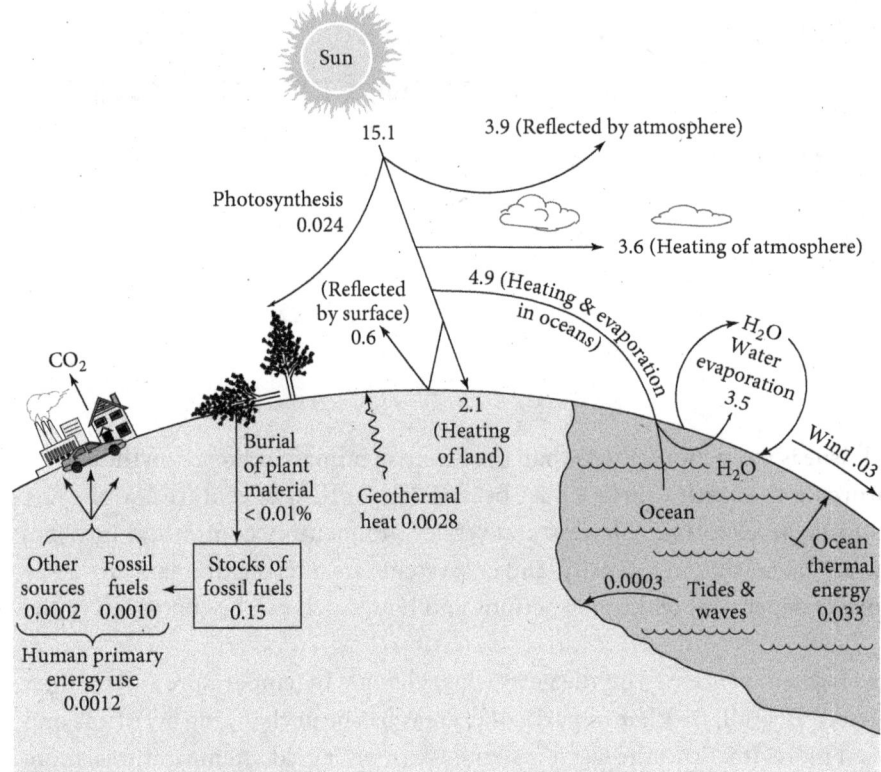

Figure 7.1 Annual energy fluxes on Earth [in 10^8 terawatt hours (TWh)].
Source: Adapted from Spiro et al., *Chemistry of the Environment, 3rd Ed.*, University Science Books, © 2012, all rights reserved.

Heating of the atmosphere $\rightarrow 3.6 \times 10^8$ TWh

Heating and evaporation in oceans $\rightarrow 4.9 \times 10^8$ TWh

Reflected by the surface $\rightarrow 0.6 \times 10^8$ TWh

Adding these together:

$$3.9 + 0.024 + 2.1 + 3.6 + 4.9 + 0.6 = 15.1 \times 10^8 \text{ TWh}$$

QUESTION 2

How much of the energy from the sun is reflected? Provide both the value and the percent.

Q2 ANSWER According to Figure 7.1, reflection of the sun's energy happens in the atmosphere (3.9×10^8 TWh) and on the surface (0.6×10^8 TWh):

$$3.9 \times 10^8 \text{ TWh} + 0.6 \times 10^8 \text{ TWh} = 4.5 \times 10^8 \text{ TWh}$$

To calculate the percent divide by the total energy from the sun (15.1×10^8 TWh):

$$\frac{\text{reflected energy}}{\text{total energy}} \times 100\% = \frac{4.5 \times 10^8 \text{ TWh}}{15.1 \times 10^8 \text{ TWh}} \times 100\% = 29.8\% \text{ or } \sim 30\% \text{ as mentioned in the text}$$

*QUESTION 3

What percent of the sun's radiation is used for photosynthesis?

QUESTION 4

If a 100-W light bulb is on for 1,000 hr annually (roughly 3 hr day^{-1}) how many such light bulbs could the sun's energy light?

Q4 ANSWER Divide the energy from the sun by the energy from 1 light bulb after converting TWh to Wh:

$$1 \text{ TWh} = 10^{12} \text{ Wh}$$

$$15.1 \times 10^8 \text{ TWh} \times \frac{10^{12} \text{ Wh}}{1 \text{ TWh}} = 15.1 \times 10^{20} \text{ Wh}$$

At 1,000 hr yr^{-1} each 100W light bulb uses:

$$1,000 \text{ hr} \times 100 \text{ W} = 1 \times 10^5 \text{ Wh}$$

$$\frac{15.1 \times 10^{20} \text{ Wh}}{1 \times 10^5 \frac{\text{Wh}}{\text{light bulb}}} = 1.51 \times 10^{16} \text{ light bulbs}$$

*QUESTION 5

How many light bulbs could geothermal energy light if a 100-W light bulb is on for 1,000 hr annually?

*QUESTION 6

How many light bulbs (at 60W) could the energy reflected by the atmosphere light for a year?

Most of the sun's rays are absorbed in the tropics, while the earth's outgoing radiation is more uniformly distributed. Consequently, there is a constant movement of energy from the equator to the poles, through the atmosphere and the oceans. For this reason, climate change is most pronounced in the polar regions. The retreat of the arctic sea ice and the melting of the permafrost provide dramatic evidence of global warming (Fig. 7.2).

---

* Answers to starred questions can be found at the end of the book.

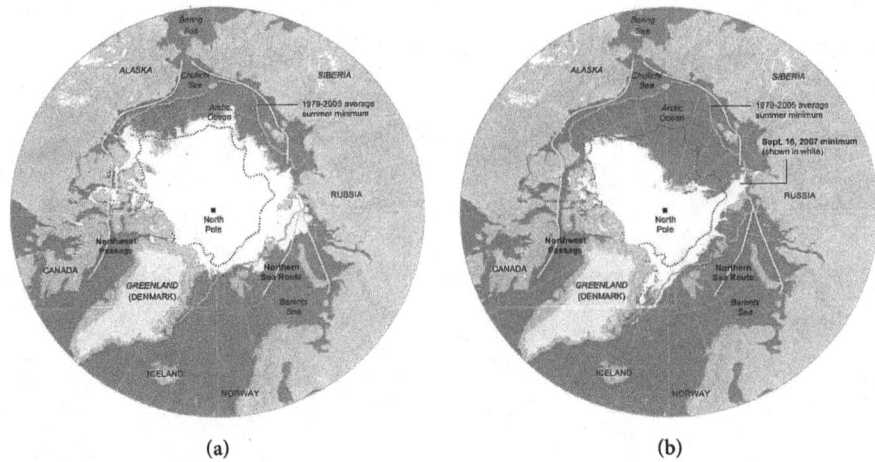

(a)   (b)

Figure 7.2 Arctic sea ice in (a) 2003 versus (b) 2007. The average sea ice between 1979 and 2005 is designated by the dotted line.
Sources: Image courtesy of the National Snow and Ice Data Center, University of Colorado, Boulder, CO, USA. Adapted from Spiro et al., *Chemistry of the Environment, 3rd Ed.*, University Science Books, © 2012, all rights reserved.

QUESTION 7

Using Figure 7.2, estimate the percent of the decrease in sea ice from 2003 to 2007.

Q7 ANSWER Answers will vary depending on the technique used. The ice shown in Figure 7.2(a) can almost be assumed to be a square and the one shown in Figure 7.2(b) is similar to a triangle. Rough estimates can be made by using these geometric figures. Answers should range between 30% and 40% less ice in 2007. All answers should be in the range of 25%–50%. Another method that can be used is breaking the figure into small squares and counting the differences. The upper right corner of ice is not present in 2007 and as well as a significant portion of the ice in the bottom right outside the average line in 2003.

*QUESTION 8

Estimate how much more ice was present in 2003 than the average line indicates.

*QUESTION 9

Estimate how much less ice was present in 2007 than the average line indicates.

QUESTION 10

Assuming the density of ice is 0.92 g mL^{-1} and the density of water is 1.00 g mL^{-1}, how much volume would 2.0L of ice occupy when melted?

Q10 ANSWER First calculate the mass of ice in 2.0L of ice:

$$2.0 \text{ L ice} \times \frac{1{,}000 \text{ mL ice}}{1 \text{ L ice}} \times \frac{0.92 \text{ g ice}}{1 \text{ mL ice}} = 1.8 \times 10^3 \text{ g ice}$$

When ice changes phases to water, no mass is lost, only the volume changes. Since the mass of ice is equal to the mass of liquid water, divide by the density of water to calculate the volume of water:

$$1.8 \times 10^3 \text{ g water} \times \frac{1 \text{ mL water}}{1 \text{ g water}} \times \frac{1 \text{ L}}{1,000 \text{ mL}} = 1.8 \text{ L liquid water}$$

Liquid water has a smaller volume than ice, which expands on freezing. When floating ice melts, however, it does not change the ocean level, because it displaces an equal mass of water.

*QUESTION 11

There are 25,000 L less ice in the Arctic versus the previous year. Calculate the change in volume when this volume of ice melts.

7.2 Radiation Balance and the Greenhouse Effect

The sun provides the earth with an enormous input of energy every day. The earth rids itself of energy at the same rate and thereby maintains a steady state, with a constant average temperature. The earth loses energy by emitting infrared (IR) radiation.

Figure 7.3 shows the blackbody spectra at different temperatures. A blackbody absorbs all the incident radiation, being nonreflective and opaque. It reradiates the absorbed energy with a smooth spectrum that depends only on temperature. The actual spectra of the sun and the earth are somewhat bumpy because specific atomic and molecular transitions contribute to the emissions.

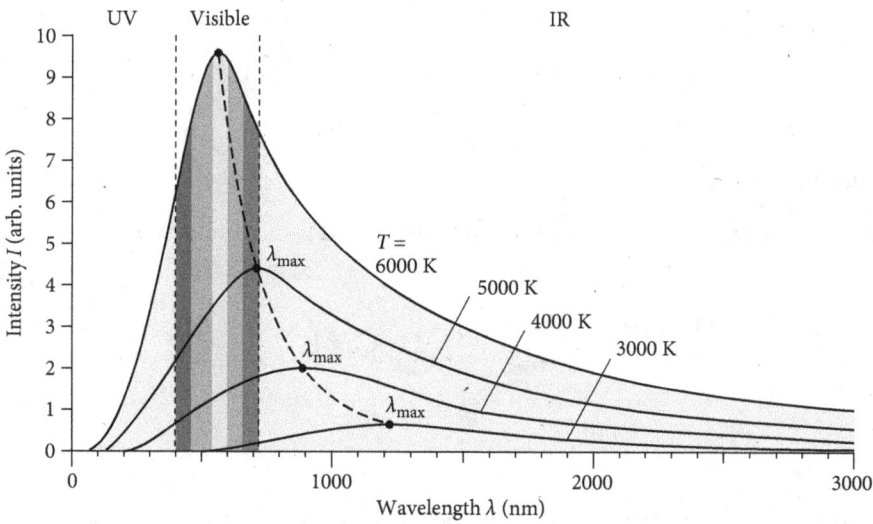

Figure 7.3 Computed blackbody emission spectra.
Source: Adapted from Spiro et al., *Chemistry of the Environment, 3rd Ed.*, University Science Books, © 2012, all rights reserved.

QUESTION 12

According to Figure 7.3, what is the wavelength range for visible light?

Q12 ANSWER Visible light begins at 400 nm and ends at ~700 nm.

*QUESTION 13

According to Figure 7.3, UV light has a wavelength lesser than what value? IR has a wavelength greater than what value?

QUESTION 14

According to the plots in Figure 7.3, which temperatures have their greatest wavelength (λ_{max}) emission in the visible spectrum?

Q14 ANSWER λ_{max} is in the visible spectrum for $T = 6{,}000K$ and $T = 5{,}000K$.

*QUESTION 15

According to the plots in Figure 7.3, which temperatures have their greater wavelength (λ_{max}) emission in the IR spectrum?

For a blackbody, the peak wavelength of radiation is inversely proportional to the absolute temperature (Wein's law):

$$\lambda_{peak} = \frac{2.9 \times 10^6 \text{ (nm K)}}{T} \tag{7.1}$$

QUESTION 16

If λ_{peak} for the sun is 483 nm, calculate T.

Q16 ANSWER Rearrange (7.1) to solve for T (multiply both sides by T and then divide by λ_{peak}. Then solve for 483 nm:

$$T = \frac{2.9 \times 10^6 \text{ (nm K)}}{\lambda_{peak}} = \frac{2.9 \times 10^6 \text{ (nm K)}}{483 \text{ nm}} = 6.0 \times 10^3 \text{ K}$$

*QUESTION 17

Earth has a λ_{peak} of ~10,000 nm. What temperature in Kelvin does this correspond to?

While the earth receives radiation mainly in the visible region, characteristic of the high temperature at the surface of the sun, it gives off radiation in the IR region, which corresponds to the much longer wavelengths characteristic of the earth's cooler surface temperature. All the energy that the earth absorbs from the sun must eventually be reemitted, so we can calculate the earth's steady-state temperature by setting its radiation rate equal to the rate at which it absorbs energy from the sun. The flux of solar energy directed at the earth (S_0) is 1,370 watts per square meter (W m^{-2})(1 W = 1 J sec^{-1}).

The solar rays strike the earth as if the planet were a disk of area πr^2 (the standard formula for the area of a circle), r being the earth's radius. Since the earth radiates from its entire surface, whose area is $4\pi r^2$ (the standard formula for the area of a sphere), we need to divide S_0 by the ratio $4\pi r^2/\pi r^2 = 4$. Moreover, not all of the sun's rays are absorbed; the fraction reflected back to space, the albedo (a), is close to 0.3. The average rate at which solar radiation is absorbed and reemitted is therefore:

$$S = \frac{(1-a)S_0}{4} \tag{7.2}$$

QUESTION 18

Given that $S_0 = 1{,}370$ W m^{-2} and $a = 0.3$ for the sun incident on the earth, calculate S.

Q18 ANSWER Use (7.2) to calculate this:

$$S = \frac{(1-a)S_0}{4} = \frac{(1-0.3)(1{,}370 \text{ W m}^{-2})}{4} = \frac{959 \text{ W m}^{-2}}{4} = 240 \text{ W m}^{-2}$$

This rate can be used to calculate the temperature according to the Stefan–Boltzmann law, which states that the rate at which a blackbody radiates energy is proportional to the fourth power of its absolute temperature:

$$S = kT^4 \tag{7.3}$$

where k is the Stefan–Boltzmann constant, 5.67×10^{-8} W m^{-2} K^{-4}.

---

*QUESTION 19

Use (7.3) and the result from Q18 to calculate the temperature of the earth's surface.

QUESTION 20

Calculate the heat balance of earth in W m^{-2}, under the following components of the atmosphere:
  (a) Shortwave solar radiation incident at the top of the earth's atmosphere.
  (b) Shortwave radiation reflected by the atmosphere and earth's surface.

(c)   Shortwave radiation absorbed by the atmosphere and earth's surface.

(d)   Longwave radiation emitted from the earth's surface.

(e)   The share of the longwave radiation emitted directly to space (through the atmospheric "window") and the share absorbed by the atmosphere.

(f)   The longwave radiation emitted downward from the atmosphere to the earth's surface, and outward from the atmosphere into space.

(g)   The sensible and latent heat transfer from the earth's surface to the atmosphere.

The following information (in units of W m^{-2}) is sufficient to conduct these calculations:

- $S_0 = 1,368$ W m^{-2}.
- The total albedo is 30%, 86% of which is provided by the atmosphere and the remainder by the earth's surface.
- 24% of the incident shortwave radiation is absorbed by the atmosphere and 46% by the earth's surface.
- The earth's surface temperature is 288K.
- About 5% of the longwave radiation emitted from the earth's surface radiates directly to space through the atmospheric window.
- The radiative cooling of the atmosphere, and corresponding radiative heating of the earth's surface, is ~106 W m^{-2}.

Q20 ANSWER

(a)   Knowing S_0, the solar radiation incident at the top of the earth's atmosphere is:

$$\frac{S_0}{4} = \frac{1,368 \text{ W m}^{-2}}{4} = 342 \text{ W m}^{-2}$$

(b)   Using the solar radiation calculated in part (a), the solar radiation reflected by the earth's atmosphere is:

$$(0.3) \times (0.86) \times (342 \text{ W m}^{-2}) = 88.2 \text{ W m}^{-2}$$

Solar radiation reflected by the earth's surface is:

$$(0.3) \times (0.14) \times (342 \text{ W m}^{-2}) = 14.4 \text{ W m}^{-2}$$

(c)   Again, using the solar radiation calculated in part (a), the solar radiation absorbed by the earth's atmosphere is:

$$(0.24) \times (342 \text{ W m}^{-2}) = 82.1 \text{ W m}^{-2}$$

Solar radiation absorbed at the earth's surface (S_1) can be calculated:

$$S_1 = (0.46) \times (342 \text{ W m}^{-2}) = 157 \text{ W m}^{-2}$$

(d)   For a surface temperature of 288 K, the flux of radiative heat from the earth's surface (S_2) can be calculated from the Stefan–Boltzmann law:

$$S = kT^4$$

Hence:

$$S_2 = \left(5.67 \times 10^{-8} \frac{\text{W}}{\text{m}^2 \text{ K}^4}\right) \times (288 \text{ K})^4 = 390 \frac{\text{W}}{\text{m}^2}$$

(e) Using the longwave radiation calculated in part (d), the share of S_2 emitted directly to space is:

$$(0.05) \times \left(390\frac{W}{m^2}\right) = 19.5\frac{W}{m^2}$$

Share of S_2 absorbed by the atmosphere is:

$$(0.95) \times \left(390\frac{W}{m^2}\right) = 371\frac{W}{m^2}$$

(f) Since radiative cooling of the atmosphere (106 W m^{-2}) equals the radiative heating of the earth's surface, the following relationship holds at the earth's surface:

radiative flux in = radiative flux out + 106 W m^{-2}

This can be expressed as:

$$S_1 + S_3 = S_2 + 106 \text{ W m}^{-2}$$

where S_3 is the longwave radiation from the atmosphere absorbed by the earth's surface. We can solve for S_3 as follows:

$$S_3 = (390 + 106 - 157)\frac{W}{m^2} = 339\frac{W}{m^2}$$

Longwave radiation emitted from the atmosphere into space (S_4) can be calculated analogously as we did for the earth's surface:

radiative flux in = radiative flux out 106 W m^{-2}

$$(0.24) \times (S_0) + (0.95) \times (S_2) = S_3 + S_4 - 106\frac{W}{m^2}$$

$$(0.24) \times \left(1{,}368\frac{W}{m^2}\right) + (0.95) \times \left(390\frac{W}{m^2}\right) = 339\frac{W}{m^2} + S_4 - 106\frac{W}{m^2}$$

$$S_4 = (328 + 371 - 339 + 106)\frac{W}{m^2}$$

$$S_4 = 466\frac{W}{m^2}.$$

(g) Sensible and latent heat lost from the earth's surface = 106 W m^{-2}. This compensates for the radiative heating of the earth's surface, and the radiative cooling of the atmosphere.

*QUESTION 21

Outgoing radiation from the top of the earth's atmosphere has been measured by satellites to be 237 W m^{-2}. From this information, calculate the temperature at the top of the atmosphere. If S_0, the solar irradiance, is 1,368 W m^{-2}, calculate the earth's albedo.

*QUESTION 22

Calculate the emission by the earth's surface, given that its mean global temperature is 288K.

TABLE 7.1 Composition of Dry Air at Ground Level in Remote Continental Areas

Constituent	Formula	Concentration (by volume, ppm)
Nitrogen	N_2	780,900
Oxygen	O_2	209,400
Argon	Ar	9300
Carbon dioxide	CO_2	390
Neon	Ne	18
Helium	He	5.2
Methane	CH_4	1.7
Krypton	Kr	1.1
Hydrogen	H_2	0.5
Nitrous oxide	N_2O	0.3
Xenon	Xe	0.08
Carbon monoxide	CO	0.04–0.08
Organic vapors		0.02
Ozone	O_3	0.01–0.04

Source: Adapted from Spiro et al., *Chemistry of the Environment, 3rd Ed.*, University Science Books, © 2012, all rights reserved.

7.3 Infrared Absorption by Greenhouse Gases: Molecular Vibrations

The trapping of the earth's radiation is not a property of the atmosphere as a whole, but rather of specific molecules, those that have more than two atoms. This requirement excludes most of the atmosphere's molecules. Leaving out water vapor, which varies greatly from place to place, 99.9% of the atmosphere consists of N_2 (78%), O_2 (21%), and Ar (0.9%) (Table 7.1).

QUESTION 23

Using the information in Table 7.1, show that the percent N_2 in the earth's atmosphere is 78%.

Q23 ANSWER Using the concentration of N_2 as 780,900 convert to percent. Parts per million (ppm) by volume means how many particles out of 1 million particles are present (remember that percent means per 100):

$$\frac{780,900}{1,000,000} = \frac{x}{100}$$

Cross multiply:

$$78{,}090{,}000 = 1{,}000{,}000\,x$$

Divide both sides by 1,000,000:

$$\frac{78{,}090{,}000}{1{,}000{,}000} = \frac{1{,}000{,}000\,x}{1{,}000{,}000}$$

$$x = 78.09\% \approx 78\%$$

*QUESTION 24

Using the information in Table 7.1, show that the percent Ar in the earth's atmosphere is 0.93%.

None of the major atmospheric gases meet the two fundamental requirements for the absorption of infrared (IR) radiation:

1. When radiation is absorbed by a molecule, the molecule undergoes a quantum transition, involving the movement either of its electrons or of its nuclei; the energy of the radiation must match the energy of the molecular transition. In the IR region of the spectrum, the available transitions involve the movement of the nuclei in molecular vibrations. Argon, the third most abundant atmospheric constituent (0.9%), is transparent to IR radiation because it is monatomic; it has no vibrations.

2. Because radiation is electromagnetic, its absorption requires that the transition change the electric field within the molecule, that is, the transition must alter the molecule's dipole moment (the vector sum of atomic charges times their distances from the molecule's center of mass). This second requirement is the reason that N_2 and O_2 are unable to absorb the earth's IR radiation. Although their nuclei do vibrate along the bond joining them, and the energy of the vibration is in the IR region, the vibration does not change the dipole moment. Because the molecule is symmetrical, the dipole moment remains zero no matter how much the bond is stretched. The vibration is IR inactive as is true for all homonuclear diatomic molecules. The dipole moment is altered by vibrations of heteronuclear diatomic molecules (e.g., CO, NO, and HCl), since their atoms have different partial charges. However, these molecules do not contribute significantly to the greenhouse effect because their concentrations in the atmosphere are too low and their absorption is too weak.

The majority of the molecules and atoms in Table 7.1 have fewer than three atoms and do not have vibrations that change the dipole moment, a requirement for absorption of electromagnetic radiation. On the contrary, polyatomic molecules have many vibrations ($3n - 6$ for nonlinear molecules, where n is the number of atoms, or $3n - 5$ for linear molecules); at least some of these vibrations change the dipole moment and are IR active. All the gases that contribute significantly to the greenhouse effect are polyatomic. The two most important greenhouse molecules are H_2O and CO_2.

QUESTION 25

Calculate the number of vibrations for an H_2O molecule.

Q25 ANSWER According to the valence shell electron pair repulsion (VSEPR) theory, the water molecule is bent and thus not linear.

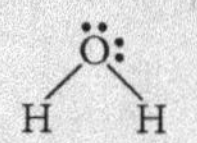

$$\overset{\overset{\displaystyle \ddot{O}:}{\diagup\diagdown}}{H\qquad H}$$

Therefore, use the equation $3n - 6$ to calculate the number of vibrations ($n = 3$ for H_2O).

$$3n - 6 = 3(3) - 6 = 9 - 6 = 3 \text{ vibrations}$$

*QUESTION 26

Calculate the number of vibrations for a CO_2 molecule.

Some of the vibrations for carbon dioxide and water are illustrated in Figure 7.4. All three vibrations of H_2O change their dipole moment. For CO_2, the symmetric stretching motion of the two O atoms leaves the dipole moment unchanged. The dipoles that each O atom generates relative to the C atom cancel one another because of the linear geometry. However, the net dipole moment is altered by the asymmetric stretch and by the bending vibration.

It might appear that H_2O and CO_2 would not be effective in heat-trapping because the earth's emissions cover a wide spectrum of wavelengths, whereas the molecular vibrational transitions correspond to specific energies. However, the molecules can undergo not only vibrations but also rotations. For each molecular vibration, IR photons can induce transitions to many different rotational levels (rates of rotation). In addition, the rotational absorptions are broadened by the gain or loss of translational energy during molecular collisions. Consequently, each vibration is associated with a broad absorption band.

IR absorption spectra are shown schematically for H_2O and CO_2 in Figure 7.5. The CO_2 bending vibration, at 14,992 nm, is responsible for the main absorption band, often called the 15-μ band (1 μ = 1 μm = 1,000 nm). In the H_2O spectrum, the bending vibration, at 6,269 nm, also produces a major absorption band, while pure rotational transitions are responsible for the broad absorption at wavelengths >20,000 nm. The stretching vibrations of both molecules are less effective at absorbing radiation, and their wavelengths are too short to absorb the earth's radiation to any significant extent.

In the top panel of Figure 7.5, the absorptions for these two molecules are added up and superimposed on the earth's emission spectrum. The combined absorption bands block most

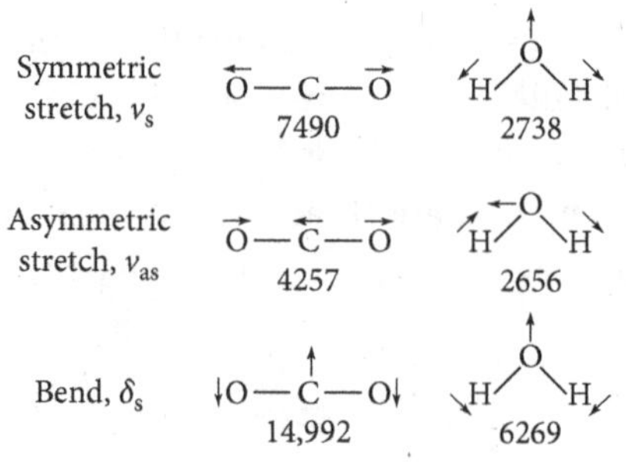

Figure 7.4 Molecular vibrations of CO_2 and H_2O and the wavelengths (nm) corresponding to their excitations. *Source:* Adapted from Spiro et al., *Chemistry of the Environment, 3rd Ed.,* University Science Books, © 2012, all rights reserved.

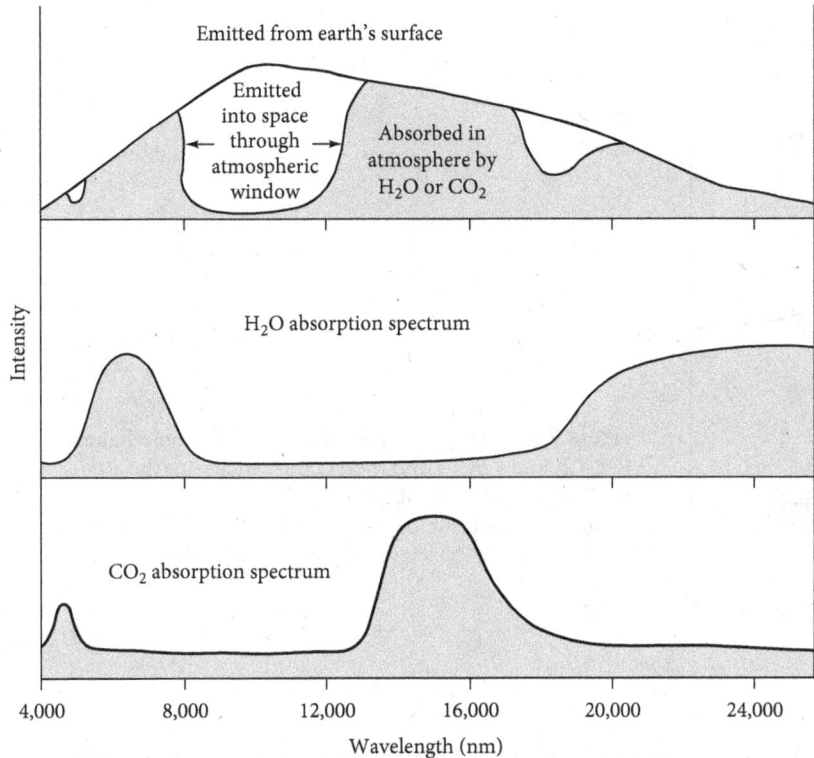

Figure 7.5 Absorption spectrum of terrestrial radiation by H_2O and CO_2.
Source: Adapted from Spiro et al., *Chemistry of the Environment, 3rd Ed.*, University Science Books, © 2012, all rights reserved.

of the terrestrial radiation. There is, however, a relatively unobstructed region of the spectrum between 8,000 and 12,000 nm through which radiation can escape. This region is called the atmospheric window.

This window can be closed by other polyatomic molecules in the atmosphere, including chlorofluorocarbons (CFCs), CH_4, N_2O, and O_3 (Fig. 7.6). The CFCs and N_2O are of major concern as destroyers of stratospheric ozone (see Chapter 10), but they are also important greenhouse gases. Their impact is significant even though their concentrations are orders of magnitude lower than that of CO_2. The reason is that the CO_2 absorptions are nearly "saturated," that is, most of the radiation emitted within the absorption bands is already absorbed. As a result, each extra CO_2 molecule contributes only a relatively small amount to the total absorption. (The same is true for each additional water molecule.) On the contrary, the contribution to overall absorption is relatively high for each extra CFC molecule precisely because the CFCs are very dilute and absorb only a small fraction of the radiation, but do so in the window region. One extra CFC molecule contributes thousands of times more to the greenhouse effect than one extra CO_2 molecule.

In order to determine the effectiveness of a molecule in absorbing radiation, the Beer–Lambert law is used:

$$T = \frac{l}{l_0} = e^{-\varepsilon l} \tag{7.4}$$

In this equation, T is the fraction of light transmitted through a global layer of absorbing gas of equivalent thickness l and ε is the absorptivity. This equation applies to any form of radiation passing through the absorbing material.

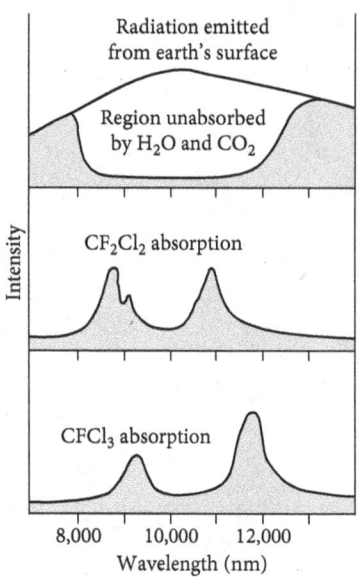

Figure 7.6 Absorption spectra of chlorofluoromethanes (CF_2Cl_2 and $CFCl_3$) and their coincidence with the atmospheric window (8,000 – 13,000 nm).
Source: Adapted from Spiro et al., *Chemistry of the Environment, 3rd Ed.,* University Science Books, © 2012, all rights reserved.

QUESTION 27

Using the Beer–Lambert law, $T = e^{-\varepsilon l}$, show graphically that the decrease in transmittance is nearly the same for each doubling of a light-absorbing gas when the concentration is low (e.g., $\varepsilon l \sim 0.001$), but the effect of doubling diminishes strongly when the concentration is high (e.g., $\varepsilon l \sim 3$).

Q27 ANSWER When $\varepsilon l = 0.001$, $T = e^{-0.001} = 0.999$, and successive doublings produce the graph below, showing a nearly linear decrease of the transmission. Each increment of gas produces almost the same effect.

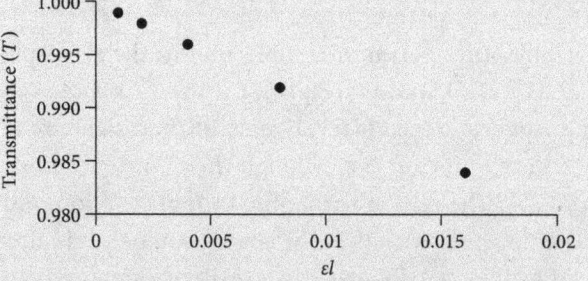

However, when $\varepsilon l = 3$, the graph shows that successive doublings produce successively smaller decreases in transmittance.

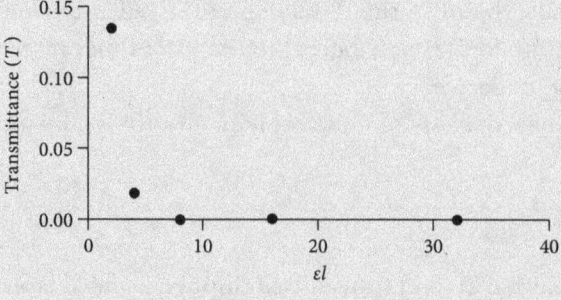

By the time $\varepsilon l = 6$, the transmittance is nearly zero, and further increases in the concentration have little effect. We can say that the absorption is "saturated."

QUESTION 28

A gas with an effective thickness of 4 cm has a transmittance of 0.56. Calculate the absorptivity.

Q28 ANSWER In order to complete this calculation, the Beer–Lambert law will be used:

$$T = e^{-\varepsilon l}$$

To solve for absorptivity, take the natural log (ln) of both sides of the equation and solve for absorptivity (ε):

$$\ln T = \ln e^{-\varepsilon l}$$

$$\ln T = -\varepsilon l$$

$$-\varepsilon = \frac{\ln T}{l}$$

$$\varepsilon = -\frac{\ln T}{l} = -\frac{\ln 0.56}{4 \text{ cm}} = 0.15 \text{ cm}^{-1}$$

*QUESTION 29

A gas with an absorptivity of 0.175 cm^{-1} has a transmittance of 0.34. Calculate the effective thickness of the gas.

*QUESTION 30

A gas with an absorptivity of 0.122 cm^{-1} has an effective thickness of 8 cm. Calculate the transmittance of the gas.

7.4 Water: Positive Feedback, Latent Heat, and Energy Flows

Water vapor is the dominant greenhouse gas, comprising on average ~4% of the molecules in the atmosphere. However, we have little control over H_2O in the atmosphere because its concentration depends mainly on the rate of evaporation of liquid H_2O. This rate depends on the vapor pressure, which increases exponentially with rising surface temperature. Consequently, water is not an independent factor in the greenhouse effect but depends on other drivers of the global temperature. Water amplifies the heat-trapping effect of other greenhouse gases. As they warm the earth's surface, the water content of the atmosphere increases, trapping more heat.

Water evaporation and precipitation, called the hydrological cycle, are also major carriers of energy in the atmosphere. While it takes 4.2J (1 cal) to heat 1g of H_2O by 1°C, much more energy is needed to vaporize the same 1g of H_2O; the energy required to vaporize a liquid is called its latent heat. At 15°C, which is the average annual global temperature, the latent heat of water is 2.46 kJ g^{-1}. The latent heat is released again when water vapor condenses into rain, which is the reason that rainfall is associated with storms; even a modest rainfall releases a huge amount of energy.

QUESTION 31

Given that evaporation of 1 mL of H_2O requires 2.46 kJ of energy at ambient temperatures, how much energy is released in a 5.0-cm rainfall over an area of 600. km^2? (For comparison, the area of Manhattan is 60 km^2.) If 1 ton (ton = metric ton = 1,000 kg) of trinitrotoluene (TNT) releases 4.18 × 10^6 kJ of energy, how many tons of TNT would be equivalent to the energy released in rainfall?

Q31 ANSWER First calculate the volume of water created by the rainfall. Recalling that 1 cm^3 = 1 mL it will be helpful to calculate the volume in cm^3.
    Convert the area from km^2 to cm^2:

$$600. \text{ km}^2 \times \frac{(100,000 \text{ cm})^2}{(1 \text{ km})^2} = 6.00 \times 10^{12} \text{ cm}^2$$

Calculate the volume by multiplying the area by depth:

$$(5.0 \text{ cm}) \times (6.00 \times 10^{12} \text{ cm}^2) = 3.0 \times 10^{13} \text{ cm}^3$$

Multiply the volume by the energy per 1 mL to obtain the total energy required:

$$3.0 \times 10^{13} \text{ cm}^3 \times \frac{1 \text{ mL}}{1 \text{ cm}^3} \times \frac{2.46 \text{ kJ}}{1 \text{ mL}} = 7.38 \times 10^{13} \text{ kJ}$$

Divide this value by the amount of energy in 1 ton of TNT to determine the number of tons of TNT this is equivalent to:

$$7.38 \times 10^{13} \text{ kJ} \times \frac{1 \text{ ton}}{4.18 \times 10^6 \text{ kJ}} = 1.8 \times 10^7 \text{ tons}$$

---

*QUESTION 32

Given that evaporation of 1.0 mL of H_2O requires 2.46 kJ of energy at ambient temperatures, calculate the volume of water in liters that should evaporate to equal the energy of 15 tons of TNT (1 ton of TNT release 4.18 10^6 kJ of energy).

*QUESTION 33

Given that evaporation of 1 mL of H_2O requires 2.46 kJ of energy at ambient temperatures, calculate the area of water in km^2 of 3.0 cm deep water that should evaporate to equal the energy of 5.0 tons of TNT (1 ton of TNT release 4.18 10^6 kJ of energy).

The combined effects of greenhouse radiation trapping, water evaporation, and cloud formation produce a complex set of energy flows through the atmosphere. Figure 7.7 shows the planet's inputs and outputs of energy in units of 10^8 TWh yr^{-1}.

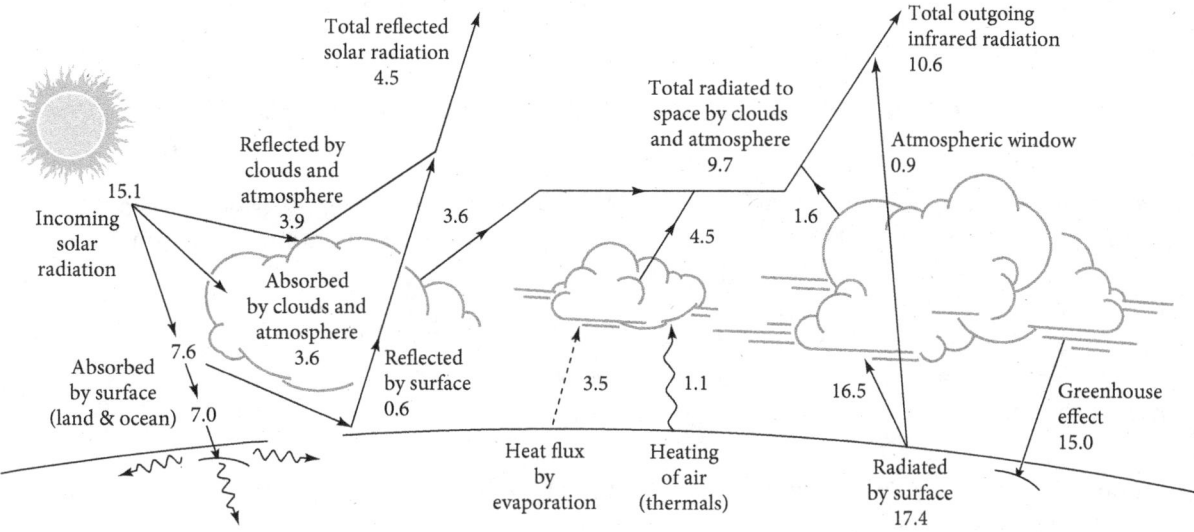

Figure 7.7 Annual heat balance of the earth (in 10^8 TWh).
Source: Adapted from Spiro et al., *Chemistry of the Environment, 3rd Ed.*, University Science Books, © 2012, all rights reserved.

---

*QUESTION 34

Which paths does the incoming solar radiation take according to Figure 7.7?

QUESTION 35

According to Figure 7.7, how much of the incoming solar radiation is reflected.

Q35 ANSWER The figure shows that the reflection of the energy comes from the clouds and atmosphere as well as by the surface. This accounts for 3.9×10^8 TWh and 0.6×10^8 TWh, respectively, of the total amount of 15.1×10^8 TWh.

For the atmosphere and clouds:

$$\text{Percent} = \frac{\text{part}}{\text{whole}} \times 100\% = \frac{3.9 \times 10^8 \text{ TWh}}{15.1 \times 10^8 \text{ TWh}} \times 100\% = 26\%$$

For the surface:

$$\text{Percent} = \frac{\text{part}}{\text{whole}} \times 100\% = \frac{0.6 \times 10^8 \text{ TWh}}{15.1 \times 10^8 \text{ TWh}} \times 100\% = 4\%$$

Adding these together gives a total of 30% for reflection.

*QUESTION 36

According to Figure 7.7, how much of the incoming solar radiation is absorbed?

QUESTION 37

Using Figure 7.7, determine where the energy absorbed by the atmosphere (total 20.1×10^8 TWh) originates.

Q37 ANSWER The figure shows that 3.6×10^8 TWh radiant from the incoming solar radiation is absorbed by the atmosphere. There are also 16.5×10^8 TWh absorbed by the atmosphere from the energy radiated by the surface of the earth:

$$16.5 \times 10^8 \text{ TWh} + 3.6 \times 10^8 \text{ TWh} = 20.1 \times 10^8 \text{ TWh}$$

*QUESTION 38

The grand total of energy gained by the atmosphere is 24.7×10^8 TWh. Use Figure 7.7 to determine the origin of all of this energy.

7.5 Greenhouse Computations: Radiative Forcings

Climate researchers deal with mathematical complexities of the greenhouse effect using global circulation models (GCMs), which apply the laws of physics to the atmosphere and the oceans, using inputs of energy from the sun to predict the dynamics of air parcels around the globe. The models have become highly sophisticated and are run on the world's largest supercomputers. However, essential aspects of the problem were captured in one of the earliest modeling efforts, now a classic, which produced a temperature profile of the atmosphere, and predicted the consequence of doubling the atmospheric CO_2 concentration (Fig. 7.8).

QUESTION 39

Using Figure 7.8, determine the temperature difference at 30 km altitude between an atmosphere with 300 ppm CO_2 and an atmosphere of 600 ppm CO_2.

Q39 ANSWER At an altitude of 30 km, the temperature is ~225K when the CO_2 concentration is 600 ppm and ~235K when the CO_2 concentration is 300 ppm. The temperature difference is approximately 10K.

*QUESTION 40

Using Figure 7.8, determine the percent temperature decrease at 40 km altitude from an atmosphere with 300 ppm CO_2 to an atmosphere of 600 ppm CO_2.

Because of the difficulty in predicting temperature changes accurately, the greenhouse effect is generally quantified as radiative forcing, the computed change in the heat balance of the earth, expressed in watts per square meter. It is defined as the change in the net irradiance (downward minus upward) at the tropopause because of an external driver of climate, such as the change in the concentration of CO_2. It is computed with models by holding all

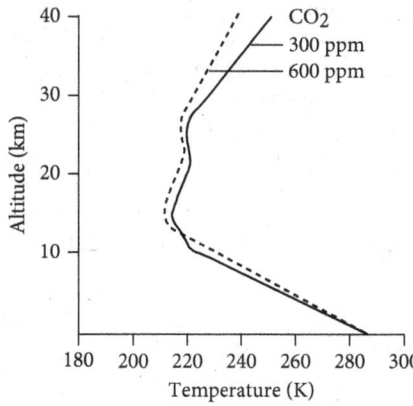

Figure 7.8 Effect of doubling the CO_2 concentration on the temperature profile of the atmosphere.
Sources: Adapted from Manabe & Wetherald, "Thermal Equilibrium of the Atmosphere With a Given Distribution of Relative Humidity," *Journal of the Atmospheric Sciences*, Vol. 24, 1967, pp. 241–259, and Spiro et al., *Chemistry of the Environment, 3rd Ed.*, University Science Books, © 2012, all rights reserved.

tropospheric properties fixed at unperturbed values, and then allowing for stratospheric temperatures to readjust to a radiative-dynamical equilibrium. The change in radiative forcing is further defined as a change relative to 1750, a preindustrial year, and is a global and annual average value.

Table 7.2 depicts the comparison of the characteristics for several important greenhouse gases, all of whose concentrations have increased since preindustrial times. Carbon dioxide is by far the most abundant of the five, followed by CH_4, N_2O, CCl_2F_2 (CFC-11), and CHF_2Cl (HCFC-22). The computed radiative forcing also decreases in this order, but is not nearly as strong as the concentrations. The radiative forcintg per molecule increases strongly in this order, reflecting increasingly strong absorption bands occurring in the atmospheric window.

QUESTION 41

According to Table 7.2, what percent of the earth's atmosphere was CO_2 in the preindustrial age and what percent of the earth's atmosphere is currently CO_2?

Q41 ANSWER Use the fact that percent is the same as part per hundred and complete a ratio.
Preindustrial: (280 ppm)

$$\frac{280}{1,000,000} = \frac{x}{100}$$

Cross multiply

$$28,000 = 1,000,000\,x$$

Current: (417 ppm)

$$\frac{417}{1,000,000} = \frac{x}{100}$$

Cross multiply

$$41,700 = 1,000,000\,x$$

$$x = 0.417\%$$

TABLE 7.2 Summary of Properties of Greenhouse Gases Affected by Human Activities

	CO_2	CH_4	N_2O	CFC-11	HCFC-22
Atmospheric concentration	ppmv[a]	ppbv[a]	ppbv[a]	pptv[a]	pptv[a]
Preindustrial (1750–1800)	~280	~700	~270	0	0
Current	417	1,892	334	222	220
Increased radiative forcing (W m^{-2})[b]	1.94	0.50	0.20	0.060	0.049
Atmospheric lifetime (years)	~100–300[c]	12.4	121	45	11.9
Per molecule ratio of radiative forcings[d] [$\Delta F(GHG)/\Delta F(GHG\ CO_2)$]	1	25	298	4,750	1,810
Major removal mechanism	I[e]	II[e]	III[e]	III[e]	III[e]

[a] The abbreviations ppmb, ppbv, and pptv are parts per millions, parts per billion, and parts per trillion by volume, respectively.

[b] Changes (since 1750) in radiative forcing represent changes in the rate per square meter at which energy is supplied to the atmosphere below the stratosphere.

[c] No single lifetime can be defined for CO_2 because of the different rates of uptake by different removal processes.

[d] Quoted forcings are for a 100-year time horizon.

[e] I = slow exchange of carbon between surface waters and deeper layers of the ocean; uptake into biomass. II = reaction with HO• radical in the troposphere. III = photolysis in the stratosphere.

Sources: Data from Working Group I of the Intergovernmental Panel on Climate Change (IPCC), *Technical Summary of the Working Group I*, IPCC, 2001, and the Carbon Dioxide Information Center, *Recent Greenhouse Gas Concentrations*, Oak Ridge National Laboratory, 2016. Adapted from Spiro et al., *Chemistry of the Environment, 3rd Ed.*, University Science Books, © 2012, all rights reserved.

---

*QUESTION 42

According to Table 7.2, what percent of the earth's atmosphere was CH_4 and what percent was N_2O in the preindustrial age and what percent of the earth's atmosphere is currently CH_4 and what percent is N_2O?

*QUESTION 43

Which compound (CH_4, CO_2, N_2O) had the highest percentage increase between preindustrial times and current times? Use the data in Table 7.2 to calculate this.

7.6 Carbon Dioxide and the Carbon Cycle

The concentration of CO_2 in the atmosphere over Hawaii has been carefully measured for the last five decades (Fig. 7.9). Every year the concentration peaks in the winter, as dormant vegetation decomposes, and hits a trough in the summer, as CO_2 is taken up by photosynthesis.

Underlying this sawtooth pattern, one can readily see a steady upward trend. This CO_2 history has become an emblem of our times, since the rising concentration correlates with the steady increase of fossil fuel consumption.

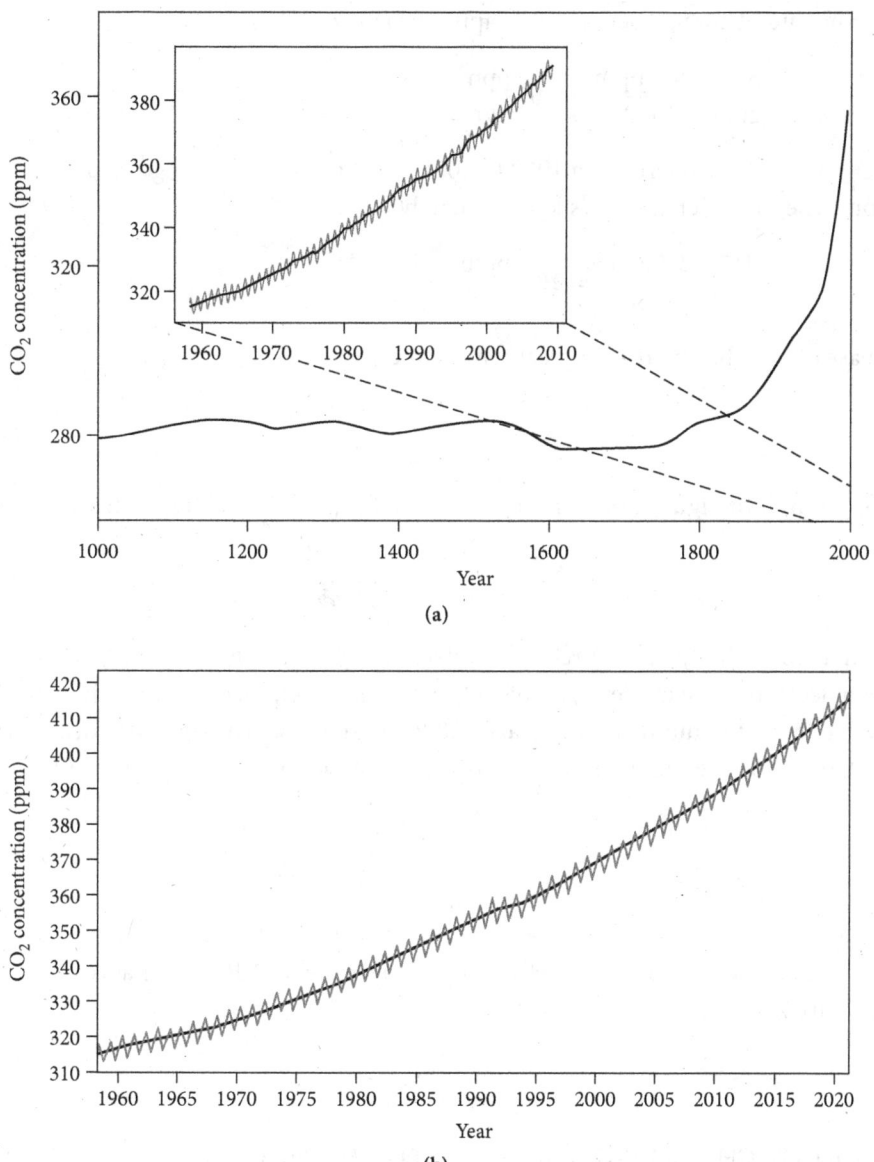

(a)

(b)

Figure 7.9 (a) Historic increase in the atmospheric concentration of CO_2. (b) The concentration of CO_2 between 1958 and 2021 from measurements taken at Mauna Loa, Hawaii. Carbon dioxide is removed from the atmosphere by photosynthesis in the summer, producing the observed annual oscillations in concentration.
Sources: Adapted from Sarmiento & Gruber, "Sinks for Anthropogenic Carbon," *Physics Today,* Vol. 55, 2002, pp. 30 –36; inset: Keeling et al., Scripps Institution of Oceanography, La Jolla, CA (prior to 1974); National Oceanic and Atmospheric Administration, Washington, DC (since 1974). Available at https:// keelingcurve.ucsd.edu/.

QUESTION 44

Compare the average rate of increase in CO_2 concentration from the year 1000 to 2000 with the increase from 1960 to 2021.

Q44 ANSWER From the year 1000 to 2000, the concentration of CO_2 in the atmosphere changed from about 280 to 350 ppm. The rate of change per year was approximately:

$$\frac{350 - 280 \text{ ppm}}{2000 - 1000 \text{ yr}} = 0.07 \frac{\text{ppm}}{\text{yr}}$$

The concentration of CO_2 in the atmosphere from 1960 to 2021 (inset of Fig. 7.9) changed from approximately 315 to 417 ppm. The rate of change was approximately:

$$\frac{417 - 315 \text{ ppm}}{2021 - 1960 \text{ yr}} = 1.67 \frac{\text{ppm}}{\text{yr}}$$

The average rate of increase of CO_2 has increased greatly more recently.

*QUESTION 45

Calculate the slope of the inset graph in Figure 7.9(b) from 2011 to 2021. Use this to extrapolate the CO_2 concentration in 2050.

When viewed against the historical record, as determined from the composition of air bubbles trapped in ice cores, the modern rise of CO_2 is dramatic (Fig. 7.10). The preindustrial level of 280 ppm had been maintained for the last 10,000 yr. The modern rise is also dramatic for CH_4 and N_2O, the two greenhouse gases with the greatest radiative forcing after CO_2.

*QUESTION 46

What is the rate of change in concentrations for CO_2, CH_4, and N_2O over the past 10,000 yr according to the graphs shown in Figure 7.10?

*QUESTION 47

What is the percent increase in CO_2, CH_4, and N_2O from 1800 to 2000? Use Figure 7.10.

Considering the global carbon cycle illustrated in Figure 7.11, which gives estimates of the pools of carbon residing in the air, soil, biomass, and oceans, and of the annual fluxes (the units are gigatons, 1 Gt = 1,015g), a perspective can be gained on atmospheric CO_2. These fluxes are very large, making it hard to account fully for the fate of the relatively small anthropogenic contribution. Yet, it is this contribution that is driving the accumulation of CO_2 in

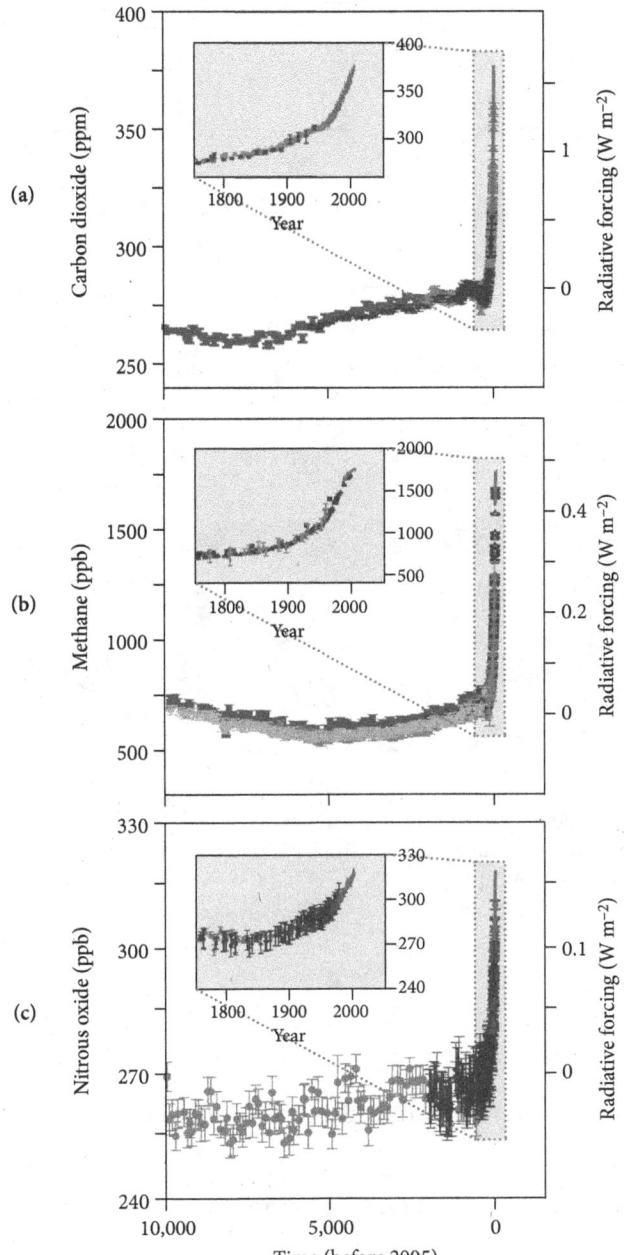

Figure 7.10 Estimated concentrations of major greenhouse gases, over the past 10,000 yr, based on ice core gas bubbles. (a) Concentrations of CO_2, (b) CH_4, and (c) N_2O.
Sources: Adapted from a report of Working Group I of the Intergovernmental Panel on Climate Change, *Summary for Policymakers*, 2007, available at http://www.ipcc.ch/ipccreports/ar4-wg1.htm. Adapted from Spiro et al., *Chemistry of the Environment, 3rd Ed.*, University Science Books, © 2012, all rights reserved.

the atmosphere, and with it, the greenhouse warming. There is a large exchange between the air and the biosphere, associated with the cycle of respiration and photosynthesis (in the figure, GPP = green plant productivity). Another large exchange is associated with CO_2 absorption and outgassing from the oceans.

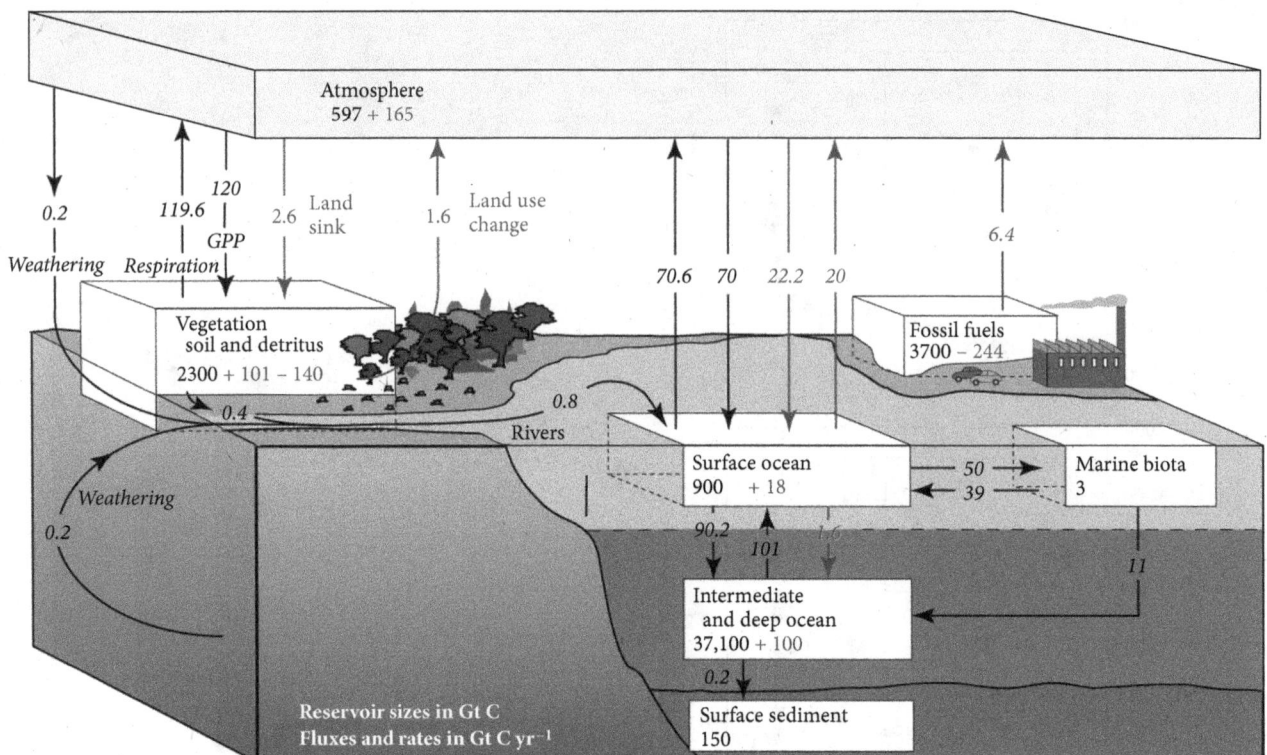

Figure 7.11 The global carbon cycle of the 1990s showing pools and main annual fluxes, in Gt C yr^{-1}. Estimates of preindustrial values are in black, while estimates of the additional anthropogenic contributions are in gray.
Sources: Adapted from a report of Working Group I of the Intergovernmental Panel on Climate Change, *Summary for Policymakers*, 2007, available at http://www.ipcc.ch/ipccreports/ar4-wg1.htm. Adapted from Spiro et al., *Chemistry of the Environment, 3rd Ed.*, University Science Books, © 2012, all rights reserved.

QUESTION 48

According to Figure 7.11, how many Gt of carbon are added to the atmosphere annually? How many Gt of carbon dioxide is this equivalent to?

Q48 ANSWER Looking at the figure, add the values with arrows pointing toward the atmosphere:

$$119.6 \text{ Gt} + 1.6 \text{ Gt} + 70.6 \text{ Gt} + 20 \text{ Gt} + 6.4 \text{ Gt} = 218.2 \text{ Gt C yr}^{-1}$$

Using the mass of carbon (12.01 g mol^{-1}) and the molar mass of CO$_2$ (12.01 g mol^{-1} + 32.00 g mol^{-1} = 44.01 g mol^{-1}) calculate the ratio of the mass of carbon to the total mass of carbon dioxide:

$$\frac{12.01 \frac{g}{mol} \text{ C}}{44.01 \frac{g}{mol} \text{ CO}_2} = 0.273 \frac{\text{C}}{\text{CO}_2}$$

Calculate the amount of CO$_2$ by dividing by the ratio of carbon to carbon dioxide:

$$218.2 \frac{\text{Gt C}}{\text{yr}} \times \frac{\text{CO}_2}{0.273 \text{ C}} = 799 \text{ Gt CO}_2$$

*QUESTION 49

According to Figure 7.11, how many Gt of carbon are removed from the atmosphere annually? How many Gt of carbon dioxide is this equivalent to?

7.7 Reservoirs, Flows, and Residence Times

It is often helpful to represent environmental processes in terms of reservoirs and flows, introduced in Chapter 1, as Figure 7.11 does for the carbon cycle. A reservoir is a region of the environment that holds a significant amount of the material under consideration and is physically or chemically separated from other reservoirs. These amounts can be estimated by multiplying the concentration of the material, obtained by sampling, with the size of the reservoir.

To understand the dynamics of the environment, one needs to understand how much material is transferred from one reservoir to another per unit of time (i.e., the flow). These numbers are generally much harder to determine, and a great deal of ingenuity must go into making flow estimates. If the flows and reservoirs are known, one can calculate residence times, assuming that the system is in a steady state (i.e., that the amounts do not change with time, or at least not rapidly). A steady state implies that inflows and outflows are in balance for each reservoir. The residence time is the average length of time the material spends in a reservoir and is just the ratio of the reservoir size to the inflow or outflow rate.

QUESTION 50

Figure 7.11 indicates that the atmosphere contains 762 Gt C (597 Gt C + 165 Gt C) as CO_2. Based on its sinks and sources, calculate the residence time for atmospheric C.

Q50 ANSWER From Q48, we calculated that the total C entering the atmosphere is:

$$119.6 \text{ Gt} + 1.6 \text{ Gt} + 70.6 \text{ Gt} + 20 \text{ Gt} + 6.4 \text{ Gt} = 218.2 \text{ Gt C yr}^{-1}$$

And from Q49 we calculated the total C leaving the atmosphere as:

$$0.2 \text{ Gt} + 120 \text{ Gt} + 2.6 \text{ Gt} + 70 \text{ Gt} + 22.2 \text{ Gt} = 215 \text{ Gt C yr}^{-1}$$

The residence time for C in the atmosphere is the reservoir divided between either the inflow:

$$\frac{762 \text{ Gt C}}{218.2 \frac{\text{Gt C}}{\text{yr}}} = 3.50 \text{ yr}$$

or outflow:

$$\frac{762 \text{ Gt C}}{215 \frac{\text{Gt C}}{\text{yr}}} = 3.54 \text{ yr}$$

So, approximately 3.5 yr.

*QUESTION 51

The reduced carbon in sedimentary rocks turns over much more slowly. Since this carbon is estimated to accumulate (and erode) at a rate of 0.2 Gt C yr^{-1}, and the total size of the reservoir is estimated to be 10 million Gt C, what is the residence time?

QUESTION 52

The ocean is the largest sink for atmospheric CO_2 on the planet. From Figure 7.11, calculate the residence time of carbon in the intermediate and deep ocean.

Q52 ANSWER The reservoir of the intermediate and deep ocean contains 37,200 Gt C, but it has two sources of additional carbon (marine biota and surface ocean) and two sources of removal (surface ocean and surface sediment).

The overall removal of C yr^{-1} is:

$$101 \text{ Gt C yr}^{-1} + 0.2 \text{ Gt C yr}^{-1} = 101.2 \text{ Gt C yr}^{-1}$$

Overall, the addition of C yr^{-1} is:

$$90.2 \text{ Gt C yr}^{-1} + 1.6 \text{ Gt C yr}^{-1} + 11 \text{ Gt C yr}^{-1} = 102.8 \text{ Gt C yr}^{-1}$$

Net addition of C yr^{-1} into the intermediate and deep ocean is:

$$102.8 \text{ Gt C yr}^{-1} - 101.2 \text{ Gt C yr}^{-1} = 1.6 \text{ Gt C yr}^{-1}$$

The residence time for CO_2 in the ocean is based on either the inflow:

$$\frac{37,200 \text{ Gt C}}{102.8 \text{ Gt C yr}^{-1}} = 362 \text{ yr}$$

or outflow:

$$\frac{37,200 \text{ Gt C}}{101.2 \text{ Gt C yr}^{-1}} = 368 \text{ yr}$$

So, approximately 365 yr.

*QUESTION 53

From Figure 7.11, calculate the residence time of carbon in the surface ocean.

7.8 Albedo, Particles, and Clouds

The albedo is a critical factor in the earth's radiation balance. It directly determines the fraction of the solar radiation that is absorbed by the earth–air system. Even a slight change in the average albedo can measurably affect global temperature. The reflectivity of solar radiation varies greatly from place to place (Fig. 7.12). The darkest regions are the oceans, which constitute ~70% of the earth's total area. The brightest parts of the globe are the snow-covered polar areas, with albedos as high as 80%. The melting of polar ice in the Arctic creates more open ocean and low-

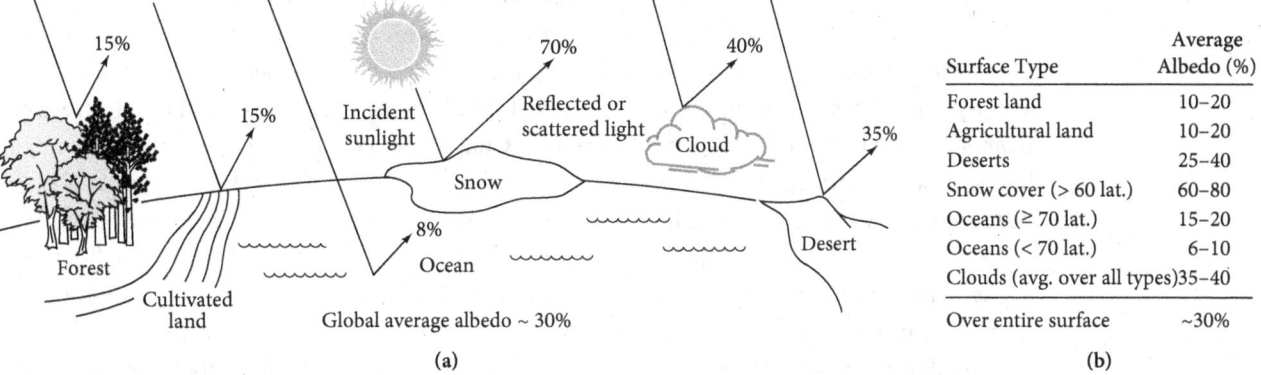

Surface Type	Average Albedo (%)
Forest land	10–20
Agricultural land	10–20
Deserts	25–40
Snow cover (> 60 lat.)	60–80
Oceans (≥ 70 lat.)	15–20
Oceans (< 70 lat.)	6–10
Clouds (avg. over all types)	35–40
Over entire surface	~30%

(a)   (b)

Figure 7.12 Variation in albedo with cloud cover and type of surface on the earth.
Source: Adapted from Spiro et al., *Chemistry of the Environment, 3rd Ed.*, University Science Books, © 2012, all rights reserved.

ers the albedo to 10%–20%. This causes ocean warming, which in turn causes more ice to melt, which is considered a positive feedback. The "positive" means that it helps the cycle of warming continue; however, this is negative because of the detrimental effects of climate change.

QUESTION 54

How many times more albedo is present in snow cover than in deserts?

Q54 ANSWER Figure 7.12 shows that deserts have 25%–40% albedo and snow cover has 60%–80% albedo.

Compare the extrema from each range. Divide the percent albedo in the snow by the percent albedo in the dessert:

$$\frac{60\%}{25\%} = 2.4 \text{ times}$$

$$\frac{80\%}{25\%} = 3.2 \text{ times}$$

$$\frac{60\%}{40\%} = 1.5 \text{ times}$$

$$\frac{80\%}{40\%} = 2.0 \text{ times}$$

The albedo is between 1.5 and 3.2 times greater in the snow-covered areas than in the desert.

*QUESTION 55

How many times more albedo is present in deserts than in forests?

The surface albedo can be affected by human activities. For example, the deposition of black carbon particles (soot) on snow decreases its albedo. There is concern that mountain snowpack will melt faster from this form of pollution. On the other hand, the albedo is increased when deserts expand as a result of deforestation and erosion.

The global albedo is dominated by clouds, and by particles and molecules in the air (65%). Humans contribute to the load of particles in the air, which can have both direct and indirect effects on albedo. Sunlight is reflected by light particles and absorbed by dark particles. The atmospheric aerosol is dominated by light particles of sulfuric acid and sulfate salts, resulting from sulfur oxidation in the atmosphere (see Chapter 9). However, black carbon from combustion can significantly heat polluted air over localities, and even whole regions. A brown haze frequently hangs over much of India, where it is believed to play a role in periodic disruptions of the monsoon rains. Because black carbon contributes to global warming, it has been argued that reductions in black carbon emissions would have immediate benefits for both human health and slowing climate change.

The dominant aerosol effect is to increase cloudiness because small particles induce the formation of cloud droplets by acting as condensation nuclei. The combined direct and indirect cooling effects of atmospheric aerosol are estimated to offset a substantial fraction of the warming that would have been expected from the increase in greenhouse gases. Ironically, measures currently being taken to reduce pollution from particle emissions will contribute to future warming.

The importance of condensation nuclei for cloud formation rests on the high surface tension of small droplets of water, which induces evaporation. The free energy required to condense water molecules into a liquid droplet is:

$$\Delta G = -NRT \ln\left(\frac{p}{p_0}\right) + 4\pi r^2 \gamma \tag{7.5}$$

Here $NRT \ln(p/p_0)$ is the liquid–vapor equilibrium term for N mol of water [R is the gas constant ($8.314\ \mathrm{J\ mol^{-1}\ K^{-1}}$) and T is the absolute temperature]. The ratio (p/p_0) is the ratio of the partial pressure of water molecules to the equilibrium vapor pressure of liquid water and must exceed unity for condensation to occur. How far it must exceed unity depends on the second term, $4\pi r^2 \gamma$, in which γ is the surface tension ($72.8\ \mathrm{erg\ cm^{-2}}$ at 20°C for H_2O), and r is the droplet radius. The erg is a unit of energy and is the work done by 1 dyne (dyn) of force acting through a 1 cm distance. One erg is equal to 1×10^{-7} J.

The number of moles in the droplet is also related to its radius:

$$N = \frac{\left(\frac{4\pi}{3}\right)r^3 \rho}{M} \tag{7.6}$$

where $(4\pi/3)r^3$ is the volume of the drop, ρ is its density ($1.00\ \mathrm{g\ cm^{-3}}$), and M is the gram molecular weight ($18.0\ \mathrm{g\ mol^{-1}}$). Thus, ΔG for a droplet is the result of two opposing terms that depend differently on r. Figure 7.13 shows a plot of ΔG against r for $p/p_0 = 1.001$ (100.1% relative humidity). The curve goes through a maximum, which defines a critical radius, $r_c = 1.08\ \mu$m. Droplets larger than this radius will accumulate more water molecules and become stable; droplets $<1.08\ \mu$m will evaporate. A 1-μm drop contains 0.23×10^{-12} mol of H_2O [Eq. (7.6)] or 1.38×10^{11} molecules. It is very improbable that this many molecules can come together simultaneously to form a growing droplet. Consequently, water vapor at 100.1% humidity in clean air does not condense. The critical droplet radius depends on the extent to which p/p_0 exceeds unity, that is, the extent to which the air is "supersaturated" with water. This dependence is obtained by differentiating (5) and setting $d(\Delta G)/dr$ equal to zero:

$$r_c = \frac{2M\gamma}{\rho RT \ln\dfrac{p}{p_0}} \tag{7.7}$$

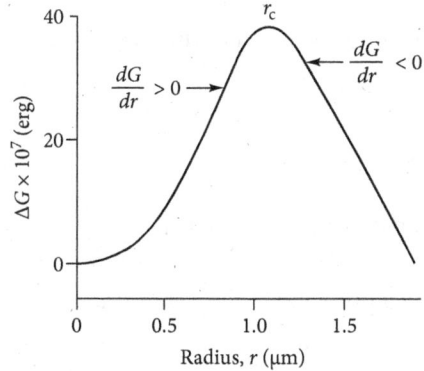

Figure 7.13 Variation of ΔG with drop size at $p/p_0 = 1.001$ ($T = 293$ K).
Source: Adapted from *Spiro et al., Chemistry of the Environment, 3rd Ed.,* University Science Books, © 2012, all rights reserved.

Thus, r_c decreases slowly as p/p_0 increases. Clean air can be supersaturated to a high degree without precipitation. In the real atmosphere, however, condensation actually does occur in the range of 100.1%–101% relative humidity, corresponding to $r_c = 1.08 - 0.108$ μm, thanks to the presence of condensation nuclei. An aerosol particle in this size range provides a surface for the accumulation of water molecules, lowering the surface tension sufficiently to permit droplet growth. There is a trade-off between droplet size and the number of condensation nuclei. A given amount of water vapor can form a small number of large drops or a large number of small ones. An excess of condensation nuclei can produce droplets that are too small to fall as rain. The fogs that often hover over cities probably reflect a large number of condensation nuclei in polluted air. An increase in the number of atmospheric particles is likely to increase the cloud cover and, therefore, the albedo.

QUESTION 56

Some scientists have developed the idea of injecting the atmosphere with small particles as "seeds" to act as cloud condensation nuclei in the hopes of providing precipitation to drought-prone areas. Using NaCl particles on the order of 0.5 μm in diameter has been proposed. From modeling studies, it has been calculated that water condenses on particles of this size at a supersaturation of 1.004%. How does the size of the particles compare with the minimum size of a pure water droplet at this humidity?

Q56 ANSWER The pure water droplet critical radius is obtained from (7.7):

$$r_c = \frac{2\,M\gamma}{\rho RT \ln \frac{p}{p_0}}$$

for a relative humidity, $(p/p_0) = 1.004$.

Using the water surface tension, γ, given above, 72.8 erg cm^{-2} or 72.8 dyn cm^{-1} at 20°C, and the molecular weight of water, 18.00 g mol^{-1}, we obtain:

$$r_c = \frac{2(18.00 \text{ g mol}^{-1})(72.8 \text{ dyn cm}^{-1})}{(1.00 \text{ g cm}^{-3})(8.314 \text{ J mol}^{-1} \text{ K}^{-1})(293 \text{ K}) \ln (1.004)} = 269.5 \text{ dyn cm}^2 \text{ J}^{-1}$$

Converting dynes cm to joules:

$$269.5 \frac{\text{dyn cm}^2}{\text{J}} \times \frac{\text{erg}}{\text{dyn cm}} \times 10^{-7} \frac{\text{J}}{\text{erg}} = 2.70 \times 10^{-5} \text{ cm} \times \frac{10^4 \, \mu\text{m}}{1 \text{ cm}} = 0.270 \, \mu\text{m}$$

The critical radius of the H_2O under these conditions is 0.270 μm, about one-half of the size of the NaCl particles.

*QUESTION 57

What if particles of 0.75 μm in diameter are used for seeding raindrops? From modeling studies, it has been calculated that water condenses on particles of this size at a supersaturation of 1.008%. How does the size of the particles compare with the minimum size of a pure water droplet at this humidity?

*QUESTION 58

Calculate the critical radius for water droplet formation in the atmosphere at 20°C and 101% relative humidity.

*QUESTION 59

What relative humidity must exist to form pure water droplets of 0.5 μm in diameter?

7.9 Global Warming Impacts

The expected warming of the earth will likely produce a cascade of impacts, including:

Storms and Droughts: Warming accelerates the hydrological cycle, increasing the rate of water evaporation and precipitation. Coastal areas experience more intense storms, with attendant flooding as the oceans warm up. At the same time, the interiors of continents become drier as soil moisture diminishes. Droughts increase and deserts expand, putting added pressure on already strained water supplies. Melting of glaciers and snowpack further threaten water supplies, as rivers swell and dry up earlier in the annual cycle.

Coastal Inundation: Much of the world's population lives in low-lying coastal areas, which will become permanently flooded as sea levels rise. The forced movement of people will impact social structures, particularly in impoverished areas of the world.

Agriculture: Growing zones will shift to higher latitudes, dislocating farms. Water scarcity will affect crop yields.

Health: Dislocation, floods, droughts, and heatwaves will all impact human health. The ranges of tropical disease insect vectors will expand.

Ecosystems: Shifting climate zones will lead to species extinction. Particularly vulnerable are the ocean corals, which grow slowly and tolerate a narrow temperature range. Even a 1°C–2°C warming is expected to produce significant coral bleaching. The ocean's biota will also be affected by increasing seawater acidification from CO_2 uptake (see Chapter 12).

The average temperature at the earth's surface is now 1°C warmer than in preindustrial times. The effects are already apparent; stronger storms and flooding, especially in coastal areas, increasing droughts and wildfires, failing crops, and bleaching coral reefs. All of these will intensify as the temperature continues to rise.

Recognizing the dangers, a series of international conferences have set targets for reducing greenhouse gas emissions, culminating in the Paris Accord of 2016, in which most of the world's nations agreed to aim for sufficient emission reductions to keep the temperature increase to 2°C or less.

7.10 Conclusions

The climate is changing over time and scientists continually collect data to assess the causes of the changes. Climate science requires the analysis of large sets of data, interpretation of graphs, and creating models to predict how the climate change will affect the earth. The skills practiced in this chapter allow scientists to make scientific conclusions. Scientists expect the changes in climate to cause the following effects: storms and droughts, coastal inundation, shifting of agricultural zones, health of living beings, and changes in ecosystems. Countries throughout the world are working to create international agreements to help combat this global issue.

8

FREE RADICAL CHEMISTRY—NITROGEN OXIDES, OZONE, AND COMBUSTION

8.1 Free Radicals

Free radicals are atoms or molecules having an unpaired electron. The chemical symbol for a radical is typically written with a single dot to symbolize that unpaired electron. Most free radicals are highly reactive, resulting in short atmospheric lifetimes. Since electron pairing is the basis of the covalent bond, free radicals can always gain stability by pairing the lone electron with an electron on another atom. A free radical can form a new bond at the expense of an existing bond in another molecule, but the result is the formation of another free radical because there is still an unpaired electron. The new radical can proceed to attack another molecule, thus creating a chain reaction driven by the instability of free radicals. The termination of these chain reactions occurs when a pair of free radicals finds each other and creates a stable bond from their previously unpaired electrons. Free radicals, being highly reactive, are present at very low concentrations; thus, they are much more likely to encounter stable molecules than other radicals.

QUESTION 1

What is a free radical? Draw an example and comment on the reactivity of free radicals.

Q1 ANSWER Free radicals are atoms or molecules having an unpaired electron, e.g., Br•. Most free radicals are highly reactive and only have a fleeting existence.

One way to know if a molecule has unpaired electrons is to draw its Lewis dot structure. As a reminder, Lewis dot structures explicitly show the valence electrons in a molecule as either covalent bonds (lines connecting atoms) or lone pair electrons (dots). To draw Lewis dot structures, first draw a skeleton of the molecule, putting the least-electronegative atom in the center. Second, use the periodic table to calculate the number of valence electrons in each molecule. Connect the atoms by covalent bonds and place the remaining electrons around each atom as lone pairs. Remember that each atom can only be surrounded by eight electrons (the octet rule), except hydrogen, which only has one electron to share with its neighbor. The covalent bond is an equal sharing of two electrons by both atoms. When there are more places to put electrons than valence electrons available, start adding double (or triple) bonds.

Sometimes there are multiple ways to draw a structure following the rules stated above; these are resonance structures. In order to decide which structure is most plausible, calculate the formal charge on each atom. An atom's formal charge is the electrical difference between the valence electrons in an isolated atom (from the periodic table) and the number of electrons assigned to that atom in the drawn Lewis structure. Break the covalent bond in half, and each atom will be assigned one electron for the purpose of calculating the formal charge. Ideally, all formal charges should be zero, but if a molecule has a charge, be sure that negative charges are assigned to the most electronegative atoms. Elements with high electronegativity have a greater tendency to attract electrons. Ignoring the noble gases in group 8A because they do not attract electrons, electronegativity generally increases as we move up and to the right on the periodic table (F is the most electronegative atom, followed by O and Cl).

QUESTION 2

Draw Lewis dot structures for the following molecules. Indicate whether each molecule is a free radical, or not.

(i)   OH (OH is one of the most important reactants in the atmosphere).

(ii) NO

(iii) N_2O

Q2 ANSWER

(i)   O has six valence electrons and H has one. Total valence electrons: $6 + 1 = 7$. Since there are only two atoms, simply connect them with a covalent bond and place the remaining electrons as lone pairs around the atoms.

$$\cdot \ddot{\underset{..}{O}}-H$$

H can only share its one electron in the covalent bond, so there are no lone pair electrons around it. O cannot reach an octet because there are only a total of seven electrons. This last unpaired electron means that OH· is a free radical.

(ii) O has six valence electrons and N has five. Total valence electrons: $5 + 6 = 11$. Since there are only two atoms, simply connect them with a covalent bond and place the remaining electrons as lone pairs around the atoms. A total of 14 electrons are needed for both N and O to have complete octets, but only 11 electrons are available. Thus, try a double bond (instead of a single bond) between the atoms.

$$\cdot\ddot{N}=\ddot{\underset{..}{O}} \qquad \ddot{\underset{..}{N}}=\ddot{O}\cdot$$
$$\textit{Structure 1} \qquad \textit{Structure 2}$$

Either way, there is an unpaired electron, so NO· is a free radical.

Notice that there are multiple ways to arrange the bonds and lone pair electrons; these are resonance structures. Use formal charges to decide the best structure.

Formal charges: valence electrons – electrons assigned

Structure 1: N: $5 - 5$ (2 from bonds, 3 lone) $= 0$

O: $6 - 6 = 0$

Structure 2: N: $5 - 6 = -1$

O: $6 - 5 = +1$

Structure 1 is favorable since all atoms have a zero formal charge.

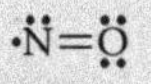

$$\cdot \ddot{N}=\ddot{O}$$

(iii) O has six valence electrons and N has five, but there are two nitrogen atoms. Total valence electrons: $6 + 2(5) = 16$

Because there are three atoms, which atom should go in the center of our structure? Nitrogen is less electronegative; it will go in the center. As in part (ii), there are not enough electrons to satisfy each atom's octet, so introduce more bonds. In fact, there are so few, the structure needs two additional bonds. But, again, there are multiple ways to arrange these resonance structures.

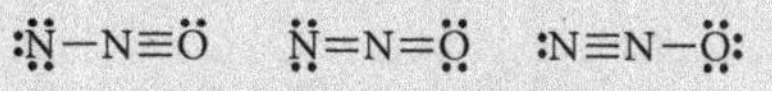

$$:\ddot{N}-N\equiv\ddot{O} \quad \ddot{N}=N=\ddot{O} \quad :N\equiv N-\ddot{O}:$$
    Structure 1   Structure 2   Structure 3

No matter how the electrons are arranged, they are all paired, so N_2O is not a free radical. Use formal charges to decide the best structure.

Structure 1: N: $5 - 7 = -2$
         N: $5 - 4 = +1$
         O: $6 - 5 = +1$

Structure 2: N: $5 - 6 = -1$
         N: $5 - 4 = +1$
         O: $6 - 6 = 0$

Structure 3: N: $5 - 5 = 0$
         N: $5 - 4 = +1$
         O: $6 - 7 = -1$

Structure 1 is the least favorable because it has a large charge (-2) and the most electronegative atom (O) has a positive formal charge. Structure 3 is most favorable because it allocates the negative charge on the most electronegative atom.

$$:N\equiv N-\ddot{O}:$$

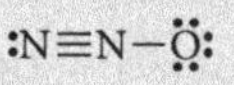

TABLE 8.1 Some Average Bond Dissociation Enthalpies

Bond	Enthalpy (kJ mol^{-1})	Bond	Enthalpy (kJ mol^{-1})
H—H	432	C≡O	1071
O=O	494	C—C	347
O—H	460	C=C	611
C—H	410	C⋯C^a	519
C—O	360	N=O	623
C=O	799	N≡N	941

^{a}Aromatic. 1.5 bond order.

Source: Adapted from Spiro et al., *Chemistry of the Environment, 3rd Ed.,* University Science Books, © 2012, all rights reserved.

QUESTION 3

Using Table 8.1, compare the amount of energy needed to dissociate OH· and NO· into individual atoms.

Q3 ANSWER A stronger bond requires more energy to break it. Table 8.1 lists the bond dissociation enthalpies. To dissociate OH·, an O–H single bond must be broken; this requires 460 kJ/mol of energy. To dissociate NO·, a N=O double bond must be broken; this requires 623 kJ/mol of energy. Thus, it takes more energy to break the NO· bond than the OH· bond.

QUESTION 4

Given your answers above, which molecule (OH· or NO·) is the most reactive? Which is the least reactive? Why?

Q4 ANSWER Given the answers above, the most reactive molecule would be OH·. It is a free radical; it is only held together by a single bond; and it requires less energy to break than NO·. The least reactive is NO· because it contains a double bond.

*QUESTION 5

Draw Lewis structures and compare the reactivity of NO_2 and HNO_3.

8.2 Nitrogen Oxides, Free Energy, and Equilibrium

The nitrogen cycle is very important to the existence of life on Earth, but it also fuels tropospheric ozone formation. Most of the nitrogen present on Earth is in the form of N_2 gas in the atmosphere. The triple bond connecting the two nitrogen atoms is so strong that breaking it requires considerable energy. Dissociating this bond can only be done with high-temperature combustion or lightning, and in the presence of oxygen, NO· is formed:

$$N_2 + O_2 \rightarrow 2\,NO· \tag{8.1}$$

Unfortunately, when the temperature is lowered, the reverse reaction is not favored because kinetics intervenes. Outside the combustion zone, the rate for NO· decomposition slows down and another reaction interferes:

$$2\,NO· + O_2 \rightarrow 2\,NO_2 \tag{8.2}$$

Note that both NO· and NO_2· are free radicals (11 and 17 valence electrons, respectively), but for some reason, convention dictates that we do not always put the single dot next to these compounds to indicate as such.

A spontaneous reaction is one that occurs under certain given conditions. We express the spontaneity of a reaction using the change in free energy (ΔG). The change in standard free energy for a reaction (ΔG°_{rxn}) is calculated by subtracting the standard free energy of formation of the reactants from that of the products, taking into account mole ratios:

$$\Delta G^\circ_{rxn} = \sum n\Delta G^\circ_f(\text{products}) - \sum m\Delta G^\circ_f(\text{reactants}) \tag{8.3}$$

where n and m are stoichiometric coefficients.

---

* Answers to starred questions can be found at the end of the book.

Standard free energy of formation values can be found in Table 8.2 for several compounds. These values are relative to the element in their stable form at 1 atmospheres (atm) and 25°C, which are given the value zero, by convention. If the overall change in free energy for a reaction (ΔG°_{rxn}) is negative, the reaction is spontaneous because there is energy available to do work.

TABLE 8.2 Standard (25°C) Enthalpies (ΔH°) and Free Energies (ΔG°) of Formation for Some Atmospheric Compounds

Compound[a]	ΔH° kJ mol^{-1}	ΔG° kJ mol^{-1}
O (g)	249.2	230.1
O_3 (g)	142.2	163.4
CO_2 (g)	−393.4	−394.3
CO (g)	−110.5	−137.2
NO (g)	90.4	86.7
NO_2 (g)	33.8	51.8
HNO_3 (aq)	−206.5	−110.5
SO_2 (g)	−296.8	−300.3
SO_3 (g)	−395.1	−370.3
H_2SO_4 (aq)	−907.3	−741.8
H_2O (g)	−241.8	−228.6
H_2O (l)	−285.7	−237.2

[a]Aqueous = aq, gas = g, and liquid = l.

Source: Adapted from Spiro et al., *Chemistry of the Environment, 3rd Ed.,* University Science Books, © 2012, all rights reserved.

QUESTION 6

Given the reaction (8.2) above: $2\,NO\,(g) + O_2\,(g) \rightarrow 2\,NO_2\,(g)$, calculate the change in standard free energy for this reaction.

Q6 ANSWER From the values presented in Table 8.2, for this reaction:

$$\Delta G^\circ_{rxn} = \left[2\text{ mol}\left(51.8\frac{kJ}{mol}\right)\right] - \left[2\text{ mol}\left(86.7\frac{kJ}{mol}\right) + 1\text{ mol}\left(0\frac{kJ}{mol}\right)\right] = -69.8\text{ kJ}$$

QUESTION 7

Is this reaction spontaneous? How do you know?

Q7 ANSWER Yes, this reaction is spontaneous in the forward direction because is negative.

The free energy change of a reaction is related to its equilibrium constant, K_{eq}, by the expression:

$$\Delta G = -RT \ln K_{eq} \tag{8.4}$$

where R is the gas constant, 0.008314 kJ K^{-1}, and T is the absolute temperature (Kelvin = °C + 273). The equilibrium constant is a unitless quantity equal to the equilibrium ratio of product concentrations divided by reactant concentrations when each is raised to a power equal to the number of moles in the reaction.

For the equation $aA + bB \rightarrow cC + dD$, the equilibrium constant, K, would be:

$$K_{eq} = \frac{[C]^c [D]^d}{[A]^a [B]^b} \tag{8.5}$$

QUESTION 8

Based on (8.2), if there is 2.5×10^{-16} atm of NO, what is the calculated equilibrium concentration of NO_2 at sea level and 25°C?

Q8 ANSWER The equilibrium constant expression can be used to calculate equilibrium concentrations. For this reaction, the equilibrium constant is:

$$K_{eq} = \frac{[NO_2]^2}{[NO]^2 [O_2]}$$

The NO concentration is given in the question. The oxygen concentration can be assumed because the total air pressure at sea level is 1 atm and 21% of it is oxygen; thus $[O_2] = 0.21$ atm. But that still leaves two unknowns: NO_2 concentration (which is the answer to the question) and the equilibrium constant (K_{eq}). There is a relationship between the equilibrium constant and free energy change of a reaction in (8.4):

$$\Delta G = -RT \ln K_{eq}$$

$$-69.8 \text{ kJ} = -(0.008314 \text{ kJ K}^{-1})(25 + 273)K \ln(K_{eq})$$

$$-69.8 \text{ kJ} = (-2.4776 \text{ kJ}) \ln(K_{eq})$$

$$28.17 = \ln(K_{eq})$$

$$K_{eq} = 1.72 \times 10^{12}$$

A very large equilibrium constant (K_{eq}) means that the product side of the reaction is favored. From the equilibrium expression, isolate the NO_2 concentration term:

$$[NO_2]^2 = K_{eq}[NO]^2[O_2]$$

$$[NO_2] = \sqrt{K_{eq}[NO]^2[O_2]}$$

$$[NO_2] = \sqrt{(1.72 \times 10^{12})(2.5 \times 10^{-16})^2(0.21)}$$

$$[NO_2] = 1.5 \times 10^{-10} \text{ atm}$$

*QUESTION 9

Is reaction (8.1) $N_2 + O_2 \rightarrow 2\ NO$ spontaneous? What is the equilibrium constant for this reaction at 25°C? Which side of the reaction is favored at that temperature?

*QUESTION 10

Using the thermodynamic data in Table 8.2, calculate the equilibrium concentration ratio of O_3 to O_2 at sea level and 25°C.

8.3 Photochemistry

Atmospheric chemistry is largely governed by the free radicals that are produced by solar radiation. Breaking bonds using the energy from light is called photolysis. We usually abbreviate "light" or "photon" with "hv" [Planck's constant (h) multiplied by frequency (v)].

QUESTION 11

Write the reaction for the photolysis of NO_2 to NO and O.

Q11 ANSWER $NO_2\ (g) + hv \rightarrow NO\ (g) + O\ (g)$

For a bond to break, the absorbed photon energy must equal or exceed the bond energy. The amount of energy from a photon of light is related to its frequency:

$$E = hv \qquad (8.6)$$

This relationship was introduced in Chapter 4 in the discussion of photovoltaic energy. Because the speed of light relates to frequency and wavelength:

$$c = \lambda v \qquad (8.7)$$

then energy is also related to the wavelength of light:

$$E = \frac{hc}{\lambda} \qquad (8.8)$$

where h is Planck's constant (6.626×10^{-34} J sec) and c is the speed of light (2.998×10^8 m sec^{-1}).

Standard enthalpy is the heat released or absorbed, under standard conditions (1 atm of pressure and 25°C) when the compound is formed from its elements. Negative values mean that heat is released by the system (exothermic), while positive values mean that heat is absorbed (endothermic). To a good approximation, the change in enthalpy of a reaction is equal to the energy required to drive a dissociation reaction. The change in standard enthalpy is calculated by subtracting the standard enthalpy of formation of the reactants from that of the products, taking into account mole ratios, analogous to the calculation of ΔG°_{rxn}:

$$\Delta H^\circ_{rxn} = \sum n \Delta H^\circ_f (\text{products}) - \sum m \Delta H^\circ_f (\text{reactants}) \qquad (8.9)$$

where n and m are stoichiometric coefficients. The standard enthalpy of formation values can be found in Table 8.2, but again, any element in its stable form at 1 atm and 25°C is zero.

QUESTION 12

Given that the standard enthalpy of formation for atomic oxygen is 249.2 kJ/mol, confirm that wavelengths less than ~400 nm will photolyze NO_2.

Q12 ANSWER For the dissociation of NO_2 above [NO_2 (g) + $h\nu \rightarrow$ NO (g) + O (g)]:

$$\Delta H^\circ_{rxn} = \left[1 \text{ mol}\left(249.2\frac{kJ}{mol}\right) + 1 \text{ mol}\left(90.4\frac{kJ}{mol}\right)\right] - \left[1 \text{ mol}\left(33.8\frac{kJ}{mol}\right)\right] = 305.8 \text{ kJ}$$

This answer shows that the reaction is endothermic, which makes sense since the light provides the energy to drive the reaction. This is the amount of energy needed to photolyze 1 mole (mol) of NO_2 molecules (305.8 kJ/mol).

From the relationship between energy and wavelength, we can calculate the maximum wavelength required to generate that amount of energy:

$$\lambda = \frac{hc}{E} = \left[\frac{\left(6.626 \times 10^{-34} \text{ J sec}\right)\left(2.998 \times 10^8 \frac{m}{sec}\right)}{(305.8 \text{ kJ})}\right]\left[\frac{1 \text{ kJ}}{1,000 \text{ J}}\right]\left[\frac{1 \times 10^9 \text{ nm}}{1 \text{ m}}\right] = 6.496 \times 10^{-22} \text{ nm}$$

Because the wavelength is the distance traveled by a single photon, not a mole of photons, there is one last step to this problem:

$$6.496 \times 10^{-22} \text{ nm}\left[\frac{6.022 \times 10^{23} \text{ photons}}{1 \text{ mol}}\right] = 391.2 \text{ nm}$$

This proves that wavelengths less than 400 nm will photolyze NO_2.

QUESTION 13

What type of light photolyzes NO_2?

Q13 ANSWER Looking at an electromagnetic spectrum (like that in Fig. 8.1), ultraviolet light has a wavelength of 391.2 nm.

QUESTION 14

What is the maximum wavelength that will photolyze molecular oxygen (O_2)? Does this light reach the surface of the earth (see Fig. 8.1)? Why or why not?

Q14 ANSWER Photolysis of molecular oxygen: O_2 (g) + $h\nu \rightarrow$ O (g) + O (g):

$$\Delta H^\circ_{rxn} = \left[2 \text{ mol}\left(249.2\frac{kJ}{mol}\right)\right] - \left[1 \text{ mol}\left(0\frac{kJ}{mol}\right)\right] = 498.4 \text{ kJ (endothermic)}$$

$$\lambda = \frac{hc}{E} = \left[\frac{\left(6.626 \times 10^{-34} \text{ J sec}\right)\left(2.998 \times 10^8 \frac{m}{sec}\right)}{(498.4 \text{ kJ})}\right]\left[\frac{1 \text{ kJ}}{1,000 \text{ J}}\right]\left[\frac{1 \times 10^9 \text{ nm}}{1 \text{ m}}\right]\left[\frac{6.022 \times 10^{23} \text{ photons}}{1 \text{ mol}}\right] = 240.0 \text{ nm}$$

Figure 8.1 shows that 240 nm lies in the UV-C range, and this wavelength is absorbed by ozone in the stratosphere, so it does not reach the surface of the Earth.

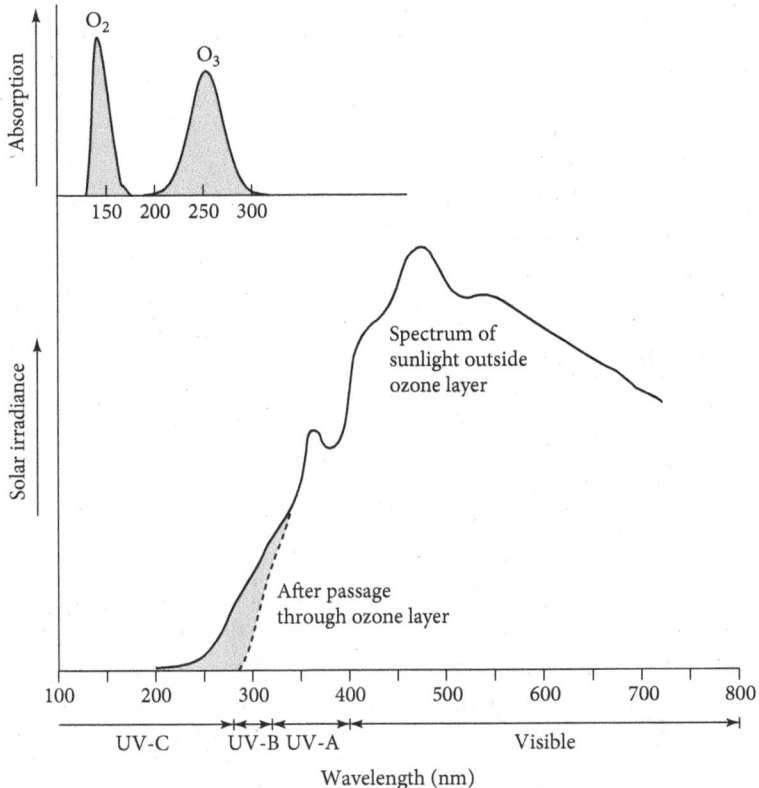

Figure 8.1 The absorption spectra of O_2 and O_3, and their effect on the spectrum of incoming solar radiation. *Source*: Adapted from Spiro et al., *Chemistry of the Environment, 3rd Ed.*, University Science Books, © 2012, all rights reserved.

---

*QUESTION 15

What is the maximum wavelength that will photolyze ozone (O_3)? What type of light dissociates ozone?

8.4 Photochemical Smog

Photochemical smog is a mixture of pollutant gases and particles. It can form wherever large quantities of automobile and industrial exhausts are photolyzed by sunlight. It is characterized by the accumulation of brown, hazy fumes, containing ozone, other oxidants, and particulate matter, with harmful health effects. Los Angeles is the classic example location for photochemical smog.

Figure 8.2 shows the time course of key atmospheric ingredients for a classic, smoggy day in Los Angeles. The concentration of hydrocarbons peaks during the early morning rush hour. The concentration of NO reaches a peak at the same time and then begins to decrease as the NO_2 concentration increases. Subsequently, the concentration of oxidants rises and the hydrocarbon concentration falls. This sequence of events results from the complex interplay of oxygen, hydrocarbons, nitrogen oxides, and sunlight. The effect is exacerbated because the emission in Los Angeles is trapped by surrounding mountains and cool breezes flowing from the sea. Hot air flowing over the mountains forms an inversion layer, preventing upward mixing, while plentiful sunshine drives photochemical reactions in the stagnant city air.

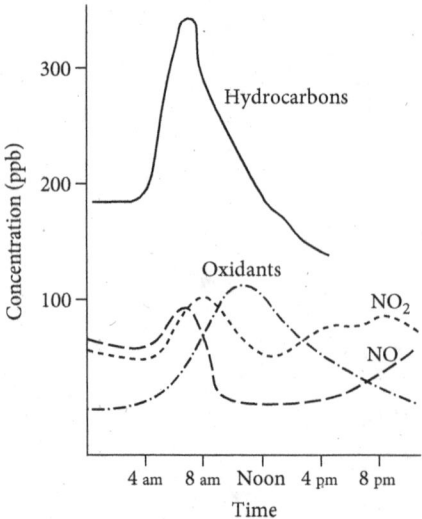

Figure 8.2 Example of a concentration-time profile of smog-forming chemicals in Los Angeles air. *Source:* Adapted from Spiro et al., *Chemistry of the Environment, 3rd Ed.,* University Science Books, © 2012, all rights reserved.

QUESTION 16

Consider the atmospheric concentrations on a typical smoggy day in Los Angeles in Figure 8.2. Is the $(NO_2)/(NO)$ ratio, say at 8 a.m., greater or less than the equilibrium value? Comment on the discrepancy.

Q16 ANSWER From the answer to Q8, we know that if $[NO] = 2.5 \times 10^{-16}$ atm, then $[NO_2] = 1.5 \times 10^{-10}$ atm at equilibrium, so $[NO_2]/[NO] = 6.0 \times 10^6$. But at 8 a.m., the concentrations are equal, so the actual $[NO_2]/[NO] = 1$, or about a million times smaller than the equilibrium ratio. The direct reaction of NO with O_2 is very slow, and the chemistry is dominated by radicals and sunlight.

8.4.1 NO_x and Ozone

$NO_2\cdot$ absorbs sunlight in the blue region of the spectrum, thus giving smoggy air a brown tint. As seen in the example above, when $NO_2\cdot$ is photolyzed, it dissociates into $NO\cdot$ and O:

$$NO_2\cdot + h\nu \rightarrow NO\cdot + O \tag{8.10}$$

The O atoms can then react immediately with abundant O_2 molecules to produce ozone (O_3). This reaction requires a third body (M) to stabilize the energetic intermediate and dissipate the excess energy as heat. M is any inert molecule (in the atmosphere it is likely N_2 or O_2):

$$O + O_2 + M \rightarrow O_3 + M \tag{8.11}$$

But since each molecule of ozone requires an oxygen atom from NO_2 dissociation, this mechanism cannot build up ozone concentration to levels greater than NO_2 itself. Moreover, the $NO\cdot$ produced can react with the ozone to restore the $NO_2\cdot$:

$$NO\cdot + O_3 \rightarrow NO_2\cdot + O_2 \tag{8.12}$$

Adding the three reactions above produces a "null" cycle, in which O_3 is neither net produced nor consumed.

Hydrocarbons are needed to have a net increase in ozone production. Hydrocarbons, represented by R–H, where "R" is any carbon-containing group, react with OH radicals. A hydrogen atom is abstracted from the hydrocarbon molecule to produce hydrocarbon radicals and water:

$$R–H + OH\cdot \rightarrow R\cdot + H_2O \tag{8.13}$$

Hydrocarbon radicals react immediately with O_2 molecules in the surrounding air to produce peroxyl radicals:

$$R\cdot + O_2 \rightarrow ROO\cdot \tag{8.14}$$

The resulting ROO· has a weakened O–O bond and can readily donate an O atom to NO·, regenerating $NO_2\cdot$, and producing an alkoxy radical:

$$ROO\cdot + NO \rightarrow NO_2 + RO\cdot \tag{8.15}$$

This mechanism allows ozone to buildup and far exceed levels of NO· that are emitted from the combustion sources, because this regenerated $NO_2\cdot$ can subsequently photolyze to O and produce additional ozone through the above reactions (8.10) and (8.11).

QUESTION 17

Why are hydrocarbons (abbreviated as RH) necessary to get a *net* production of ozone?

Q17 ANSWER Without hydrocarbons, there would be a net null cycle, where ozone is both produced and consumed:

$$NO_2 + hv \rightarrow NO + O$$

$$O + O_2 + M \rightarrow O_3 + M$$

$$\underline{NO + O_3 \rightarrow NO_2 + O_2}$$

NULL

Hydrocarbons provide an alternative pathway by which NO· can be converted to $NO_2\cdot$ without using ozone itself. Once $NO_2\cdot$ is present, it is photolyzed and the atomic oxygen radical easily reacts with O_2 to create ozone:

$$NO_2\cdot + hv \rightarrow NO\cdot + O \tag{8.10}$$

$$O + O_2 + M \rightarrow O_3 + M \tag{8.11}$$

$$NO\cdot + ROO\cdot \rightarrow RO\cdot + NO_2\cdot \tag{8.15}$$

$$NO_2\cdot + hv \rightarrow NO\cdot + O \tag{8.10}$$

$$O + O_2 + M \rightarrow O_3 + M \tag{8.11}$$

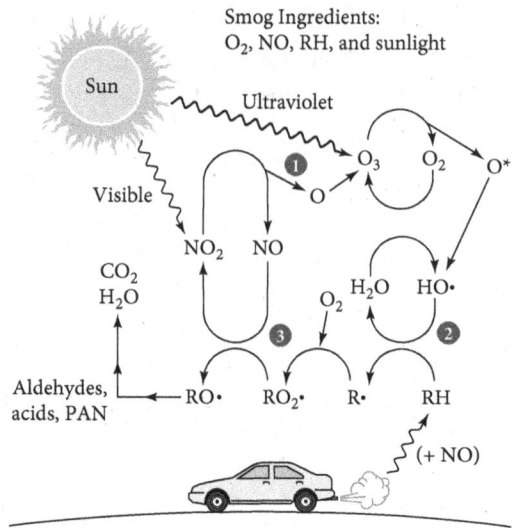

Figure 8.3 Smog formation from oxygen, nitric oxides, hydrocarbons, and sunlight.
Source: Adapted from Spiro et al., *Chemistry of the Environment, 3rd Edition,* University Science Books, © 2012, all rights reserved.

There is a net production of ozone:

The ROO· is from:

$$R\text{–}H + OH\cdot \rightarrow R\cdot + H_2O \tag{8.13}$$

$$R\cdot + O_2 \rightarrow ROO\cdot \tag{8.14}$$

Because NO and NO_2 are constantly cycling back and forth, they are often grouped together in the "NO_x" family. The generic recipe for photochemical smog is NO_x + RH + sunlight in the presence of oxygen (Fig. 8.3).

8.4.2 *Nitric Acid*

One way to end the $NO_2\cdot$ free radical cycle is to form nitric acid via:

$$4\,NO_2\cdot + O_2 + 2\,H_2O \rightarrow 4\,HNO_3 \tag{8.16}$$

Nitric acid molecules, being hydroscopic, are absorbed in raindrops and rained out of the atmosphere.

QUESTION 18

Is the conversion of NO_2 to HNO_3 endo- or exothermic (Table 8.2)? Calculate the enthalpy and free energy changes for reaction (8.16) in the aqueous phase.

Q18 ANSWER Enthalpy:

$$\Delta H^\circ_{rxn} = \left[4 \text{ mol}\left(-206.5\,\frac{\text{kJ}}{\text{mol}}\right)\right] - \left[4 \text{ mol}\left(33.8\,\frac{\text{kJ}}{\text{mol}}\right) + 2 \text{ mol}\left(-285.7\,\frac{\text{kJ}}{\text{mol}}\right)\right] = -389.8 \text{ kJ}$$

Therefore, the reaction is exothermic.

Free energy:

$$\Delta G^\circ_{rxn} = \left[4\,mol\left(-110.5\frac{kJ}{mol}\right)\right] - \left[4\,mol\left(51.8\frac{kJ}{mol}\right) + 2\,mol\left(-237.2\frac{kJ}{mol}\right)\right] = -174.8\,kJ$$

The reaction is also spontaneous, as ΔG°_{rxn} is <0.

QUESTION 19

What volume [in microliters (μL)] of nitric acid would be produced from 1.5×10^{-10} atm NO_2. via the above reaction ($4\,NO_2 + O_2 + 2\,H_2O \rightarrow 4\,HNO_3$) in a closed room with dimensions of 4 m × 5 m × 2 m at 20.0°C? Assume ideal gas behavior and an excess of water and oxygen. The density of nitric acid is 1.51 g mL^{-1}.

Q19 ANSWER First, make a plan to solve this problem. The known variables are (1) partial pressure of the limiting reagent (NO_2); (2) the volume and temperature of the room; and (3) the density of nitric acid.

The partial pressure of NO_2 is related to the moles of NO_2 through the ideal gas law. $PV = nRT$, where the volume and temperature are constant for all gases in the room. Thus, the partial pressure of NO_2 is related to the moles of NO_2:

$$P_{NO_2}V = n_{NO_2}RT$$

Solve for moles by rearranging the equation:

$$n_{NO_2} = \frac{P_{NO_2}V}{RT}$$

Remember that there are specific units in the ideal gas law, dictated by the ideal gas constant, R. R is equal to 0.08206 (L atm)/(mol K); thus, volume must be in liters, pressure in atmospheres, and temperature in Kelvin.

Temperature is given in Celsius, so 20.0°C + 273 = 293 K

Volume is simply the volume of the room, but it must be in liters:

$$4\,m \times 5\,m \times 2\,m = 40\,m^3$$

$$40\,m^3\left(\frac{100\,cm^3}{1\,m}\right)^3\left(\frac{1\,mL}{1\,cm^3}\right)\left(\frac{1\,L}{1{,}000\,mL}\right) = 40{,}000\,L$$

Plugging in the values:

$$n_{NO_2} = \frac{P_{NO_2}V}{RT} = \frac{(1.5 \times 10^{-10}\,atm)(40.000\,L)}{\left(0.08206\,\frac{L\,atm}{mol\,K}\right)(293\,K)} = 2.495 \times 10^{-7}\,mol$$

The moles of NO_2 are related to the moles of nitric acid because of stoichiometry. The moles of nitric acid can be converted to mass via molar mass, and then to volume using density. Convert moles of NO_2 to volume of HNO_3:

$$2.495 \times 10^{-7}\,mol\,NO_2\left(\frac{4\,mol\,HNO_3}{4\,mol\,NO_2}\right)\left(\frac{63.01\,g\,HNO_3}{1\,mol\,HNO_3}\right)\left(\frac{mL\,HNO_3}{1.51\,g\,HNO_3}\right)\left(\frac{1\,L}{1{,}000\,mL}\right)\left(\frac{1 \times 10^6\,\mu L}{1\,L}\right)$$

$$= 0.010\,\mu L\,HNO_3$$

Typically, atmospheric constituents are reported as a mixing ratio in units of parts per million (ppm) or parts per billion (ppb) by volume. This is a ratio where the "part" is the atmospheric constituent and the "whole" is the air itself. This ratio is unitless, so the part and the whole must be in the same "moles" or "molecules" unit.

QUESTION 20

Convert 0.298 μg m^{-3} of HNO$_3$ to units of ppb if the total atmospheric pressure is 0.93 atm and the temperature is 25.0°C. Assume ideal gas behavior.

Q20 ANSWER The "part" is micrograms of nitric acid. The "whole" is cubic meters of air.

To achieve ppb, or parts per billion, both micrograms and cubic meters must be converted to moles. (The ratio also works if both are converted to molecules.) Micrograms of nitric acid are related to moles through molar mass. Moles of air can be calculated from the ideal gas law since the temperature and pressure are given.

$$\text{HNO}_3\text{: } 0.298\ \mu g\left(\frac{1g}{1\times 10^6\ \mu g}\right)\left(\frac{1\ mol}{63.0129\ g}\right) = 4.729 \times 10^{-9}\ \text{mol HNO}_3$$

$$\text{Air: } 1\ m^3\left(\frac{100\ cm^3}{1\ m}\right)^3\left(\frac{1\ mL}{1\ cm^3}\right)\left(\frac{1\ L}{1{,}000\ mL}\right) = 1{,}000\ L$$

$$n = \frac{PV}{RT} = \frac{(0.93\ \text{atm})(1{,}000\ \text{L})}{\left(0.08206\ \frac{\text{L atm}}{\text{mol K}}\right)(298\ \text{K})} = 38.03\ \text{mol air}$$

$$\text{Ratio: } \frac{4.729\times 10^{-9}\ \text{mol HNO}_3}{38.03\ \text{mol air}} = 1.244 \times 10^{-10}$$

One billion is 10^9.

$$(1.244 \times 10^{-10}) \times (1 \times 10^9) = 0.12\ \text{ppb}$$

---

*QUESTION 21

Convert 0.392 μg m^{-3} of HNO$_3$ to units of ppb if the total atmospheric pressure is 0.95 atm and the temperature is 20.0°C. Assume ideal gas behavior.

8.4.3 Hydrocarbon Oxidation

Hydrocarbons are essential to the net production of tropospheric ozone. Hydroxyl radicals would like to react with H atoms because of the stability of the resulting water molecule. The water O–H bonds are stronger than the C–H bonds of organic compounds. Consequently, OH· can attack any organic compound with a C–H bond (and many other compounds as well). The hydroxyl radical has been called "nature's vacuum cleaner."

However, not all C–H bonds are the same. The stronger the C–H bond, the slower the reaction with OH·. This is why different fuel formulations will have differing effects on air pollution. Olefins (hydrocarbons containing C=C double bonds) play a special role in smog formation because they are highly reactive with OH· radicals (see the next section of this chapter).

Eventually, all of the carbon in the hydrocarbons will be fully oxidized to carbon dioxide (CO_2). Along the way, many oxidized hydrocarbons are produced that accompany ozone production.

First, a quick review of the typical reaction scheme of an alkane (R–H) as described previously:

$$R\text{–}H + OH\cdot \rightarrow R\cdot + H_2O \tag{8.13}$$

$$R\cdot + O_2 \rightarrow ROO\cdot \tag{8.14}$$

$$ROO\cdot + NO\cdot \rightarrow NO_2\cdot + RO\cdot \tag{8.15}$$

Because the alkoxy molecule is not the most oxidized form of carbon, the mechanism continues. Alkoxy radicals can produce carbonyls if the O-bearing C atom has a H atom attached. Upon encountering an O_2 molecule, this H atom can transfer to the O_2, thereby creating a carbonyl:

$$R'CHO\cdot + O_2 \rightarrow HO_2\cdot + R'C{=}O \tag{8.17}$$

QUESTION 22

Usually radicals in the air react with oxygen to produce peroxyl radicals, but sometimes H atom transfer to O_2 occurs instead, even though the O–H bond of HOO· is relatively weak. Suggest why this happens in the case of reaction (8.17).

Q22 ANSWER In this case, the product of the reaction is stabilized by the C=O bond, thereby favoring the reaction.

Like other peroxy radicals, the hydroperoxy radical ($HO_2\cdot$) can transfer an O atom to a NO radical, producing another OH radical:

$$HO_2\cdot + NO\cdot \rightarrow NO_2\cdot + OH\cdot \tag{8.18}$$

Carbonyls are subject to photolysis where UV rays split the R–C bond and produce two new radicals, alkyl and acyl:

$$R_2C{=}O + hv \rightarrow R\cdot + RCO\cdot \tag{8.19}$$

Both radicals react with O_2. The resulting peroxyl radical from the hydrocarbon radical can then transfer an O atom to NO·, further feeding the smog cycle:

$$R\cdot + O_2 \rightarrow ROO\cdot \tag{8.14}$$

$$ROO\cdot + NO\cdot \rightarrow NO_2\cdot + RO\cdot \tag{8.15}$$

The acyl radical is capable of transferring its R group to O_2 because the stability of the resulting product molecule, that is, the triple-bonded C≡O:

$$RCO\cdot + O_2 \rightarrow ROO\cdot + CO \tag{8.20}$$

Once more, the accompanying peroxyl radical can transfer an O atom to a NO radical, leaving an alkoxy (or, if R=H, a OH·) radical, and further promoting smog formation.

Figure 8.4 shows a flowchart summary of the radical pathways in smog formation.

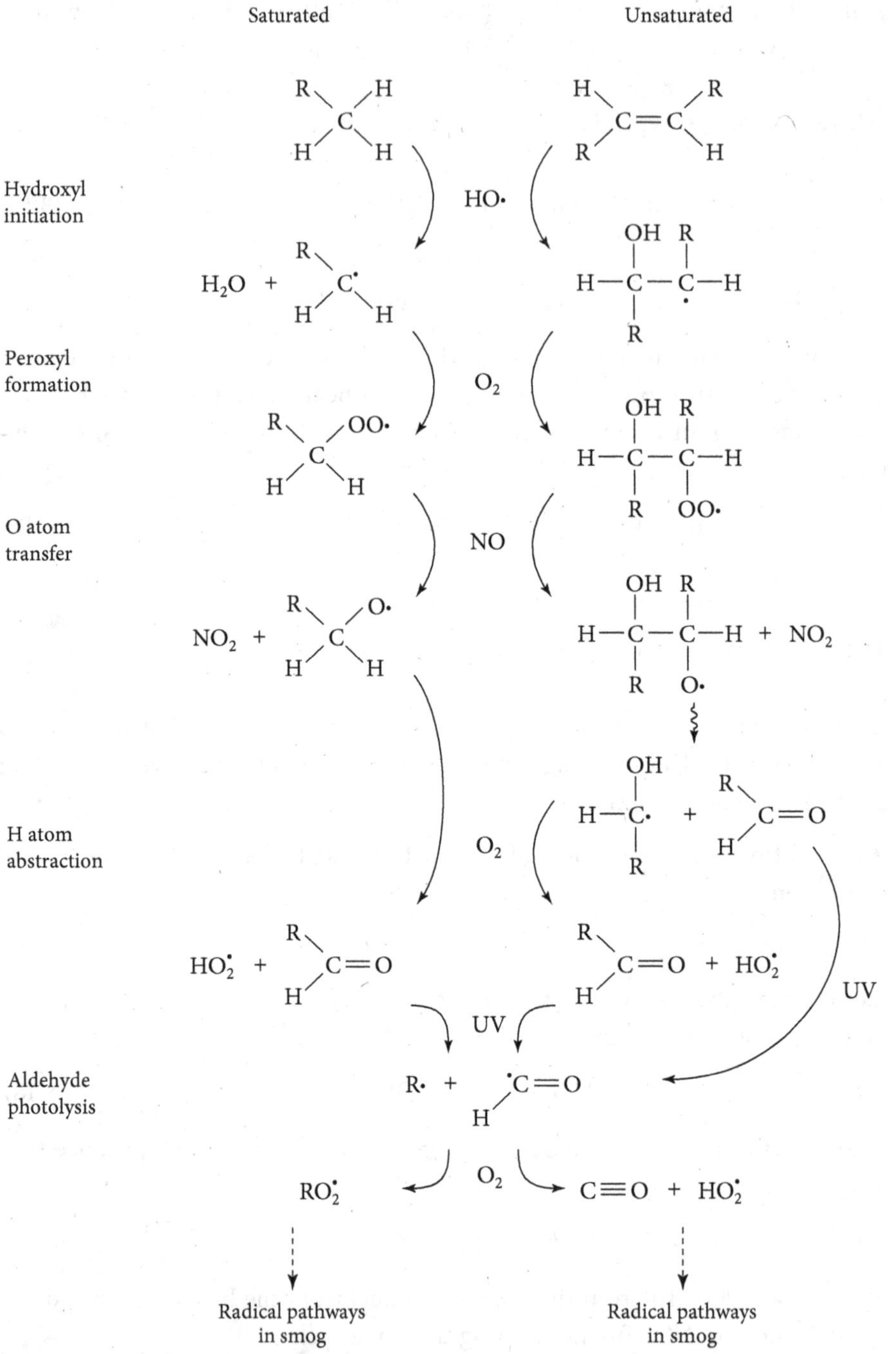

Figure 8.4 Overview of free radical pathways in smog formation.
Source: Adapted from Spiro et al., *Chemistry of the Environment, 3rd Ed.*, University Science Books, © 2012, all rights reserved.

While carbon monoxide is a stable molecule, it is still not fully oxidized. With an oxidation number of +4, the carbon in carbon dioxide (CO_2) is in its most oxidized, and most stable, form. CO, like other stable molecules, reacts with the hydroxyl radical:

$$CO + OH\cdot \rightarrow HOCO\cdot \qquad (8.21)$$

The resulting radical is also capable of donating its H atom to O_2. The reaction proceeds because of the strong C=O bonds in the product:

$$HOCO\cdot + O_2 \rightarrow CO_2 + HO_2\cdot \tag{8.22}$$

The carbon is now fully oxidized, but the resulting HO_2 radical behaves like other peroxyl radicals and promotes further smog formation:

$$HO_2\cdot + NO\cdot \rightarrow NO_2\cdot + OH\cdot \tag{8.18}$$

QUESTION 23

Using a stoichiometric analysis of the photochemical smog scheme laid out above, determine how many ozone molecules can be produced from the complete oxidation of the simplest hydrocarbon, methane. "Complete oxidation" of carbon means the production of CO_2. Assume an excess of NO_x and do not forget about HO_2 reactions and NO_2 photolysis.

Q23 ANSWER

$$CH_4 + OH\cdot \rightarrow CH_3\cdot + H_2O$$

$$CH_3\cdot + O_2 \rightarrow CH_3OO\cdot$$

$$CH_3OO\cdot + NO \rightarrow NO_2 + CH_3O\cdot$$

$$CH_3O\cdot + O_2 \rightarrow HO_2\cdot + CH_2O$$

$$CH_2O + hv \rightarrow H\cdot + CHO\cdot$$

$$H\cdot + O_2 \rightarrow HO_2\cdot$$

$$CHO\cdot + O_2 \rightarrow CO + HO_2\cdot$$

$$CO + OH\cdot \rightarrow HOCO\cdot$$

$$HOCO\cdot + O_2 \rightarrow CO_2 + HO_2\cdot$$

$$CH_4 + 2\,OH\cdot + 5\,O_2 + NO \rightarrow NO_2 + H_2O + CO_2 + 4\,HO_2\cdot$$

Each $HO_2\cdot$, as a peroxyl radical, can react with NO (due to the excess of NO_x):

$$4\,HO_2\cdot + 4\,NO \rightarrow 4\,NO_2 + 4\,OH\cdot$$

$$CH_4 + 5\,O_2 + 5\,NO \rightarrow 5\,NO_2 + H_2O + CO_2 + 2\,OH\cdot$$

Each NO_2 is photolyzed, which ultimately results in the production of one ozone molecule:

$$5\,NO_2 + hv \rightarrow 5\,NO + 5\,O$$

$$5\,O + 5\,O_2 \rightarrow 5\,O_3$$

Thus, the complete oxidation of one methane molecule, in excess NO_x, results in five ozone molecules.

QUESTION 24

Use Figure 8.2 to describe the following atmospheric chemistry phenomena.
  (i)   Why does the NO concentration increase before the NO_2 concentration?
  (ii) Why does the peak hydrocarbon concentration peak with that of NO in the urban basin?

Q24 ANSWER
  (i)   NO is directly emitted from combustion sources and then becomes NO_2 in the atmosphere.
  (ii) While the "background" of hydrocarbons is due to natural plant emissions, like isoprene and pinenes (which is why the concentrations never dip below ~100 ppb), the morning peak hydrocarbon concentration is due to rush-hour traffic. Hydrocarbons, as well as NO, are a by-product of the internal combustion engine in vehicles.

*QUESTION 25

Use Figure 8.2 to describe why the oxidant concentrations peak later in the day.

8.4.4 Bond Strength and Photochemical Activity

Not all hydrocarbons are equally reactive with the OH radical. They vary in dissociation energies (Table 8.3) and, therefore, in their reactivity. The stronger the C–H bond, the slower the reaction with the OH radical.

TABLE 8.3 Dissociation Energy of the C–H Bond Related to Substituents on the Carbon Atom[a]

Compound	Bond	Energy (kJ mol^{-1})
Water	HO—H	460
Methane	H_3C—H	427
Ethane	H_3CH_2C—H	406
Isopropane	$(H_3C)_2HC$—H	393
Tertiary butane	$(H_3C)_3C$—H	381
Trifluoromethane	F_3C—H	446
Chloroform	Cl_3C—H	401
Methanol	HOH_2C—H	393
Ethylene	H_2CHC—H	444
Benzene	H_5C_5C—H	427
Toluene	$H_5C_6H_2C$—H	326

[a] All are lower than the dissociation energy of the water 0–H bond, formed in the reaction with the HO• radical.

Source: Adapted from Spiro et al., *Chemistry of the Environment,*
3rd Edition, University Science Books, © 2012, all rights reserved.

Olefins are highly reactive with $OH\cdot$. The radical readily attacks the double bond and adds itself to the molecule, leaving the unpaired electron on the other carbon:

$$R_2C = CR_2 + OH\cdot \rightarrow R_2C(OH) - C(\cdot)R_2 \tag{8.23}$$

The new radical reacts with surrounding O_2 to form a peroxy radical:

$$R_2C(OH) - C(\cdot)R_2 + O_2 \rightarrow R_2C(OH) - C(OO\cdot)R_2 \tag{8.24}$$

The peroxy radical reacts with $NO\cdot$ and transfers an O to form $NO_2\cdot$:

$$R_2C(OH) - C(OO\cdot)R_2 + NO\cdot \rightarrow NO_2\cdot + R_2C(OH) - C(O\cdot)R_2 \tag{8.25}$$

This time the alkoxy radical product is unstable and it splits at the central C–C bond, producing a smaller radical and a carbonyl compound:

$$R_2C(OH) - C(O\cdot)R_2 \rightarrow R_2C(OH)\cdot + R_2C=O \tag{8.26}$$

Although olefins constitute a small fraction of refined petroleum products, they play a disproportional role in smog formation because of their high reactivity; thus, their content is regulated. However, olefins are also present in hydrocarbons emitted by vegetation. Deciduous trees emit isoprene, while conifers emit monoterpenes, all molecules with C=C bonds.

---

*QUESTION 26

Using the photochemical smog scheme laid out above, how many ozone molecules can be produced from the complete oxidation of ethene. "Complete oxidation" of carbon means the production of CO_2. Assume an excess of NO_x and do not forget about HO_2 reactions and NO_2 photolysis.

8.5 Combustion: Gasoline and Formulations

Much of the discussion in any environmental science textbook deals with combustion. On the surface, combustion is a simple process by which a hydrocarbon releases the energy tied up in its bonds by reacting with oxygen:

$$\text{Hydrocarbon} + O_2 \rightarrow CO_2 + H_2O \tag{8.27}$$

QUESTION 27

Write a reaction to represent the complete combustion of methane.

Q27 ANSWER $CH_4 + 2O_2 \rightarrow CO_2 + 2H_2O$

---

*QUESTION 28

Write a reaction to represent the complete combustion of ethanol (C_2H_5OH).

This net effect describes complete combustion, but it does not represent reality. There are many products of incomplete combustion (e.g., CO) and partially burned or unburned organics that exist in the fuel mixtures, not to mention the vast amount of NO_x that is produced simply by having N_2 and O_2 in the air near high-temperature combustion processes (see Chapter 9 for more information about air pollution).

8.5.1 *Free Radicals, Knocking, and Octane*

Combustion is a free radical process that allows oxygen to combine very rapidly with fuel molecules. Much of urban pollution is produced by transportation, and pollution regulations have substantially changed the composition of gasoline. Compromises have been made between fuel efficiency, engine performance, and pollution reduction.

Automobile engines work by igniting a mixture of gasoline and air with a spark. The air/fuel mixture is compressed by a piston and then ignited. The spark fragments the fuel molecules, generating enough free radicals to set off a chain reaction. The force of the explosion on the piston delivers power to the drive-train. However, the compression itself heats the fuel; at sufficiently high temperatures, the fuel molecules can react with thermally activated O_2 molecules to form radicals, thereby setting off the explosion prematurely. This preignition lowers the power generated and produces extra wear on the engine. This phenomenon is called "knocking" due to the noise produced by the engine under acceleration.

The temperature required for radical generation depends on the structure of the fuel molecule. Hydrocarbons are broken down into fragments and hydrogen atoms transfer to the hot O_2 molecules. The reaction rate depends primarily on the strength of the C–H bond (just as it does for OH· radical reactions). Branched hydrocarbons are more resistant to radical formation than straight-chained hydrocarbons because methylene ($-CH_2$) groups are more susceptible to attack by the thermally activated oxygen molecules than methyl ($-CH_3$) groups. With more methylene groups, straight-chained hydrocarbons are subject to preignition at a lower temperature.

The resistance of gasoline to knocking is specified by its octane number. Gasoline is a mixture of hydrocarbons with low boiling points, most of them containing seven or eight carbon atoms. Isooctane (2,2,4-trimethylpentane) is particularly resistant to preignition due to its highly branched structure; it is assigned an octane number of 100. The zero of the octane scale is set by *n*-heptane. Table 8.4 lists the octane numbers for several components of gasoline.

QUESTION 29

Compare the octane number of two isomers that both have 7-carbon atoms: (a) 2-methylhexane and (b) 3,3-dimethylpentane.

Q29 ANSWER The 3,3-dimethylpentane (b) molecule is more highly branched with more methyl groups, which are less susceptible to reacting with thermally active oxygen molecules. 2-methylhexane (a) has more methylene groups, making it more likely to react with the thermally active oxygen. 3,3-dimethylpentane will have a higher octane number.

TABLE 8.4 Properties of Some Components of Gasoline

Component	Research Octane Number (RON)	Motor Octane Number (MON)	Vapor Pressure (psi at 100°F)	PA[a]
Butane			51	3.23
n-Pentane	62	67	15.5	4.80
n-Hexane	19	22	5.0	5.90
Methyl propane			82	2.83
2-Methylbutane	99	104	20	
2-Methylpentane	83	79	6.6	5.82
2-Methylhexane	41	42	2.2	6.85
Isooctane	100	110	1.65	3.15
1-Butene	144	126	50	24.4
1-Methylpropene	170	139	62	24.4
1-Pentene	118	109	19	35.0
Cyclohexane	110	97	3.3	8.50
Methylcyclohexane	104	84	1.6	7.87
Benzene	99	91	3.3	0.88
Toluene	124	112	1.04	5.98
meta-Xylene	145	124	0.33	22.8
Ethanol	115[b]		17	3.3
Methanol	123	93	60	1.0
Methyl *tert*-butyl ether (MTBE)	123	97	8	2.6
Ethyl *tert*-butyl ether (ETBE)	111[b]		4	8.1

[a]Photochemical activity (PA) measured as rate of reaction with HO• radicals, units, cm^3 molecule^{-1} s^{-1} × 10^{12}.

[b]Average (RON + MON).

Sources: Adapted from Seddon, "Reformulated Gasoline, Opportunities for New Catalyst Technology," 1992, *Catalysis Today*, Vol. 15, pp. 1–21, and Spiro et al., *Chemistry of the Environment, 3rd Ed.,* University Science Books, © 2012, all rights reserved.

---

*QUESTION 30

Compare the octane number of two 9-carbon hydrocarbons: (a) isopropylbenzene and (b) 1,2,3-trimethylcyclohexane.

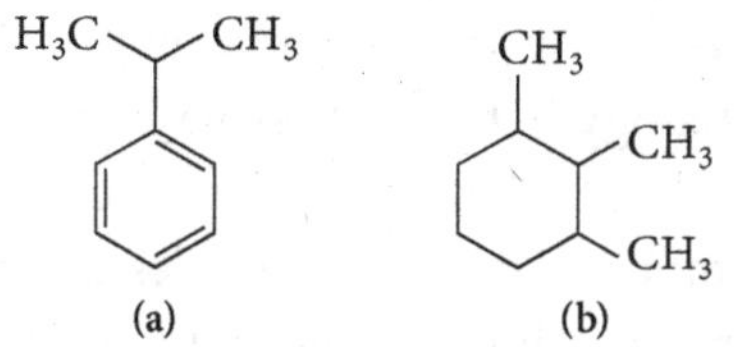

(a)   (b)

Table 8.4 also lists the vapor pressure, an important factor in controlling the volatility of gasoline blends. Generally, an increase in the number of carbon atoms will decrease the molecule's vapor pressure. To reduce evaporating hydrocarbons (i.e., volatile organic compounds [VOCs]), and adding excess emissions to the atmosphere, "summer blends" used in times of warmer temperatures have smaller amounts of smaller-sized alkanes than what is used in cooler temperatures.

QUESTION 31

How would the vapor pressure of *n*-heptane compare with that of butane, *n*-pentane, and *n*-hexane?

Q31 ANSWER *n*-heptane has seven carbons (C_7H_{16}), which makes it the largest straight-chained alkane listed in the question. This larger compound will have a lower vapor pressure; less of it will evaporate into the gas phase at a given temperature.

QUESTION 32

To combat excess hydrocarbon (i.e., VOC) emissions, would you want a higher fraction of *n*-heptane or *n*-pentane in a "summer blend"?

Q32 ANSWER Because the "summer blend" is used during times of higher ambient temperatures, we want to increase the fraction of larger hydrocarbons to reduce VOC evaporation. Thus, a higher fraction of *n*-heptane (versus *n*-pentane) would be desired.

*QUESTION 33

Rank the following hydrocarbons in order of increasing vapor pressure: 1-heptene, ethane, 1-butene, and 1-pentene.

Finally, Table 8.4 also lists the photochemical activity (PA), which is the rate of reaction with OH· radicals. The PA values can be understood based on the number of C–H bonds and their relative strengths as well as the special reactivity of C=C double bonds, as discussed above. Alkenes have high PA values. Rates are low for benzene and for compounds whose C–H bonds are mostly in methyl groups (methanol, ethanol, MTBE, and methylpropane). As the number of secondary C–H bonds increases, so does the OH· reactivity. Among the straight-chain alkanes, the rate increases from butane to pentane to hexane. High rates are also found for methyl-substituted benzenes (toluene, xylene), because of the special stabilization of methylene radicals by the aromatic ring.

Note that the analogy between thermally active O_2 molecules and OH· radicals does not extend to alkenes, or the alkylated benzenes, which have high octane numbers, despite having high PA numbers. Hot O_2 molecules abstract H atoms from saturated hydrocarbons, like the OH· radical, but they do not react rapidly with C=C bonds or with benzene substituents.

QUESTION 34

Compare the PA (i.e., reactivity with OH· radical) of *n*-heptane and 1-heptene. Which is more reactive?

Q34 ANSWER Both compounds have seven carbon atoms, but heptene is an alkene. The C=C double bond in heptene is more reactive than the C–C bonds in heptane. 1-heptene is more reactive with OH·; thus, it will have a higher PA (higher PA value on Table 8.4).

*QUESTION 35

Compare the PA of hexane and isoprene.

8.5.2 *Gasoline Additives*

It was discovered in the 1920s that knocking could be diminished if organolead compounds (i.e., tetramethyllead and tetraethyllead) were added to gasoline. These lead (Pb) additives suppress the radical chain reactions in the preignition phase. To avoid building up Pb deposits inside the engine, gasoline would also contain ethylene dichloride or ethylene dibromide. These organohalogens act as scavengers producing $PbCl_2$ or $PbBr_2$, compounds that are volatile at the high temperature of the exhaust gases, and thus removed from inside the engine and released into the atmosphere.

Lead accumulates in the soil through the deposition of aerosols. It is a persistent environmental pollutant because it does not readily degrade. With the introduction of catalytic converters in the mid-1970s, unleaded gasoline was introduced. This was necessary because lead compounds in the exhaust reacted with the rhodium and platinum catalysts in the converters, "poisoning" their surfaces and rendering them inactive. As the fraction of the car fleet with catalysts increased, so did the unleaded fraction of the gasoline supply.

QUESTION 36

Explain the data in Figure 8.5.

Q36 ANSWER Lead was introduced as a gasoline additive in the late 1920s and it was heavily used until the 1970s. Lead was removed from gasoline (giving us "unleaded" gasoline) because it interfered with the functioning of the catalytic converter. There was also a growing awareness that lead is one of the most toxic pollutants, especially to children's brain development, thus it should be removed from our environment.

An alternative to adding free radical scavengers to reduce knocking is to alter the gasoline composition by reducing the fraction of low octane components and increasing the fraction of high octane components. In the United States, the removal of lead was initially compensated by increasing the content of aromatic compounds, principally benzene, toluene, and xylenes (sometimes called BTX). Although BTX gasolines have high octane ratings, there are other concerns with these compounds. Xylenes react rapidly with OH· (see Table 8.4) and thus have a greater potential to form smog than the alkanes. Benzene, although low in PA, is a carcinogen.

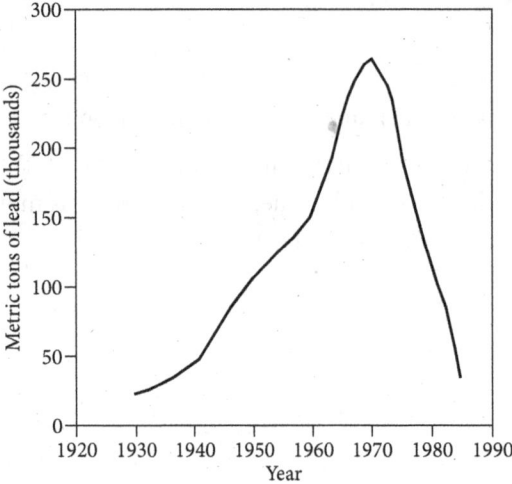

Figure 8.5 The historical consumption of lead in gasoline in the United States.
Source: Adapted from Dunlop et al., "Past Leaded Gasoline Emissions as a Nonpoint Source Tracer in Riparian Systems," 2000, *Environmental Science and Technology*, Vol. 34, No. 7, p. 1211.

Beginning in the 1980s, aromatics were replaced with "oxygenates," fuel molecules with one or more oxygen atoms. Oxygenates considered were methanol, ethanol, methyl tert-butyl ether (MTBE), and ethyl tert-butyl ether (ETBE), which all have octane ratings substantially higher than 100. Oxygenate additives also help decrease the emission of carbon monoxide (CO) during cold engine starts; because the fuel already contains O atoms, their conversion to CO_2 during fuel-rich combustion is more complete.

The oil industry first settled on MTBE, which is made in refineries, but it has become unpopular due to its propensity to contaminate groundwater through gasoline leaks. Ethanol emerged as a strong competitor for MTBE, and eventual replacement, because of the vast amount of domestic corn grown in the United States that can be used to produce ethanol.

QUESTION 37

Why would the concentration of olefins (alkenes) be restricted to a small value in gasoline while that of aromatics is allowed to be much higher, and alkanes are virtually unrestricted?

Q37 ANSWER Assuming a comparison of all similar-sized compounds, alkenes are more reactive in creating smog (higher PA value) than alkanes or aromatics because they have at least one C=C bond that is susceptible to OH· attack (to generate peroxy radicals, RO_2·). Substituted aromatics do have higher PA, while benzene itself is carcinogenic. Alkanes are quite unreactive because they do not contain C=C bonds.

*QUESTION 38

Why did the new unleaded gasoline contain higher concentrations of aromatics and olefins (also known as alkenes)?

*QUESTION 39

Other than reducing the production and emission of CO during combustion, what other advantages do ethanol and MTBE provide?

8.5.3 *Diesel*

A diesel engine works very differently from a spark-ignition engine. In diesels, air in the piston is preheated by compression, and the fuel is sprayed into the hot chamber, burning on contact. Since preignition is not an issue, the degree of compression can be very high, allowing for greater efficiency. This explains the higher fuel efficiency and thus the popularity of diesel vehicles across Europe.

The engine is built ruggedly to accommodate the higher compression forces, and since there are fewer moving parts than in a spark engine, a diesel engine wears out more slowly. Thus, diesel is generally the engine of choice for large trucks and buses.

Easy fragmentation of fuel molecules is desirable for diesel engines because it enhances the combustion of the injected fuel. Thus, straight-chained hydrocarbons are abundant in diesel fuel. Also, diesel engines work best with fuel that contains larger hydrocarbons (11–16 carbons) than that of gasoline (6–10 carbons).

While molecules at the air/fuel interface burn completely, molecules in the center of the injected plume can heat up before they have access to the O_2 molecules and, therefore, tend to decompose to solid "black carbon" particulates. Thus, diesel emissions contain much more particulate matter pollution than spark engine emissions. Diesel emissions also emit a lot of NO_x because of the high compression of the engines. An ordinary catalytic converter cannot be used to reduce the NO_x because there is too much O_2 in the exhaust, so diesel emissions require an active control catalyst.

QUESTION 40

Compare gasoline and diesel combustion characteristics, pros and cons.

Q40 ANSWER Gasoline fuel is made up of a larger fraction of smaller (lower molecular weight) hydrocarbons, thus there is higher direct fuel evaporation of VOCs. Additives must be addressed to combat preignition and volatility issues. Diesel fuel is made up of a larger fraction of large (higher molecular weight) hydrocarbons; thus, there is less evaporation of VOCs. The diesel combustion process is more efficient, but it results in large emissions of particulate matter, which has direct deleterious health effects.

8.6 Conclusions

Free radical chemistry is a key mechanism by which photochemical smog forms in the air of cities and populated regions around the world. The chemical reactivity of molecules with free radicals depends on the chemical structure of the molecule. One of the largest sources of atmospheric pollution is from combustion and the emission from automobile exhaust in particular. In order to limit smog formation in the atmosphere, the most volatile and reactive chemicals in gasoline mixtures are limited.

9

AIR POLLUTION

9.1 Pollutant

Atmospheric pollutants, or "air pollution," directly or indirectly harm animals, plants, and humans. This chapter highlights six pollutants generally viewed as needing control measures: carbon monoxide (CO), sulfur dioxide (SO_2), toxic organics, particulate matter (PM), nitrogen oxides (NO_x), and volatile organic compounds (VOCs). The first four directly harm human welfare, whereas the last two do so indirectly by generating ozone and other oxidants via photochemical smog.

Before exploring each pollutant individually, we will have a general discussion of how to handle units and kinetics of these atmospheric pollutants.

9.1.1 Gas-Phase Units and Kinetics

Because pressure, volume, and temperature are interdependent properties for gas-phase molecules, an absolute concentration unit like mass or moles per volume (e.g., molarity, M) cannot be used unless the pressure and temperature are specified. Since ambient pressure and temperature vary greatly over local, regional, and global scales, the amount of gases in the air is often described using a relative unit – a mole fraction (χ_i):

$$\chi_i = \frac{\text{mol of gas } i}{\text{total mol of air}} \tag{9.1}$$

This unitless quantity can be used to determine partial pressure (P_i) or partial moles (n_i), if the total pressure (P_T) or total moles (n_T) of air is known:

$$P_i = \chi_i P_T \tag{9.2}$$

$$n_i = \chi_i n_T \tag{9.3}$$

QUESTION 1

What is the partial pressure of nitrogen gas where the total pressure is 0.85 atm? (Hint: N_2 makes up 78% of the ambient air).

Q1 ANSWER The mole fraction for nitrogen will be 0.78 [78 moles (mol) of N_2 per 100 mol of air]. Thus, plug in the values into the partial pressure equation:

$$P_i = (0.78)(0.85 \text{ atm}) = 0.66 \text{ atm}$$

The partial pressure of N_2 would be 0.66 atmosphere (atm).

*QUESTION 2

What is the partial pressure of oxygen gas where the total pressure is 760 mm Hg? 760 mm of mercury (mmHg) is a measure of pressure and is equal to 1 atm. (Hint: O_2 makes up 21% of the ambient air).

*QUESTION 3

How many moles of O_2 are there in a sample that has a total of 1.50 mol of air?

Most air pollutants exist in very small (trace) amounts, so the mole fraction must be scaled. Common units are parts per million (ppm), parts per billion (ppb), and parts per trillion (ppt). For example:

$$\text{ppm} = \frac{\text{mol of air pollutant}}{\text{million mol of air}} = \frac{\text{mol pollutant}}{1 \times 10^6 \text{ mol air}} \tag{9.4}$$

Concentrations of a pollutant in a particular region of atmosphere depend on its emissions from sources, chemistry in the atmosphere, transport, and deposition. Chemical reactions in the atmosphere may form the pollutant or remove it. Transport of air pollution is carried away by winds from the source, and depositional loss involves both "dry" uptake onto surfaces and "wet" loss by falling raindrops. A one-box model of these processes on atmospheric species i is shown in Figure 9.1. Refer to Section 1.3.1 in Chapter 1 for an introduction to reservoirs and flows, which are involved in these types of models.

If the box (Fig. 9.1) is the entire atmosphere, then there is no additional flow into or out of the box, thus both "transport" terms would equal zero. If we assume that the concentration of the pollutant is relatively constant within the box, then the rate of production (chemical production rate plus emission rate) equals the loss rate (chemical loss rate plus deposition rate). This is the steady-state assumption and it allows scientists to estimate production or loss rates, given the other components in the equation.

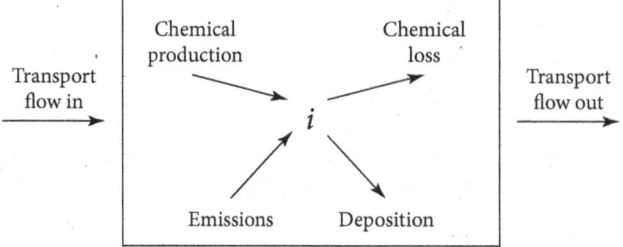

Figure 9.1 One-box model showing the production and loss of an atmospheric pollutant "i."

---

* Answers to starred questions can be found at the end of the book.

The lifetime of pollutant i is the average time a molecule of i remains in the box. It is the ratio of i's concentration in the box to its total loss rate. Lifetime is often referred to as residence time (e.g., the average amount of time i resides in the box). For i in Figure 9.1:

$$\text{lifetime} = \frac{\text{concentration of } i}{\text{flow out} + \text{deposition rate} + \text{chemical loss rate}} \tag{9.5}$$

For many reactive pollutants, chemical oxidation dominates its loss rate from the atmosphere. The most important oxidant in the atmosphere is the hydroxyl radical (OH·). Because the reaction of a pollutant with OH· is often the rate-limiting step, the step that determines a pollutant's loss rate is given by:

$$i + OH· \rightarrow \text{products}$$

This is a bimolecular reaction. The rate of this reaction could be described as a loss of reactants or a gain in products over time:

$$\text{rate} = \frac{-d[i]}{dt} = \frac{-d[OH·]}{dt} = \frac{d[\text{products}]}{dt} = k[i][OH·] \tag{9.6}$$

The rate law would then be the product of the rate constant (k) and the concentrations of reactants. The rate depends on the probability of pollutant i hitting OH· and the reaction's activation energy (E_a), which is the minimum energy required to initiate a chemical reaction. Thus, the rate constant is dependent on temperature; this is described by the Arrhenius equation:

$$k = Ae^{-\frac{E_a}{RT}} \tag{9.7}$$

where E_a is the activation energy (kJ mol^{-1}), R is the gas constant (8.314 J K^{-1} mol^{-1}), T is the temperature (K), and A is the frequency factor which represents the collision frequency. A can be treated as a constant for a given system.

Rate constants are determined under specific temperatures and ambient pressures; thus, we can use absolute units for concentrations within rate equations. For example, the rate law (9.6) is second order overall and would have the following units:

$$\text{rate} = k[i][OH·] \tag{9.6}$$

$$\frac{\text{molec}}{\text{cm}^3 \text{ sec}} = \left(\frac{\text{cm}^3}{\text{molec sec}}\right)\left(\frac{\text{molec}}{\text{cm}^3}\right)\left(\frac{\text{molec}}{\text{cm}^3}\right)$$

Rates, by definition, are changes in concentration per time, and since we can use absolute concentration units, each reactant would have units of molecules per cubic centimeter. If the lifetime is a ratio of concentration and loss rate and the loss is dominated by reaction with OH·, then the lifetime of i would be:

$$\text{lifetime} = \frac{[i]}{\text{loss rate}} = \frac{[i]}{k[i][OH·]} = \frac{1}{k[OH·]} \tag{9.8}$$

For these second-order reactions, the atmospheric lifetime of the pollutant only depends on what the pollutant is reacting with, and not on the concentration of the pollutant itself. Because we need to go back and forth between relative and absolute concentration units, it will be important to interconvert units for air pollutant concentrations, given specific pressure and temperature conditions.

QUESTION 4

Convert 15 ppm CO to units of molec cm^{-3} at 1.0 atm and 25°C.

Q4 ANSWER We saw that ppm is a mole ratio between moles of CO and millions of moles of air, but we ultimately want molecules of CO, not moles. Because of the relationship between moles and molecules, we can also think of ppm as a ratio of molecules:

$$15 \text{ ppm CO} = \frac{15 \text{ mol CO}}{1 \times 10^6 \text{ mol air}} \left(\frac{1 \text{ mol air}}{6.022 \times 10^{23} \text{ molec air}} \right) \left(\frac{6.022 \times 10^{23} \text{ molec CO}}{1 \text{ mol CO}} \right) = \frac{15 \text{ molec CO}}{1 \times 10^6 \text{ molec air}}$$

Now we just have to convert the molecules of air to a volume (cm^3). We will assume that this air sample behaves ideally and use the relationships built into the ideal gas law ($PV = nRT$) to create our own conversion factor. The volume taken up by exactly 1 mol at 1.0 atm and 25°C is:

$$V = \frac{nRT}{P} = \frac{(1 \text{ mol})\left(0.08206 \frac{\text{L atm}}{\text{mol K}}\right)(298 \text{ K})}{1.0 \text{ atm}} = 24.45 \text{ L}$$

Using this conversion factor:

$$\frac{15 \text{ molec CO}}{1 \times 10^6 \text{ molec air}} \left(\frac{6.022 \times 10^{23} \text{ molec CO}}{1 \text{ mol air}} \right) \left(\frac{1 \text{ mol air}}{24.45 \text{ L}} \right) = 3.69 \times 10^{17} \frac{\text{molec CO}}{\text{L air}}$$

Now convert liters to cubic centimeters:

$$3.69 \times 10^{17} \frac{\text{molec CO}}{\text{L air}} \left(\frac{1 \text{ L}}{1,000 \text{ mL}} \right) \left(\frac{1 \text{ mL}}{1 \text{ cm}^3} \right) = 3.7 \times 10^{14} \frac{\text{molec CO}}{\text{cm}^3}$$

QUESTION 5

Convert 125 μg m^{-3} SO_2 to units of ppb at 28°C and 1.10 atm.

Q5 ANSWER Convert μg SO_2 to moles SO_2 and m^3 air to moles of air to get a mixing ratio, then simply multiply by 1 billion. We will assume that this air sample behaves ideally and use the relationships built into the ideal gas law ($PV = nRT$) to create our own conversion factor. The volume taken up by exactly 1 mol at 1.10 atm and 28°C is:

$$V = \frac{nRT}{P} = \frac{(1 \text{ mol})\left(0.08206 \frac{\text{L atm}}{\text{mol K}}\right)(301 \text{ K})}{1.10 \text{ atm}} = 22.45 \text{ L air}$$

Using this conversion factor:

$$\frac{125 \text{ } \mu\text{g SO}_2}{m^3 \text{ air}} \left(\frac{1 \text{ m}}{100 \text{ cm}} \right)^3 \left(\frac{1 \text{ cm}^3}{1 \text{ mL}} \right) \left(\frac{1,000 \text{ mL}}{1 \text{ L}} \right) \left(\frac{22.45 \text{ L}}{1 \text{ mol air}} \right) = 2.806 \frac{\mu\text{g SO}_2}{\text{mol air}}$$

$$\frac{2.806 \text{ } \mu\text{g SO}_2}{\text{mol air}} \left(\frac{1 \text{ g SO}_2}{1 \times 10^6 \text{ } \mu\text{g SO}_2} \right) \left(\frac{1 \text{ mol SO}_2}{64.066 \text{ g SO}_2} \right) = 4.380 \times 10^{-8} \frac{\text{mol SO}_2}{\text{mol air}}$$

$$\left(4.380 \times 10^{-8} \frac{\text{mol SO}_2}{\text{mol air}} \right)(1 \times 10^9) = 43.8 \text{ ppb}$$

---

*QUESTION 6

Convert 22 ppt naphthalene to molecules per cm^3 at 34°C and 1.1 atm.

*QUESTION 7

Convert 0.736 mg CO_2 L^{-1} to ppm CO_2 at 25°C and 1.0 atm.

These ideas will be practiced in the subsequent sections.

Six air pollutants (carbon monoxide, sulfur dioxide, toxic organics, particulates, nitrogen dioxides, and VOCs) all need control measures to protect human health and the environment. The first four directly harm human welfare, whereas the last two do so indirectly, by generating ozone and other "oxidant" molecules via photochemical smog (Chapter 8).

9.2 Carbon Monoxide

Carbon monoxide (CO) emissions in the United States peaked around 1970 and have been declining since then (Fig. 9.2). By far, the major source of CO has been transportation. Since 1970, emissions have declined, despite the continued increase in vehicle travel, thanks to increasingly stringent emission control standards and improvements in energy efficiency.

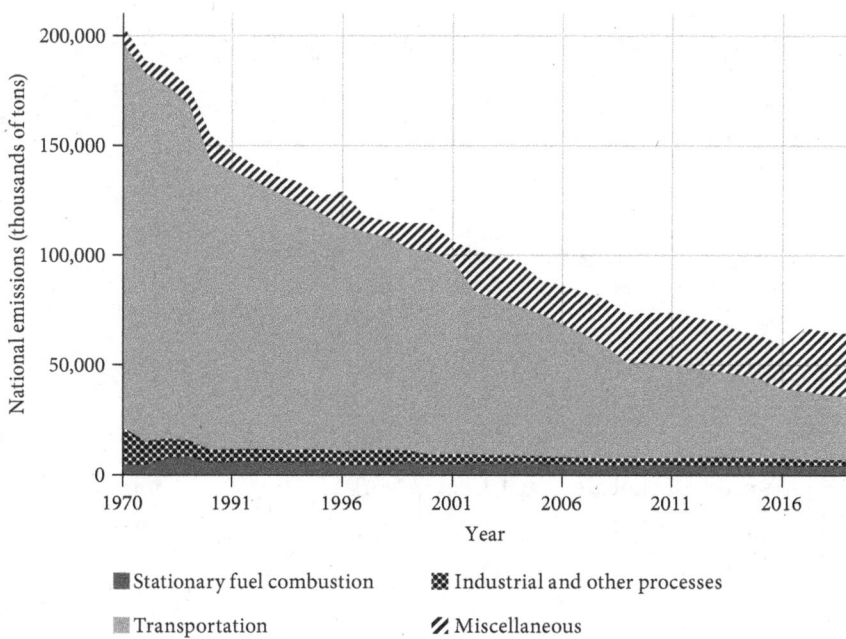

Figure 9.2 Carbon monoxide emission trends in the United States from 1970 to 2019 by source type. The CO emissions from each source are stacked (e.g., in 1970, there were 4,632 thousand tons emitted by stationary fuel combustion, 16,899 thousand tons emitted by industrial and other processes, 174,602 thousand tons emitted by transportation, and 7,909 thousand tons of CO from the miscellaneous category). *Source:* Data from the U.S. Environmental Protection Agency (EPA) National Emissions Inventories, *Air Pollution Emission Data,* average annual emissions, available at https://www.epa.gov/air-emissions -inventories/air-pollutant-emissions-trends-data.

QUESTION 8

Use the following given information to answer the questions below. Let us assume there is 150 ppb CO and a hydroxyl radical (OH·) concentration of 8.7×10^5 molec cm^{-3} in the northern hemisphere. Assume an average temperature of 30°C and pressure of 1.0 atm. A simple box model for CO is shown below.

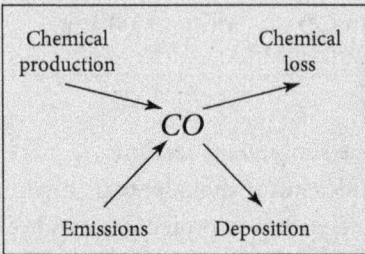

(i)   Calculate the rate of CO loss via chemistry in this box if the rate constant (k) for the reaction between CO and OH· is $5 \times 10^{-13} e^{(-300 \text{ K}/T)}$ cm^3/(molec sec).

(ii)   Calculate the lifetime (in days) of CO in this box, assuming the deposition rate is minimal and can be ignored.

(iii) Assume that the chemical production of CO is dominated by the oxidation of methane ($CH_4 + OH· \rightarrow$ products), where the rate constant for this reaction is 4.0×10^{-15} cm^3 (molec sec)$^{-1}$ and the average methane concentration is 1.7 ppm. If CO is in a steady state where the production rate equals the loss rate (and deposition is minimal), what is the rate of CO emission from combustion in the northern hemisphere?

(iv) How would the emission sources be different in the southern hemisphere?

Q8 ANSWER

(i)   The rate expression for chemical loss would be [from Eq. (9.6)]:

$$\text{rate} = k[i][OH·]$$

To use this expression, we need all of the concentration units to be in molec cm^{-3}:

$$PV = nRT$$

$$V = \frac{nRT}{P} = \frac{(1 \text{ mol})\left(0.08206 \frac{\text{L atm}}{\text{mol K}}\right)[(273 + 30) \text{ K}]}{1.0 \text{ atm}} = 24.86 \text{ L air}$$

$$\frac{150 \text{ molec CO}}{1 \times 10^9 \text{ molec air}} \left(\frac{6.022 \times 10^{23} \text{ molec air}}{1 \text{ mol air}}\right)\left(\frac{1 \text{ mol air}}{24.86 \text{ L}}\right)\left(\frac{1 \text{ L}}{1,000 \text{ mL}}\right)\left(\frac{1 \text{ mL}}{1 \text{ cm}^3}\right) = 3.63 \times 10^{12} \frac{\text{molec CO}}{\text{cm}^3 \text{ air}}$$

Now plug the values into the rate expression. Note that the rate constant has the temperature in its exponent:

$$\text{rate} = \left(5 \times 10^{-13} e^{-\frac{300K}{303K}} \frac{\text{cm}^3}{\text{molec sec}}\right)\left(3.63 \times 10^{12} \frac{\text{molec CO}}{\text{cm}^3 \text{ air}}\right)\left(8.7 \times 10^5 \frac{\text{molec OH·}}{\text{cm}^3 \text{ air}}\right) = 5.9 \times 10^5 \frac{\text{molec}}{\text{cm}^3 \text{ sec}}$$

(ii)   The lifetime of CO in the atmosphere is calculated by dividing the concentration of CO by its loss rate, but the deposition rate is minimal, so it is assumed to be zero:

$$\text{lifetime} = \frac{\text{concentration of CO}}{\text{deposition rate + chemical loss rate}} = \frac{[\text{CO}]}{\text{chemical loss rate}}$$

Since we already calculated the chemical loss rate, we can plug in the values:

$$\text{lifetime} = \frac{3.63 \times 10^{12} \frac{\text{molec CO}}{\text{cm}^3 \text{ air}}}{5.9 \times 10^5 \frac{\text{molec}}{\text{cm}^3 \text{ air}}} = 6{,}152{,}542 \text{ sec} \left(\frac{1 \text{ min}}{60 \text{ sec}}\right)\left(\frac{1 \text{ hr}}{60 \text{ min}}\right)\left(\frac{1 \text{ day}}{24 \text{ hr}}\right) = 71 \text{ days}$$

(iii)   The steady-state assumption says that the rate of production (chemical production rate plus emission rate) equals the loss rate (chemical loss rate plus deposition rate). If deposition is minimal, then chemical production plus emission equals chemical loss:

$$\text{(chemical production from CH}_4\text{) + (emission) = (chemical loss)}$$

$$\text{(rate} = k[\text{CH}_4][\text{OH·}]) + \text{(emission)} = \text{(rate} = k[\text{CO}][\text{OH·}])$$

Thus, we can solve for just the emission component:

$$\text{emission} = \text{(rate} = k[\text{CO}][\text{OH·}]) - \text{(rate} = k[\text{CH}_4][\text{OH·}])$$

Since we already have the chemical loss rate from above, we need to solve for the chemical production rate from the oxidation of methane:

$$\text{rate} = k[\text{CH}_4][\text{OH·}]$$

We need methane in the correct units:

$$\frac{1.7 \text{ molec CH}_4}{1 \times 10^6 \text{ molec air}}\left(\frac{6.022 \times 10^{23} \text{ molec air}}{1 \text{ mol air}}\right)\left(\frac{1 \text{ mol air}}{24.86 \text{ L}}\right)\left(\frac{1 \text{ L}}{1{,}000 \text{ mL}}\right)\left(\frac{1 \text{ mL}}{1 \text{ cm}^3}\right) = 4.12 \times 10^{13} \frac{\text{molec CH}_4}{\text{cm}^3 \text{ air}}$$

Then plug in the values into the rate expression:

$$\text{rate} = \left(4.0 \times 10^{-15} \frac{\text{cm}^3}{\text{molec sec}}\right)\left(4.12 \times 10^{13} \frac{\text{molec}}{\text{cm}^3 \text{ air}}\right)\left(8.7 \times 10^5 \frac{\text{molec}}{\text{cm}^3 \text{ air}}\right) = 1.4 \times 10^5 \frac{\text{molec}}{\text{cm}^3 \text{ sec}}$$

Now subtracting the two rates:

$$\text{emission} = 5.9 \times 10^6 \frac{\text{molec}}{\text{cm}^3 \text{ sec}} - 1.4 \times 10^5 \frac{\text{molec}}{\text{cm}^3 \text{ sec}} = 5.8 \times 10^6 \frac{\text{molec}}{\text{cm}^3 \text{ sec}}$$

(iv)   There is less combustion in the southern hemisphere overall, but there is a larger biomass burning component to the emissions of CO.

---

*QUESTION 9

Assuming the main loss of methane is via a chemical reaction with OH radical under the same conditions described above, calculate the lifetime of methane in the northern hemisphere.

*QUESTION 10

Now let us consider CO in the contiguous United States (instead of the entire northern hemisphere) at 25°C and 1.00 atm. Use the same box model in Q8 to answer the following questions:

(i) If the OH· radical concentration is 1×10^6 molec cm^{-3} and the rate constant (k) for the reaction between CO and OH· is $5 \times 10^{-13} e^{(-300\ K/T)}$ cm^3 (molec sec)$^{-1}$, what is the lifetime of CO? Assume that loss via deposition is negligible.

(ii) Figure 9.2 shows that there were 64,188 thousand tons of CO emitted from the United States in 2019. The area of the United States is 7.66×10^6 km^2 and most CO is within the first 1 km in altitude. If the average concentration of CO over the entire country is 1.4 ppm, what is the chemical production rate of CO?

(iii) If the oxidation of methane (CH_4 + OH·) is the main chemical production of CO in the atmosphere, what is the concentration of methane (in ppm) in the United States? Assume the OH· radical concentration is 1×10^6 molec cm^{-3} and the rate constant (k) for the reaction between methane and OH· is 4.0×10^{-15} cm^3/(molec sec).

Although CO occurs naturally in the environment at low levels (a few ppm), it is an asphyxiating poison at high levels because it displaces the O_2 bound to hemoglobin, the blood protein that transports oxygen from the lungs to the body's tissues. While the main source of anthropogenic CO is automotive transport, individuals are at greater risk of CO poisoning from malfunctioning stoves and heaters in their homes. CO is produced whenever fuel combustion is incomplete.

9.3 Sulfur Dioxide

The main source of anthropogenic sulfur dioxide (SO_2) emissions is the combustion of coal (Fig. 9.3).

Though SO_2 is a lung irritant, the most damaging health effects (for both plants and animals) in urban atmospheres is caused by sulfuric acid (H_2SO_4), which is formed from the oxidation of SO_2 in air. Sulfuric acid is a major component in acid rain. Oxidation in the gas phase is shown below, where "M" is called a "third body," which can be any inert air molecule (generally N_2 or O_2) that will remove excess energy from the unstable reaction intermediate:

$$SO_2 + OH· \xrightarrow{M} HOSO_2$$

$$HOSO_2 + O_2 \xrightarrow{M} HO_2· + SO_3$$

$$SO_3 + H_2O \rightarrow\rightarrow\rightarrow H_2SO_4$$

Sulfuric acid has a low vapor pressure, so it exists in the aqueous phase.

QUESTION 11

Figure 9.3 shows that 1,438,000 tons of SO_2 were emitted in 2019 from stationary fuel combustion. How much limestone would be needed to react with this amount of sulfur dioxide in 2019? (Hint: 1 ton is 1,000 kg.)

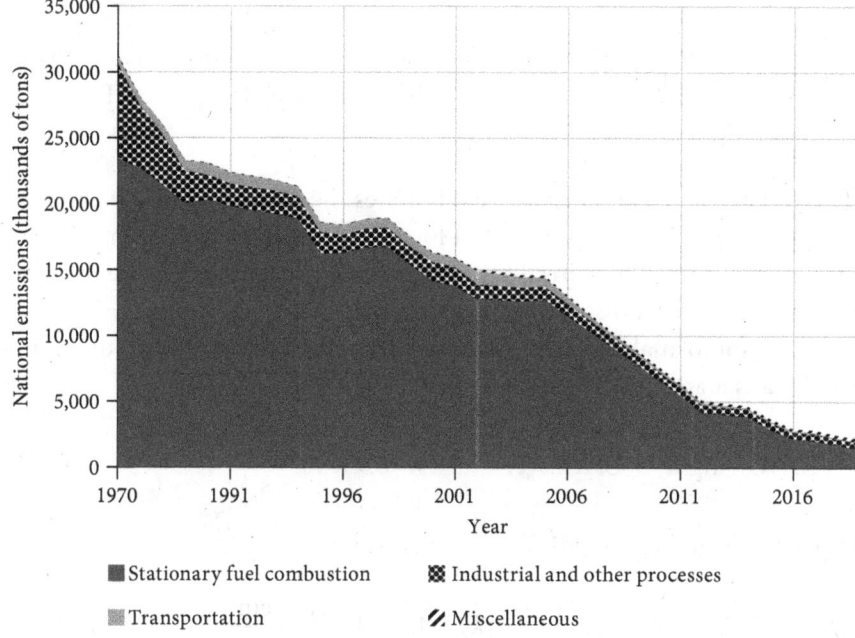

Figure 9.3 Sulfur dioxide emission trends in the United States from 1970 to 2019 by source type. *Source:* Data from the U.S. EPA National Emissions Inventories, *Air Pollution Emission Data, average annual emissions,* available at https://www.epa.gov/air-emissions-inventories/air-pollutant-emissions -trends-data.

Q11 ANSWER Use the amount of SO_2 and the stoichiometry for the following equation:

$$SO_2 + CaCO_3 + \tfrac{1}{2}O_2 \rightarrow CaSO_4 + CO_2$$

provided in the above equation. Calculated molar masses of SO_2 and $CaCO_3$ are 64.066 and 100.09 g mol^{-1}, respectively:

$$1{,}438{,}000 \frac{\text{tons } SO_2}{\text{yr}} \left(\frac{1{,}000 \text{ kg } SO_2}{1 \text{ ton } SO_2}\right)\left(\frac{1{,}000 \text{ g } SO_2}{1 \text{ kg } SO_2}\right)\left(\frac{1 \text{ mol } SO_2}{64.066 \text{ g } SO_2}\right)\left(\frac{1 \text{ mol } CaCO_3}{1 \text{ mol } SO_2}\right)\left(\frac{100.09 \text{ g } CaCO_3}{1 \text{ mol } CaCO_3}\right)$$

$$= 2.2466 \times 10^{12} \text{ g } CaCo_3$$

$$2.2466 \times 10^{12} \text{ g } CaCo_3 \left(\frac{1 \text{ kg } CaCO_3}{1{,}000 \text{ g } CaCO_3}\right)\left(\frac{1 \text{ ton } CaCO_3}{1{,}000 \text{ kg } CaCO_3}\right) = 2.247 \times 10^{12} \text{ tons } CaCO_3$$

Although limestone is relatively cheap, a great deal of it would be used for this application. In addition, the $CaSO_3$ sludge has to be disposed.

*QUESTION 12

How much $CaSO_3$ would have to be disposed if 2,035,000 tons of SO_2 are scrubbed using calcium carbonate?

9.4 Organics

Although there are countless organic compounds in the air, we will discuss here inhalable organics of particular concern: aldehydes, benzene, and polycyclic aromatic hydrocarbons (PAHs).

9.4.1 *Formaldehyde*

Formaldehyde (CH_2O) is a reactive molecule that irritates the eyes at quite low concentrations, just over 100 ppb. There is also evidence that formaldehyde is a carcinogen. A great deal of formaldehyde is emitted from industrial sources and transportation, but the greatest exposure to formaldehyde occurs indoors (see Section 9.8).

The main loss processes for formaldehyde are photolysis based on the overhead sun in the United States (the sun breaks it apart):

$$CH_2O + light \rightarrow CHO\cdot + H\cdot \qquad k_1 = 6.9 \times 10^{-5}\frac{1}{s}$$

and chemical reaction with the hydroxyl radical:

$$CH_2O + OH\cdot \rightarrow CHO\cdot + H_2O \qquad k_2 = 4.8 \times 10^{-12}\frac{cm^3}{molec\ sec}$$

The rate constants (k) are given for the individual equations above.

QUESTION 13

Compare the two loss processes for formaldehyde. Calculate the atmospheric lifetimes for formaldehyde as if each loss process was independent. Assume the OH· radical concentration is 2.0×10^6 molec cm^{-3}.

Q13 ANSWER The rate constant for photolysis is much larger, so the rate of photolysis should be faster, resulting in a shorter lifetime:

$$lifetime = \frac{[i]}{loss\ rate}$$

For photolysis:

$$lifetime = \frac{[CH_2O]}{k_1[CH_2O]} = \frac{1}{k_1} = \frac{1}{6.9 \times 10^{-5}\frac{1}{sec}} = 14{,}493\ sec\left(\frac{1\ hr}{3{,}600\ sec}\right) = 4.0\ hr$$

For the chemical reaction:

$$lifetime = \frac{[CH_2O]}{k_2[CH_2O][OH\cdot]} = \frac{1}{k_2[OH\cdot]} = \frac{1}{\left(4.8 \times 10^{-12}\frac{cm^3}{molec\ sec}\right)\left(2.0 \times 10^6\frac{molec}{cm^3}\right)}$$

$$= 104{,}167\ sec\left(\frac{1\ hr}{3{,}600\ sec}\right) = 29\ hr$$

9.4.2 *Benzene*

Benzene (C_6H_6) is derived mainly from crude oil and is widely used in the petroleum, chemical, and manufacturing industries. It is a known carcinogen and is implicated as a causative agent in human leukemia.

The main atmospheric loss process is the reaction with OH radical, but it is slow with a reaction rate constant of 1.22×10^{-12} cm^3 (molec sec)$^{-1}$.

---

*QUESTION 14

Calculate the lifetime of benzene in the atmosphere, assuming the main loss is via the reaction with OH radicals and the OH· concentration is 2.0×10^6 molec cm^{-3}.

9.4.3 *PAHs*

Compounds with C_6H_6 rings fused together (PAHs) are potent carcinogens. PAHs are formed as by-products of carbon fuel combustion. They are mainly adsorbed on soot particles in diesel exhaust and smoke from coal or wood fires. PAH reactivities vary, as does their carcinogenicity. Some examples include naphthalene, anthracene, and fluoranthene (Table 9.1).

TABLE 9.1 Example PAHs, Structures, and Reaction Rate Constants for Reaction With OH Radical

Compound	Structure	k[cm^3/(molec sec)]
Naphthalene	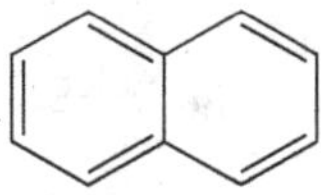	2.16×10^{-11}
Anthracene	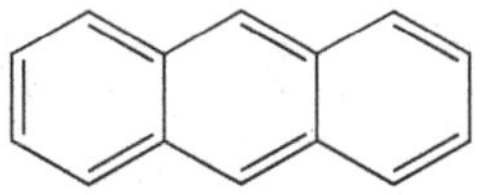	1.7×10^{-11}
Fluoranthene	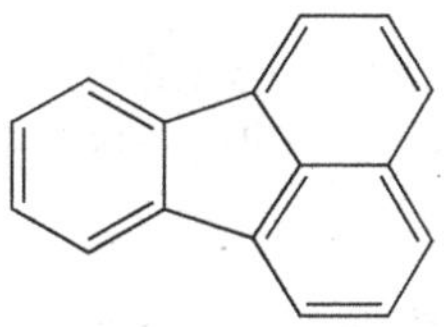	$\sim 5 \times 10^{-11}$

Source: Data from Finlayson-Pitts & Pitts, Jr., *Chemistry of the Upper and Lower Atmosphere*, Academic Press, 2000.

QUESTION 15

Calculate the lifetime of each of the three PAHs above, assuming the main loss is via reaction with OH radicals and the OH· radical concentration is 2.0×10^6 molec cm^{-3}.

TABLE 9.2 Reaction Rate Constants for the Oxidation of Phenanthrene

Oxidant	Reaction Rate Constant (k), $\left(\dfrac{cm^3}{molec\ sec}\right)$
OH·	1.3×10^{-11}
NO$_3$·	1.2×10^{-13}
O$_3$	4.0×10^{-19}

Source: Data from Finlayson-Pitts & Pitts, Jr., *Chemistry of the Upper and Lower Atmosphere*, Academic Press, 2000.

Q15 ANSWER

$$\text{Naphthalene lifetime} = \frac{1}{k[\text{OH}]} = \frac{1}{\left(2.16 \times 10^{-11}\,\frac{cm^3}{molec\ sec}\right)\left(2.0 \times 10^6\,\frac{molec}{cm^3}\right)}$$

$$= 23{,}148\ sec\left(\frac{1\ hr}{3{,}600\ sec}\right) = 6.4\ hr$$

$$\text{Anthracene lifetime} = \frac{1}{k[\text{OH}]} = \frac{1}{\left(1.7 \times 10^{-11}\,\frac{cm^3}{molec\ sec}\right)\left(2.0 \times 10^6\,\frac{molec}{cm^3}\right)}$$

$$= 29{,}412\ sec\left(\frac{1\ hr}{3{,}600\ sec}\right) = 8.2\ hr$$

$$\text{Fluoranthene lifetime} = \frac{1}{k[\text{OH}]} = \frac{1}{\left(5 \times 10^{-11}\,\frac{cm^3}{molec\ sec}\right)\left(2.0 \times 10^6\,\frac{molec}{cm^3}\right)}$$

$$= 10{,}000\ sec\left(\frac{1\ hr}{3{,}600\ sec}\right) = 3\ hr$$

*QUESTION 16

Another PAH is phenanthrene ($C_{14}H_{10}$). Assume the oxidant concentrations are $[\text{OH·}] = 2 \times 10^6$ molec cm^{-3}, $[\text{NO}_3\text{·}] = 5 \times 10^8$ molec cm^{-3}, and $[\text{O}_3] = 7 \times 10^{11}$ molec cm^{-3}.
  (i)   Calculate the atmospheric lifetime of phenanthrene.
  (ii) How much does each oxidant contribute to the overall loss?

9.5 Particulate Matter

Atmospheric particles are perhaps the most serious of air pollution health hazards. They penetrate the lungs, blocking and irritating air passages, and can have toxic effects. Small particles have the greatest health impact because they penetrate most deeply into the lungs and get trapped there. The smallest particles can even pass across the alveoli of the lungs directly into the bloodstream. The EPA regulates particles based on two size classifications: PM$_{10}$ and PM$_{2.5}$, which are PM with aerodynamic diameters ≤10 and 2.5 μm, respectively. Particles larger than 10 μm are more easily eliminated from the human nose and throat.

QUESTION 17

If an adult's breathing rate is 354 L hr^{-1}, how much fine particular matter (in mg) is inhaled by an adult each year? Assume the average PM$_{2.5}$ concentration in the air is 10 μg m^{-3}.

Q17 ANSWER Given that a person breaths 354 L hr^{-1}, we can find the volume of air taken in during 1 yr. If this volume has an average particle concentration of 10 μg m^{-3}, then we can calculate the total mass inhaled (after making sure the units cancel correctly):

$$\frac{354\ L}{hr}\left(\frac{24\ hr}{1\ day}\right)\left(\frac{365\ day}{1\ yr}\right)(1\ yr) = 3.10 \times 10^6\ L$$

$$\frac{10\ \mu g}{m^3}(3.10 \times 10^6\ L)\left(\frac{1,000\ mL}{1\ L}\right)\left(\frac{1\ cm^3}{1\ mL}\right)\left(\frac{1\ m}{100\ cm}\right)^3 = 3.10 \times 10^4\ \mu g$$

$$3.10 \times 10^4\ \mu g\left(\frac{1\ g}{1 \times 10^6\ \mu g}\right)\left(\frac{1,000\ mg}{1\ g}\right) = 31\ mg$$

*QUESTION 18

Different locations around the world have different concentrations of particular matter. For example, the World Health Organization (WHO) data show that the annual average PM$_{2.5}$ concentrations for Beijing, China; Toulouse, France; and Hilo, Hawaii are 85, 12, and 8 μg m^{-3}, respectively (data acquired from Ambient Air Pollution Database, WHO, May 2016[1]). Compare the total mass of PM$_{2.5}$ particles (in mg) inhaled in 1 yr by a child who lives in each one of these cities. Assume that each child breathes an average of 8 L min^{-1}.

*QUESTION 19

Figure 9.4 shows that there is a total of 5,686 thousand tons of PM$_{2.5}$ emitted over the United States in 2019. If the area of the United States is 7.66 $\times$ 10^6 km^2 and most of the PM is within the first 1 km in altitude,

    (i)   What is the average concentration of PM$_{2.5}$ in μg m^{-3} due to emissions?

    (ii)   The average PM$_{2.5}$ concentration across the United States in 2019 was 7.7 μg m^{-3}. [Data from EPA PM (PM$_{2.5}$ Trends)[2].] Compare your answer to the data; what are some reasons for the discrepancy?

QUESTION 20

Assume a polluted atmosphere has an average PM$_{2.5}$ concentration of 170 μg m^{-3} and the particles are 14% sulfur by weight. An average person respires 354 L hr^{-1} and retains 50% of the particles smaller than 1 μm in the lungs. How much sulfur accumulates in the lungs in 1 yr if 75% of the particulate mass is contained in particles smaller than 1 μm?

---

[1] World Health Organization, *Air Pollution*, available at https://www.who.int/gho/phe/outdoor_air _pollution/exposure/en/

[2] U.S. Environmental Protection Agency, *Particulate Matter (PM2.5) Trends*, available at https://www.epa .gov/air-trends/particulate-matter-pm25-trends

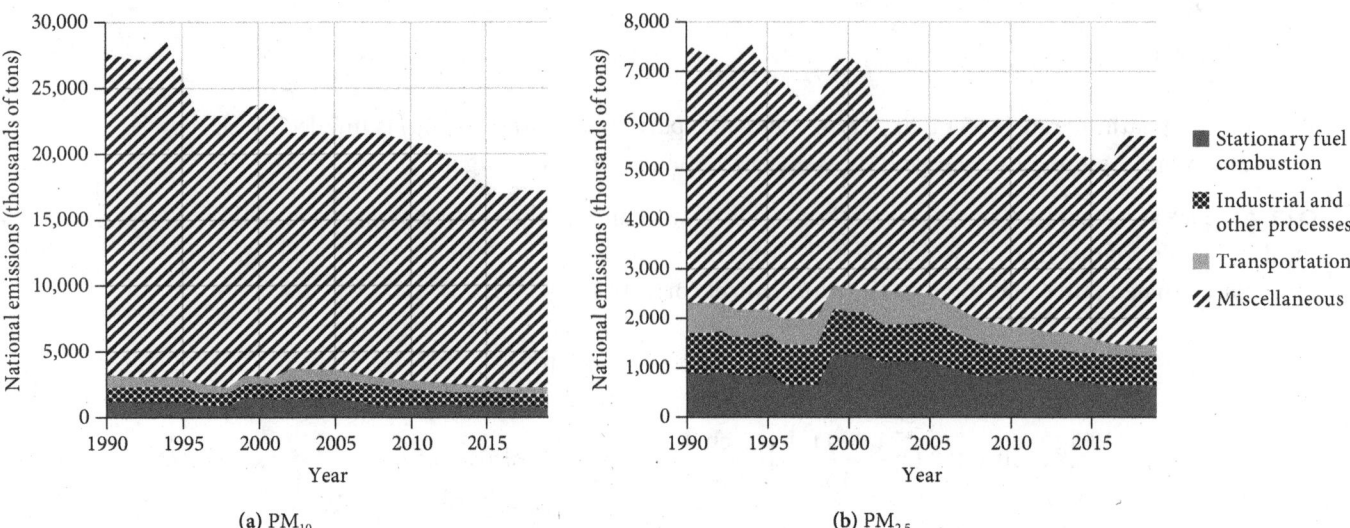

Figure 9.4 Trend in primary emissions of (a) course PM_{10} and (b) fine $PM_{2.5}$ particles in the United States. "Miscellaneous" includes road dust, fertilizers, and livestock dust.
Source: Data from the U.S. EPA National Emissions Inventories, *Air Pollution Emission Trends Data, average annual emissions,* available at https://www.epa.gov/air-emissions-inventories/air-pollutant-emissions-trends-data.

Q20 ANSWER First, find the mass of $PM_{2.5}$ inhaled in 1 yr (like we did in Q17):

$$\frac{170\mu g}{m^3}\left(\frac{1\ m}{100\ cm}\right)^3\left(\frac{1\ cm^3}{1\ mL}\right)\left(\frac{1,000\ mL}{1\ L}\right)\left(\frac{354\ L}{1\ hr}\right)\left(\frac{24\ hr}{1\ day}\right)\left(\frac{365\ day}{1\ yr}\right)(1\ yr)\left(\frac{1\ g}{1\times10^6\ \mu g}\right)\left(\frac{1,000\ mg}{1\ g}\right)=527\ mg$$

Amount of particles retained smaller than 1 μm

$$(0.50)(527\ mg)=263.5\ mg$$

Amount of mass in particles smaller than 1 μm

$$(0.75)(263.5\ mg)=197.6\ mg$$

Amount of sulfur in the particles

$$(0.14)(197.6\ mg)=27.7\ mg\ S$$

9.6 NO_x and VOCs

Nitrogen oxides (NO_x) and VOCs in outdoor air have minor direct health effects, but they are the main ingredients in the formation of photochemical smog (see Chapter 8). Controlling smog (and thus ozone) formation requires reducing emissions of NO_x and VOCs. Trends in U.S. air emissions of NO_x and VOCs are shown in Figure 9.5.

NO_x is emitted as NO due mainly to high-temperature combustion in the presence of air.

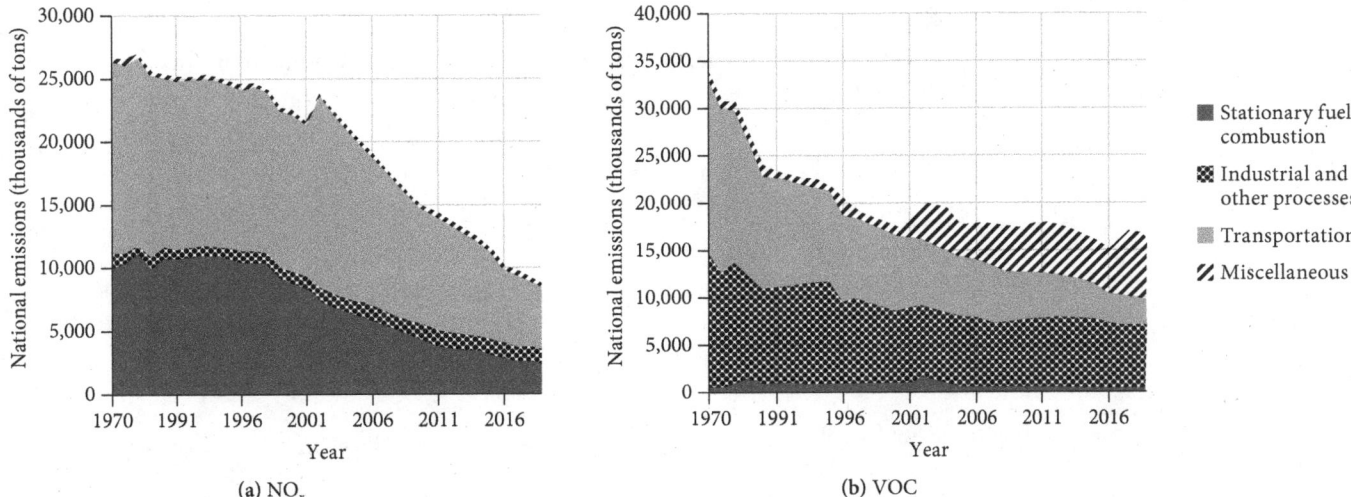

Figure 9.5 (a) Nitrogen oxide (NO$_x$) and (b) VOC emission trends in the United States, 1970–2019. "Miscellaneous" includes fertilizers, livestock waste, and newly submitted sources over time.
Source: Data from the U.S. EPA National Emissions Inventories, *Air Pollution Emission Data, average annual emissions,* available at https://www.epa.gov/air-emissions-inventories/air-pollutant-emissions-trends-data.

QUESTION 21

In the overall reaction mechanism that produces NO from N$_2$ and O$_2$, the rate-limiting step is the reaction between atomic oxygen and molecular nitrogen to produce nitric oxide and atomic nitrogen.

(i)   Write a chemical equation that describes this rate-limiting step.

(ii) Write a rate law to describe this rate-limiting (slowest) step.

(iii) If this mechanism has an activation energy of 315 kJ mol^{-1} and its rate constant is 9.7 × 10^{10} L (mol sec)$^{-1}$ at 800°C, what is the rate constant at 300°C?

Q21 ANSWER

(i)   Following the description given in the question:

$$O + N_2 \rightarrow NO\cdot + N\cdot$$

(ii) The rate law depends on the concentration of reactants:

$$\text{rate} = k[O][N_2]$$

(iii) Rate constants are temperature-dependent and they are related through the Arrhenius equation:

$$k = Ae^{-\frac{E_a}{RT}}$$

Before we even begin the mathematics, as you decrease the temperature, the rate will be slower, so we expect the rate constant to be smaller. The activation energy (E_a), frequency factor (A), and gas constant (R) will be the same at both temperatures, but the rate constant (k) will change as temperature (T) changes. Ultimately, we will solve for k_2, given T_2, but first we must solve for A:

$$A = \frac{k}{e^{\left(-\frac{E_a}{RT}\right)}}$$

$$A = \frac{9.7 \times 10^{10} \frac{L}{mol\ sec}}{e^{-\left(\frac{\left(315\frac{kJ}{mol}\right)\left(\frac{1,000\ J}{1\ kJ}\right)}{8.314\frac{J}{mol\ K}(1,073\ K)}\right)}} = 2.098 \times 10^{26} \frac{L}{mol\ sec}$$

Now use A to solve for k at 300°C:

$$k = Ae^{-\frac{E_a}{RT}}$$

$$k = \left(2.098 \times 10^{26} \frac{L}{mol\ sec}\right)e^{-\left(\frac{\left(315\frac{kJ}{mol}\right)\left(\frac{1,000\ J}{1\ kJ}\right)}{8.314\frac{J}{mol\ K}(573\ K)}\right)} = 0.0040 \frac{L}{mol\ sec}$$

*QUESTION 22

What is the rate constant for this system at 1,000°C?

9.7 Ozone

While anthropogenic gases are destroying ozone in the stratosphere (see Chapter 10), they are helping to create ozone in the troposphere via photochemical smog (see Chapter 8). Though ozone in the stratosphere protects us from the harmful effects of UV rays, ozone at the ground level is quite harmful, producing cracks in rubber, destroying plants, and causing respiratory distress and eye irritation in humans. Ozone is a strong oxidant, reacting rapidly with electron-rich molecules. Such molecules are abundant in rubber, in the photosynthetic apparatus of green plants, and in the lining of the lung's air passages.

Ozone is found throughout the world, but with an increase in the precursors (NO_x and VOCs) due to anthropogenic activity, there has been an increase in photochemical ozone across the globe (Fig. 9.6). In addition, ozone can be transported for long distances; thus, it has become a regional air pollution problem.

Los Angeles is ideally suited for smog production because the air is trapped by high mountains and onshore breezes from the ocean. Regulations have become steadily more stringent, and the air quality has improved (Fig. 9.7), despite the steadily increasing population and transportation density. California's emission regulation has benefitted the rest of the country as well because carmakers have had to improve fleet emissions in order to sell cars in California.

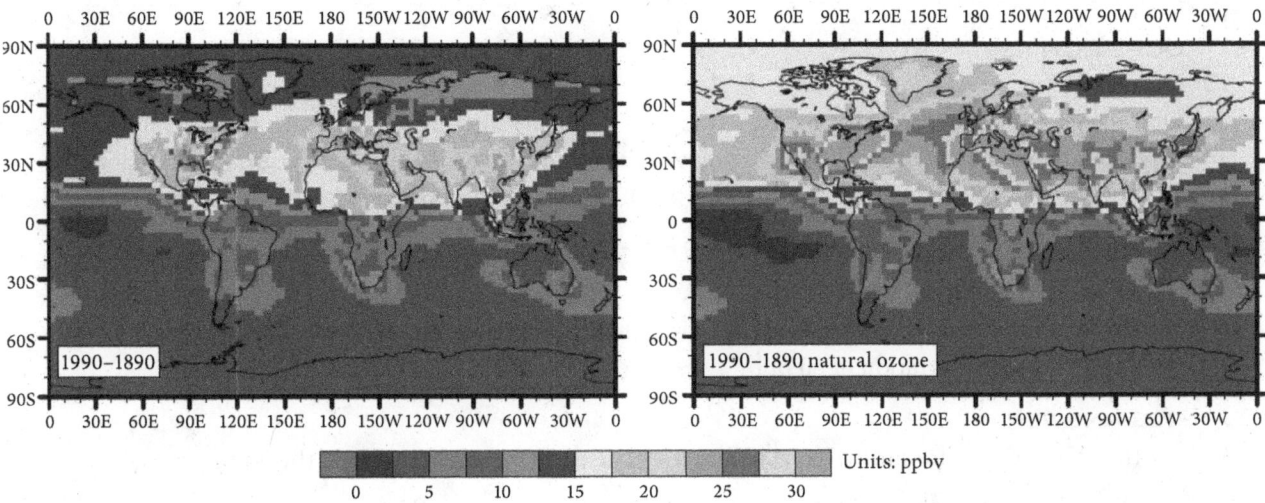

Figure 9.6 Surface ozone increase (in ppb) from 1890 to 1990 from model simulations across the Earth with (right) and without (left) anthropogenic emissions.
Source: Adapted from Lamarque et al., "Tropospheric Ozone Evolution Between 1890 and 1990," 2005, *Journal of Geophysical Research*, Vol. 110, No. D08304, https://doi.org/10.1029/2004JD005537.

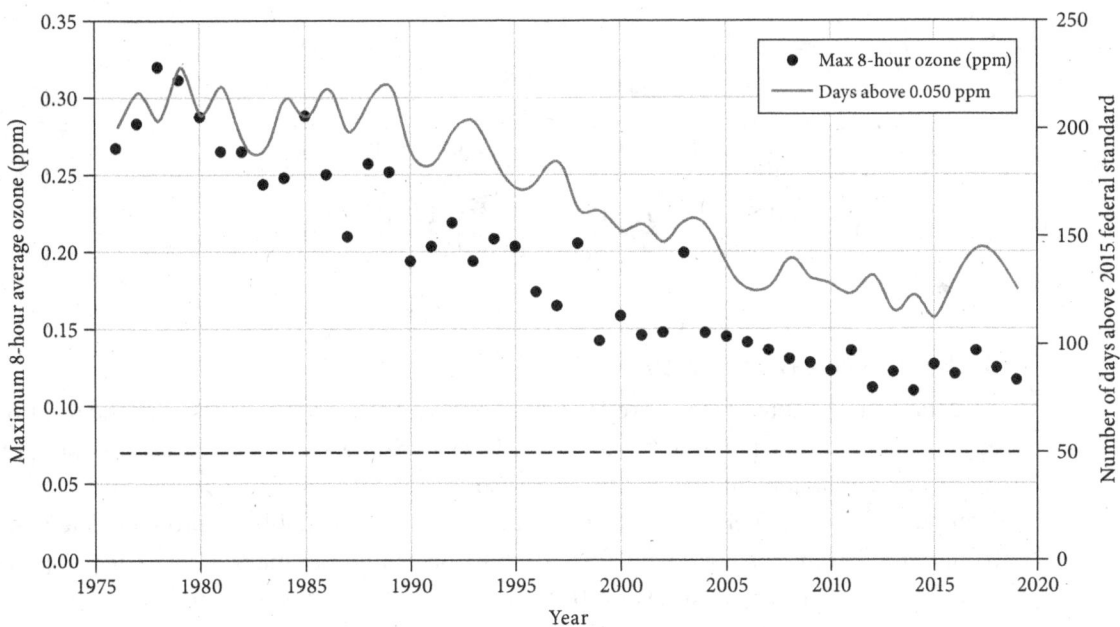

Figure 9.7 Maximum 8-hr average ozone measured in the South Coast Air Basin and the number of days the federal standard (0.070 ppm) was exceeded in each year. The California standard is also 0.070 ppm, which is indicated on the chart.
Source: Data from the South Coast Air Quality Management District (AQMD), *Air Quality Investigations,* available at http://www.aqmd.gov/home/air-quality/air-quality-data-studies/historic-ozone-air-quality-trends.

QUESTION 23

One way to eliminate ozone from an air sample is to chemically react it away. Ambient air flows through a filter coated with an aqueous solution of sodium thiosulfate. The thiosulfate ion reacts with ozone to produce the tetrathionate ion ($S_4O_6^{2-}$) and molecular oxygen:

$$S_2O_3^{2-} + O_3 + H^+ \rightarrow S_4O_6^{2-} + O_2 + H_2O$$

(i)   Balance this equation.

(ii) How much thiosulfate is needed to react with a 20.0-L sample of air at 1.0 atm and 25°C with an ozone concentration of 60 ppb?

Q23 ANSWER

(i)   You need more sulfur, oxygen, and hydrogen on the reactant side of the equation:

$$2\,S_2O_3^{2-} + O_3 + 2\,H^+ \rightarrow S_4O_6^{2-} + O_2 + H_2O$$

(ii) All we need is the stoichiometry from the balanced equation to relate the amount of ozone to the amount of thiosulfate, but this relationship only works in moles. First, let us figure out how many moles of ozone are present in the air sample. Because ppb is a mole ratio, we also need to know how many moles of air are present in the sample.

$$n_{O_3} = \chi_{O_3} n_T$$

$$n_T = \frac{PV}{RT} = \frac{(1.0 \text{ atm})(20.0 \text{ L})}{\left(0.08206\,\dfrac{\text{L atm}}{\text{mol K}}\right)(298 \text{ K})} = 0.8179 \text{ mol air}$$

$$n_{O_3} = \left(\frac{60 \text{ mol } O_3}{1 \times 10^9 \text{ mol air}}\right)(0.8179 \text{ mol air}) = 4.907 \times 10^{-8} \text{ mol } O_3$$

$$4.907 \times 10^{-8} \text{ mol } O_3\left(\frac{2 \text{ mol } S_2O_3^{2-}}{1 \text{ mol } O_3}\right)\left(\frac{112.132 \text{ g } S_2O_3^{2-}}{1 \text{ mol } S_2O_3^{2-}}\right)\left(\frac{1 \times 10^6\,\mu g}{1 \text{ g}}\right) = 11\,\mu g\, S_2O_3^{2-}$$

9.7.1 Controlling Ozone Precursors

As detailed in Chapter 8, a net production of tropospheric ozone occurs with the reaction of NO_x and VOCs in the presence of sunlight. Limiting ozone formation, as a secondary pollutant, requires emission controls for the precursors.

Combustion of any fuel in air will produce NO_x as an inevitable by-product. The main contributors are transportation and fuel combustion from stationary sources (e.g., power plants and space heating). Combustion from transport also accounts for much of the atmospheric CO and hydrocarbons, especially in urban areas, because automotive exhaust contains substantial quantities of unburned gases. In addition, some of the volatile automotive fuel escapes before combustion due to evaporation. On the other hand, substantial quantities of VOCs are emitted by vegetation. Plants release a variety of hydrocarbons; in particular, the terpenes have a C=C double bond; thus, they react rapidly with the hydroxyl radical (OH·) (see Chapter 8). In some areas, vegetation is a major contributor to the reactive hydrocarbons responsible for smog. In most places, the key to ozone reduction is control of NO· emissions; hydrocarbons are generally too abundant to be brought low enough to be the limiting reagent.

Because automotive transportation was a large source of NO_x and VOCs, the catalytic converter is quite an effective tool for reducing emissions. The three-way catalytic converter, so named because it reduces emissions of hydrocarbons, CO, and NO·, uses reduction–oxidation chemistry to remove pollutants from the exhaust.

The reduction chamber includes the following reactions:

$$Hydrocarbon + H_2O \rightarrow H_2 + CO$$

$$2NO\cdot + 2H_2 \rightarrow N_2 + 2H_2O$$

The oxidation chamber includes the following reactions:

$$2CO + O_2 \rightarrow 2CO_2$$

$$Hydrocarbon + 2O_2 \rightarrow CO_2 + 2H_2O$$

QUESTION 24

In the first stage, in the reduction chamber, NO reacts with H_2 to yield N_2 and water. Hydrogen comes from the reaction of unburnt hydrocarbons. Identify the element that is being reduced in this single reaction.

Q24 ANSWER Nitrogen is reduced. It goes from an oxidation state of +2 (because O is –2) in NO· on the reactant side to an oxidation state of 0 due to the stability of N_2. Moving from +2 to zero is a gain of electrons, which is reduction.

---

*QUESTION 25

In the second stage, in the oxidation chamber, unburnt hydrocarbons and CO mix with air to form carbon dioxide. Identify the element being oxidized.

9.8 Indoor Air Pollution

In many cases, individual exposure to air pollutants is greater indoors than outdoors. Because of the enclosed environment indoors, pollutant levels can easily build without proper dilution or ventilation. In addition, many people spend up to 90% of their time in a building or transportation vehicle. This section will focus on the two most common, and preventable, indoor pollutants—formaldehyde and carbon monoxide—the health effects of each have been mentioned above (see Sections 9.2 and 9.4.1).

9.8.1 *Formaldehyde*

Formaldehyde ($H_2C{=}O$) is a stable product formed from the oxidation of hydrocarbons (see Chapter 8). Even in a clean environment, there can be ~10 ppb of formaldehyde present. Indoor concentrations are typically in the 5–20 ppb range, but it can get very high, ~1,000 ppb near emission hotspots. Sources of formaldehyde indoors include cigarette smoke and synthetic materials. These materials, such as foam insulation and particleboard, have resins formed by combining formaldehyde with other organic substances. In addition, formaldehyde itself is used in the dyeing and gluing of carpets. Formaldehyde evaporates easily; its emission increases with increasing temperature and relative humidity. The release of formaldehyde then decreases with the age of the product.

QUESTION 26

The detection threshold for formaldehyde is 100 ppb. Would we detect it in a 4 m × 6 m × 3 m room with new carpet (4 m × 6 m) if the pollutant off-gases at a rate of 57.2 μg (m^2 hr)$^{-1}$ over the first 24 hr? The ambient temperature and pressure in the room are 27°C and 1.0 atm, respectively. Assume there is no ventilation so that formaldehyde can accumulate over this time period.

Q26 ANSWER First we must determine how much formaldehyde is released in the 24-hr period then we have to convert the units into a ppb mixing ratio to compare it to the detection threshold:

$$\frac{57.2 \, \mu g}{m^2 \, hr}(24 \, hr)(4 \, m \times 6 \, m) = 32{,}947 \, \mu g \text{ released}$$

In order to find the mixing ratio, convert the amount of formaldehyde released into moles:

$$32{,}947 \, \mu g \left(\frac{1 \, g}{1 \times 10^6 \, \mu g}\right)\left(\frac{mol}{30.03 \, g}\right) = 1.097 \times 10^{-3} \, mol$$

Now determine how many moles of air are in the room. If we can calculate the volume in liters, then we can use the ideal gas law to relate the volume to moles at the temperature and pressure that were given:

$$4 \, m \times 6 \, m \times 3 \, m = 72 \, m^3$$

$$72 \, m^3 \left(\frac{100 \, cm}{1 \, m}\right)^3\left(\frac{1 \, mL}{1 \, cm^3}\right)\left(\frac{1 \, L}{1{,}000 \, mL}\right) = 72{,}000 \, L \text{ air}$$

$$n_T = \frac{PV}{RT} = \frac{(1.0 \, atm)(72{,}000 \, L)}{\left(0.08206 \frac{L \, atm}{mol \, K}\right)(300 \, K)} = 2{,}925 \, mol \text{ air}$$

Now make the ratio of moles:

$$\frac{1.097 \times 10^{-3} \, mol \text{ formaldehyde}}{2{,}925 \, mol \text{ air}} = 3.75 \times 10^{-7}$$

Convert to ppb for comparison:

$$(3.75 \times 10^{-7})(1 \times 10^9) = 375 \text{ ppb}$$

YES, we can detect formaldehyde under these conditions (375 ppb > 100 ppb).

*QUESTION 27

What mass of formaldehyde must be released from building materials to produce 0.30 ppm in a room with the dimensions of 5 m × 7 m × 3 m at 27°C and 1.0 atm?

9.8.2 *Carbon Monoxide*

Carbon monoxide is a product of incomplete combustion of carbon-containing fuels such as wood, gasoline, kerosene, or gas (used in stoves and heaters). Cigarettes are also a source of CO. Average outdoor levels of CO are 1–20 ppm, while indoor levels can easily be 50–90 ppm. There are commercial carbon monoxide detectors to monitor indoor levels of the pollutant.

QUESTION 28

A concentration of 800 ppm of CO is considered lethal to humans. Calculate the minimum mass of CO in grams that would create a lethal environment in a closed room that is 17.6 m × 8.80 m × 2.64 m. The temperature and pressure are 20.0°C and 1.00 atm, respectively.

Q28 ANSWER Using the mixing ratio of 800 ppm, we can figure out the amount of CO present:

$$n_{CO} = \chi_{CO} n_r$$

$$800 \text{ ppm} = \frac{800 \text{ mol CO}}{1 \times 10^6 \text{ mol air}}$$

The moles of air total can be calculated from the volume of the room:

$$17.6 \text{ m} \times 8.8 \text{ m} \times 2.64 \text{ m} = 409 \text{ m}^3$$

$$409 \text{ m}^3 \left(\frac{100 \text{ cm}}{1 \text{ m}}\right)^3 \left(\frac{1 \text{ mL}}{1 \text{ cm}^3}\right) \left(\frac{1 \text{ L}}{1,000 \text{ mL}}\right) = 408,883 \text{ L air}$$

$$n_T = \frac{PV}{RT} = \frac{(1.00 \text{ atm})(408,883 \text{ L})}{\left(0.08206 \frac{\text{L atm}}{\text{mol K}}\right)(293 \text{ K})} = 17,006 \text{ mol air}$$

Now find moles of CO and then use molar mass to find grams:

$$\frac{800 \text{ mol CO}}{1 \times 10^6 \text{ mol air}} (17,006 \text{ mol air}) = 13.6 \text{ mol CO} \left(\frac{28.01 \text{ g}}{1 \text{ mol}}\right) = 381 \text{ g CO}$$

*QUESTION 29

Fazlzadeh et al. measured CO inside and outside of 68 cafés in Iran. The average indoor concentration was 24.8 ppm, while the average outdoor concentration was 2.7 ppm.[3]

(i)   If the average café had a volume of 394m³, what is the mass of carbon monoxide inside a café at 25°C and 1.00 atm?

(ii)   If a person breathes about 354 L hr⁻¹, how many CO molecules are inhaled in a 2-hr period? What is the mass of CO inhaled during that time period? [Assume the same temperature and pressure as in part (i).]

9.9 Conclusions

Understanding general sources and concentrations of air pollutants in the atmosphere is important to ameliorate both the health and environmental consequences of the pollution. Here we focused on six criteria pollutants—CO, SO_2, toxic organics, PM, NO_x, and VOCs—because they cause the most health and environmental harm and need to be controlled. Knowing how to calculate the concentrations of these pollutants in the air and think about the transport of pollutants through the atmosphere with a one-box model can help us predict which areas in the world the air pollutants will impact most and how long they might stay in the atmosphere. Some of the air pollutants are more localized in their impact, whereas others are stable enough to move to different regions and around the world.

---

[3] Fazlzadeh et al., *Atmospheric Pollution Research*, Vol. 6, No. 4, 2015, pp. 550–555.

10

STRATOSPHERIC OZONE SHIELD

10.1 Background About the Stratospheric Ozone Shield

The stratospheric ozone (O_3) layer screens out the highest energy ultraviolet (UV) photons, preventing damage to the plants and animals below. Because this high-energy radiation is so harmful to biological molecules, life on Earth could not have evolved as it did without the protective ozone layer. Recent human activities, such as the injection of nitric oxide (NO) into the stratosphere via high-altitude aviation and sources of chlorine radicals (Cl) via chloro-fluorocarbons (CFCs) used as propellants and refrigerants, have threatened this protective layer with catalytic destruction cycles. The resulting stratospheric ozone layer destruction is dangerous because it allows more high-energy UV light to penetrate the atmosphere to the surface, where it can cause DNA damage in humans, animals, and plants. The basic chemical mechanisms of catalytic ozone destruction were elucidated in the 1970s, and governments acted quickly to enact an international treaty, the Montreal Protocol, to ban the use of CFCs and later other ozone-depleting chemicals. A major impetus for this rapid international political action was the discovery of a dramatic hole in the ozone layer over Antarctica in the austral spring of 1985. The extent and magnitude of this Antarctic ozone hole were not predicted with known gas-phase catalytic loss mechanisms, as its formation was hastened by meteorological conditions enabling additional chemical reactions on the surface of polar stratospheric clouds to enhance the ozone loss. In the sections that follow, we first describe the temperature and pressure structure of the atmosphere and chemical reactions that enable O_3 to form, and then discuss the loss processes of O_3.

10.2 Structure of the Atmosphere

The surface pressure of the atmosphere is a consequence of the weight of all of the gases of the atmosphere pressing down on the earth. We first consider the total mass of the atmosphere.

QUESTION 1

With its average composition of 78% N_2, 21% O_2, <1% Ar, and other trace gases, the average molecular weight of air is 29 g mol^{-1}. Based on a total 1.8×10^{20} moles (mol) of gas in the earth's atmosphere, what does the atmosphere weigh?

Q1 ANSWER Use the molecular weight as a conversion factor to convert from total moles to total grams:

$$1.8 \times 10^{20} \text{ mol air} \times \frac{29 \text{ g air}}{\text{mol air}} = 5.2 \times 10^{21} \text{ g air}$$

*QUESTION 2

If the composition of the atmosphere was only N_2 and did not include any other gases, what would the atmosphere weigh? What is the percent difference between assuming only nitrogen gas in the atmosphere and accounting for the true mixture?

QUESTION 3

Now that we know the mass of the atmosphere, use the mixture of gases from Q1 to convert this to a surface pressure. How good is this approximation of total moles and molecular weight? Useful information: Earth has a radius of 6,378 km and the observed average surface pressure is 14 psi (pounds per square inch); 2.2 lb = 1 kg, 2.54 cm = 1 in, surface area = $4\pi r^2$.

Q3 ANSWER Surface pressure can be expressed as weight per unit surface area (e.g., 14 psi), so we need to first determine the earth's surface area and then divide the total mass by this. Then, through unit conversions, we can compare the result to 14 psi.

For a sphere of radius r, surface area = $4\pi r^2$

$$\frac{\text{mass}}{\text{area}} = \frac{5.2 \times 10^{21} \text{ g air}}{4\pi(6{,}378 \text{ km})^2} \times \frac{\dfrac{2.2 \text{ lb}}{1{,}000 \text{ g}}}{\left(\dfrac{1 \text{ in}}{0.0254 \text{ m}}\right)^2 \times \left(\dfrac{1{,}000 \text{ m}}{1 \text{ km}}\right)^2} = 14.4 \text{ psi}$$

This is a pretty good approximation of the total mass of the atmosphere and agrees with observation within 0.4 psi.

*QUESTION 4

Now calculate the surface pressure assuming only nitrogen gas from Q2. Useful information: Earth has a radius 6,378 km and observed average surface pressure is 14 psi; 2.2 lb = 1 kg, 2.54 cm = 1 in, surface area = $4\pi r^2$.

10.2.1 *Total Ozone: Dobson Units*

The stratospheric ozone layer occurs in the lower stratosphere, with the highest O_3 concentrations between 20 and 30 km (as we will explore further below). However, if all the ozone layers were compressed down to sea level and spread evenly around the Earth at 1 atm pressure and 25°C temperature, this layer would be only 3-mm thick. This effective thickness is reported in Dobson units (DUs), named after the pioneering experimentalist in the 1920s who developed an instrument to measure total ozone column. One DU is equal to 0.01-mm-thick

---

* Answers to starred questions can be found at the end of the book.

O_3 layer at standard temperature and pressure. A typical average ozone column around the globe (i.e., not in the Antarctic ozone hole) is about 300 DU.

The number of moles of a gas (n) is related to volume (V) via the gas law:

$$PV = nRT \qquad (10.1)$$

where R, the gas constant, is 0.08206 L atm mol^{-1} K^{-1}, when P, V, and T are expressed in liters (L), atmospheres (atm), and K ($= 273 + °C$), respectively. Under standard conditions ($P = 1$ atm, $T = 298$ K), the volume per mole of gas is 22.4 L. The volume of a thin band around the earth is the product of its thickness, 3 mm, times the earth's area, $4\pi r^2$, where r is the earth's radius, which is 6,378 km given above in Q1.

QUESTION 5

How many moles of ozone does the stratosphere hold, assuming the average column of 300 DU is spread around the entire globe?

Q5 ANSWER To use the ideal gas law, we need to obtain the volume in liters, so we will convert length into centimeters, and divide the resulting volume by 1,000 cm^3 L^{-1}. At standard T and P, n is proportional to V, so moles of ozone can be obtained by dividing the calculated volume by 22.4 L, the volume of 1 mol of a gas. Putting all this together we have:

$$V = 4\pi r^2 \times \text{thickness} = 4\pi(6{,}378 \text{ km})^2 \times 3.00 \text{ mm} \times \frac{0.1 \text{ cm}}{\text{mm}} \times \left(\frac{1 \times 10^5 \text{ cm}}{\text{km}}\right)^2 \times \frac{\text{L}}{1{,}000 \text{ cm}^3} = 1.53 \times 10^{15} \text{ L}$$

$$n = \frac{V}{22.4 \text{ L mol}^{-1}} = \frac{1.53 \times 10^{15} \text{ L}}{22.4 \text{ L mol}^{-1}} = 6.83 \times 10^{13} \text{ mol ozone}$$

---

*QUESTION 6

How many moles of ozone does the stratosphere hold if the average column of 100 DU (the column of ozone measured in the ozone hole) is spread around the earth?

10.2.2 Scale Height of the Atmosphere

The atmospheric mass is not distributed equally with height; rather, the pressure of the atmosphere decreases exponentially with altitude (Fig. 10.1), a consequence of the balance between buoyant forces and the gravitational pull of the Earth's atmosphere toward the surface. The temperature structure of the atmosphere is more complex, with a minimum at the tropopause and maximum at the stratopause, due to a combination of heating of the surface and chemical reactions that generate heat at higher altitudes (Fig. 10.1).

The exponential decay of pressure with height has a "scale height" of 8 km. This means the exponential decay of pressure in the Earth's atmosphere can be described by the following equation:

$$P(z) = P(0)e^{-\frac{z}{H}} \qquad (10.2)$$

where $P(0)$ is the surface pressure, $P(z)$ is the pressure at altitude z, and H is the scale height (8.0 km). Because the number density (the number of molecules of air per unit volume of air)

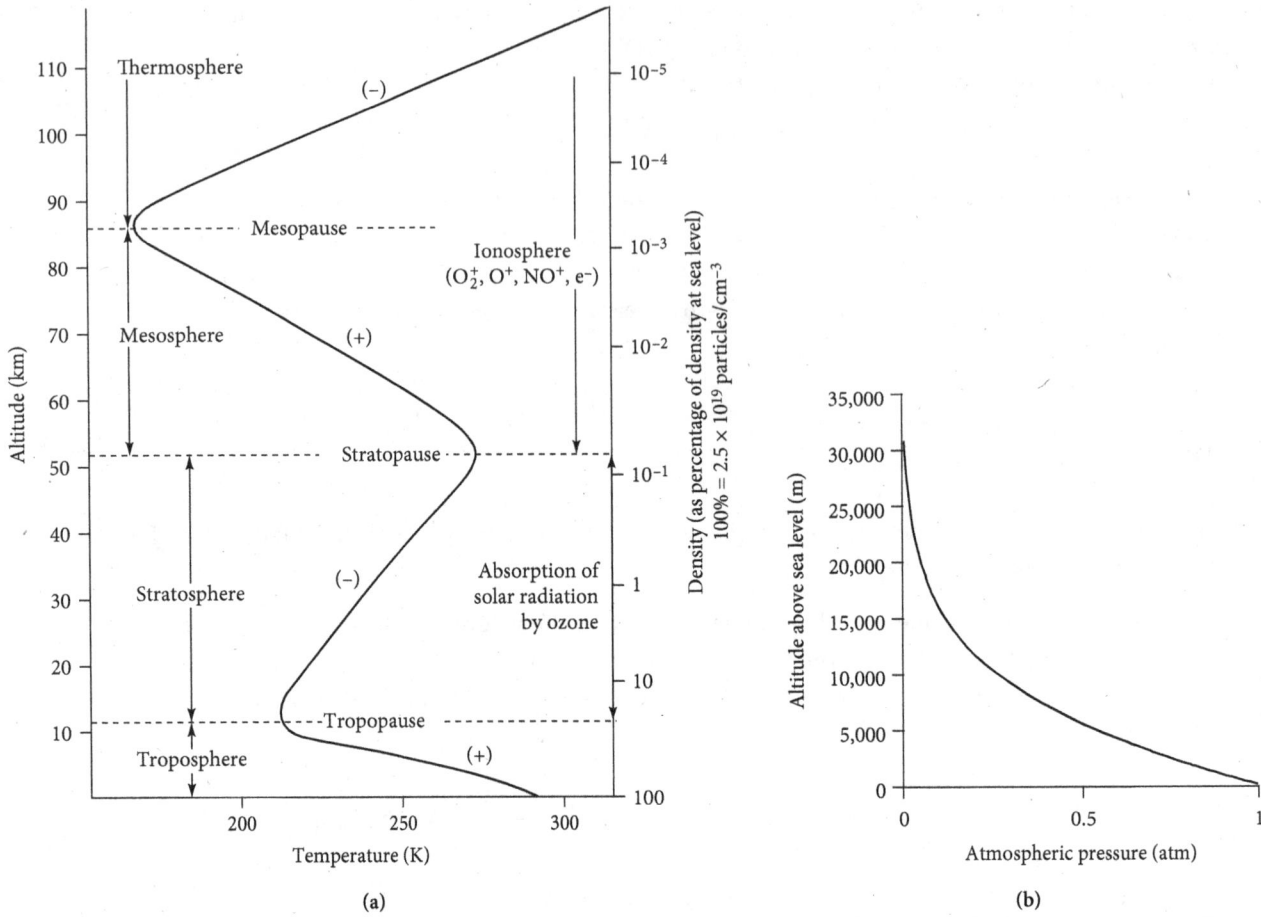

Figure 10.1 Temperature and pressure structure of the atmosphere.
Source: Adapted from Spiro et al., *Chemistry of the Environment, 3rd Ed.*, University Science Books,
© 2012, all rights reserved.

is directly proportional to pressure, to a first approximation, the same equation can be used to determine how total gas concentration [M] decreases with altitude. Use this equation, and the ideal gas law equation of $PV = nRT$ [see Eq. (10.1)], to answer the following questions.

QUESTION 7

First, we will calculate the surface total number density of air (in units of molecules per cubic centimeter, molec cm^{-3}) using the ideal gas law. For average surface temperature and pressure, use 298 K and 1 atm.

Q7 ANSWER Rearrange $PV = nRT$ to solve for number density and use Avogadro's number to convert from moles to molecules:

$$\frac{n}{V} = \frac{P}{RT} = \frac{1\ atm}{\left(0.08206\ \frac{L\ atm}{mol\ K}\right)(298\ K)} \times \frac{6.023 \times 10^{23}\ molec}{1\ mol} \times \frac{1\ L}{1 \times 10^3\ cm^3} = 2.46 \times 10^{19}\ molec\ cm^{-3}$$

*QUESTION 8

At 10 km in altitude, the temperature is approximately 215 K and the pressure is 0.15 atm. Calculate the total number density of air under these conditions using the ideal gas law.

*QUESTION 9

In a colder climate, the lower surface temperature can lead to a higher surface concentration. At $P = 1.0$ atm and $T = -5.0°C$, what is the number density of air?

QUESTION 10

What is the number density of O_2 gas at $z = 0.0$, 10.0, 20.0, 30.0, 40.0, and 50.0 km, assuming a constant fraction of 21% and using the above calculated surface total number density of air of $[M] = 2.46 \times 10^{19}$ molec cm^{-3}?

Q10 ANSWER Use the scale height equation to determine the total concentration of air at each altitude, then multiply by 0.21 to calculate the $[O_2]$ at each altitude:

At surface:

$$[O_2](0 \text{ km}) = 0.21 \times [M] = 0.21 \times 2.46 \times 10^{19} \text{ molec cm}^{-3} = 5.17 \times 10^{18} \text{ molec cm}^{-3}$$

$$[M](z) = [M](0)e^{-\frac{z}{H}}$$

$$[M](10 \text{ km}) = (2.46 \times 10^{19} \text{ molec cm}^{-3})e^{-\frac{10.0\text{km}}{8.0\text{km}}}$$

$$[M](10 \text{ km}) = 7.05 \times 10^{18} \text{ molec cm}^{-3}$$

$$[O_2](10 \text{ km}) = 0.21 \times (7.05 \times 10^{18} \text{ molec cm}^{-3}) = 1.5 \times 10^{18} \text{ molec cm}^{-3}$$

Similarly:

$$[M](20 \text{ km}) = (2.46 \times 10^{19} \text{ molec cm}^{-3})e^{-\frac{20.0\text{km}}{8.0\text{km}}}$$

$$[M](20 \text{ km}) = 2.02 \times 10^{18} \text{ molec cm}^{-3}$$

$$[O_2](20 \text{ km}) = 0.21 \times (2.02 \times 10^{18} \text{ molec cm}^{-3}) = 4.2 \times 10^{17} \text{ molec cm}^{-3}$$

$$[O_2](30 \text{ km}) = 1.2 \times 10^{17} \text{ molec cm}^{-3}$$

$$[O_2](40 \text{ km}) = 3.5 \times 10^{16} \text{ molec cm}^{-3}$$

$$[O_2](50 \text{ km}) = 1.0 \times 10^{16} \text{ molec cm}^{-3}$$

*QUESTION 11

What is the number density of N_2 gas at $z = 0.0$, 10.0, 20.0, 30.0, 40.0, and 50.0 km, assuming a constant fraction of 78% and using the above calculated surface total number density of air of $[M] = 2.46 \times 10^{19}$ molec cm^{-3}?

To understand the chemistry occurring in a particular region of the atmosphere, we typically want concentration or number densities of gases to determine reaction kinetics. However, to assess the bigger picture mass balance questions, we often wish to describe atmospheric gases in terms of the total mass. We often need to convert between concentrations and total mass units; an example of this follows.

QUESTION 12

How much carbon? The atmosphere currently contains 418 parts per million by volume (ppmv) CO_2. How many gigatons of carbon is this? Note: You can again use the fact that the atmosphere contains total 1.8×10^{20} mol of air.

Q12 ANSWER Use the atomic weight of carbon to convert between moles of CO_2 and mass of C, and use the total number of moles in the atmosphere to scale up the mixing ratio to a total carbon burden:

$$\frac{418 \text{ mol } CO_2}{10^6 \text{ mol air}} \times \frac{1 \text{ mol C}}{1 \text{ mol } CO_2} \times \frac{12.011 \text{ g C}}{\text{mol}} \times \frac{1 \text{ Pg}}{10^{15} \text{ g}} \times \frac{1 \text{ Gt}}{1 \text{ Pg}} \times 10^{20} \text{ mol air} = 904 \text{ Gt C}$$

*QUESTION 13

Currently, the atmosphere contains approximately 1,892 parts per billion by volume (ppbv) methane. How many gigatons of carbon is this? Note: You can again use the fact that the atmosphere contains total 1.8×10^{20} mol of air.

10.3 Transmission of UV Light Through Ozone

The attenuation of light by absorbing gases (such as ozone) can be described by the Beer–Lambert law, which shows the exponential decay of transmitted light with length through the sample, at a given wavelength:

$$T = \frac{l}{l(0)} = e^{-\varepsilon l} \tag{10.3}$$

where ε is the absorptivity or extinction coefficient (units: inverse length) and l is the path length of light through the sample. The effect of a small decrease in ozone layer thickness can be obtained by differentiating the above equation, and then dividing by (10.3) to obtain:

$$\frac{dT}{T} = -\varepsilon dl = -\varepsilon l \frac{dl}{l} \tag{10.4}$$

so that a fractional decrease in layer thickness can be readily converted to the fractional increase in transmitted radiation, with proportionality constant εl.

An alternative but an equivalent formulation of the Beer–Lambert law can be written in terms of the column density of gas (c, units molec cm^{-2}) and absorption cross section (σ, units cm^2):

$$T = \frac{l}{l(0)} = e^{-\sigma c} \tag{10.5}$$

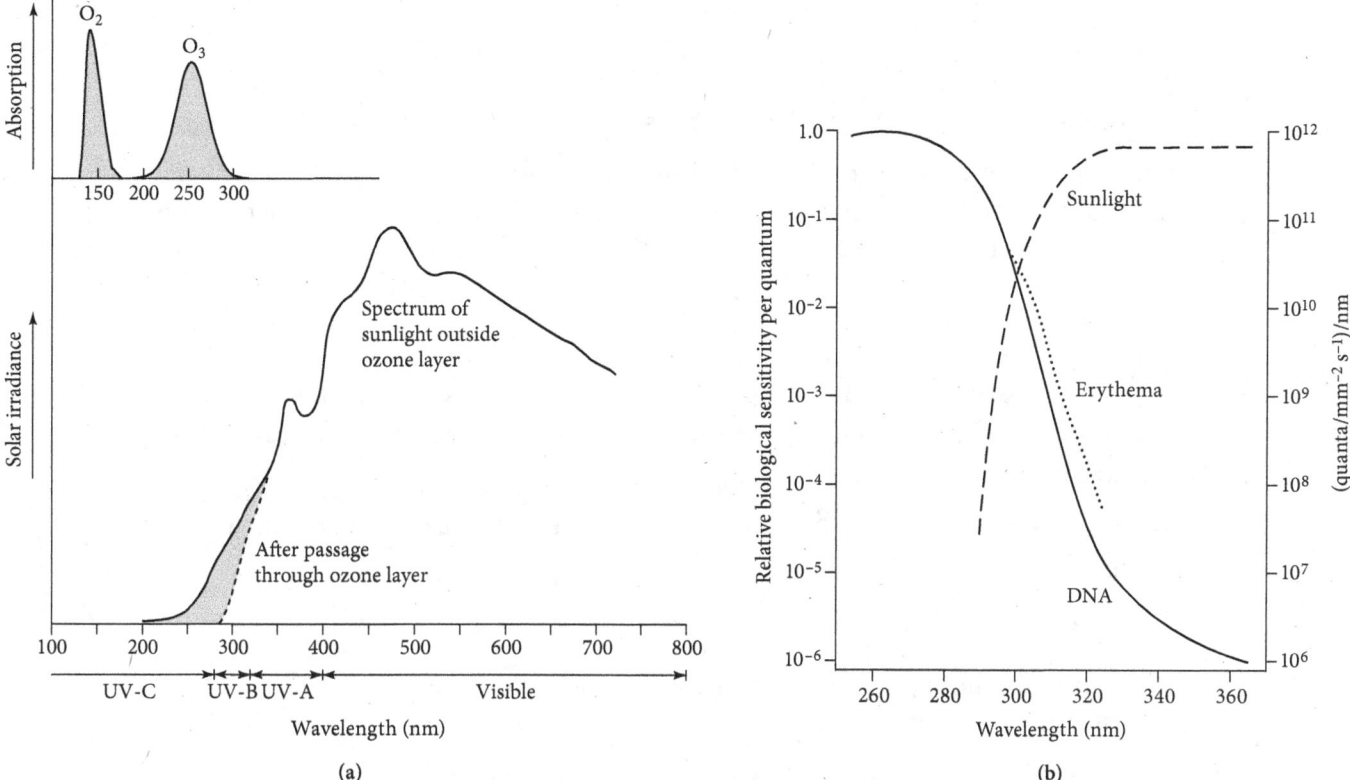

Figure 10.2 (a) The absorption spectra of O_2 and O_3 and their effect on the sunlight that penetrates the Earth's atmosphere to the surface. (b) Spectrum of sunburn sensitivity and DNA UV absorption at constant light intensity. *Source:* Adapted from Spiro et al., *Chemistry of the Environment, 3rd Ed.,* University Science Books, © 2012, all rights reserved.

The wavelength-dependent absorption of UV light by O_2 and O_3 is shown in Figure 10.2, alongside a comparison of the spectral shape of sunburn sensitivity relative to penetrating sunlight.

Use the above relationships and figures to answer the following questions:

QUESTION 14

What fraction of 340-nm light is transmitted [σ(340 nm)= 2×10^{-21} cm²] through a typical ozone column of 8×10^{18} molec cm⁻²?

Q14 ANSWER Use the column density formula:

$$T = e^{-\sigma c} = e^{-(2 \times 10^{-21} \text{ cm}^2)(8 \times 10^{18} \text{ molec cm}^{-2})} = 0.98$$

98% of 340-nm light is transmitted through the typical ozone layer.

*QUESTION 15

What fraction of 320-nm light is transmitted [σ(320 nm) = 2.7×10^{-20} cm²] through an ozone column of 2.6×10^{19} molec cm⁻²?

QUESTION 16

How does stratospheric ozone shield the earth's surface from harmful UV radiation?

Q16 ANSWER Stratospheric ozone shields the earth's surface from harmful UV radiation by absorption of solar UV radiation. The peak absorption occurs at 260 nm, with high absorption spanning from 200 to 300 nm.

QUESTION 17

Calculate the fractional increase in transmission (dT/T) for a 1% decrease in the thickness (l) of the ozone layer for 310 and 295 nm. (Assume $l = 0.34$ cm and $\varepsilon = 3$ and 18 cm^{-1} for 310 and 295 nm, respectively.)

Q17 ANSWER Here, use the differential form of the extinction coefficient equation:

$$\frac{dT}{T} = -\varepsilon l \frac{dl}{l}$$

with $l = 0.34$ cm and (dl/l) = -0.01 in each case for the 1% decrease in thickness.
     At 310 nm, $\varepsilon = 3$ cm^{-1} and:

$$\frac{dT}{T} = -\varepsilon l \frac{dl}{l} = -(3 \text{ cm}^{-1})(0.34 \text{ cm})(-0.01) = 0.01$$

This would cause a 1% increase in transmission of 310-nm light.
     At 295 nm, $\varepsilon = 18$ cm^{-1} and:

$$\frac{dT}{T} = -\varepsilon l \frac{dl}{l} = -(18 \text{ cm}^{-1})(0.34 \text{ cm})(-0.01) = 0.06$$

This would cause a 6% increase in transmission of 295-nm light.

*QUESTION 18

Calculate the fractional increase in transmission (dT/T) for a 1% decrease in the thickness (l) of the ozone layer for 285 nm. (Assume $l = 0.34$ cm and $\varepsilon = 3$, 18, and 56 cm^{-1} for 310, 295, and 285 nm, respectively.)

QUESTION 19

Will the calculated increase in transmission at 285 nm cause more damage to biological tissue than the calculated increase in transmission at 295 nm? Explain your reasoning.

Q19 ANSWER The curve of relative sensitivity to sunburn in Figure 10.2(b) shows that sensitivity at 285 nm is about 10 times higher than at 295 nm, and UV transmission increases 3.2 times more at 285 than 295 nm (19% versus 6%) so damage would be much greater at 285 nm.

QUESTION 20

What column density (molec cm^{-2}) of ozone would be required to remove 99.9% of sunlight at 310 nm [O_3 absorption cross section σ(310 nm) $= 9 \times 10^{-20}$ cm^2]? Convert this column density to DUs, and compare it with the average ozone column. Is there enough O_3 in the atmosphere to cut out this much light at 310 nm? (1 DU $= 2.69 \times 10^{16}$ molec cm^{-2}).

Q20 ANSWER To cut out 99.9% of the light means a transmission of $T = 0.001$:

$$T = \frac{l}{l(0)} = e^{-\sigma c}$$

$$0.001 = e^{-\sigma c}$$

$$\ln(0.001) = \ln(e^{-\sigma c})$$

$$\ln(0.001) = -\sigma c$$

$$\frac{\ln(0.001)}{-9 \times 10^{-20} \text{ cm}^2} = c = 7.7 \times 10^{19} \text{ molec cm}^{-2}$$

Now convert to DUs:

$$7.7 \times 10^{19} \text{ molec cm}^{-2} \times \frac{1 \text{ DU}}{2.69 \times 10^{16} \text{ molec cm}^{-2}} = 2{,}900 \text{ DU}$$

As mentioned in the text above and confirmed by NASA's OzoneWatch website (https://ozonewatch.gsfc.nasa.gov/), the global average O_3 column is 300 DU. No, there is not enough O_3 in the atmosphere to cut out this much light.

*QUESTION 21

Calculate the column density (molec cm^{-2}) of ozone required to remove 99.9% of sunlight at 320 nm and compare this to the column density calculated for 310 nm above. The O_3 absorption cross section σ(320 nm) $= 2.7 \times 10^{-20}$ cm^2. Convert this column density to DUs and compare it with the average ozone column. Is there enough O_3 in the atmosphere to cut out this much light at 320 nm? (1 DU $= 2.69 \times 10^{16}$ molec cm^{-2}).

10.4 Reactions that Create and Destroy Ozone

Ozone (O_3) is formed by the reaction of atomic oxygen (O radical), which is produced photochemically (by photolysis of O_2 in the stratosphere), with molecular oxygen (O_2), in a pressure-dependent reaction. This reaction is pressure-dependent because excess energy must be carried away by a third collision partner, abbreviated as the generic molecule "M." Because of this, a peak in ozone concentration occurs in the stratosphere, where pressure is high enough to stabilize the $O + O_2 + M$ reaction product O_3 (pressure decreases with altitude), but photolysis of O_2 is also high enough to produce a significant concentration of O radical (photolysis increases with altitude). Stratospheric ozone is catalytically destroyed by radical cycles involving the radical families HO_x (OH + HO_2) and NO_x (NO + NO_2), as well as the halides bromine (Br) and chlorine (Cl), which are elevated in the stratosphere due to human production and use of CFCs.

10.4.1 Chapman Mechanism

The ozone–oxygen cycle of photochemical reactions controls ozone steady-state concentration in the stratosphere. It is also often referred to as the Chapman mechanism, named after chemist Sydney Chapman, who determined the underlying mechanism in the 1930s. The reactions and rate constants are listed below, with equations for determining the temperature-dependent chemical rate constants and Table 10.1 for the photochemical rate constants, since these vary with photon flux. Temperatures are also included in the table to assist in calculation of the reaction rates:

$$O_2 + hv \rightarrow 2\,O \tag{10.6a}$$

Rate $a = k_a[O_2]$; where k_a is a photochemical rate constant

$$O + O_2 + M \rightarrow O_3 + M \tag{10.6b}$$

Rate $b = k_b[O][O_2][M]$

$$O_3 + hv \rightarrow O_2 + O \tag{10.6c}$$

Rate $c = k_c[O_3]$; another photochemical rate constant

$$O + O_3 \rightarrow 2\,O_2 \tag{10.6d}$$

Rate $d = k_d[O][O_3]$.
Reaction b is a termolecular reaction with rate:

$$k_b(\text{cm}^6\ \text{molec}^{-2}\ \text{sec}^{-1}) = 6.0 \times 10^{-34} \left(\frac{T}{300\ \text{K}}\right)^{-2.3}$$

Reaction d has a more straightforward Arrhenius form[1] of its temperature dependence:

$$k_d(\text{cm}^3\ \text{molec}^{-1}\ \text{sec}^{-1}) = 4.0 \times 10^{-11}\, e^{\left(\frac{-2{,}060\,\text{K}}{T}\right)}\ \frac{\text{cm}^3}{\text{molec sec}}$$

The two photochemical rate constants can be obtained in Table 10.1, along with temperatures at each altitude.

---

[1] The generic form of an Arrhenius rate constant is $k = A \exp(-E_a/RT)$, where A is the frequency factor, describing the likelihood of a collision being reactive and E_a is the activation barrier for the reaction. R is the ideal gas constant and T is the temperature in Kelvin. In the above equations, these have been condensed into a simple temperature-dependent equation to enable calculation of the rate constant at various altitudes (and thus differing temperatures).

TABLE 10.1 Temperatures and Photochemical
Rate Constants at Each 5-km Altitude

Height (km)	$T(K)$	k_a (sec^{-1})	k_c (sec^{-1})
0	288	1.00×10^{-20}	6.00×10^{-04}
5	262	1.00×10^{-20}	6.00×10^{-04}
10	220	1.00×10^{-20}	6.00×10^{-04}
15	218	1.00×10^{-14}	6.00×10^{-04}
20	218	4.00×10^{-13}	6.00×10^{-04}
25	222	9.00×10^{-12}	6.00×10^{-04}
30	226	4.00×10^{-11}	6.00×10^{-04}
35	234	1.00×10^{-10}	7.00×10^{-04}
40	250	2.00×10^{-10}	8.00×10^{-04}
45	260	5.00×10^{-10}	1.00×10^{-03}
50	270	6.00×10^{-10}	1.80×10^{-03}

Source: Data from NASA Panel for Data Evaluation, *Chemical Kinetics and Photochemical Data for Use in Atmospheric Studies: Evaluation Number 19,* 2019, JPL Publication 19-5, available at https://jpldataeval.jpl.nasa.gov/pdf /NASA-JPL%20Evaluation%2019-5.pdf.

QUESTION 22

Calculate the rate of formation of O for reaction (10.6a) at the high stratospheric level of 50 km if $[O_2] = 2 \times 10^{16}$ molec cm^{-3}. How many molecules per cm^3 O are produced by this reaction per hour?

Q22 ANSWER The rate law for this reaction is Rate = $k_a[O_2]$
We need to look in Table 10.1 to find the rate constant, k_a, at 50 km:

$$k_a = 6.00 \times 10^{-10} \text{ sec}^{-1}$$

Putting this information together:

$$\text{rate} = k_a[O_2] = \left(6.00 \times 10^{-10} \frac{1}{\text{sec}}\right) \times \left(2 \times 10^{16} \frac{\text{molec}}{\text{cm}^3}\right) = 1.2 \times 10^7 \frac{\text{molec}}{\text{cm}^3 \text{ sec}}$$

This means that 1.2×10^7 molec of oxygen react per second per cm^3. Because 1 molec of O_2 forms two atoms of O:

$$2 \times \left(1.2 \times 10^7 \frac{\text{molec}}{\text{cm}^3 \text{ sec}}\right) = 2.4 \times 10^7 \frac{\text{molec}}{\text{cm}^3 \text{ sec}} \text{ of O are formed}$$

Converting to per hour:

$$2.4 \times 10^7 \frac{\text{molec}}{\text{cm}^3 \text{ sec}} \times \frac{3,600 \text{ sec}}{\text{hr}} = 8.6 \times 10^{10} \frac{\text{molec}}{\text{cm}^3 \text{ sec}} \text{ of O formed}$$

*QUESTION 23

Calculate the rate of formation of O for reaction (10.6c) above at the mid-stratospheric level of 25 km if the $[O_3] = 4 \times 10^{12}$ molec cm^{-3}. How many molec cm^{-3} O is produced by this reaction per hour?

QUESTION 24

Write the differential rate law expression for the above reaction d and determine the rate of O_3 consumption by this reaction in the mid-stratospheric ozone layer ($z = 25$ km) at $[O] = 8 \times 10^{10}$ molec cm^{-3} and $[O_3] = 4 \times 10^{12}$ molec cm^{-3}. How many molec cm^{-3} O_2 are produced by this reaction per hour?

Q24 ANSWER The rate law is rate $= k_d[O][O_3]$. To calculate the rate constant:

$$k_d(cm^3\ molec^{-1}\ sec^{-1}) = 4.0 \times 10^{-11}\ e^{\left(\frac{-2,060\,K}{T}\right)} \frac{cm^3}{molec\ sec}$$

we need an estimate of the temperature. From Table 10.1, at 25 km the temperature is about 220 K. Thus:

$$rate = 4.0 \times 10^{-11}\ e^{\left(\frac{-2.060\,K}{T}\right)} \frac{cm^3}{molec\ sec}[O][O_3]$$

$$rate = 4.0 \times 10^{-11}\ e^{\left(\frac{-2.060\,K}{T}\right)} \frac{cm^3}{molec\ sec}\left[8 \times 10^{10}\frac{molec}{cm^3}\right]\left[4 \times 10^{12}\frac{molec}{cm^3}\right]$$

$$= 1.1 \times 10^9 \frac{molec}{cm^3\ sec}\ \text{destruction rate of } O_3 \text{ by reaction } d$$

O_2 is produced at twice this rate, given the reaction stoichiometry, and converting to per hour:

$$O_2 \text{ production rate: } 2 \times 1.1 \times 10^9 \frac{molec}{cm^3\ sec} \times \frac{3,600\ sec}{hr} = 7.9 \times 10^{12}\frac{molec}{cm^3\ sec}$$

10.4.2 *Using Steady-State Approximations to Determine Ozone Vertical Profiles*

Steady-state approximations (meaning concentrations do not change as a function of time) for [O] and [O_3] based on the above-mentioned Chapman mechanism can be used to obtain an expression for the ratio of O_3 to O_2 as a function of the rate constants and overall air concentration ([M], which can be determined by the scale height expression).

We first calculate steady-state expressions for both [O] and [O_3]. For example:

$$\frac{d[O_3]}{dt} = \text{production of } O_3 - \text{loss of } O_3 = 0$$

To derive this relationship, we must use the Chapman mechanism in Section 4.1:

$$\text{production of O} - \text{loss of O} = 0 \tag{10.7}$$

$$2(\text{Rate 6a}) + \text{Rate 6c} - \text{Rate 6b} - \text{Rate 6d} = 0 \tag{10.8}$$

$$\text{production of } O_3 - \text{loss of } O_3 = 0 \tag{10.9}$$

$$\text{Rate 6b} - \text{Rate 6c} - \text{Rate 6d} = 0 \tag{10.10}$$

Next, we do some algebra to cancel terms. Start by adding (10.8) and (10.10):

$$2(\text{Rate 6a}) + \text{Rate 6c} - \text{Rate 6b} - \text{Rate 6d} = 0$$

$$+ \text{Rate 6b} - \text{Rate 6c} - \text{Rate 6d} = 0$$

$$\overline{2(\text{Rate 6a}) - 2(\text{Rate 6d}) = 0} \tag{10.11}$$

Now we substitute the rate laws from the Chapman mechanism into (10.11):

$$2 k_a[O_2] - 2 k_d[O][O_3] = 0 \tag{10.12}$$

Next, we subtract (10.8) and (10.10):

$$2(\text{Rate 6a}) + \text{Rate 6c} - \text{Rate 6b} - \text{Rate 6d} = 0$$

$$- (\text{Rate 6b} - \text{Rate 6c} - \text{Rate 6d}) = 0$$

$$\overline{2(\text{Rate 6a}) - 2(\text{Rate 6b}) + 2(\text{Rate 6c}) = 0} \tag{10.13}$$

Now we substitute the rate laws from the Chapman mechanism into (10.13)

$$2 k_a[O_2] - 2 k_b[O][O_2][M] + 2 k_c[O_3] = 0 \tag{10.14}$$

The next step is to solve both (10.12) and (10.14) for the [O] and equate. For (10.12):

$$2 k_a[O_2] - 2 k_d[O][O_3] = 0$$

$$2 k_a[O_2] = 2 k_d[O][O_3]$$

$$[O] = \frac{2 k_a[O_2]}{2 k_d[O_3]} \tag{10.15}$$

For (10.14):

$$2 k_a[O_2] - 2 k_b[O][O_2][M] + 2 k_c[O_3] = 0$$

$$2 k_a[O_2] + 2 k_c[O_3] = 2 k_b[O][O_2][M]$$

$$[O] = \frac{2 k_a[O_2] + 2 k_c[O_3]}{2 k_b[O_2][M]} \tag{10.16}$$

Now we can set (10.15) and (10.16) equal to each other, since they are both equal to [O]:

$$\frac{2 k_a[O_2]}{2 k_d[O_3]} = \frac{2 k_a[O_2] + 2 k_c[O_3]}{2 k_b[O_2][M]}$$

Note that Rate a is always much slower than Rate c, so we can drop the $k_a[O_2]$ term where they are added together:

$$\frac{2 k_a[O_2]}{2 k_d[O_3]} = \frac{2 k_c[O_3]}{2 k_b[O_2][M]}$$

Then, we rearrange to give a ratio of O_3 to O_2 to give the following relationship:

$$\frac{[O_3]^2}{[O_2]^2} = \frac{4 k_a k_b[M]}{4 k_c k_d}$$

Then, we cancel the 4s in the numerator and denominator and take the square root of both sides to give:

$$\frac{O_3}{O_2} = \left(\frac{k_a k_b [M]}{k_c k_d}\right)^{\frac{1}{2}}$$

(10.17)

QUESTION 25

Use the relationship derived by (10.17) to calculate the concentration (molec cm^{-3}) of O$_3$ at 0, 10, 20, 30, 40, and 50 km.

Q25 ANSWER Plugging in temperatures from Table 10.1, we can calculate the rate constants at each altitude.

Example at 0 km:

$$k_b(\text{cm}^6 \text{ molec}^{-2} \text{ sec}^{-1}) = 6.0 \times 10^{-34}\left(\frac{T}{300 \text{ K}}\right)^{-2.3}$$

$$= k_b(\text{cm}^6 \text{ molec}^{-2} \text{ sec}^{-1}) = 6.0 \times 10^{-34}\left(\frac{288 \text{ K}}{300 \text{ K}}\right)^{-2.3}$$

$$= 6.59 \times 10^{-34} \text{ cm}^6 \text{ molec}^{-2} \text{ sec}^{-1}$$

$$k_d(\text{cm}^3 \text{ molec}^{-1} \text{ sec}^{-1}) = 4.0 \times 10^{-11} e^{\left(\frac{-2.060 \text{ K}}{T}\right)} \frac{\text{cm}^3}{\text{molec sec}}$$

$$= k_d(\text{cm}^3 \text{ molec}^{-1} \text{ sec}^{-1}) = 4.0 \times 10^{-11} e^{\left(\frac{-2.060 \text{ K}}{288 \text{ K}}\right)} \frac{\text{cm}^3}{\text{molec sec}}$$

$$= 3.13 \times 10^{-14} \text{ cm}^3 \text{ molec}^{-1} \text{ sec}^{-1}$$

$$\frac{O_3}{O_2} = \left(\frac{k_a k_b [M]}{k_c k_d}\right)^{\frac{1}{2}} = \left(\frac{(1.00 \times 10^{-20} \text{ sec}^{-1})(6.59 \times 10^{-34} \text{ cm}^6 \text{ molec}^{-2} \text{ sec}^{-1})(2.46 \times 10^{19} \text{ molec cm}^{-3})}{(6.00 \times 10^{-4} \text{ sec}^{-1})(3.13 \times 10^{-14} \text{ cm}^3 \text{ molec}^{-1} \text{ sec}^{-1})}\right)^{\frac{1}{2}}$$

$$= 2.94 \times 10^{-9}$$

Then, use the scale height expression to calculate [M] at each altitude, the resulting [O$_3$]/[O$_2$] ratio, and use the uniform 21% O$_2$ to calculate [O$_2$] = 0.21 × [M]:

$$[O_2](0 \text{ km}) = 0.21 \times [M] = 0.21 \times 2.46 \times 10^{19} \text{ molec cm}^{-3} = 5.17 \times 10^{18} \text{ molec cm}^{-3}$$

$$[M](z) = [M](0)e^{-\frac{z}{H}}$$

$$[O_2](0 \text{ km}) = (5.17 \times 10^{18} \text{ molec cm}^{-3})e^{-\frac{0 \text{ km}}{8 \text{ km}}} = 5.17 \times 10^{18} \text{ molec cm}^{-3}$$

Finally, multiplying the [O$_3$]/[O$_2$] ratio by [O$_2$]:

$$[O_3] = (2.94 \times 10^{-9}) \times (5.17 \times 10^{18} \text{ molec cm}^{-3}) = 1.52 \times 10^{10} \text{ molec cm}^{-3}$$

Then, we can determine a number density concentration of O_3 at each altitude:

Height (km)	[M] (# cm^{-3})	[O$_2$] (# cm^{-3})	[O$_3$]/[O$_2$]	[O$_3$] (# cm^{-3})
0	2.46×10^{19}	5.17×10^{18}	2.94×10^{-9}	1.52×10^{10}
10	7.05×10^{18}	4.24×10^{17}	6.47×10^{-9}	2.75×10^9
20	2.02×10^{18}	3.48×10^{16}	2.31×10^{-5}	8.05×10^{11}
30	5.79×10^{17}	2.86×10^{15}	1.00×10^{-4}	2.87×10^{11}
40	1.66×10^{17}	2.35×10^{14}	5.99×10^{-5}	1.40×10^{10}
50	4.75×10^{16}	1.93×10^{13}	2.50×10^{-5}	4.80×10^8

*QUESTION 26

Use the relationship derived by (10.17) to calculate the concentration (molec cm^{-3}) of O_3 at t 5, 15, 25, 35, and 45 km.

QUESTION 27

Your above calculations showed a peak in ozone concentration at a 30-km altitude. See answers to Q25 and Q26 for a table of ozone concentrations at all altitudes (see table above). Based on the

TABLE 10.2 Calculated Temperature Dependent Rate Constants

Height (km)	k_b(cm^6 molec^{-2} sec^{-1})	k_d(cm^3 molec^{-1} sec^{-1})
0	6.59×10^{-34}	3.13×10^{-14}
5	8.19×10^{-34}	1.54×10^{-14}
10	1.22×10^{-33}	3.43×10^{-15}
15	1.25×10^{-33}	3.15×10^{-15}
20	1.25×10^{-33}	3.15×10^{-15}
25	1.20×10^{-33}	3.73×10^{-15}
30	1.15×10^{-33}	4.40×10^{-15}
35	1.06×10^{-33}	6.01×10^{-15}
40	9.13×10^{-34}	1.06×10^{-14}
45	8.34×10^{-34}	1.45×10^{-14}
50	7.65×10^{-34}	1.94×10^{-14}

Source: NASA Panel for Data Evaluation, *Chemical Kinetics and Photochemical Data for Use in Atmospheric Studies: Evaluation Number 19,* 2019, JPL Publication 19-5, available at https://jpldataeval.jpl.nasa.gov/pdf/NASA-JPL%20Evaluation%2019-5.pdf

expression for the ozone/oxygen ratio (reproduced below), explain qualitatively why this peak occurs where it does:

$$\frac{O_3}{O_2} = \left(\frac{k_a k_b [M]}{k_c k_d}\right)^{\frac{1}{2}}$$

Q27 ANSWER At the highest altitudes, photolysis is maximized, increasing the rates of reactions a and c. The rate of reaction b, however, is maximized at the surface, where $[M]$ is the largest. So the peak at 30 km represents the optimum overlap of rapid photolysis producing O atoms, but with still enough air concentration to allow the stabilization to occur in the three-body ozone formation reaction.

10.5 Catalytic Destruction of Ozone

Catalytic destruction of ozone happens generically through the below two-step mechanism, resulting in the below overall reaction (10.18c):

$$X + O_3 \rightarrow XO + O_2 \tag{10.18a}$$

$$XO + O \rightarrow X + O_2 \tag{10.18b}$$

Overall:

$$O + O_3 \rightarrow 2O_2 \tag{10.18c}$$

The catalyst "X" in this mechanism can be various radical species, such as Cl, Br, NO, or HO, which have differing photolytic sources in the stratosphere.

To produce the radical catalysts for ozone destruction, a bond must be broken to liberate the radical. The relationship between the photon wavelength (λ) and the energy (E) of the bond that it will break is:

$$\lambda = \frac{hc}{E} \tag{10.19}$$

This equation was first introduced in Chapter 4 in relation to photovoltaic energy.

QUESTION 28

One of the catalysts of ozone destruction is the chlorine radical (Cl), of which a source to the stratosphere is CFC-11 (CCl_3F). Write the reaction that occurs in the stratosphere to introduce the catalyst. A typical carbon–chlorine bond has a bond dissociation energy (BDE) of 397 kJ mol^{-1}. What wavelength of light is required to initiate this reaction?

Q28 ANSWER Cl is photolytically produced by breaking the C–Cl bond:

$$CCl_3F + h\nu \rightarrow Cl + CCl_2F$$

To determine the wavelength of light required for this reaction, we first convert the given BDE of 397 kJ mol^{-1} to Joules per molecule required to break the bond:

$$\frac{397\ kJ}{mol} \times \frac{1{,}000\frac{J}{kJ}}{6.023 \times 10^{23}\frac{molec}{mol}} = 6.59 \times 10^{-19}\ J\ molec^{-1}$$

and then use:

$$\lambda = \frac{hc}{E}$$

$$\lambda = \frac{(6.626 \times 10^{-34}\ J\ sec)(3.00 \times 10^{8}\ m\ sec^{-1})}{6.59 \times 10^{-19}\ J} = 3.02 \times 10^{-7}\ m \times \frac{10^{9}\ nm}{m} = 302\ nm$$

*QUESTION 29

One of the catalysts of ozone destruction is the bromine radical (Br), of which a source to the stratosphere is bromotrifluoromethane (CF_3Br), a halon used for firefighting purposes. Write the reaction that occurs in the stratosphere to introduce the catalyst. A typical carbon–bromine bond has a BDE of 280 kJ mol^{-1}. What wavelength of light is required to initiate this reaction?

QUESTION 30

Now write the catalytic cycle for O_3 destruction initiated by the chlorine radical catalyst. Write the rate law for each step. What is the intermediate in this mechanism? Write the overall reaction. What is one possible sink reaction that could remove this stratospheric ozone destruction catalyst?

Q30 ANSWER The catalytic cycle is:

$$Cl + O_3 \rightarrow ClO + O_2 \qquad\qquad Rate = k[Cl][O_3]$$

$$ClO + O \rightarrow Cl + O_2 \qquad\qquad Rate = k[ClO][O]$$

Overall:

$$O + O_3 \rightarrow 2O_2$$

The intermediate in this cycle is ClO. One reaction that could remove Cl from the stratosphere is the reaction with methane: $Cl + CH_4 \rightarrow HCl + CH_3$

*QUESTION 31

Now write the catalytic cycle for O_3 destruction initiated by the bromine radical catalyst. Write the rate law for each step. What is the intermediate in this mechanism? Write the overall reaction. What is one possible sink reaction that could remove this stratospheric ozone destruction catalyst?

The hydroxyl (HO·) radical accounts for nearly one-half of total O_3 destruction in the lower stratosphere (16–20-km altitude). The HO· radical is capable of accepting an O atom from O_3 to produce HO_2· [reaction (10.20a)], which in turn can react with O atoms, regenerating HO· [reaction (10.20b)]:

$$HO· + O_3 \rightarrow HO_2· + O_2 \qquad\qquad (10.20a)$$

$$HO_2\cdot + O \rightarrow HO\cdot + O_2 \tag{10.20b}$$

The combined reactions lead to the ozone destruction reaction (10.18c).

The HO· radical is formed in the atmosphere by the reaction of water with electronically excited oxygen (*O) atoms:

$$*O + H_2O \rightarrow 2\,OH\cdot \tag{10.21}$$

with *O constituting 2% of the O atoms generated from the photolysis of O_3. The same *O atoms can also react with CH_4:

$$*O + CH_4 \rightarrow OH\cdot + CH_3\cdot \tag{10.22}$$

and CH_3· reacts with O_2, entering a cycle of free radical reactions, which produce additional HO· radicals. Both H_2O and CH_4 drift up to the stratosphere from the troposphere and increase as a result of human activity. Methane emissions are increasing and global warming is enhancing the rate of water evaporation from the earth's surface (see Chapter 7). Consequently, the stratospheric production rate of HO· radicals is also increasing.

Although NO· is produced abundantly in the lower atmosphere by combustion and lightning, almost all of it is oxidized to NO_2· and converted to nitric acid (HNO_3) in the troposphere (see Chapter 8):

$$HO\cdot + NO_2\cdot \rightarrow HONO_2 \tag{10.23}$$

after which it is rained out before reaching the stratosphere. On the other hand, nitrous oxide (N_2O), a potent greenhouse gas (see Chapter 7), although much less abundant, is also much less reactive; it does eventually reach the stratosphere. Above 30 km, most of the N_2O is photolyzed by UV photons to produce N_2 and *O atoms:

$$N_2O + h\nu(UV) \rightarrow N_2 + *O \tag{10.24}$$

A small percentage, 10% or less, of the N_2O molecules react with *O atoms to produce NO·:

$$N_2O + *O \rightarrow 2\,NO\cdot \tag{10.25}$$

This is the main source of NO· in the stratosphere.

Here NO· can act as X in ozone-destroying reactions and (10.18a) and (10.18b), cycling through NO_2 in the process:

$$NO\cdot + O_3 \rightarrow NO_2\cdot + O_2 \tag{10.26}$$

$$NO_2\cdot + O \rightarrow NO\cdot + O_2 \tag{10.27}$$

However, NO_2· also reacts with other carriers in O_3 destruction chains, HO· and ClO·. The reaction with HO· produces HNO_3:

$$HO\cdot + NO_2\cdot \rightarrow HONO_2 \tag{10.28}$$

while the reaction with ClO· produces an analogous molecule, chlorine nitrate:

$$ClO\cdot + NO_2\cdot \rightarrow ClONO_2 \tag{10.29}$$

Nitric acid and chlorine nitrate do not participate in O_3 destruction directly. Rather, they are reservoir molecules; they sequester the chain-carrying species HO· and ClO· in less reactive forms, releasing them in response to UV light:

$$HONO_2 + h\nu(UV) \rightarrow HO\cdot + NO_2\cdot \qquad (10.30)$$

$$ClONO_2 + h\nu(UV) \rightarrow ClO\cdot + NO_2\cdot \qquad (10.31)$$

Although HO· and ClO· remain available for O_3 destruction, binding to NO_2 reduces their concentration significantly and thereby reduces O_3 destruction. Furthermore, the reservoir molecules can also be rained out when stratospheric and wet tropospheric air mix in the upper troposphere.

Thus, NO· has a two-sided effect, on the one hand providing another catalytic chain mechanism for O_3 destruction, but, on the other, inhibiting two other major mechanisms for O_3 destruction. Which of these effects dominates depends on the altitude.

Above an altitude of 25 km, the net effect of nitrogen oxides is to reduce O_3 concentrations via reactions (10.26) and (10.27); in the middle and upper stratosphere, nitrogen oxides account for >50% of total O_3 destruction. However, in the lower stratosphere, the overall effect of nitrogen oxides is to protect O_3 from destruction via reactions (10.28) and (10.29). This protective role is declining in importance as CFC and halon restrictions reduce chlorine and bromine concentrations in the stratosphere. Indeed, the declining halogen content is making NO· the dominant ozone destroyer, after HO·. As N_2O emissions increase (see Chapter 7), the relative importance of the NO· ozone pathway will continue to increase.

QUESTION 32

Early observations of ClO, HCl, and $ClONO_2$ in the mid-stratosphere could not be reconciled with the accepted kinetic models of halogen chemistry until laboratory studies (motivated by these observations) showed that HCl is formed as a minor product channel in the reaction of OH + ClO, in the so-called four-center elimination reaction:

$$OH\cdot + ClO\cdot \rightarrow HO_2\cdot + Cl\cdot \qquad k = 6.7 \times 10^{-12} \, e^{\frac{270\,K}{T}} \, cm^3 \, molec^{-1} \, sec^{-1}$$

$$OH\cdot + ClO\cdot \rightarrow O_2 + HCl \qquad k = 9.7 \times 10^{-14} \, e^{\frac{320\,K}{T}} \, cm^3 \, molec^{-1} \, sec^{-1}$$

Recalling the Arrhenius rate constant definition (see the above discussion on Arrhenius rate expressions), why is the "A-factor" for the channel leading to HCl so much lower than the channel leading to HO_2?

Q32 ANSWER The A-factor in an Arrhenius rate constant indicates the likelihood of a collision going to reaction products. Thus, a smaller A-factor suggests greater geometric constraints on what collisions are likely to be successful. Here, it is much more likely for the OH and ClO to collide in a way that allows HO_2 + Cl to form than O_2 + HCl. The former requires a transition state with the atoms strung along in a line, while the latter requires a four-center transition state, with the very particular alignment of the OH and ClO as they approach one another:

$$\begin{array}{ccc} & & O \text{----} O \\ H \text{----} O \text{----} O \text{----} Cl \quad \text{versus} & & | \quad\ | \\ & & H \text{----} Cl \end{array}$$

QUESTION 33

In the mid-stratosphere, the partitioning between atomic chlorine and ClO is determined by the following reactions among combustion-produced nitrogen oxides and oxygen species:

$$ClO\cdot + NO\cdot \rightarrow Cl\cdot + NO_2\cdot \qquad k_1 = 6.4 \times 10^{-12}\, e^{\frac{290\,K}{T}}\ cm^3\ molec^{-1}\ sec^{-1}$$

$$ClO\cdot + O\cdot \rightarrow Cl\cdot + O_2 \qquad k_2 = 3.0 \times 10^{-11}\, e^{\frac{70\,K}{T}}\ cm^3\ molec^{-1}\ sec^{-1}$$

$$Cl\cdot + O_3 \rightarrow ClO\cdot + O_2 \qquad k_3 = 2.9 \times 10^{-11}\, e^{\frac{-260\,K}{T}}\ cm^3\ molec^{-1}\ sec^{-1}$$

Write down an expression for the ratio [Cl]/[ClO] based on these reactions assuming a steady state for either Cl or ClO, and assuming Chapman mechanism's steady state for [O] (reproduced below). This expression will allow us to solve for this ratio for conditions typical in the daytime stratosphere at different altitudes. First calculate the ratio at 35 km:

$$[O] = \frac{k_c[O_3]}{k_b[M][O_2]}$$

$$35\ km: [NO] = 2.0 \times 10^9\ molec\ cm^{-3}$$

$$[M] = 2 \times 10^{17}\ molec\ cm^{-3}$$

$$[O_3] = 1 \times 10^{12}\ molec\ cm^{-3}$$

$$k_c = 1 \times 10^{-3}\ sec^{-1}$$

$$T = 235\ K$$

Q33 ANSWER First, write a steady-state expression for [Cl] in the above mechanism:

$$\frac{d[Cl]}{dt} = k_1[ClO][NO] + k_2[ClO][O] - k_3[Cl][O3] = 0$$

Now divide by [ClO]:

$$0 = k_1[NO] + k_2[O] - k_3[O_3]\frac{[Cl]}{[ClO]}$$

And solve for the ratio we are interested in:

$$\frac{[Cl]}{[ClO]} = \frac{k_1[NO] + k_2[O]}{k_3[O_3]}$$

Separate terms:

$$\frac{[Cl]}{[ClO]} = \frac{k_1[NO]}{k_3[O_3]} + \frac{k_2[O]}{k_3[O_3]} = \frac{[O]}{[O_3]}\left(\frac{k_1[NO]}{k_3[O]} + \frac{k_2}{k_3}\right)$$

And plug in $[O]/[O_3]$ (10.17):

$$\frac{[O]}{[O_3]} = \frac{k_c}{k_b[M][O_2]}$$

$$\frac{[Cl]}{[ClO]} = \frac{k_c}{k_b[M][O_2]}\left(\frac{k_1[NO]}{k_3[O]} + \frac{k_2}{k_3}\right)$$

Insert [O] into the above equation and simplify:

$$[O] = \frac{k_c[O_3]}{k_b[M][O_2]}$$

$$\frac{[Cl]}{[ClO]} = \frac{k_c}{k_b[M][O_2]}\left(\frac{k_1[NO]}{k_3[O]} + \frac{k_2}{k_3}\right)$$

$$= \frac{k_c}{k_b[M][O_2]}\left(\frac{k_1 k_b[M][O_2][NO]}{k_3 k_c[O_3]} + \frac{k_2}{k_3}\right) = \frac{k_1[NO]}{k_3[O_3]} + \frac{k_2 k_c}{k_3 k_b[M][O_2]}$$

$[O_2] = 0.21[M]$, so:

$$\frac{[Cl]}{[ClO]} = \frac{k_1[NO]}{k_3[O_3]} + \frac{k_2 k_c}{0.21 k_3 k_b[M]^2}$$

Calculate the rate constants at 235 K and 35 km:

$$k_b(\text{cm}^6\ \text{molec}^{-2}\ \text{sec}^{-1}) = 6.0 \times 10^{-34}\left(\frac{T}{300\ \text{K}}\right)^{-2.3} = 6.0 \times 10^{-34}\left(\frac{235\ \text{K}}{300\ \text{K}}\right)^{-2.3}$$

$$= 1.05 \times 10^{-33}\ \text{cm}^6\ \text{molec}^{-2}\ \text{sec}^{-1}$$

$$k_1 = 6.4 \times 10^{-12}\, e^{\frac{290\ \text{K}}{235\ \text{K}}}\ \text{cm}^3\ \text{molec}^{-1}\ \text{sec}^{-1} = 2.20 \times 10^{-11}\ \text{cm}^3\ \text{molec}^{-1}\ \text{sec}^{-1}$$

$$k_2 = 3.0 \times 10^{-11}\, e^{\frac{70\ \text{K}}{235\ \text{K}}}\ \text{cm}^3\ \text{molec}^{-1}\ \text{sec}^{-1} = 4.04 \times 10^{-11}\ \text{cm}^3\ \text{molec}^{-1}\ \text{sec}^{-1}$$

$$k_3 = 2.9 \times 10^{-11}\, e^{\frac{-260\ \text{K}}{235\ \text{K}}}\ \text{cm}^3\ \text{molec}^{-1}\ \text{sec}^{-1} = 9.59 \times 10^{-12}\ \text{cm}^3\ \text{molec}^{-1}\ \text{sec}^{-1}$$

$$\frac{[Cl]}{[ClO]} = \frac{k_1[NO]}{k_3[O_3]} + \frac{k_2 k_c}{0.21 k_3 k_b[M]^2}$$

$$= \frac{(2.20 \times 10^{-11}\ \text{cm}^3\ \text{molec}^{-1}\ \text{sec}^{-1})(2.0 \times 10^9\ \text{molec}^{-1}\ \text{cm}^{-3})}{(9.59 \times 10^{-12}\ \text{cm}^3\ \text{molec}^{-1}\ \text{sec}^{-1})(1.0 \times 10^{12}\ \text{molec}^{-1}\ \text{cm}^{-3})}$$

$$+ \frac{(4.04 \times 10^{-11}\ \text{cm}^3\ \text{molec}^{-1}\ \text{sec}^{-1})(1.0 \times 10^{-3}\ \text{sec}^{-1})}{(0.21)(9.59 \times 10^{-12}\ \text{cm}^3\ \text{molec}^{-1}\ \text{sec}^{-1})(1.05 \times 10^{-33}\ \text{cm}^6\ \text{molec}^{-2}\ \text{sec}^{-1})(2.0 \times 10^{17}\ \text{molec}\ \text{cm}^{-3})^2}$$

$$= 4.59 \times 10^{-3} + 4.78 \times 10^{-4} = 5.1 \times 10^{-3}$$

*QUESTION 34

Now calculate the ratio of [Cl]/[ClO] at 20 km, using the equations derived in Q33 and the following information:

$$[NO] = 5.0 \times 10^8 \text{ molec cm}^{-3}$$

$$[M] = 2.0 \times 10^{18} \text{ molec cm}^{-3}$$

$$[O_3] = 3.0 \times 10^{12} \text{ molec cm}^{-3}$$

$$k_c = 5.0 \times 10^{-4} \text{ sec}^{-1}$$

$$T = 220 \text{ K}$$

10.6 Polar Ozone Destruction

The unique phenomenon of the Antarctic ozone hole, with its rapid onset and breakup, and isolated area of dramatic O_3 depletion, can only be explained by adding additional chemistry and meteorology to the simple Cl catalyzed depletion chemistry described above. In particular, the storage of catalyst reservoirs on the surface of polar stratospheric clouds over the dark winter primes the polar vortex for rapid depletion when the sun returns in austral spring. This is shown schematically in Figure 10.3.

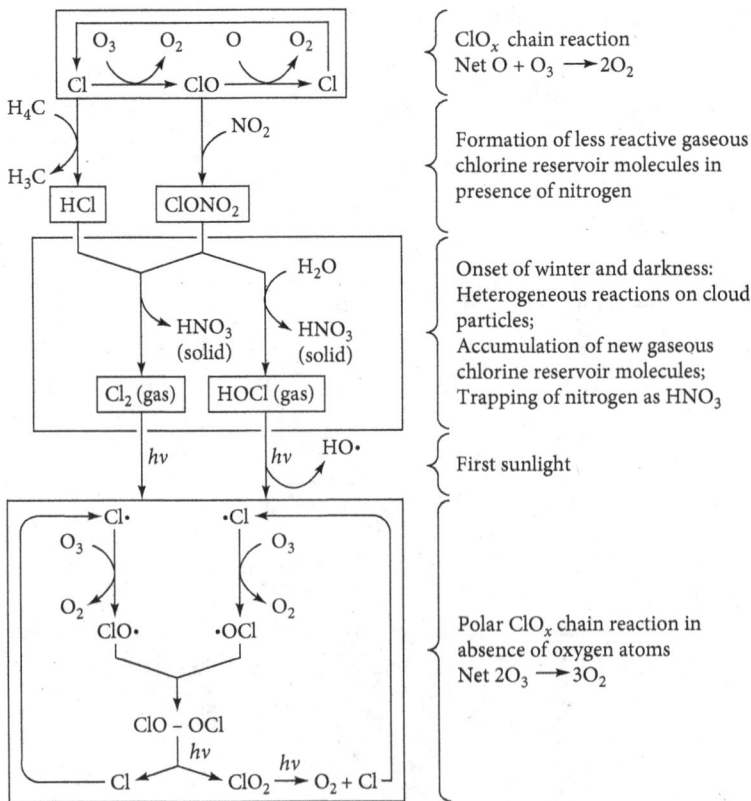

Figure 10.3 Reactions responsible for the Antarctic O_3 hole.
Source: Adapted from Spiro et al., *Chemistry of the Environment, 3rd Ed.,* University Science Books, © 2012, all rights reserved.

QUESTION 35

Write the catalytic cycle for ozone destruction via ClOOCl, which does not require the O atom.

Q35 ANSWER Replace the ClO + O reaction in the Cl initiate cycle above with the ClOOCl chemistry, which requires running the first reaction twice:

$$2 \times (Cl + O_3 \rightarrow ClO + O_2)$$

$$ClO + ClO \rightarrow ClOOCl$$

$$ClOOCl + h\nu \rightarrow 2\,Cl + O_2$$

Overall: $2\,O_3 \rightarrow 3\,O_2$

*QUESTION 36

An additional key reaction of the ClO intermediate in the mechanism in Q35, under certain conditions is to form the ClOOCl dimer, which can cycle back to Cl without requiring appreciable O atom concentrations. ClOOCl is an example of a "reservoir molecule," a molecule that is formed by radical + radical reactions (e.g., ClO + NO_2 to produce $ClONO_2$), and serves as a temporary storage for radicals, until they are broken back apart, usually by photolysis, to re-form the radicals:

$$ClO + ClO + M \rightarrow ClOOCl + M$$

$$ClOOCl + h\nu \rightarrow 2Cl + O_2$$

Write the rate law for first step in this mechanism. At typical lower stratosphere (220 K, [M] = 2.0×10^{18} molec cm^{-3}) with a ClO concentration of 2×10^7 molec cm^{-3}, what is the rate of ClOOCl formation? This dimer formation is a three-body association reaction with $k = 2.2 \times 10^{-32} (T/300)^{-3.1}$ cm^6 $molec^{-2}$ sec^{-1}. How does this mechanism explain the observed rapid onset of the Antarctic ozone hole?

QUESTION 37

What role do heterogeneous denitrification reactions (forming nitric acid in clouds) play in the polar vortex O_3 destruction chemistry (see Fig. 10.3)?

Q37 ANSWER As HNO_3 is heterogeneously formed on cloud droplets (middle box in Fig. 10.3), the NOx reservoir from $ClONO_2$ is physically separated from the Cl reservoir, which builds up in the gas phase as HOCl. Thus, while HNO_3 is trapped in ice particles, the Cl reservoir is primed to be activated at first sunlight.

*QUESTION 38

Write a reaction for release of O_3-destroying catalysts from their reservoir molecules.

QUESTION 39

What causes the breakup of the polar vortex and annual recovery of the Antarctic ozone hole?

Q39 ANSWER As the polar vortex warms, its isolation as a distinct vortex "reactor" wanes and eventually outside air can mix in, replenishing the stratospheric ozone layer. Second, the HNO_3 ice particles can melt and NO_2 can sequester and slow the ClOx chain reaction:

$$NO_2 + ClO \rightarrow ClONO_2$$

QUESTION 40

Assuming an ozone hole approximately the dimensions of Antarctica (area $= 1.4 \times 10^7$ km^2) in which the ozone column falls to 100 DU, how many molecules of O_3 have been destroyed? What fraction of total global ozone is this (see problem Q4)?

Q40 ANSWER The average ozone column without the ozone hole is 300 DU, so this represents a difference of 200 DU over an area of 1.4×10^7 km^2. This is equivalent to:

$$200 \text{ DU} \times \frac{2.69 \times 10^{16} \text{ molec cm}^{-2}}{1 \text{ DU}} \times 1.4 \times 10^7 \text{ km}^2 \times \left(\frac{100{,}000 \text{ cm}}{1 \text{ km}}\right)^2 = 7.53 \times 10^{35} \text{ molec}$$

From Q4, we know that global ozone is:

$$6.85 \times 10^{13} \text{ mol} \times \frac{6.022 \times 10^{23} \text{ molec}}{\text{mol}} = 4.13 \times 10^{37} \text{ molec } O_3$$

$$\text{Fraction is: } \frac{7.53 \times 10^{35} \text{ molec}}{4.13 \times 10^{37}} \times 100 = 1.8\%$$

1.8% of global ozone is lost under these ozone hole conditions.

10.7 The Montreal Protocol and CFC Substitutes

Upon recognition of the ozone hole crisis, industries requiring blowing agents and refrigerants began to seek alternative molecules with lower global warming potentials (GWPs) and ozone destruction potentials (ODPs). These data for a subset of CFCs and their substitutes are shown in Table 10.3.

Refer to Table 10.3, answer the following questions.

QUESTION 41

HCFCs were the first proposed replacement for CFCs. What can you glean about the likely reason for their selection based on the table? What chemical attribute is responsible for the key differences you see here?

TABLE 10.3 CFCs and Their Substitutes

Trade Name	Chemical Formula	Market	Atmospheric Lifetime (year)	100 Year GWP[a]	ODP[b]
CFC-11	CCl_3F	Blowing agent	50	4000	1.0
CFC-12	CCl_2F_2	Refrigerant	102	8500	1.0
CFC-113	CCl_2FCClF_2	Cleaning agent	85	5000	0.8
HCFC-22	CHF_2Cl	Refrigerant, blowing agent	12.1	1700	0.055
HCFC-14lb	CH_3CFCl_2	Blowing agent	9.4	630	0.11
HCFC-123	CF_3CHCl_2	Blowing agent	1.4	93	0.02
HFC-134a	CH_2FCF_3	Refrigerant, blowing agent	14.6	650	0.0
HFC-23	CHF_3	Fire extinguisher	260	11,700	0.0
HFC-227ea	C_3HF_7	Fire extinguisher	36.5	2900	0.0
HFC-245fa	$C_3H_3F_5$	Blowing agent	6.6	790	0.0

[a]Global warming potential = GWP. It is a measure of the degree of radiative forcing (see Chapter 7) of a given molecule compared to a molecule of CO_2, which is assigned a GWP value of 1.

[b]Ozone depletion potential = ODP. lf is the ratio of the impact on O_3 of a chemical compared to the impact of a similar mass of $CFCL_3$ (CFC-11), which is assigned the value of 1.

Sources: IPCC Special Report on Emission Scenarios, Section 5.4.3. Halocarbons and Other Halogenated Compounds, available at https://archive.ipcc.ch/ipccreports/sres/emission/index.php?idp=123; *Montreal Protocol on Substances That Deplete the Ozone Layer,* available at https://ozone.unep.org/treaties/montreal-protocol-substances-deplete-ozone-layer/text; adapted from Spiro et al., *Chemistry of the Environment, 3rd Ed.,* University Science Books, © 2012, all rights reserved.

Q41 ANSWER The HCFCs have smaller ODPs, primarily due to their shorter lifetimes. They have shorter lifetimes because the H–C bonds present in these molecules (and absent in the CFCs) enable them to be reactively lost to OH oxidation in the lower atmosphere before making their way up to the stratosphere.

*QUESTION 42

HFCs seem to have the best ODPs, all exactly = 0! Why are they not the CFC substitute of choice?

10.8 Conclusions

Ozone is maintained in steady state by a balance of radical reactions initiated by sunlight and requiring pressure stabilization to produce ozone. Catalytic loss cycles can throw this steady state out of balance, by initiating chain reactions that rapidly deplete ozone. There is a particularly strong depletion in the austral springtime, due to a combination of chemical and meteorological factors. Ozone is an example of a science policy success story, in which the time between scientific understanding of the problem and international political action is very brief.

PART IV

HYDROSPHERE AND LITHOSPHERE

11

WATER RESOURCES

Earth is quite literally a watery world. All living things depend absolutely on a supply of water. The biochemical reactions of every living cell take place in aqueous solution, and water is the transport medium for the nutrients a cell requires, and for the waste products it excretes. Water is abundant on the planet's surface, but ~97% of the earth's supply of water is in the oceans, where it is too salty to be used by humans or other land creatures. Thus, only 3% of the water on earth, including the glaciers, is potentially available for human consumption. Every day, however, the sun's rays distill a large quantity of water into the atmosphere that falls back to the surface as rain or snow. Proportionately more rain falls on land than on the oceans, providing a continual supply of freshwater. Water is often treated as if it were free and, in a sense, it is free. It is a by-product of the enormous flux of solar energy on earth. The hydrologic cycle accounts for about one-half of the solar energy absorbed by the earth's surface.

This chapter examines the hydrosphere by surveying the water resources and their locations. Topics include freshwater supplies, crop irrigation, groundwater, and water patterns in the United States, and ocean circulation.

11.1 Global Perspective

The global movement and storage of water are illustrated in Figure 11.1. Every year, approximately 111,000 km^3 of water falls on land, and around 71,000 km^3 returns to the atmosphere via evaporation from wet surfaces and transpiration from plants; these two processes are collectively called evapotranspiration. The remainder, about 40,000 km^3, is the runoff, which eventually reaches the oceans. If the runoff were divided evenly, it could provide each person with approximately 5,150 m^3 a year of freshwater (2020 population). But, of course, water access is not divided evenly.

QUESTION 1

If all of the precipitation falling from the sky were available for use by people, how much water is available per person (in m^3) if there are 7.8 billion people in the world (2020 data)?

Q1 ANSWER Divide the total precipitation by the population to determine the water per person:

$$\frac{111,000 \text{ km}^3}{7,800,000,000 \text{ people}} = 1.4 \times 10^{-5} \frac{\text{km}^3}{\text{person}}$$

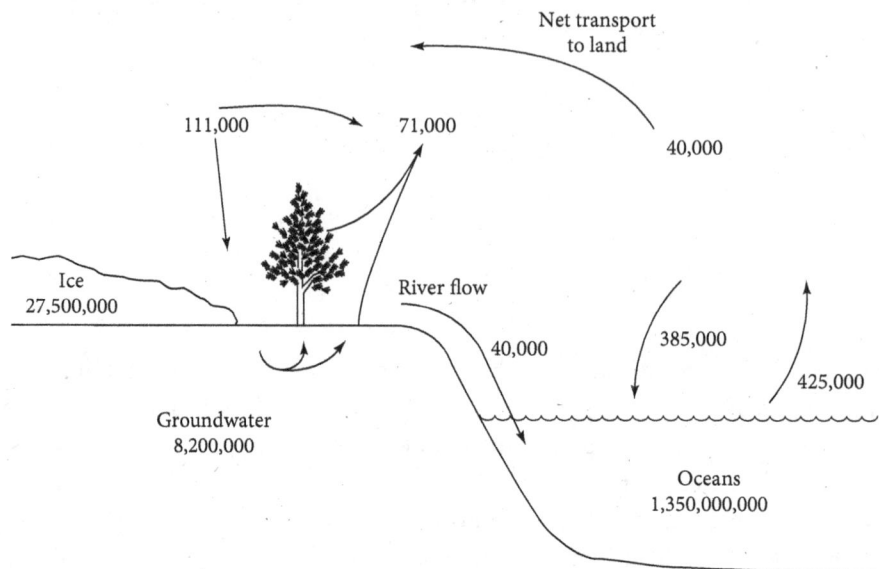

Atmosphere
13,000

Net transport
to land

111,000   71,000

40,000

Ice
27,500,000

River flow

385,000   425,000

40,000

Groundwater
8,200,000

Oceans
1,350,000,000

Figure 11.1 The global water cycle. The numbers are in cubic kilometers (km³) for the water reservoirs and cubic kilometers per year (km³ yr⁻¹) for the flows.
Sources: Adapted from Schlesinger, "The Global Water Cycle," in *Biogeochemistry: An Analysis of Global Change*, Academic Press, 1991, and Spiro et al., *Chemistry of the Environment, 3rd Ed.*, University Science Books, © 2012, all rights reserved.

Now convert to m³:

$$1.4 \times 10^{-5} \frac{km^3}{person} \times \left(\frac{1,000 \text{ m}}{1 \text{ km}}\right)^3 = 1.4 \times 10^4 \text{ m}^3$$

*QUESTION 2

What is the water available per person (in m³) in 2020 given a water runoff of 40,000 km³ and a population of 7.8 billion people.

QUESTION 3

If the water available per person is 5,100 m³ and a bath is 150 L, how many baths could a person take with this amount of water?

Q3 ANSWER First convert 150 L to m³ by converting to mL, then cm³, and finally m³:

$$150 \text{ L} \times \frac{1,000 \text{ mL}}{1 \text{ L}} \times \frac{1 \text{ cm}^3}{1 \text{ mL}} \times \left(\frac{1 \text{ m}}{100 \text{ cm}}\right)^3 = 0.15 \text{ m}^3$$

---

* Answers to starred questions can be found at the end of the book.

Divide the amount of water by the water in a bath to determine the number of baths:

$$\frac{5,100 \text{ m}^3}{0.15 \text{ m}^3} = 3.4 \times 10^4 \text{ baths}$$

This equates to 93 baths a day.

QUESTION 4

If the water available per person is 5,100 m^3 and you drink 2.7 L day^{-1}, how many days would it take a person to drink all of the water?

Q4 ANSWER First convert 2.7 L to m^3 by converting to mL, then cm^3 and finally m^3.

$$\frac{2.7 \text{ L}}{1 \text{ day}} \times \frac{1,000 \text{ mL}}{1 \text{ L}} \times \frac{1 \text{ cm}^3}{1 \text{ mL}} \times \left(\frac{1 \text{ m}}{100 \text{ cm}}\right)^3 = 0.0027 \frac{\text{m}^3}{\text{day}}$$

Divide the amount of water by the water a person can drink in a day to determine the number of days it would take to drink the amount of water.

$$\frac{5,100 \text{ m}^3}{0.0027 \frac{\text{m}^3}{\text{day}}} = 1.9 \times 10^6 \text{ days}$$

This equates to 5,175 yr of drinking 2.7 L of water a day.

*QUESTION 5

If the water available per person is 5,100 m^3 and you drink 3.5L a day, how many days would it take a person to drink all of the water?

QUESTION 6

If a person drinks 3.0 L of water a day, what percent of the 5,100 m^3 amount of water available to them would they drink in a year?

Q6 ANSWER First calculate the amount of water that the person would drink in a year:

$$3.0 \frac{\text{L}}{\text{day}} \times \frac{365 \text{ day}}{1 \text{ yr}} = 1,095 \text{ L}$$

Next convert L to m^3

$$\frac{1,095 \text{ L}}{1 \text{ yr}} \times \frac{1,000 \text{ mL}}{1 \text{ L}} \times \frac{1 \text{ cm}^3}{1 \text{ mL}} \times \left(\frac{1 \text{ m}}{100 \text{ cm}}\right)^3 = 1.095 \frac{\text{m}^3}{\text{day}}$$

Divide by the total amount available per year and multiply by 100%:

$$\frac{1.095 \text{ m}^3}{5,100 \text{ m}^3} \times 100\% = 0.021\%$$

*QUESTION 7

If a person bathes once a day in 175 L of water, what percent of the 5,100 m³ amount of water would be available to them for bathing in a year?

QUESTION 8

According to Figure 11.1, what percent of the water in reservoirs on the earth is in the oceans?

Q8 ANSWER First add the water from the ice, groundwater, and oceans:

$$27,500,000 \text{ km}^3 + 8,200,000 \text{ km}^3 + 1,350,000,000 \text{ km}^3 = 1,385,700,000 \text{ km}^3$$

Divide the water in the oceans by the total amount and multiply by 100%:

$$\frac{1,350,000,000 \text{ km}^3}{1,385,700,000 \text{ km}^3} \times 100\% = 97.4\%$$

*QUESTION 9

According to Figure 11.1, what percent of the water on earth is in ice?

*QUESTION 10

According to Figure 11.1, what percent of the water on earth is in groundwater?

Although the calculations above show there is a large amount of water per person on earth, water use is not universally divided among locations, and therefore available to people. Figure 11.2 shows the use of water on earth.

QUESTION 11

According to Figure 11.2, what percent of water is in saltwater form?

Q11 ANSWER Use the water drop visual in the upper left corner of the figure to answer this question. There are 100 drops in total and 97 of those drops are colored for saltwater. Therefore, approximately 97% of the water is saltwater.

*QUESTION 12

How much more water is used per capita in the United States than the global average?

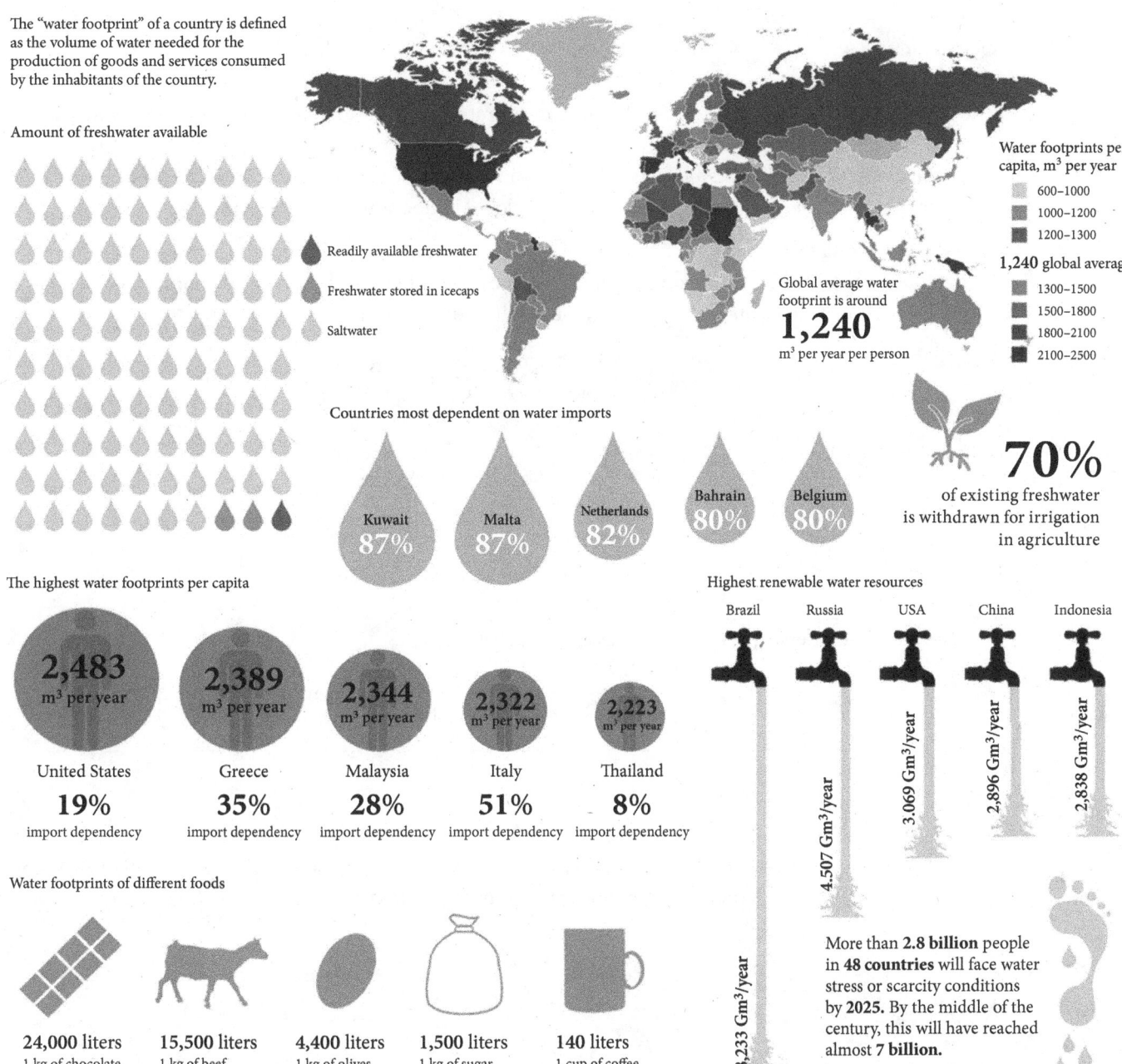

The "water footprint" of a country is defined as the volume of water needed for the production of goods and services consumed by the inhabitants of the country.

Amount of freshwater available

Readily available freshwater

Freshwater stored in icecaps

Saltwater

Water footprints per capita, m³ per year

600–1000
1000–1200
1200–1300

1,240 global average

1300–1500
1500–1800
1800–2100
2100–2500

Global average water footprint is around **1,240** m³ per year per person

Countries most dependent on water imports

Kuwait **87%**   Malta **87%**   Netherlands **82%**   Bahrain **80%**   Belgium **80%**

70% of existing freshwater is withdrawn for irrigation in agriculture

The highest water footprints per capita

2,483 m³ per year — United States — **19%** import dependency

2,389 m³ per year — Greece — **35%** import dependency

2,344 m³ per year — Malaysia — **28%** import dependency

2,322 m³ per year — Italy — **51%** import dependency

2,223 m³ per year — Thailand — **8%** import dependency

Highest renewable water resources

Brazil — 8,233 Gm³/year
Russia — 4,507 Gm³/year
USA — 3,069 Gm³/year
China — 2,896 Gm³/year
Indonesia — 2,838 Gm³/year

Water footprints of different foods

24,000 liters 1 kg of chocolate

15,500 liters 1 kg of beef

4,400 liters 1 kg of olives

1,500 liters 1 kg of sugar

140 liters 1 cup of coffee

More than **2.8 billion** people in **48 countries** will face water stress or scarcity conditions by **2025**. By the middle of the century, this will have reached almost **7 billion**.

Figure 11.2 Water footprint and availability.
Source: Adapted from WaterFootprint.org and WWF, *The Global Water Footprint,* available at https://aquadoc.typepad.com/.a/6a00d8341bf80a53ef0133f22eef77970b-pi.

Many people around the world are chronically short of water for personal needs. In numerous locations, freshwater aquifers are being drained faster than they can be replenished. Local reservoirs might be insufficient, especially in times of drought. Water resources are further strained by the increasing population. For example, unless the water supply in Africa and Asia increases significantly, the expected increases in population are predicted to place both continents under "water stress."

QUESTION 13

The global hydrological cycle is driven by solar evaporation of water on land and sea. This is how water is distributed to different locations around the world. Calculate the solar energy required to drive the global hydrological cycle using the data in Figure 11.1, assuming that heat of evaporation of water (both salt and fresh) at 15°C (the average global temperature) is 44.3 kJ mol^{-1}.

Q13 ANSWER Globally, a total of 496,000 km^3 of water (425,000 km^3 from oceans and 71,000 km^3 from land, represented by the arrows moving into the atmosphere in the figure) evaporates per year. Moles of water in 1 km^3 can be calculated (assume the density of water is 1 g mL^{-1}):

$$\text{grams water} = 1 \text{ km}^3 \times \left(\frac{100{,}000 \text{ cm}}{\text{km}}\right)^3 \times \frac{\text{mL}}{\text{cm}^3} \times 1\frac{\text{g}}{\text{mL}} = 1 \times 10^{15} \text{ g water}$$

$$\text{mol water} = (1 \times 10^{15} \text{ g}) \times \frac{1 \text{ mol}}{18.00 \text{ g}} = 5.56 \times 10^{13} \text{ mol water in 1 km}^3$$

Globally, total moles of water evaporated are:

$$5.56 \times 10^{13} \frac{\text{mol}}{\text{km}^3} \times 496{,}000 \text{ km}^3 = 2.76 \times 10^{19} \text{ mol water}$$

The global solar energy expended in evaporation of water is:

$$(2.76 \times 10^{19} \text{ mol water}) \times \left(44.3 \frac{\text{kJ}}{\text{mol}}\right) = 1.2 \times 10^{21} \text{ kJ}$$

11.2 Irrigation

Worldwide, agriculture accounts for the large majority of water use, approximately 70%, and agricultural demand is growing as the population continues to increase (Fig. 11.3).

QUESTION 14

Using Figure 11.3, calculate the rate of change in water use projected from 1900 to 2025 in terms of agricultural use.

Q14 ANSWER According to the figure, the water use in 1900 was about 550 km^3 yr^{-1} and is projected to be approximately 3,300 km^3 yr^{-1} in the year 2025.

To calculate the rate of change, divide the difference in the amount of water used by the total amount of years:

$$\frac{3{,}200\frac{\text{km}^3}{\text{yr}} - 550\frac{\text{km}^3}{\text{yr}}}{2025 - 1900} = \frac{2{,}650\frac{\text{km}^3}{\text{yr}}}{125} = 21\frac{\text{km}^3}{\text{yr}}$$

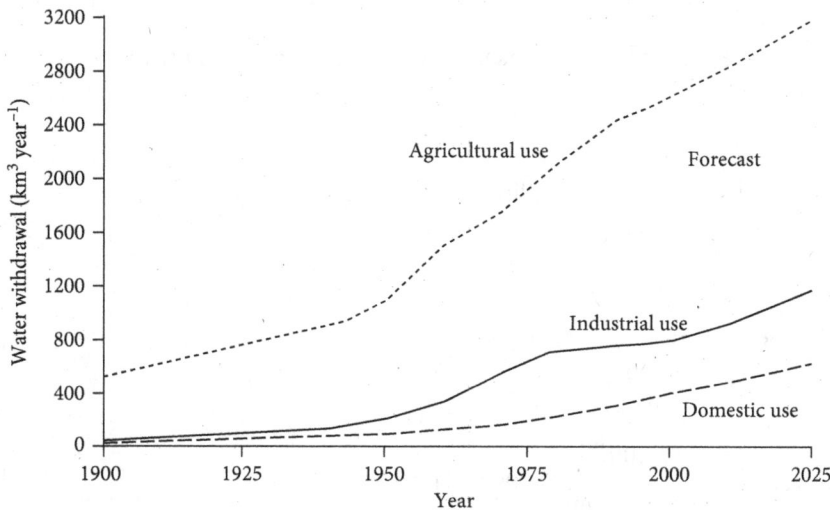

Figure 11.3 Global water use, 1900–2025.
Sources: Adapted from Shiklomanov, Ed., *World Water Resources at the Beginning of the 21st Century*, State Hydrological Institute/UNESCO and Spiro et al., *Chemistry of the Environment, 3rd Ed.*, University Science Books, © 2012, all rights reserved.

---

*QUESTION 15

Using Figure 11.3, calculate the rate of change in water use projected from 1900 to 2025 in terms of industrial use.

*QUESTION 16

Using Figure 11.3, calculate the rate of change in water use projected from 1900 to 2025 in terms of domestic use.

QUESTION 17

Calculate the percent agricultural, industrial, and domestic use in the year 2000.

Q17 ANSWER Using the graph, the amount of water withdrawal in 2000 for each of the uses is as follows:

Industrial use: 800 km^3 yr^{-1}

Domestic use: 400 km^3 yr^{-1}

Agricultural use: 2,600 km^3 yr^{-1}

Add the values to get the total amount:

$$800\frac{km^3}{yr} + 400\frac{km^3}{yr} + 2,600\frac{km^3}{yr} = 3,800\frac{km^3}{yr}$$

In order to calculate percent, take the amount for each use, divide by the total amount, and multiply by 100%:

$$\text{Industrial} = \frac{800\,\frac{km^3}{yr}}{3{,}800\,\frac{km^3}{yr}} \times 100\% = 20\%$$

$$\text{Domestic} = \frac{400\,\frac{km^3}{yr}}{3{,}800\,\frac{km^3}{yr}} \times 100\% = 10\%$$

$$\text{Agricultural} = \frac{2{,}600\,\frac{km^3}{yr}}{3{,}800\,\frac{km^3}{yr}} \times 100\% = 68\%$$

*QUESTION 18

Calculate the percent agricultural, industrial, and domestic use projected in the year 2025.

Water use in irrigation is dependent on the plant type as well. Figure 11.4 shows the amount of water needed for the transpiration of some plants that humans depend on for food. Transpiration is the movement of water through a plant followed by evaporation into the atmosphere.

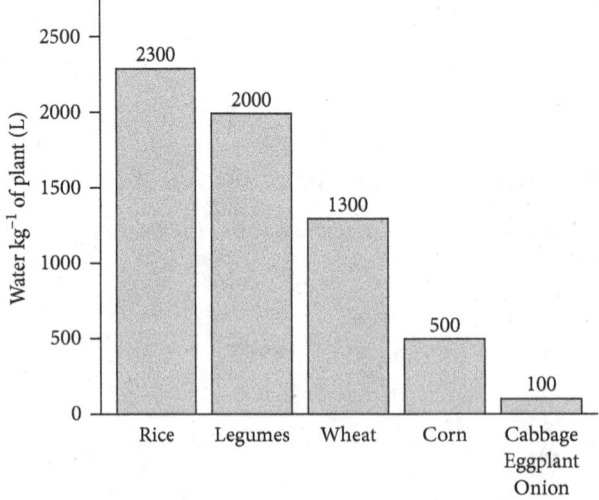

Figure 11.4 Transpiration of water from a variety of crops.
Sources: Adapted from Smil, "Water News: Bad, Good and Virtual," *American Scientist*, Vol. 96, No. 5, 2008, pp. 399–407, and Spiro et al., *Chemistry of the Environment, 3rd Ed.,* University Science Books, © 2012, all rights reserved.

QUESTION 19

According to Figure 11.4, how much more water transpires from rice than corn? Provide both relative and absolute amounts.

Q19 ANSWER The absolute amount is:

$$2{,}300\,\frac{\text{L H}_2\text{O}}{\text{kg plant}} - 500\,\frac{\text{L H}_2\text{O}}{\text{kg plant}} = 1{,}800\,\frac{\text{L H}_2\text{O}}{\text{kg plant}}$$

In order to calculate the relative amount, divide the amount of water for rice by the amount for corn:

$$\frac{2{,}300\,\frac{\text{L H}_2\text{O}}{\text{kg plant}}}{500\,\frac{\text{L H}_2\text{O}}{\text{kg plant}}} = 4.6 \text{ times}$$

*QUESTION 20

What is the percent increase in the amount of water that transpires between cabbage/eggplant/onion plants and wheat?

QUESTION 21

Consider a corn yield of 7,400 kg ha^{-1} (equivalent to 120 bushels acre^{-1}). If 25 kg (one bushel) of corn consumes ~20 m^3 of water during the growing season, what is the ratio of the mass of corn to the mass of water consumed? Where does most of the water end up? Assuming a rainfall of 30 cm yr^{-1}, calculate the minimum quantity of irrigation water required per hectare to grow the corn (1 ha = 1.0 × 10^4 m^2). If the efficiency of irrigation water delivery is only 50%, calculate how much water will need to be used to provide sufficient water.

Q21 ANSWER Given that 25 kg of corn takes up and transpires 20 m^3 of water, the weight of water is:

$$(20 \text{ m}^3 \text{ water}) \times \left(\frac{100 \text{ cm}}{\text{m}}\right)^3 \times \left(\frac{\text{mL}}{\text{cm}^3}\right) \times \left(1.00\,\frac{\text{g}}{\text{mL}}\right) = 2.0 \times 10^7 \text{ g water} = 2.0 \times 10^4 \text{ kg water}$$

The ratio of the mass of corn to the mass of water is:

$$\frac{25 \text{ kg corn}}{2.0 \times 10^4 \text{ kg water}} = 0.0013$$

Most of the water is lost through evapotranspiration.

The volume of 30 cm (0.30 m) rainfall on a hectare of land is:

$$(1.0 \times 10^4 \text{ m}^2) \times (0.30 \text{ m}) = 3.0 \times 10^3 \text{ m}^3$$

The mass of the rainfall is:

$$(3.0 \times 10^3 \text{ m}^3 \text{ water}) \times \left(\frac{100 \text{ cm}}{\text{m}}\right)^3 \times \left(\frac{\text{mL}}{\text{cm}^3}\right) \times \left(1\,\frac{\text{g}}{\text{mL}}\right) = 3.0 \times 10^9 \text{ g water} = 3.0 \times 10^6 \text{ kg water}$$

The weight of water consumed (W_{water}) in the production of 7,400 kg of corn from 1 hectare (ha) of land is:

$$W_{water} = (7.4 \times 10^3 \text{ kg corn}) \times \left(\frac{2.0 \times 10^4 \text{ kg water}}{25 \text{ kg corn}}\right) = 5.92 \times 10^6 \text{ kg water}$$

Minimum water irrigation:

$$5.92 \times 10^6 \text{ kg water needed} - 3.0 \times 10^6 \text{ kg water from rainfall} =$$
$$2.92 \times 10^6 \text{ kg water from irrigation}$$

The minimum amount of irrigation water required to grow the corn is 2.92×10^6 kg. If efficiency of delivering irrigation water to the plant were 50%, the amount of water required would be 5.84×10^6 kg.

*QUESTION 22

A farm yields 4,500 kg ha^{-1} of rice. If 55 kg of rice consumes ~184 m^3 of water during the growing season, what is the ratio of the mass of rice to the mass of water consumed? Where does most of the water end up? Assuming a rainfall of 15 cm yr^{-1}, calculate the minimum quantity of irrigation water required per hectare to grow the corn (1 ha = 1.0×10^4 m^2).

11.3 "Virtual Water" and the Issue of Meat Production

Many products require water for their production, and trade in these products, therefore, represents a transfer of "virtual" water (the volume required in the production) from the producing to the consuming locale. The numbers are particularly large for agricultural products. The real standout for "virtual" water is meat. Because the food value of plants is transferred to animal flesh at only ~10% efficiency, the virtual water content of meat is an order of magnitude higher than plants (Fig. 11.5).

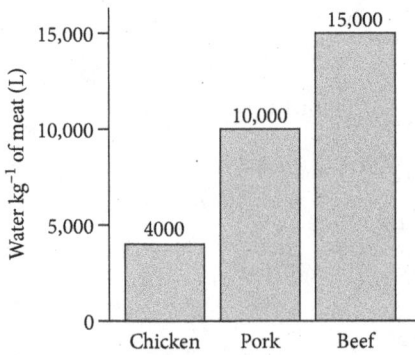

Figure 11.5 Liters of water used per kilogram of meat produced.
Sources: Adapted from Smil, "Water News: Bad, Good and Virtual," *American Scientist*, Vol. 96, No. 5, 2008, pp. 399–407, and Spiro et al., *Chemistry of the Environment, 3rd Ed.*, University Science Books, © 2012, all rights reserved.

*QUESTION 23

According to Figure 11.4, how much more water is used in beef than in chicken? Find absolute and relative values.

*QUESTION 24

What is the percent increase in the amount of water that is used between pork and chicken?

11.4 Water Resources in the United States

To gain a more detailed perspective on water resources, we examine the pattern of distribution and use in the continental United States. A schematic of water flows is shown in Figure 11.6. As described in the global water cycle, about two-thirds of the precipitation is returned to the atmosphere via evapotranspiration; the balance goes to runoff, amounting to ~2,000 km^3 yr^{-1}. Most of the water is in the east of the country and flows to the Atlantic and Gulf coasts. One-quarter of the annual runoff (468 km^3) is withdrawn for various uses, of which three-quarters (338 km^3) is subsequently returned to the stream flow and the remainder is "consumed," mostly through evapotranspiration after or during use; eventually, of course, all the water is returned to the global cycle.

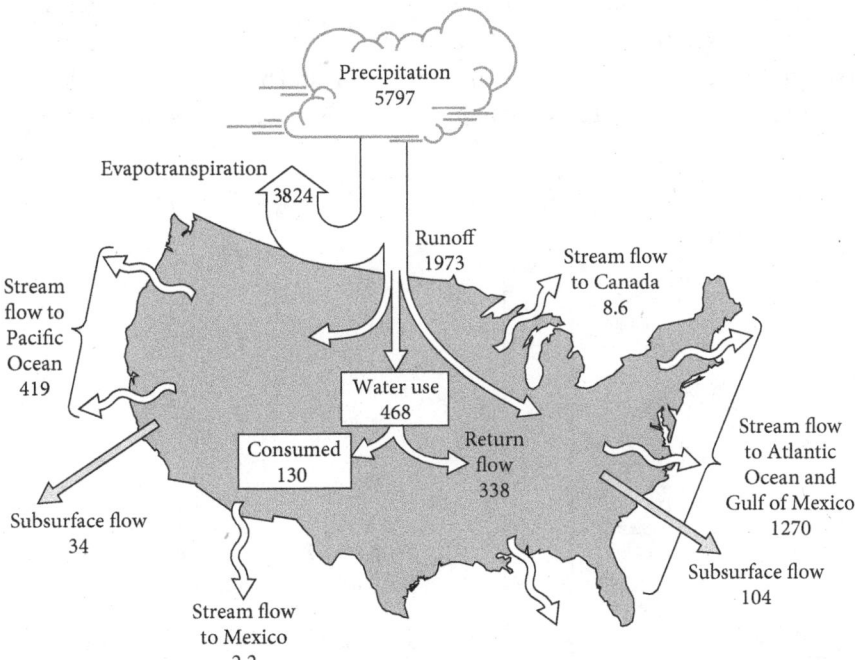

Figure 11.6 Annual water flows in the United States. The numbers are in units of cubic kilometers per year (km^3 yr^{-1}).
Sources: Adapted from the U.S. Water Resources Council, *The Nation's Water Resources, 1975–2000: Second National Water Assessment, Vol. 1: Summary* (Stock number 052-045-00051-7), 1978, Superintendent of Documents, and Spiro et al., *Chemistry of the Environment, 3rd Ed.,* University Science Books, © 2012, all rights reserved.

QUESTION 25

What percent of the water from precipitation in the United States goes through the evapotranspiration process?

Q25 ANSWER Divide the amount of water undergoing evapotranspiration by the amount of water from precipitation and multiply by 100%:

$$\frac{3{,}824\frac{km^3}{yr}}{5{,}797\frac{km^3}{yr}} \times 100\% = 65.97\%$$

*QUESTION 26

What percent of the water from precipitation in the United States goes through streamflow?

*QUESTION 27

What percent of the water from precipitation in the United States goes through subsurface flow?

The pattern of water utilization is shown in Figure 11.7. Of the 542 km^3 of freshwater used in 2005, 80% was from surface supplies and the rest was from groundwater, which is eventually recharged from the surface. Water usage is dominated by irrigation and by the cooling of electric generators. The remaining water is divided among industry and mining, domestic, and livestock

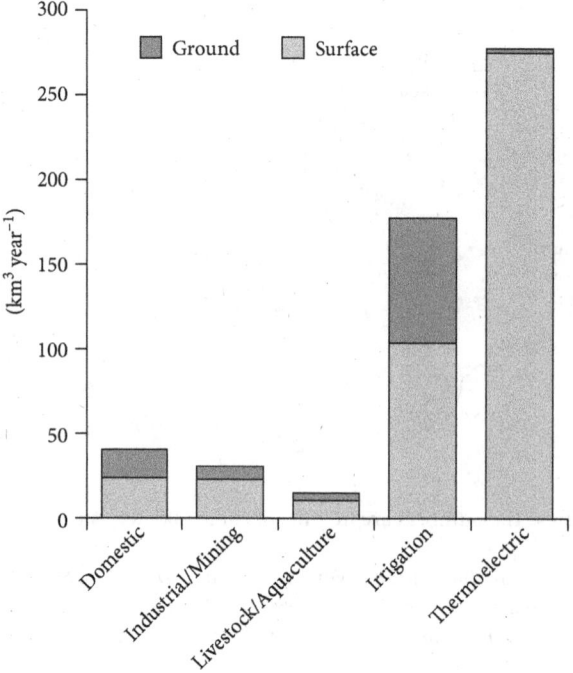

Figure 11.7 Sources and uses of water in the United States in 2005. Total use was 542 km^3 yr^{-1}.
Sources: Adapted from Kenny et al., "Estimated Use of Water in the United States in 2005," *U.S. Geological Survey Circular*, Vol. 1344, 2009, p. 52, and Spiro et al., *Chemistry of the Environment, 3rd Ed.*, University Science Books, © 2012, all rights reserved.

and aquaculture. Major industrial users are the steel, chemical, and petroleum industries; the mining uses include water extraction of minerals and fossil fuels, and milling and related activities. Most of the industrial and mining water is obtained directly from surface and groundwater, but ~20% of it comes from public supplies, which also provide most of the domestic and commercial water. Domestic uses include drinking, food preparation, bathing, washing clothes and dishes, flushing toilets, and watering lawns and gardens. The total domestic uses were about three times higher than commercial uses (hotels, restaurants, office buildings, and so on).

QUESTION 28

Using Figure 11.7, calculate the percent of groundwater and surface water from the total water used for irrigation.

Q28 ANSWER From the figure, approximately 100 km³ yr⁻¹ is used from surface water for irrigation and approximately 80 km³ yr⁻¹ is used from groundwater. First determine the total amount of water used and then divide the surface and groundwater amounts by the total amount. Then multiply by 100% to get the percent used:

$$\text{Total} = 100\frac{km^3}{yr} + 80\frac{km^3}{yr} = 180\frac{km^3}{yr}$$

$$\text{Groundwater: } \frac{80\frac{km^3}{yr}}{100\frac{km^3}{yr}} \times 100\% = 40\%$$

$$\text{Surface water: } \frac{100\frac{km^3}{yr}}{180\frac{km^3}{yr}} \times 100\% = 60\%$$

*QUESTION 29

Using Figure 11.7, calculate the percent of groundwater and surface water from the total water used for domestic use.

*QUESTION 30

Using Figure 11.7, determine the percent water used for each of the five listed uses from the total water used.

11.5 Conclusions

Although a lot of water is cycled on and through the earth, the amount of water available in different areas is not equitably distributed. Water is also not used equally around the world. Understanding the use of water and its location is important for scientists, environmentalists, government officials, and citizens. It is important to monitor the water supply and find ways to ensure the population receives fresh water.

12

WATER AS SOLVENT: ACIDS AND BASES

12.1 Water: Hydrogen Bonds and Unique Properties

One characteristic that makes Earth uniquely suitable for the evolution of life is its surface temperature that, over most of the planet, lies within the liquid range of water. One of the most commonplace substances, water, has some remarkable properties, such as its unusually high melting and boiling points, and the ease with which it dissolves ionic compounds or polar molecules. Specifically, compared to any neighboring hydride (a compound formed between hydrogen and another atom close to oxygen in the periodic table), water has by far the highest melting and boiling points. Fundamentally, this is due to the unique molecular structure of water, in particular its ability to form a three-dimensional (3D) network of hydrogen bonds (or H bonds). A hydrogen bond is the strong electrostatic interaction between a H atom bonded with an electronegative atom (one that has a strong attraction for electrons), X, and a lone pair of electrons on another electronegative atom, Y, as shown below by the dotted line between H and Y:

$$X - H \cdots :Y -$$

The elements (X or Y) that most commonly form H bonds are N, O, and F, and to a weaker extent P, S, and Cl, due to the difference in electronegativity among the different atoms. From N to O to F, the atoms are increasingly electronegative because of their growing nuclear charge; therefore, they can effectively donate H bonds (by pulling the bonding electrons toward X and away from H) as well as accept H bonds (by retaining lone pairs of electrons on Y). In nature, the most important H bonds are those with N and/or O as X and Y; for example, they occur at key points in biological molecules such as deoxyribonucleic acids (DNAs).

Let us examine water's unique structure and properties with two sets of comparisons in the next two problems.

QUESTION 1

H_2S boils at $-61°C$, H_2Se at $-42°C$, and H_2Te at $-2°C$. Based on this trend in group VI of the periodic table, at what temperature would one expect H_2O to boil? Explain any anomaly in the trend.

Q1 ANSWER According to the trend, water would be expected to boil around $-70°C$ or lower. Its actual boiling point, $100°C$, is much higher. This anomaly is primarily due to a water molecule's ability to donate two hydrogen bonds, one from each of its H atoms, while simultaneously accepting

---

two H bonds, one to each of the electron lone pairs on the O atom. Note that the two H–O bonds and two electron lone pairs form a tetrahedral structure, which can be extended in space. This 3D network of H bonds greatly enhances the intermolecular force between liquid water molecules, thus requiring more energy (and thus, higher temperature) to release molecules into the gas phase. This network of H bonds also accounts for the very high melting point of ice. Figure 12.1 shows the 3D network of H bonds in ice.

*QUESTION 2

As seen in Table 12.1, H_2O has the highest melting point and boiling point, compared to those of the hydrides of other elements in the same period. Explain the anomaly in the trend.

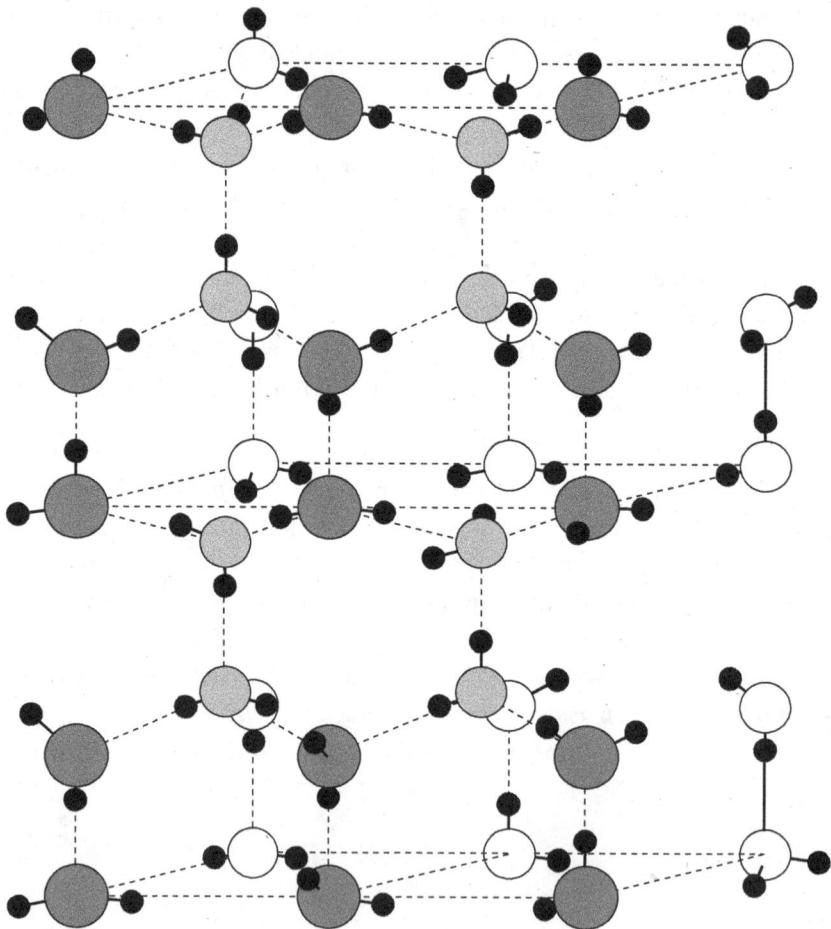

Figure 12.1 The crystal structure of ice showing the network of H bonds. The larger circles represent O atoms and the smaller circles represent H atoms. O atoms with the same color shading indicate that they lie in the same plane.
Source: Adapted from Spiro et al., *Chemistry of the Environment, 3rd Ed.,* University Science Books, © 2012, all rights reserved.

---

* Answers to starred questions can be found at the end of the book.

TABLE 12.1 Melting and Boiling Points for Water and
Neighboring First-Row-Element Hydrides

Temperature	CH_4	NH_3	H_2O	HF
Melting point (°C)	−182	−78	0	−83
Boiling point (°C)	−164	−33	100	20

Source: Adapted from Spiro et al., *Chemistry of the Environment, 3rd Ed.,* University
Science Books, © 2012, all rights reserved.

12.2 Ions, Autoionization, and pH

As mentioned earlier, another unique property of water is the ease with which it dissolves ionic
compounds. Many salts dissolve readily in water because water molecules strongly solvate
both cations and anions. The strong solvation forces stem from water's large dipole moment
and H bonding ability. Anions interact with the positive end of the molecule via H bonds,
while cations interact with the negative end via the formation of coordinate bonds with the
oxygen. Figure 12.2 shows these interactions using the example of the solvation of NaCl.

Interestingly, water can self-ionize because the ions (H^+ and OH^-) are stabilized by other
water molecules. This autoionization reaction can be fully written as:

$$H_2O + H_2O \leftrightharpoons H_3O^+(aq) + OH^-(aq) \tag{12.1}$$

For simplification, it is often written as:

$$H_2O \leftrightharpoons H^+(aq) + OH^-(aq) \tag{12.2}$$

The position of equilibrium of reactions (12.1) and (12.2) lies far to the left, resulting in
a very small equilibrium constant value. The ion-product constant for pure water, K_w, which
equals $[H^+][OH^-]$ [or strictly speaking $[H_3O^+][OH^-]$, since H^+ is hydrated in water as shown
in reaction (12.1)], is 1.00×10^{-14} at 25°C. Equilibrium constants do change somewhat with
temperature. We will explore this temperature dependence of water autoionization equilibrium in the next few problems.

In solution chemistry, it is standard to express the $[H^+]$ in its negative 10-based logarithm
(symbolized by "p") as pH:

$$pH = -\log[H^+] \tag{12.3}$$

Similarly:

$$pOH = -\log[OH^-] \tag{12.4}$$

Since $K_w = [H^+][OH^-]$, we can take the "negative logarithm" on both sides:

$$-\log(K_w) = -\log([H^+][OH^-]) = -\log[H^+] - \log[OH^-]$$

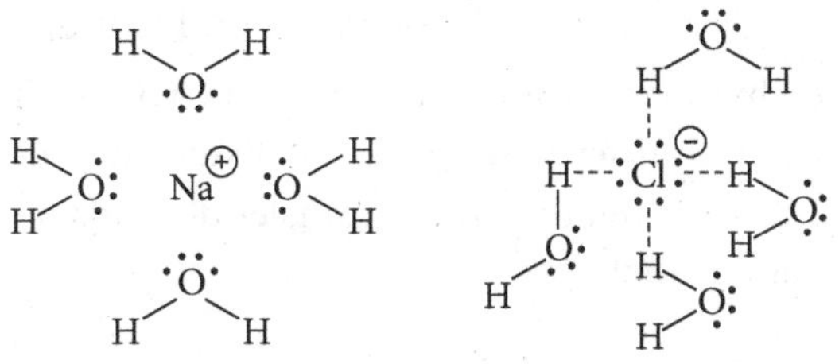

Figure 12.2 Solvation of ions from NaCl in
water.
Source: Adapted from Spiro et al., *Chemistry
of the Environment, 3rd Ed.,* University
Science Books, © 2012, all rights reserved.

To obtain the following relationship:

$$pK_w = pH + pOH \qquad (12.5)$$

Therefore, $pH = pK_w - pOH$, which can be used to calculate the pH value where $[OH^-]$ is given.

QUESTION 3

At 25°C, show that water has a pH of 7.00. K_w for pure water is 2.93×10^{-15} and 1.47×10^{-14} at 10°C and 30°C, respectively. Calculate the pH of water at each temperature. Is water neutral at each of these temperatures?

Q3 ANSWER Since for pure water, regardless of the temperature, $[H^+] = [OH^-]$, $K_w = [H^+][OH^-] = [H^+]^2$. Using each given K_w value, $[H^+]$ can be calculated by the square root of K_w:

At 10°C: $[H^+] = \sqrt{2.93 \times 10^{-15}} = 5.41 \times 10^{-8}$ M:

$$pH = -\log [5.41 \times 10^{-8}] = 7.27$$

At 25°C: $[H^+] = \sqrt{1.00 \times 10^{-14}} = 1.00 \times 10^{-7}$ M:

$$pH = -\log [1.00 \times 10^{-7}] = 7.00$$

At 30°C: $[H^+] = \sqrt{1.47 \times 10^{-14}} = 1.21 \times 10^{-7}$ M:

$$pH = -\log [1.21 \times 10^{-7}] = 6.92$$

Even though the pH values at nonstandard temperatures are not 7.00 (the value commonly recognized as the neutral pH, but strictly speaking only true at 25°C, the standard temperature), water is still neutral, regardless of the temperature because H^+ and OH^- are equal in molar concentrations. However, these calculations show that the pH value for neutral water decreases slightly with temperature increase and vice versa.

---

*QUESTION 4

For 1.00 M HCl (strong acid) and NaOH (strong base), calculate the pH at 10°C, 25°C, and 30°C. You may use the K_w values given in Q3. Explain the implication of your results.

12.3 Acid and Base Dissociation Constants, Conjugate Acid–Base Pairs, and Buffers

Strong acids transfer protons completely to water, but many acidic substances, such as weak acids, hold the proton to some extent, and the proton transfer to water, or the dissociation of the acid, is incomplete. For a generalized acid (HA), the extent of proton transfer depends on the equilibrium constant (K_a) for the acid dissociation reaction (12.6a):

$$HA\,(aq) \rightleftharpoons H^+\,(aq) + A^-\,(aq) \tag{12.6a}$$

$$\text{At equilibrium, } K_a = \frac{[H^+][A^-]}{[HA]} \tag{12.6b}$$

K_a is also often referred to as acid dissociation constant for a weak acid HA. The smaller the K_a value, the lesser the extent of proton transfer from the acid to water, and the smaller the fraction of the acid that dissociates.

If H^+ does not dissociate from HA completely, then the anion (A^-) will have some tendency to remove a proton from water to form HA, as shown below:

$$A^-\,(aq) + H_2O \rightleftharpoons HA\,(aq) + OH^-\,(aq) \tag{12.7a}$$

Thus, if a salt of A^- is added to water, its hydrolysis, or the base dissociation reaction (12.7a), will proceed to some extent. Since OH^- is produced, A^- acts as a base. Examples of A^- include acetate ($C_2H_3O_2^-$), formate (HCO_2^-), fluoride (F^-), and nitrite (NO_2^-), to name a few.

Assuming the equilibrium in (12.7a) does not lie completely to the right, A^- is a weak base. The base dissociation constant, or basicity constant, is:

$$K_b = \frac{[HA][OH^-]}{[A^-]} \tag{12.7b}$$

The water concentration, $[H_2O]$, being constant because the concentration is so large, is not included in the equilibrium constant K_b.

The acid and base dissociation reactions (12.6a) and (12.7a) are linked by the autoionization reaction (12.2). This can be seen by adding reactions (12.6a) and (12.7a), which yields reaction (12.2). When reactions are added, the equilibrium constants multiply; that is:

$$K_a K_b = K_w \tag{12.8}$$

Thus, K_b of base A^- can be derived from K_a of the acid HA, and vice versa, according to (12.8). A weak acid (HA) and its anion (A^-) constitute a conjugate acid–base pair. The two are of the same molecular entity, differing only by one proton. For example, acetic acid (CH_3COOH) and the acetate ion (CH_3COO^-) constitute a conjugate acid–base pair, as seen in:

$$CH_3COOH\,(aq) \rightleftharpoons H^+\,(aq) + CH_3COO^-\,(aq)$$

Similarly, the ammonium ion (NH_4^+) and ammonia (NH_3) constitute a conjugate acid–base pair, as observed in:

$$NH_3\,(aq) + H_2O \rightleftharpoons NH_4^+\,(aq) + OH^-\,(aq)$$

An important characteristic of a conjugate acid–base pair is that a mixture of the acid and its conjugate base has a pH that is close to the negative logarithm of the acid dissociation constant (pK_a). This result can be derived by rearranging the equilibrium expression (12.6b) and taking negative log on both sides:

$$[H^+] = \frac{K_a[HA]}{[A^-]}$$

$$-\log[H^+] = -\log[K_a] - \log\frac{[HA]}{[A^-]}$$

Thus:

$$pH = pK_a + \log\frac{[A^-]}{[HA]} \qquad (12.9)$$

If $[A-] = [HA]$, or the concentration of the weak acid and its conjugate base are equal, then the pH is exactly pK_a because log 1 = 0. Moreover, the pH will be close to pK_a, as long as $[A-]/[HA]$ is not far from unity. Even if this ratio reaches a value of 10 or 0.1, the pH will deviate from pK_a by no more than one numerical unit. Thus, the pH is buffered against large changes. A mixture of HA and A– constitutes a buffer solution, because it resists large pH changes when small amounts of other acids or bases are added to the solution. Equation (12.9) is called the Henderson–Hasselbalch (H-H) equation, which is frequently used in buffer-related calculations, including in calculations related to titrations below.

QUESTION 5

Calculate the pH of a buffer solution of formic acid (HCOOH) with a pK_a of 3.75 if the concentration of the acid is 1.5 M and the formate ion is 1.25 M in solution.

Q5 ANSWER Use the H–H equation to calculate the pH:

$$pH = pK_a + \log\frac{[A^-]}{[HA]} = 3.75 + \log\frac{1.25 \text{ M}}{1.5 \text{ M}} = 3.67$$

*QUESTION 6

Benzoic acid (C_6H_5COOH) is a weak acid with a pK_a of 4.20. Calculate the pH of a buffer solution of 0.75 M benzoic acid and 1.25 M benzoate ion.

QUESTION 7

Calculate the ratio of molarities of PO_4^{3-} and HPO_4^{2-} ions in a buffer solution with a pH of 11.0. pK_a for $HPO_4^- \rightleftharpoons H^+ + PO_4^{3-}$ is 12.32.

Q7 ANSWER We can use the H–H equation to calculate the ratio of molarities

$$pH = pK_a + \log\frac{[A^-]}{[HA]} = 12.32 + \log\frac{[PO_4^{3-}]}{[HPO_4^{2-}]} = 11.0$$

$$\frac{[PO_4^{3-}]}{[HPO_4^{2-}]} = 0.048$$

*QUESTION 8

A solution of acetic acid has a pH of 7.00. Calculate the ratio of conjugate base to acid. pK_a of acetic acid is 4.75.

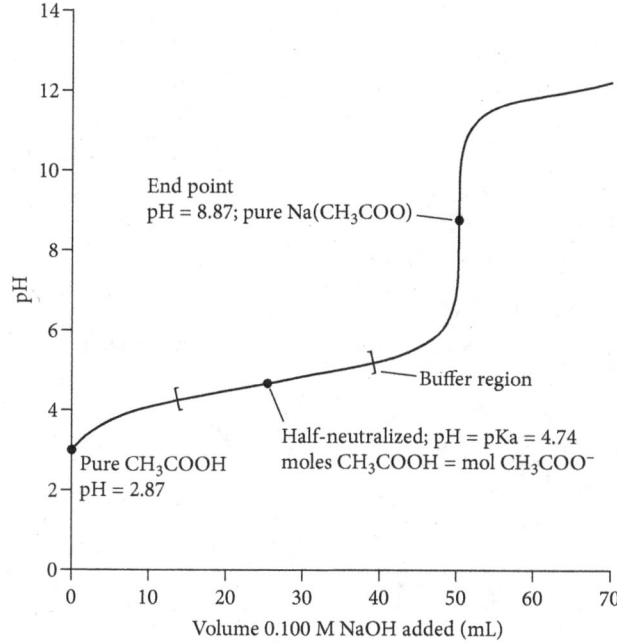

Figure 12.3 The change in pH during the titration of a weak acid (50.00 mL of 0.100 M acetic acid) with a strong base (0.100 M NaOH). *Source:* Adapted from Spiro et al., *Chemistry of the Environment, 3rd Ed.,* University Science Books, © 2012, all rights reserved.

12.4 Weak Acids and Bases: Titration

Titration is a common laboratory method to determine quantitatively the concentration of an identified analyte. A reagent, often termed the titrant, is prepared as a standard solution of known concentration and reacts with the solution of the analyte until the reaction is just complete according to the stoichiometry (i.e., reaching the equivalence point or instant of equal moles of titrant and analyte). Identifying the equivalence point is often done with the help of an indicator that changes color at the completion of the reaction. The analyte concentration can then be deduced based on the titration data. Common types include acid-base, redox, precipitation, and complexometric titrations. These analyses can be used to evaluate environmental samples for acid–base properties, redox potential, and metal concentrations.

Figure 12.3 shows the titration curve for a 0.100 M acetic acid solution (50.0 mL, with the acid dissociation constant $K_a = 1.78 \times 10^{-5}$ at 25°C), to which successive amounts of a strong base (NaOH) are added. As the titration progresses, the dominant acid–base equilibrium actually changes, which is examined in stepwise detail in Q9.

QUESTION 9

Calculate the pH of the solution when the following volumes of 0.100 M NaOH are added to 50.00 mL of 0.100 M acetic acid (HA) solution:

(a) 0 mL
(b) 5.00 mL (10% to equivalence point)
(c) 25.00 mL (50% to equivalence point)
(d) 45.00 mL (90% to equivalence point)
(e) 50.00 mL (equivalence point)

(f)   55.00 mL (10% excess of the base titrant)

(g)   70.00 mL (40% excess of the base titrant)

(h)   Generate the approximate shape of the titration curve from the results in parts (a)–(g) and compare with Figure 12.3.

Acetic acid has an acid dissociation constant of 1.78×10^{-5} at 25°C.

Q9 ANSWER Refer to Figure 12.3 as you work through the calculation.

(a)   When no NaOH is added, only 0.100 M acetic acid (HA) is present. We lay out the Initial, Change and Equilibrium concentrations in the original acetic acid solution (this "ICE" approach is frequently used in acid–base calculations):

	$HA\,(aq) \rightleftharpoons H^+\,(aq) + A^-\,(aq)^*$		
Initial Concentration (M)	0.100	0	0
Change (M)	$-x$	$+x$	$+x$
Equilibrium Concentration (M)	$0.100 - x$	$+x$	$+x$

*Strictly speaking, this reaction proceeds as $HA\,(aq) + H_2O \rightleftharpoons H_3O^+\,(aq) + A^-\,(aq)$ in the solution, and $[H_3O^+]$ is the hydrated proton and is used to calculate the pH in some other books. The results will be the same if using H^+ in the above-implied equation in the ICE table, and $pH = -\log[H^+]$ as defined earlier.

Note that in this particular calculation, the volume of the solution is fixed, so the change in the molar concentration of each reactant and product is proportional to that of the corresponding mole, obeying the stoichiometric ratio. The decrease in the reactant amount is indicated by a "–" sign, whereas the increase in the product amount is indicated by a "+" sign. The "Equilibrium Concentration" is the addition of the "Initial Concentration" with "Change."

At equilibrium:

$$K_a = \frac{[H^+][A^-]}{[HA]} \tag{12.6b}$$

Thus, according to the above ICE table:

$$\frac{x^2}{0.100 - x} = 1.78 \times 10^{-5}$$

Since K_a is very small, indicating that acetic acid ionizes only to a small extent, we can assume that x is far smaller than 0.100 M. We can then make the approximation and check at the end of the calculation to see if our assumption is correct:

$$0.100 - x \approx 0.100$$

Otherwise we would need to solve the quadratic equation as discussed below.

Now the above equation becomes:

$$\frac{x^2}{0.100} = 1.78 \times 10^{-5}$$

$$x^2 = 0.100 \times 1.78 \times 10^{-5} = 1.78 \times 10^{-6}$$

$$x = 1.33 \times 10^{-3}\,M$$

Since $x = [H^+]$, as seen in the above ICE table:

$$pH = -\log[H^+] = -\log(1.33 \times 10^{-3}) = 2.87$$

The above assumption can be examined by calculating the percent dissociation of the acid:

$$\frac{x}{0.100} \times 100\% = \frac{1.33 \times 10^{-3}}{0.100} \times 100\% = 1.33\%$$

So x is far smaller than 5% of 0.100, so the approximation made above is valid. Typically, if K_a is less than the initial [HA] by at least two orders of magnitude, the assumption that [H$^+$] is far smaller than the initial [HA] is correct. It is good practice to double-check the assumption and make sure that the percent dissociation is ≤5%. This is a useful technique to avoid solving the quadratic formula while largely maintaining the mathematical accuracy of the answer. We shall use this simplification technique frequently in acid–base equilibrium calculations.

However, if x turns out to be larger than 5% of the value from which it is subtracted, then the quadratic formula needs to be solved to obtain the most accurate answer. The quadratic equation will be used in later calculation.

(b) When 5.00 mL of 0.100 M NaOH is added to 50.00 mL of 0.100 M HA, the former would titrate 5.00 mL of HA according to reaction (12.7a), leaving 45.00 mL of HA unreacted:

$$HA + OH^- \rightarrow A^- + H_2O \tag{12.7a}$$

So, 10% (= 5.00/50.00 mL) of HA (based on moles) is converted to A$^-$, with 90% of HA remaining unreacted. This is now a buffer solution containing an acid (HA) and its conjugate base (A$^-$) in mole (as well as concentration) is a ratio of 9:1. As discussed earlier, the H–H equation (12.9) is used to calculate the pH of a buffer:

$$pH = pK_a + \log\frac{[A^-]}{[HA]} = -\log(1.78 \times 10^{-5}) + \log\frac{1}{9} = 3.80$$

(c) When 25.00 mL of NaOH is added to 50.00 mL of HA, 50% of HA is converted to A$^-$ with 50% of HA remaining unreacted, according to reaction (12.7a). The moles of HA is equal to that of A$^-$(ratio 1:1). This is also a buffer solution. Using the H-H equation:

$$pH = pK_a + \log\frac{[A^-]}{[HA]} = -\log(1.78 \times 10^{-5}) + \log\frac{1}{1} = 4.75$$

Note that pH = pK_a at this half-equivalence, or half-neutralized, point.

(d) When 45.00 mL of NaOH is added to 50.00 mL of HA, 90% of HA is converted to A$^-$, with 10% of HA remaining unreacted, according to reaction (12.7a). This is again a buffer solution. Using the H–H equation:

$$pH = pK_a + \log\frac{[A^-]}{[HA]} = 4.75 + \log\frac{9}{1} = 5.70$$

Note in scenarios (b)–(d), NaOH is the limiting reactant in the acid–base titration.

(e)   When 50.00 mL of NaOH is added to 50.00 mL of HA, all of HA is converted to A^-, so this is an equivalence point. The total volume is 100.00 mL and the total conjugate base concentration C_{A^-} is:

$$\text{mol } C_{A^-} = 0.100\frac{\text{mol}}{\text{L}} \times \frac{1\text{ L}}{1,000\text{ mL}} \times 50.00\text{ mL} = 5.00 \times 10^{-3}\text{ mol } A^-$$

$$[C_{A^-}] = \frac{5.00 \times 10^{-3}\text{ mol } A^-}{(50.00 + 50.00\text{ mL})} \times \frac{1,000\text{ mL}}{1\text{ L}} = 5.00 \times 10^{-2}\text{ mol}$$

A^- will reach equilibrium with water, HA, and OH^- as shown in the ICE table below:

	$A^-\,(aq) + H_2O \rightleftharpoons HA\,(aq) + OH^-\,(aq)$		
Initial Concentration (M)	0.0500	0	0
Change (M)	$-x$	$+x$	$+x$
Equilibrium Concentration (M)	$0.0500 - x$	$+x$	$+x$

We can derive the equilibrium constant for A^-, the conjugate base of HA, according to (12.8):

$$K_b = \frac{K_w}{K_a} = \frac{1.00 \times 10^{-14}}{1.78 \times 10^{-5}} = 5.62 \times 10^{-10}$$

At equilibrium:

$$K_b = \frac{[HA][OH^-]}{[A^-]}$$

According to the above ICE table,

$$K_b = \frac{x^2}{0.0500 - x} = 5.62 \times 10^{-10}$$

Assuming x is far smaller than 0.0500, we can make the following approximation:

$$\frac{x^2}{0.0500} = 5.62 \times 10^{-10}$$

$$x = 5.30 \times 10^{-6}\text{ M}$$

(Verify: x is less than 5% of 0.050 M, the above assumption is correct)
Thus:

$$pOH = -\log[OH^-] = -\log(x) = -\log(5.30 \times 10^{-6}\text{ M}) = 5.28$$

$$pH = 14.00 - pOH = 14.00 - 5.28 = 8.72$$

(f)   When 55.00 mL NaOH is added, the excess NaOH (HA now being the limiting reactant) is the dominant base, and OH^- generated from A^- reacting with water can be ignored because it is so small.
   First, we need to subtract the number of moles of OH^- reacted away from the total number of moles of OH^- added to the solution, and then divide by the total volume of the solution:

$$\text{mol } OH^- = \left(55.00\text{ mL} \times \frac{1\text{ L}}{1,000\text{ mL}} \times \frac{0.100\text{ mol}}{\text{L}}\right) - \left(50.00\text{ mL} \times \frac{1\text{ L}}{1,000\text{ mL}} \times \frac{0.100\text{ mol}}{\text{L}}\right)$$

$$= 5.00 \times 10^{-4}\text{ mol } OH^-$$

$$[OH^-] = \frac{5.00 \times 10^{-4} \text{ mol OH}^-}{55.00 + 50.00 \text{ mL}} \times \frac{1,000 \text{ mL}}{1 \text{ L}} = 4.76 \times 10^{-3} \text{ M OH}^-$$

Thus, pOH = $-\log(4.76 \times 10^{-3}$ M$) = 2.32$ and pH = $14.00 -$ pOH = $14.00 - 2.32 = 11.68$

(g)   When 70.00 mL NaOH is added, the excess NaOH is again the dominant base; thus:

$$OH^- = \left(70.00 \text{ mL} \times \frac{1 \text{ L}}{1,000 \text{ mL}} \times \frac{0.100 \text{ mol}}{L}\right) - \left(50.00 \text{ mL} \times \frac{1 \text{ L}}{1,000 \text{ mL}} \times \frac{0.100 \text{ mol}}{L}\right)$$

$$= 2.00 \times 10^{-3} \text{ mol OH}^-$$

$$[OH^-] = \frac{2.00 \times 10^{-3} \text{ mol OH}^-}{70.00 + 50.00 \text{ mL}} \times \frac{1,000 \text{ mL}}{1 \text{ L}} = 1.67 \times 10^{-2} \text{ M OH}^-$$

Thus, pOH = $-\log[OH^-] = -\log(1.67 \times 10^{-2}$ M$) = 1.78$:

$$pH = 14.00 - pOH = 14.00 - 1.78 = 12.22$$

(h)   When these pH values are plotted against the respective volumes of the added NaOH, the approximate titration curve is produced. More data points would produce a titration curve with better accuracy. This approaches the actual titration curve shown in Figure 12.3.

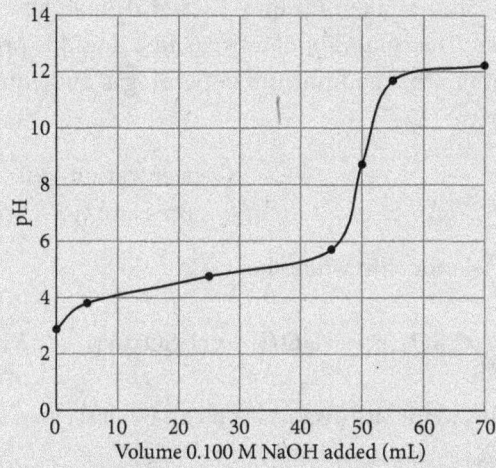

In Q9, we calculated that [H$^+$] equals 1.33×10^{-3} M in the 0.100 M solution of acetic acid, with a pH of 2.88. The percent dissociation (%Dis) is the mole percent of the initial acid, HA (with concentration C_{HA}), that is dissociated into H$^+$ and A$^-$ ions. Thus, for a 0.1 M solution of acetic acid:

$$\%Dis = \frac{[H^+]}{C_{HA}} \times 100\% = \frac{1.33 \times 10^{-3} \text{ M}}{0.100 \text{ M}} \times 100\% = 1.33\%$$

*QUESTION 10

Calculate the percent dissociation of a 0.0100 M solution of acetic acid. How does this percent compare with that in the 0.100 M acetic acid? What is the general trend for dilute versus concentrated solutions?

*QUESTION 11

Calculate the pH of the solution where the following volumes of 0.100 M of HCl are added to 50.00 mL of 0.100 M of ammonia (NH_3). Ammonia has a base dissociation constant 1.8×10^{-5} at 25°C.

(a)   0 mL

(b)   5.00 mL (10% to the equivalence point)

(c)   25.00 mL (50% to the equivalence point)

(d)   45.00 mL (90% to the equivalence point)

(e)   50.00 mL (equivalence point)

(f)   55.00 mL (10% excess of the acid titrant)

(g)   70.00 mL (40% excess of the acid titrant)

(h)   Generate the approximate shape of the titration curve from these results and compare with that in Q9.

12.5 Acid Rain, Atmospheric CO_2, and Climate Change

Acid–base equilibrium has important applications for many environmental issues. For example, gaseous SO_2 can be taken up into a water droplet and form sulfurous acid (H_2SO_3) in the aqueous phase. The solubility of any gas in a liquid is proportional to the pressure of the gas over the solution; this equilibrium between gas and aqueous phases can be described by Henry's law constant (K_H):

$$K_H = \frac{\text{Concentration (aq)}}{\text{Pressure (g)}} \tag{12.10}$$

For SO_2, the relationship would be:

$$SO_2\,(g) + H_2O\,(l) \leftrightarrow H_2SO_3\,(aq) \qquad K_H = 1.24\,\frac{M}{atm}$$

Once an acid is made, its dissociation can be described by the equilibrium acid dissociation constant or K_a as explained above:

$$H_2SO_3\,(aq) \leftrightarrow H^+\,(aq) + HSO_3^-\,(aq) \qquad K_{a1} = 1.32 \times 10^{-2}$$

QUESTION 12

Calculate the pH of rainwater if there is 75 ppb (EPA standard) SO_2 in the atmosphere that dissolves into the raindrops. Assume sulfur is the only contributor to pH.

Q12 ANSWER First, we have to figure out how much H_2SO_3 acid will be made at equilibrium from dissolving 75 ppb SO_2. The relationship between SO_2 and H_2SO_3 is described by Henry's law. All equilibrium constants are ratios of the concentrations of the products divided by the concentrations of reactants at equilibrium. As noted by the units of Henry's law constant, the concentration units will be the aqueous molarity of the products divided by the gas pressure of the reactant:

$$K_H = \frac{[H_2SO_3]}{P_{SO_2}} = 1.24 \frac{M}{atm}$$

To use this equation, we have to have the amount of SO_2 in units of atm. Assuming the total pressure (P_T) is 1 atm, a 75 ppb mixing ratio would give a partial pressure (P_i) of 75×10^{-9} atm. Plugging all of this in, we get:

$$[H_2SO_3] = \left(1.24 \frac{M}{atm}\right)(75 \times 10^{-9} \text{ atm}) = 9.3 \times 10^{-8} \text{ M}$$

Next, we have to use the acid dissociation relationship to determine the amount of H^+ ions in the water. The amount of products is unknown, but since there is 1 mol of each according to the stoichiometry of the equation, they will have the same equilibrium amount:

$$K_{a1} = \frac{[H^+][HSO_3^-]}{[H_2SO_3]} = \frac{x^2}{[H_2SO_3]} = \frac{x^2}{9.3 \times 10^{-8}} = 1.3 \times 10^{-2}$$

$$x = \sqrt{(9.315 \times 10^{-8})(1.32 \times 10^{-2})} = 3.5 \times 10^{-5} = [H^+]$$

Since pH $= -\log[H^+]$:

$$pH = -\log(3.5 \times 10^{-5}) = 4.46$$

Aqueous oxidation of sulfur continues until it is fully oxidized (oxidation number +6) as in SO_4^{-2} or H_2SO_4. Much of the "particulate matter" in the atmosphere includes sulfate particles formed in the atmosphere as a consequence of SO_2 oxidation.

---

*QUESTION 13

Assuming the acidity only comes from SO_2, what is the concentration of dissolved SO_2 gas (in ppb) needed to produce a raindrop with a pH of 4.05.

Different control measures can be used to reduce the amount of sulfuric acid in urban air. Coal-fired power plants can reduce SO_2 emissions or limit the amount of sulfur in the fuel itself. Most plants currently remove SO_2 from the effluent in the stack by installing chemical scrubbers where the gas passes through a slurry of limestone (calcium carbonate), converting the SO_2 to calcium sulfite ($CaSO_3$):

$$CaCO_3 + SO_2 \rightarrow CaSO_3 + CO_2$$

The acidification of rainwater can also occur due to the equilibrium of gaseous carbon dioxide in the atmosphere, its aqueous form (carbonic acid), and the subsequent acid dissociation. With ongoing climate change and, in particular, the rapidly increasing mixing ratio of CO_2 in the atmosphere owing to anthropogenic emissions, it is worthy to examine the acidity of rain under various future CO_2 scenarios.

QUESTION 14

The current global average atmospheric CO_2 concentration recently reached 419 ppm (parts per million)[1]. Future greenhouse gas concentrations in the atmosphere depend greatly on emission controls and other scenarios. According to the Representative Concentration Pathways (RCPs) in the Intergovernmental Panel on Climate Change (IPCC) Fifth Assessment Report (AR5), globally averaged CO_2 concentration may reach 538 ppm by 2100 in one of the emission stabilization scenarios (RCP4.5)[2]. Calculate the pH of rainwater at this atmospheric CO_2 level (at 25°C at sea level), ignoring other types of acids in the atmosphere in parts (a) and (b) below.

(a)   Will the natural rainwater be more or less acidic in the future atmosphere scenario of higher concentrations of CO_2?

(b)   Verify any assumptions made in this calculation and compare with results obtained using the quadratic formula.

(c)   List two other types of possible acids in the atmosphere and their sources and impacts on rainwater. If 0.010 grams (g) of each acid eventually dissolves into every 1 liter (L) of rainwater in an industrial region, calculate the pH of each type of acid rain.

Q14 ANSWER

(a)   Aqueous rain is in equilibrium with atmospheric CO_2, forming carbonic acid:

$$CO_2\,(g) + H_2O\,(l) \rightleftharpoons H_2CO_3\,(aq)$$

If the CO_2 concentration equals 538 ppm, the mixing ratio of CO_2 in the atmosphere is:

$$\frac{538\ CO_2\ \text{molecules}}{1 \times 10^6\ \text{air molecules}} = 538 \times 10^{-6}$$

Since the atmospheric pressure is 1 atm (at 25°C, at sea level), the CO_2 partial pressure is:

$$P_{CO_2} = (538 \times 10^{-6}) \times 1\ \text{atm} = 5.38 \times 10^{-4}\ \text{atm}$$

Using Henry's law constant (K) for CO_2 at 25°C ($K_H = 3.2 \times 10^{-2}$ M atm^{-1}), the initially dissolved carbonic acid concentration in rainwater is:

$$[H_2CO_3] = P_{CO_2} \times K_H = (5.38 \times 10^{-4}\ \text{atm}) \times (3.2 \times 10^{-2}\ \text{M atm}^{-1}) = 1.70 \times 10^{-5}\ \text{M}$$

Rainwater is, thus, a diluted solution of carbonic acid in equilibrium with its conjugate base HCO_3^-. K_a of carbonic acid is 4.3×10^{-7}. Using the above ICE approach:

	$H_2CO_3\,(aq) \rightleftharpoons H^+\,(aq) + HCO_3^-\,(aq)$		
Initial Concentration (M)	1.70×10^{-5}	0	0
Change (M)	$-x$	$+x$	$+x$
Equilibrium Concentration (M)	$1.70 \times 10^{-5} - x$	$+x$	$+x$

[1] Global Monitoring Laboratory, National Oceanic & Atmospheric Administration, *Trends in Atmospheric Carbon Dioxide,* available at https://www.ipcc.ch/report/ar5/syr/.

[2] IPCC, *AR5 Synthesis Report: Climate Change 2014,* available at https://www.ipcc.ch/report/ar5/syr/.

Set up the equilibrium as:

$$K_a = \frac{x^2}{1.70 \times 10^{-5} - x} = 4.3 \times 10^{-7}$$

Assuming x is far smaller compared to 1.70×10^{-5}, the equation becomes:

$$K_a = \frac{x^2}{1.70 \times 10^{-5}} = 4.3 \times 10^{-7}$$

$$x = 2.7 \times 10^{-6} \text{ M}$$

$$\text{pH} = -\log[\text{H}^+] = -\log(x) = -\log(2.7 \times 10^{-6}) = 5.57$$

This is 0.07 pH unit lower than that of the present-day rainwater (pH = 5.63, with CO_2 at ~415 ppm) due to the elevated CO_2 concentration (538 ppm). In the future, with a higher concentration of CO_2 in the atmosphere, the rainwater will be more acidic due to the shift in this equilibrium reaction alone.

(b)   Now we will compare the calculated results by using the quadratic formula versus using the x as a small approximation to calculate the pH of weak acid solutions.

First, we verify the assumption in part (a) above:

$$\frac{x^2}{1.70 \times 10^{-5}} \times 100\% = \frac{2.7 \times 10^{-6}}{1.70 \times 10^{-5}} \times 100\% = 16\%$$

The assumption that [H$^+$] is far smaller than the initial [H$_2$CO$_3$] is not valid. Indeed, K_a (4.3×10^{-7}) is within two orders of magnitude of the initial [H$_2$CO$_3$] (1.70×10^{-5} M), giving rise to the relatively higher x in Q14 compared to the examples in Q9. Let us use the quadratic formula and compare it with the result from using the simplified formula.

When $K_a \ll$ [H$_2$CO$_3$] is not valid, one can still set up the ICE table and the equilibrium relationship as we did earlier.

	$H_2CO_3\,(aq) \rightleftharpoons H^+\,(aq) + HCO_3^-\,(aq)$		
Initial Concentration (M)	1.70×10^{-5}	0	0
Change (M)	$-x$	$+x$	$+x$
Equilibrium Concentration (M)	$1.70 \times 10^{-5} - x$	$+x$	$+x$

The equilibrium is:

$$K_a = \frac{x^2}{1.70 \times 10^{-5} - x} = 4.3 \times 10^{-7}$$

Now combine the terms to simplify the equation:

$$x^2 = 7.3 \times 10^{-12} - 4.3 \times 10^{-7}x$$

$$x^2 + 4.3 \times 10^{-7}x - 7.3 \times 10^{-12} = 0$$

The general quadratic equation is:

$$ax^2 + bx + c = 0$$

So, in our case, $a = 1$, $b = 4.3 \times 10^{-7}$, and $c = -7.3 \times 10^{-12}$

Now use the quadratic equation to solve for x:

$$x = \frac{-b \pm \sqrt{b^2 - 4ac}}{2a}$$

There are two possible answers for the quadratic equation:

$$x = \frac{-b + \sqrt{b^2 - 4ac}}{2a} = \frac{-4.3 \times 10^{-7} + \sqrt{(4.3 \times 10^{-7})^2 - 4(1)(-7.3 \times 10^{-12})}}{2(1)}$$

$$x = 2.5 \times 10^{-6}$$

$$x = \frac{-b - \sqrt{b^2 - 4ac}}{2a} = \frac{-4.3 \times 10^{-7} - \sqrt{(4.3 \times 10^{-7})^2 - 4(1)(-7.3 \times 10^{-12})}}{2(1)}$$

$$x = -2.9 \times 10^{-6}$$

A negative value does not make sense for concentration, so we can ignore that answer. Solving for this quadratic equation, we get $x = [H^+] = 2.5 \times 10^{-6}$ M:

$$pH = -\log[H^+] = -\log(2.5 \times 10^{-6} \text{ M}) = 5.60$$

which is slightly lower (by 0.03 pH unit) than the pH value of the present-day rainwater from CO_2 dissolution (pH = 5.63, with CO_2 at ~415 ppm).

In comparison, the x is small approximation [Q9 part (a) above] gives a result of $[H^+] = 2.92 \times 10^{-6}$ M, that is, about 9% higher than that when solving the quadratic formula (2.68×10^{-6} M). The calculated pH values using these two methods (5.53 versus 5.57, respectively) are within 1%. Thus, in this case, the error using the simplified formula remains small. Overall, in cases where it is difficult to judge whether $K_a \ll$ initial acid concentration is strictly valid, one should calculate with caution and can always rely on the quadratic formula to calculate the most accurate answer.

(c)   The contributions from other acidic constituents of the atmosphere, particularly HNO_3 and H_2SO_4, must often be considered, in addition to the acidifying effect of CO_2. These acids can both form naturally. For example, HNO_3 derives from NO produced in lightning and forest fires, while H_2SO_4 derives from volcanoes and biogenic sulfur compounds (Chapter 9). At natural background concentrations, these acids rarely influence rainwater pH significantly.

In polluted areas, however, the concentrations of these acids can be much higher and can reduce the pH of rainwater substantially over extended regions, producing what is commonly known as acid rain. It is not uncommon in polluted areas to find the pH of rainwater in the range of pH 3.5–5 (see solutions below for such examples). Moreover, acid rain can fall quite far from the sources of pollution, due to long-range atmospheric transport. In particular, acid rain is a pressing problem for areas downwind of coal-fired power plants with often high SO_2 and NO_x emissions, although emissions of SO_2 have been reduced in recent decades in some countries. While acid rain is a phenomenon in the hydrosphere, it depends on conditions in the atmosphere, such as the extent of acidic emissions and the prevailing weather patterns. This is an example of where the different spheres of the earth are closely interconnected.

If 0.010 g of sulfuric acid dissolves into 1 L of rainwater:

$$0.010 \text{ g of } H_2SO_4 \times \frac{1 \text{ mol}}{98.00 \text{ g } H_2SO_4} = 0.00010 \text{ mol } H_2SO_4$$

Since only the first proton in H_2SO_4 dissociates completely, whereas the second proton dissociates much less, we approximate that 0.00010 mol of H^+ dissociates and dissolves into every 1 L of rainwater:

$$[H^+] = \frac{0.00010 \text{ mol}}{1.00 \text{ L}} = 0.00010 \text{ M}$$

$$pH = -\log[H^+] = -\log(0.00010 \text{ M}) = 4.00$$

If 0.010 g of nitric acid dissolves into 1 L of rainwater:

$$0.010 \text{ g of HNO}_3 \times \frac{1 \text{ mol}}{63.00 \text{ g HNO}_3} = 0.00016 \text{ mol HNO}_3$$

Since only one proton in HNO_3 dissociates completely, we calculate that 0.00016 mol of H^+ dissociates and dissolves into every 1 L of rainwater:

$$[H^+] = \frac{0.00016 \text{ mol}}{1.00 \text{ L}} = 0.00016 \text{ M}$$

$$pH = -\log[H^+] = -\log(0.00016 \text{ M}) = 3.80$$

Therefore, acid rain affected by sulfuric acid and nitric acid could have pH values of 4.00 and 3.80, respectively, in these scenarios.

*QUESTION 15

Let us also explore the extremely high CO_2 scenario the earth may face in the future. In the extreme, albeit not unlikely, scenario of the atmospheric CO_2 reaching 800 ppm, roughly doubling that of today's level, calculate the pH of rainwater at 25°C at sea level, ignoring other potential sources of acids in the air. With respect to rising CO_2 levels, how much should we be concerned about the enhanced acidity of rain as compared to climate warming and other associated environmental impacts?

12.6 Polyprotic Acids, Conjugate Bases of Polyprotic Acids, and Alkalinity

So far, we have dealt with the gas–liquid equilibrium of CO_2 and the ionization of the first proton of an acid. In solution, H_2CO_3 actually has two ionizable protons; they dissociate successively, with very different K_a values:

$$H_2CO_3 \text{ (aq)} \rightleftharpoons H^+ \text{ (aq)} + HCO_3^- \text{ (aq)} \qquad K_{a1} = 4.3 \times 10^{-7} \qquad (12.11)$$

$$HCO_3^- \text{ (aq)} \rightleftharpoons H^+ \text{ (aq)} + CO_3^- \text{ (aq)} \qquad K_{a2} = 4.7 \times 10^{-11} \qquad (12.12)$$

Let us explore the acidity and basicity of solutions with respect to different types of molecular and ionic species derived from carbonic acid. Other polyprotic acids, such as phosphoric acid, can be analyzed similarly. These acids are all found in the environment, so it is important to understand how they react in different conditions.

QUESTION 16

Calculate the pH of a buffer solution with equal molar concentrations of H_2CO_3 and HCO_3^-.

Q16 ANSWER This is a buffer solution, H_2CO_3 is the acid and HCO_3^- is the conjugate base. Using the H–H equation (12.9):

$$pH = pK_a + \log\frac{[HCO_3^-]}{[H_2CO_3]} = -\log(4.3 \times 10^{-7}) - \log(1) = 6.37$$

*QUESTION 17

Calculate the pH of a buffer having an equal molar concentration of HCO_3^- and CO_3^{2-}.

*QUESTION 18

Calculate the pH of a solution of 0.050 M carbonate ions (CO_3^{2-}).

When an ion, such as bicarbonate (HCO_3^-) has both a conjugate acid (H_2CO_3) and a conjugate base (CO_3^{2-}), then a different method for calculating pH must be used. The proton dissociated from HCO_3^- [reaction (12.11)] is taken up by another HCO_3^- via the reverse reaction (12.12) which becomes:

$$H^+(aq) + HCO_3^-(aq) \rightleftharpoons H_2CO_3(aq) \qquad K'_{a1} = \frac{1}{K_{a1}} = \frac{1}{4.3 \times 10^{-7}} = 2.3 \times 10^6$$

$$HCO_3^-(aq) \rightleftharpoons H^+(aq) + CO_3^{2-}(aq) \qquad K_{a2} = 4.7 \times 10^{-11}$$

Adding the above two equations, we obtain:

$$2\,HCO_3^-(aq) \rightleftharpoons H_2CO_3(aq) + CO_3^{2-}(aq)$$

The equilibrium constant for the combined reaction is then the multiplication of those for the two equations that were combined:

$$K'_{a1}K_{a2} = \frac{K_{a2}}{K_{a1}} = \frac{[H_2CO_3][CO_3^{2-}]}{[HCO_3^-]^2}$$

For a solution initially containing only the bicarbonate ion, we see that $[H_2CO_3] = [CO_3^{2-}]$ from the combined equation. Therefore:

$$\frac{K_{a2}}{K_{a1}} = \frac{[H_2CO_3]^2}{[HCO_3^-]^2}$$

Or:

$$\frac{[H_2CO_3]}{[HCO_3^-]} = \sqrt{\frac{K_{a2}}{K_{a1}}}$$

To calculate the pH, we can go back to (12.11) and write:

$$K_{a1} = \frac{[HCO_3^-][H^+]}{[H_2CO_3]}$$

Rearrange and insert the equation relating K_{a1} and K_{a2}, we get:

$$[H^+] = K_{a1}\frac{[H_2CO_3]}{[HCO_3^-]} = K_{a1}\sqrt{\frac{K_{a2}}{K_{a1}}}$$

$$[H^+]^2 = K_{a1}^2 \times \frac{K_{a2}}{K_{a1}} = K_{a1}K_{a2}$$

Take the log of both sides and multiply through by -1:

$$-2\log[H^+] = -\log K_{a1} - \log K_{a2}$$

To give:

$$pH = \frac{1}{2}(pK_{a1} + pK_{a2}) \qquad (12.13)$$

From this relationship, we see that the pH of the monoprotic form of any diprotic acid is just the mean of the two bracketing pK_a values, largely independent of the molar concentration.

QUESTION 19

Calculate the pH of a solution of 0.050 M bicarbonate ions (HCO_3^-).

Q19 ANSWER For 0.050 M bicarbonate, its pH $= \frac{1}{2}(-\log(4.3 \times 10^{-7}) - \log(4.7 \times 10^{-11})) = 8.35$

---

*QUESTION 20

There can also be more than two ionizable protons. For example, phosphoric acid has three:

$$H_3PO_4\,(aq) \rightleftharpoons H^+\,(aq) + H_2PO_4^-\,(aq) \qquad pK_{a1} = 2.1$$

$$H_2PO_4^-\,(aq) \rightleftharpoons H^+\,(aq) + HPO_4^{2-}\,(aq) \qquad pK_{a2} = 7.2$$

$$HPO_4^{2-}\,(aq) \rightleftharpoons H^+\,(aq) + PO_4^{3-}\,(aq) \qquad pK_{a3} = 12.4$$

Calculate the pH for 0.050 M of $H_2PO_4^-$ and HPO_4^{2-}.

QUESTION 21

What is the dominant form of the phosphate ion in (1) a lake with pH 5.0 and (2) seawater with pH 8.1, and what is the ratio of this form to the next most abundant form?

$$H_3PO_4 \rightleftharpoons H^+ + H_2PO_4^- \qquad pK_{a1} = 2.1$$

$$H_2PO_4^- \rightleftharpoons H^+ + HPO_4^{2-} \qquad pK_{a2} = 7.2$$

$$HPO_4^{2-} \rightleftharpoons H^+ + PO_4^{3-} \qquad pK_{a3} = 12.4$$

Q21 ANSWER The lake pH is higher than pK_{a1}, but lower than pK_{a2}, so most of the phosphate will be present as $H_2PO_4^-$. For the seawater, HPO_4^{2-} dominates because the pH is between pK_{a2} and pK_{a3}. Because the lake pH is closer to pK_{a2} than pK_{a1}, the next most abundant form is HPO_4^{2-}, while for the seawater $H_2PO_4^-$ is the next most abundant form, since the pH is closer to pK_{a2} than to pK_{a3}. The ratios can be obtained from the equilibrium expression for the second ionization:

$$\frac{[HPO_4^{2-}]}{[H_2PO_4^-]} = \frac{K_{a2}}{[H^+]}$$

At pH = 5.0:

$$\frac{[HPO_4^{2-}]}{[H_2PO_4^-]} = \frac{10^{-7.2}}{10^{-5.0}} = 10^{-2.2} = 0.0063$$

$$\frac{[HPO_4^{2-}]}{[H_2PO_4^-]} = 158$$

At pH = 8.1:

$$\frac{[HPO_4^{2-}]}{[H_2PO_4^-]} = \frac{10^{-7.2}}{10^{-8.1}} = 10^{0.9} = 7.9$$

*QUESTION 22

Calculate the concentration of all forms of carbonate in the surface waters of the oceans, with the pH at around 8.1 after acidification since the industrial revolution. The ocean is also in equilibrium with the atmospheric CO_2, of which the mixing ratio is around 415 ppm as of January 2021 (NOAA).

Most natural bodies of water contain more bases than acids, mainly because they are in contact with basic minerals (see Chapter 13). The alkalinity (Alk) is the total amount of bases, which can be determined by titration with a standard acid, and is reported in equivalents (of H^+) per liter (equiv L^{-1}). The alkalinity provides a mass balance, summing the concentrations of the bases, multiplied by the number of protons taken up. For example, if the bases present are carbonates, then:

$$Alk = [HCO_3^-] + 2[CO_3^{2-}]$$

For phosphates:

$$Alk = [H_2PO_4^-] + 2[HPO_4^{2-}] + 3[PO_4^{3-}]$$

QUESTION 23

Using the concentration of atmospheric CO_2 and ocean surface pH given in Q22, calculate the alkalinity of ocean surface waters, assuming that only carbonate and bicarbonate are contributing to the alkalinity.

Q23 ANSWER From Q22, $[HCO_3^-]$ and $[CO_3^{2-}]$ are already calculated using Henry's law constant and the two-stage equilibrium of carbonic acid [Eqs. (12.11) and (12.12)]:

$$[HCO_3^-] = 7.09 \times 10^{-4} \, M; \; [CO_3^{2-}] = 4.19 \times 10^{-6} \, M$$

Thus:

$$\text{Alk} = [HCO_3^-] + 2[CO_3^{2-}] = 7.09 \times 10^{-4} \, M + 2(4.19 \times 10^{-6} \, M) = 7.17 \times 10^{-4} \, \text{equiv/L}$$

12.7 Conclusions

Acids and bases are found in many of earth's systems, such as raindrops, lakes, and the ocean. Sources of acidic inputs include anthropogenic CO_2, SO_2, and NO_x emissions through the burning of fossil fuels. NO_x also derives from natural sources such as lightning. Analyzing environmental samples with acid–base titrations or calculating the pH of buffer solutions is important to learn about the state of the system and monitor changes over time.

13

WATER AND THE LITHOSPHERE

13.1 Weathering and Solubilization Mechanisms

13.1.1 *Ionic Solids and the Solubility Product*

Although water is generally an excellent solvent for ions, many ionic compounds are only sparingly soluble because the ion–water forces are outweighed by the forces that hold the ions together, particularly when the ions can arrange themselves in an energetically favorable way in a crystalline lattice. The energies stabilizing a lattice are generally maximal when the positive and negative ions have equal size and/or charge. For example, lithium fluoride is less soluble than lithium iodide because fluoride is closer in size to the small lithium-ion than iodide; on the contrary, cesium iodide is less soluble than cesium fluoride because the cesium ion is closer in size to iodide. Similarly, both sodium carbonate and calcium nitrate are highly soluble because the positive and negative ions differ in charge, but calcium carbonate is sparingly soluble because both the cation and anion are doubly charged. Therefore, they have large lattice energy. Magnesium carbonate is more soluble than calcium carbonate, while sodium silicates are more soluble than potassium silicates, in both cases because the smaller cations do not fit the lattice and interact more strongly with water. These differing solubilities account for Mg and Na being much more abundant in seawater than Ca and K, respectively, despite the similar abundance of these pairs of metals in the earth's crust (Table 13.1).

The solubility of a sparingly soluble salt is governed by the equilibrium constant for the dissolution reaction, called the solubility product (K_{sp}). For example, barium sulfate ($BaSO_4$) dissolves with a K_{sp} of 10^{-10}:

$$BaSO_4 \rightarrow Ba^{2+} + SO_4^{2-}$$

$$K_{sp} = [Ba^{2+}][SO_4^{2-}] = 10^{-10} \ M^2 \tag{13.1}$$

(Barium sulfate does not appear in (13.1) because the effective concentration of a solid phase is constant.) This equation is valid as long as there is solid barium sulfate in equilibrium with the solution. If the concentration product of Ba^{2+} and SO_4^{2-} exceeds the value of K_{sp}, barium sulfate will precipitate. It actually does precipitate on dead phytoplankton in the ocean, because their decomposition produces enough sulfate that the barium ions in the ocean are sufficient to exceed the solubility product.

TABLE 13.1 Major Elements in the Earth's Crust and Their Abundance in Seawater

Element	Symbol	Crustal Average[a] (% by weight)	Seawater[b] (ppm by weight)
Oxygen	O	46.7	883,000
Silicon	Si	27.7	2.9
Aluminum	Al	8.1	0.001
Iron	Fe	5.1	0.003
Calcium	Ca	3.7	411
Sodium	Na	2.8	10,800
Potassium	K	2.6	392
Magnesium	Mg	2.1	1,290
Titanium	Ti	0.62	0.001
Hydrogen	H	0.14	110,000
Phosphorus	P	0.13	0.09
Carbon	C	0.09	28.0
Manganese	Mn	0.09	0.0004
Fluorine	F	0.08	13
Sulfur	S	0.05	904
Barium	Ba	0.05	0.02
Chlorine	Cl	0.05	19,400
Chromium	Cr	0.04	0.0002
Strontium	Sr	0.04	8.1
Zirconium	Zr	0.03	0.00003

[a]Sources for abundance of elements in the earth's crust: "List of Periodic Table Elements Sorted by Abundance in Earth's Crust," *Israel Science and Technology Homepage*, 2011, available at http://www.science.co.il/PTelements.asp?s=Earth; "Elements, Terrestrial Abundance," *The Internet Encyclopedia of Science*, 2011, available at https://www.daviddarling .info/encyclopedia/E/elterr.html; *Abundance in the Earth's Crust: Periodicity, Web Elements Chemistry*, 2010, available at https://www.webelements.com/periodicity/abund_crust/.
[b]Source for abundance of elements in seawater: Turekian, *Oceans*, Prentice-Hall, 1968.

Source: Adapted from Spiro et al., *Chemistry of the Environment, 3rd Ed.*, University Science Books, © 2012, all rights reserved.

QUESTION 1

What is the solubility of barium sulfate ($BaSO_4$) in (a) pure water or (b) 1 mM sodium sulfate? The K_{sp} for $BaSO_4$ is 1.1×10^{-10} M^2.

Q1 ANSWER

(a)   When pure water is equilibrated with barium sulfate, the concentration of barium matches the concentration of sulfate in the solution:

$$[Ba^{2+}] = [SO_4^{2-}] = K_{sp}^{1/2} = (1.1 \times 10^{-10}\ M^2)^{1/2} = 1.0 \times 10^{-5}\ M$$

Thus, the $BaSO_4$ solubility is 1.0×10^{-5} M.

(b)   When either Ba^{2+} or SO_4^{2-} is present in excess, the solubility is the concentration of its partner, which decreases in inverse proportion to the excess concentration, following (13.1). This is due to Le Chatelier's principle that says if we have excess SO_4^{2-}, the reaction shifts to more solids (to the left) to restore solubility equilibrium.

Thus, when $[SO_4^{2-}] = 1 \times 10^{-3}$ M

The solubility is:

$$[Ba^{2+}] = \frac{K_{sp}}{[SO_4^{2-}]} = \frac{1.1 \times 10^{-10}\ M^2}{1 \times 10^{-3}\ M} = 1 \times 10^{-7}\ M$$

Actually, $[SO_4^{2-}]$ is slightly larger than the initial sodium sulfate concentration, since the dissolution of each Ba^{2+} ion is accompanied by one SO_4^{2-} ion. Consequently, the concentration of sulfate ion is:

$$[SO_4^{2-}]_{actual} = [SO_4^{2-}] + [Ba^{2+}] = [SO_4^{2-}] + \frac{K_{sp}}{[SO_4^{2-}]} = 1 \times 10^{-3}\ M + 1 \times 10^{-7}\ M \sim 1 \times 10^{-3}\ M$$

However, since K_{sp} is so small, the second term is negligible.

*QUESTION 2

Calculate the solubility of magnesium carbonate ($MgCO_3$) in 50 mM magnesium chloride. K_{sp} for $MgCO_3$ is $4.0 \times 10^{-5}\ M^2$.

QUESTION 3

Calculate the solubility of $Cu(OH)_2$ in pure water. K_{sp} for $Cu(OH)_2$ is 2.2×10^{-20}.

Q3 ANSWER   Consider the dissociation of copper hydroxide in water:

$$Cu(OH)_2 \rightleftharpoons Cu^{2+} + 2\,OH^-$$

		Cu^{2+}	OH^-
Initial (M)		0	0
Change (M)	$-s$	$+s$	$+2s$
Equilibrium		s	$+2s$

There are 2 mol of OH^- for every 1 mol of $Cu(OH)_2$ dissolved, which is why the change is $2s$.

* Answers to starred questions can be found at the end of the book.

The relationship between K_{sp} and concentration of ions is as follows. The concentration of the hydroxide ion is squared because 2 mol of hydroxide is formed when 1 mol of copper hydroxide dissolves. Furthermore, we can put in the relationship we determined from the above equilibrium table:

$$K_{sp} = [Cu^{2+}][OH^-]^2 = (s)(2s)^2 = 4s^3 = 2.2 \times 10^{-20}$$

$$s^3 = \frac{2.2 \times 10^{-20}}{4} = 5.5 \times 10^{-21}$$

$$s = 1.8 \times 10^{-7} \text{ M}$$

*QUESTION 4

Calculate the solubility of $Fe(OH)_3$ in pure water. K_{sp} for $Fe(OH)_3$ is 2.6×10^{-39}.

13.1.2 *Solubility and Basicity*

If one of the ions of a sparingly soluble salt is an acid or a base, then the solubility increases because the ion is partially converted to its basic or acid form, thereby pulling the solubility equilibrium toward more dissolution. For example, K_{sp} of calcium carbonate is 5.0×10^{-9}, and if the only reaction were:

$$CaCO_3 \rightleftharpoons Ca^{2+} + CO_3^{2-} \tag{13.2}$$

then the solubility would be 7.1×10^{-5}. However, the solubility equilibrium is shifted to the right because the basic carbonate ion reacts further with water (see Chapter 12):

$$CO_3^{2-} + H_2O \rightleftharpoons HCO_3^- + OH^- \qquad K_b = 2.1 \times 10^{-4} \tag{13.3}$$

and is largely converted to bicarbonate.

QUESTION 5

Calculate the soltubility of calcium carbonate in pure water and the pH of the resulting solution. The Ks for $CaCO_3$ and HCO_3^- are 5.0×10^{-9} and 2.1×10^{-4}, respectively.

Q5 ANSWER The first step is to add the two equations (13.2) and (13.3) together:

$$CaCO_3 \rightleftharpoons Ca^{2+} + CO_3^{2-}$$

$$+ CO_3^{2-} + H_2O \rightleftharpoons HCO_3^- + OH^-$$

$$\overline{CaCO_3 + H_2O \rightarrow Ca^{2+} + HCO_3^- + OH^-}$$

Then, multiply the equilibrium constants together, which you do when you add the corresponding chemical equations together (because the free energies add, and they are proportional to the log of the equilibrium constants):

$$K = (5.0 \times 10^{-9}) \times (2.1 \times 10^{-4}) = 1.1 \times 10^{-12}$$

Then, calculate the product concentrations of this combined reaction, all of which are the same as the solubility:

$$K_{sp} = [Ca^{2+}][HCO_3^-][OH^-] = s^3 = 1.1 \times 10^{-12}$$

$$s = 1.0 \times 10^{-4} \text{ M}$$

Since $[OH^-]$ is equal to the solubility, the pH is given by:

$$pH = 14.0 - pOH = 14.0 - (-\log 1.0 \times 10^{-4} \text{ M}) = 10.0$$

*QUESTION 6

Calculate the solubility of magnesium carbonate ($MgCO_3$) in pure water. Calculate the pH of the resulting solution. The Ks for $MgCO_3$ and HCO_3^- are 1.6×10^{-8} and 2.1×10^{-4}, respectively.

The solubility is increased further when the solution is exposed to the atmosphere because the carbonate ion then reacts with the acidic CO_2:

$$CO_3^{2-} + H_2CO_3 \rightleftharpoons 2\,HCO_3^- \qquad K = \frac{K_{a1}}{K_{a2}} = \frac{4.3 \times 10^{-7}}{4.7 \times 10^{-11}} = 9.1 \times 10^3 \qquad (13.4)$$

The equilibrium constant is large, so the equilibrium lies far to the right, and the $CaCO_3$ solubility is increased considerably. [Reaction (13.4) is obtained by subtracting the two successive dissociation reactions of H_2CO_3 (see Chapter 12), so the equilibrium constant is obtained by dividing the two acidity constants.]

QUESTION 7

Calculate $[Ca^{2+}]$, $[HCO_3^-]$ and pH when air-saturated water is in equilibrium with calcium carbonate ($CaCO_3$). (See Q22 in Chapter 12 for additional information.)

Q7 ANSWER In equilibrium with the atmosphere, H_2O contains H_2CO_3, which dissolves $CaCO_3$:

$$CaCO_3 + H_2CO_3 \rightleftharpoons Ca^{2+} + 2\,HCO_3^- \qquad (13.5)$$

The equilibrium constant is:

$$K'_{sp} = \frac{[Ca^{2+}][HCO_3^-]^2}{[H_2CO_3]} = \frac{K_{sp}K_{a1}}{K_{a2}} = (5.0 \times 10^{-9}) \times (9.1 \times 10^3) = 4.6 \times 10^{-5}$$

Since reaction (13.5) is the sum of reactions (13.2) and (13.4), the equilibrium constant is obtained by multiplying the two equilibrium constants.

From the stoichiometry of reaction (13.5), we see that $[HCO_3^-] = 2[Ca^{2+}]$, and consequently:

$$K'_{sp} = \frac{4[Ca^{2+}]^3}{H_2CO_3}$$

But $[H_2CO_3]$ is fixed by the atmospheric CO_2 concentration at 1.31×10^{-5} M (see Q22 in Chapter 12). Therefore:

$$[Ca^{2+}] = \left[\frac{K'_{sp}[H_2CO_3]}{4}\right]^{\frac{1}{3}} = \left[\frac{(4.6 \times 10^{-5}) \times (1.31 \times 10^{-5})}{4}\right]^{\frac{1}{3}} = 5.3 \times 10^{-4} \text{ M}$$

Note that this is five times the solubility without CO_2 (Q5).
Then,

$$[HCO_3^-] = 2[Ca^{2+}] = 1.1 \times 10^{-3} \text{ M}$$

and the pH can be obtained from the H_2CO_3 acidity constant:

$$pH = pK_{a1} + \log\frac{[HCO_3^-]}{[H_2CO_3]} = 6.3 + \log\frac{1.1 \times 10^{-3}}{1.31 \times 10^{-5}} = 8.2$$

This equation is a form of the Henderson–Hasselbach equation; refer to Section 3 in Chapter 12 for more details about these calculations.

This is close to the pH of seawater, 8.1, and is much lower than in the absence of CO_2 (Q5).

QUESTION 8

How much Fe^{2+} could be present in H_2O containing 1.0×10^{-2} M HCO_3^-, without causing precipitation of $FeCO_3$ ($K_{sp} = 2.0 \times 10^{-11}$)?

Q8 ANSWER At equilibrium with solid $FeCO_3$:

$$K_{sp} = [Fe^{2+}][CO_3^{2-}] = 2.0 \times 10^{-11}$$

$$\text{or } [Fe^{2+}] = \frac{2.0 \times 10^{-11}}{[CO_3^{2-}]}$$

The $[Fe^{2+}]$ must be less than this, if there is no precipitation. If the water contains 1.0×10^{-2} M HCO_3^-, then the main acid–base equilibrium is:

$$2\,HCO_3^- \rightleftharpoons H_2CO_3 + CO_3^{2-}$$

$$K = \frac{K_{HCO_3^-}}{K_{H_2CO_3}} = \frac{[H_2CO_3][CO_3^{2-}]}{[HCO_3^-]^2} = \frac{K_{a2}}{K_{a1}} = \frac{4.7 \times 10^{-11}}{4.3 \times 10^{-7}} = 1.1 \times 10^{-4}$$

Since H_2CO_3 and CO_3^{2-} are formed in equal amounts:

$$[CO_3^{2-}]^2 = [1.1 \times 10^{-4}][HCO_3^-]^2 = [1.1 \times 10^{-4}][1.0 \times 10^{-2}]^2 = 1.1 \times 10^{-8}$$

$$[CO_3^{2-}] = 1.0 \times 10^{-4}$$

Consequently, if we put the concentration of CO_3^{2-} into the concentration of Fe^{2+} relationship, the concentration of Fe^{2+} must be lower than this value:

$$[Fe^{2+}] < \frac{2.0 \times 10^{-11}}{1.0 \times 10^{-4}} = 1.9 \times 10^{-7} \text{ M}$$

---

*QUESTION 9

Calculate the solubility of silver (II) carbonate, $AgCO_3$ ($K_{sp} = 6.2 \times 10^{-12}$), in H_2O. Then, assume that most of the carbonate ions produced react with water to form bicarbonate ions. How would your results differ based on the two different assumptions?

The chemistry of the inorganic carbon cycle is determined by the precipitation and dissolution of calcium carbonate. The pH of the ocean is 8.1, and CO_2 dissolves readily since the carbonic acid is converted to bicarbonate. The ocean is, therefore, a vast store of inorganic carbon ($39,000 \times 10^{15}$ g; Fig. 13.5).

The values of $[HCO_3^-]$ and pH calculated in Q5 are very close to those observed for seawater, ~2 mM and 8.1, respectively. Thus, the oceans seem to behave like a bowl of H_2O, lined with $CaCO_3$ and in contact with air.

A small fraction of this C is present as CO_3^{2-}, ~0.4%. However, its concentration is sufficient that when Ca^{2+} enters the ocean, as a result of the weathering of terrestrial rocks, calcium carbonate precipitates. This process is inherently slow because nucleation of $CaCO_3$ requires a high degree of supersaturation, but sea creatures can accelerate this nucleation in forming $CaCO_3$ shells. Most of the world's $CaCO_3$ forms as seashells. Magnesium is also involved, although to a lesser extent, because some seashells contain dolomite, $MgCa(CO_3)_2$. The $MgCO_3$ by itself is too soluble to precipitate in the ocean. For simplicity, we will limit the reactions that follow Ca, but keep in mind that Mg is sometimes also involved.

When $CaCO_3$ [or $MgCa(CO_3)_2$] precipitates:

$$Ca^{2+} + CO_3^{2-} \rightarrow CaCO_3 \tag{13.6}$$

The carbonate acid–base equilibrium readjusts to restore the CO_3^{2-} concentration, via the reaction:

$$2\,HCO_3^- \rightleftharpoons CO_3^{2-} + H_2CO_3 \tag{13.7}$$

(The proton released to form CO_3^{2-} is taken up by a second HCO_3^-.)
The sum of these two reactions is:

$$Ca^{2+} + 2\,HCO_3^- \rightarrow CaCO_3 + H_2CO_3 \tag{13.8}$$

We see that precipitation of each mole of $CaCO_3$ produces 1 mol of H_2CO_3, thereby adding CO_2 to the atmospheric reservoir. However, if the source of Ca (or Mg) is the weathering of terrestrial carbonate minerals, then reaction (13.8) has been run in reverse on land, and the net effect on global CO_2 is zero. Weathering of terrestrial carbonates consumes the same amount of CO_2 as is released by the precipitation of carbonates in the ocean.

An ominous consequence of this simple chemistry is that sea creatures may have difficulty making their calcium carbonate shells as the atmospheric CO_2 rises. The rate of shell production depends on the carbonate concentration, which decreases as CO_2 rises and the pH falls. In fact, this relationship between increased atmospheric CO_2 concentrations and a decrease in ocean pH has already been observed in Hawaii and elsewhere around the world, where the atmospheric CO_2 concentration has long been monitored (see Chapter 7). Experiments in controlled environments have demonstrated that adding CO_2 to seawater slows the rate at which coral and reef-building algae secrete $CaCO_3$.

The geological record indicates that ocean pH held constant at 8.2 during the last 600,000 yr, but has dropped to 8.1 since the 1800s. If the atmospheric CO_2 concentration doubles, $[H^+]$ will be halved, and the pH will drop further to 7.8 [recall that log (2) = 0.3]. The disruption to ocean life could be extensive. The growth of coral reefs is already affected by ocean acidification.

13.2 Weathering of Silicate Minerals

The situation is quite different when silicate minerals are weathered because silicate acts as a base but is not a source of carbonate to the pool. Silicate chemistry is complex, but the essence of the acid–base process is that the equivalent of one SiO_3^{2-} unit consumes two protons and is converted to $Si(OH)_4$:

$$SiO_3^{2-} + 2\,H^+ + H_2O \rightarrow Si(OH)_4 \tag{13.9}$$

If the protons are provided by CO_2 (i.e., H_2CO_3), and if the silicate is the source of the Ca^{2+} (or Mg^{2+}) ions that later precipitate carbonate in the ocean, then the silicate weathering contribution to the carbon cycle can be represented as:

$$CaSiO_3 + 2\,H_2CO_3 + H_2O \rightarrow Si(OH)_4 + Ca^{2+} + 2\,HCO_3^- \tag{13.10}$$

When this reaction is added to the carbonate precipitation reaction (13.8), the resultant is:

$$CaSiO_3 + H_2CO_3 + H_2O \rightarrow Si(OH)_4 + CaCO_3 \tag{13.11}$$

Thus, if silicate, rather than carbonate, is the source of the Ca^{2+} that subsequently produces carbonate precipitation, then 1 mol of H_2CO_3 (or equivalently of CO_2) is consumed for every mole of carbonate precipitated. Consequently, silicate weathering draws down the atmospheric reservoir of CO_2.

In fact, weathering provides negative feedback on fluctuations in the CO_2 level because of the operation of the greenhouse effect. The weathering rate increases with temperature and moisture.

QUESTION 10

Does the weathering rate depend on the CO_2 concentration itself?

Q10 ANSWER Yes. Higher CO_2 increases $[H_2CO_3]$ in rainwater, which increases the rate of reaction (13.11).

If the CO_2 rises, so does the earth's surface temperature and the flow of water through the geological cycle. The weathering rate increases, and so does the flow of CO_2 to the carbonate sediments, thereby diminishing the atmospheric CO_2 concentration. This mechanism acts to stabilize CO_2 levels over geologic time.

If reaction (13.11) operated only in the forward direction, then CO_2 would slowly, but surely, disappear from the atmosphere. However, the reaction is eventually reversed by the operation of plate tectonics. As the earth's crust is subducted at the plate boundaries, the increase in temperature and pressure reverses reaction (13.11); the volatile components, H_2O and CO_2, escape through volcanoes, leaving the silicate rock behind. This process completes the inorganic carbon cycle, whose reactions are shown later in the chapter in Figure 13.5.

13.3 Ion Exchange: Clays and Soils

Many solids have ions that are loosely held at fixed charge sites. These solids can be exchanged with ions that are free in the solution. The exchanging ions may be positively charged (cations) or negatively charged (anions):

$$R^-M^+ + M'^+ \rightarrow R^-M'^+ + M^+ \qquad \text{Cation exchange} \qquad (13.12)$$

$$R^+X^- + X'^- \rightarrow R^+X'^- + X^- \qquad \text{Anion exchange} \qquad (13.13)$$

In these reactions, R represents a fixed charge site. It attracts ions of the opposite charge, the strength of attraction being proportional to the charge/radius ratio. (Multiple-charged ions can occupy more than one ion exchange site.) This reaction is the underlying mechanism of ion exchange resins, with organic polymers having numerous covalently attached charged groups.

Because of their abundance in the earth's crust, silicates are the major component of soils (63% on average) and are responsible for most of the ion exchange capacity. Silica itself (as found naturally in quartz) has no ion-exchange sites because of its closed 3D network of oxygen atoms. However, the substitution of metal ions in the silica framework leaves uncompensated negative charges because the charge on the metal ions (+3, +2, or +1) is less than the charge on Si (+4) (counting the oxide ions as –2). These uncompensated charges are balanced by mobile cations, giving the silicate mineral cation exchange character. Of special importance are the clays, which result from the weathering of primary silicate minerals, and are, therefore, abundant in soils.

All of these substitutions of cations in soil (Al^{3+} and Si^{4+}) with lower charges (e.g., Fe^{3+}, Mg^{2+}) produce excess negative charge from the oxygen atoms, which is balanced by the adsorption of other cations, commonly Na^+, K^+, Mg^{2+}, or Ca^{2+}. These are the base cations, so-called because their oxides are strong bases. They are readily exchanged for other cations in solution. The order of exchange depends on the affinity of the cations for anionic sites on the clay compared to their attraction for water molecules. Generally, aluminum cations are the most difficult to exchange and sodium cations the least difficult, with other ions between, in the order $Al^{3+} > H^+ > Ca^{2+} > Mg^{2+} > K^+ > NH_4^+ > Na^+$. Because protons are more tightly held than most other cations, if the soil solution is acidic, the protons will exchange with the adsorbed cations. The exchange of protons for other ions increases both the solution's pH and its base cation concentration. Thus, like limestone, clays neutralize the acids in soil water while increasing the concentration of base cations.

One of the most commonly measured properties of soil is the cation-exchange capacity (CEC). This is a quantitative assessment of how clay or soil interacts with cations. A standard method for measuring the CEC of a sample is to extract the soil with 1M ammonium chloride at pH = 4.5. The sample is then filtered, and the concentration of calcium, magnesium, potassium, and sodium are determined by flame atomic absorption spectroscopy. The concentration of the hydronium ion is determined by titration.

QUESTION 12

Calculate the CEC if you extract a 2.50 g sample of soil with 250. mL of ammonium chloride, and obtained the following results:

$[Ca^{2+}]$	35.8 $\mu g\ mL^{-1}$
$[Mg^{2+}]$	16.3 $\mu g\ mL^{-1}$
$[K^+]$	not detected
$[Na^+]$	6.5 $\mu g\ mL^{-1}$
$[H_3O^+]$	8.60 $mmol\ L^{-1}$

Q12 ANSWER The first step is to calculate the total positive charge for these cations. Because the only cations we measure were present in the exchange sites of the soil, the positive charge is equivalent to the number of negative sites in the 2.50 g of soil.

For calcium, the concentration of positive charge due to this +2 charged cation:

$$2 \times 35.8 \frac{\mu g}{mL} \times \frac{mol}{40.1\ g} \times \frac{1 \times 10^6\ \mu mol}{1\ mol} \times \frac{g}{1 \times 10^6\ \mu g} = 1.79 \frac{\mu mol}{mL}$$

Now we can take into account the volume of ammonium chloride used to extract the 2.50g of soil sample. The convention is to report the results in cmol (10^{-2} mol) of positive charge per kg of soil:

$$1.79 \frac{\mu mol}{mL} \times \frac{250.\ mL}{2.50\ g\ soil} \times \frac{1,000\ g}{kg\ soil} \times \frac{10^{-4}\ cmol}{\mu mol} = 17.9 \frac{cmol}{kg\ soil}$$

Similar calculations can be done for the other ions:

$$Mg^{2+}: 2 \times 16.3 \frac{\mu g}{mL} \times \frac{mol}{24.31\ g} \times \frac{1 \times 10^6\ \mu mol}{1\ mol} \times \frac{g}{1 \times 10^6\ \mu g} = 1.34 \frac{\mu mol}{mL}$$

$$1.34 \frac{\mu mol}{mL} \times \frac{250.\ mL}{2.50\ g\ soil} \times \frac{1,000\ g}{kg\ soil} \times \frac{10^{-4}\ cmol}{\mu mol} = 13.4 \frac{cmol}{kg\ soil}$$

$$Na^+: 1 \times 6.5 \frac{\mu g}{mL} \times \frac{mol}{23.00\ g} \times \frac{1 \times 10^6\ \mu mol}{1\ mol} \times \frac{g}{1 \times 10^6\ \mu g} = 0.28 \frac{\mu mol}{mL}$$

$$0.28 \frac{\mu mol}{mL} \times \frac{250.\ mL}{2.50\ g\ soil} \times \frac{1,000\ g}{kg\ soil} \times \frac{10^{-4}\ cmol}{\mu mol} = 2.8 \frac{cmol}{kg\ soil}$$

To calculate the concentration of H_3O^+:

$$8.60\frac{mmol}{L} \times \frac{L}{1{,}000\ mL} \times \frac{250.\ mL}{2.50\ g\ soil} \times \frac{1{,}000\ g}{kg\ soil} \times \frac{10^{-1}\ cmol}{mmol} = 86.0\frac{cmol}{kg\ soil}$$

The CEC is calculated by adding the individual values calculated:

$$CEC = (17.9 + 13.4 + 2.8 + 86.0)\frac{cmol}{kg\ soil} = 120.1\frac{cmol}{kg\ soil}$$

*QUESTION 13

Calculate the CEC if you extract a 1.75g sample of soil with 200. mL of ammonium chloride, and obtained the following results:

$[Ca^{2+}]$	20.1 μg mL^{-1}
$[Mg^{2+}]$	13.7 μg mL^{-1}
$[K^+]$	8.5 μg mL^{-1}
$[Na^+]$	not detected
$[H_3O^+]$	3.60 mmol L^{-1}

13.4 Effects of Freshwater Acidification

13.4.1 Soil Neutralization

Figure 13.1 shows the acid–base reactions that occur after rain falls to the ground. Initially, the pH can diminish, because topsoil contains large amounts of CO_2 due to bacterial decomposition of the organic matter, up to 100 times the concentration of CO_2 in the atmosphere. In addition, plants exude a variety of organic acids, and the decay of plant matter produces other acids *en route* to the eventual conversion of the organic molecules to CO_2. The pH values in the topsoil are frequently <5.

As the water percolates into the mineral layers, neutralization reactions come into play. In soils containing limestone (calcareous soils), neutralization raises the pH toward 8.4. The effectiveness of this reaction is reflected in the widespread practice of liming (mixing crushed limestone into the soil) agricultural fields, as well as lawns and gardens in order to raise the pH of acidic soils.

When limestone is absent, neutralization occurs via the exchange of protons for the base cations on ion exchange sites of clay particles and of the humus. The number of exchangeable cation sites is the CEC, measured in units of acid equivalents (equiv) per square meter of soil or cmol of acid equivalents per kg of soil. The fraction of the sites occupied by base cations, rather than by H$^+$, is called the base saturation (β). These ion-exchange sites consist of anionic oxygen atoms bound to the silicate lattice (or organic acid anions in the case of humus) and are much less basic than the carbonate ion. Consequently, the pH rises much less in noncalcareous than in calcareous soils; typically, the pH in noncalcareous soils is ~5.5.

Because the ion exchange sites are on the particle surface, the CEC is limited and is much less than the neutralizing capacity of calcareous soils. However, the CEC is slowly replenished by silicate weathering reactions, such as reaction (13.2), which release additional base cations

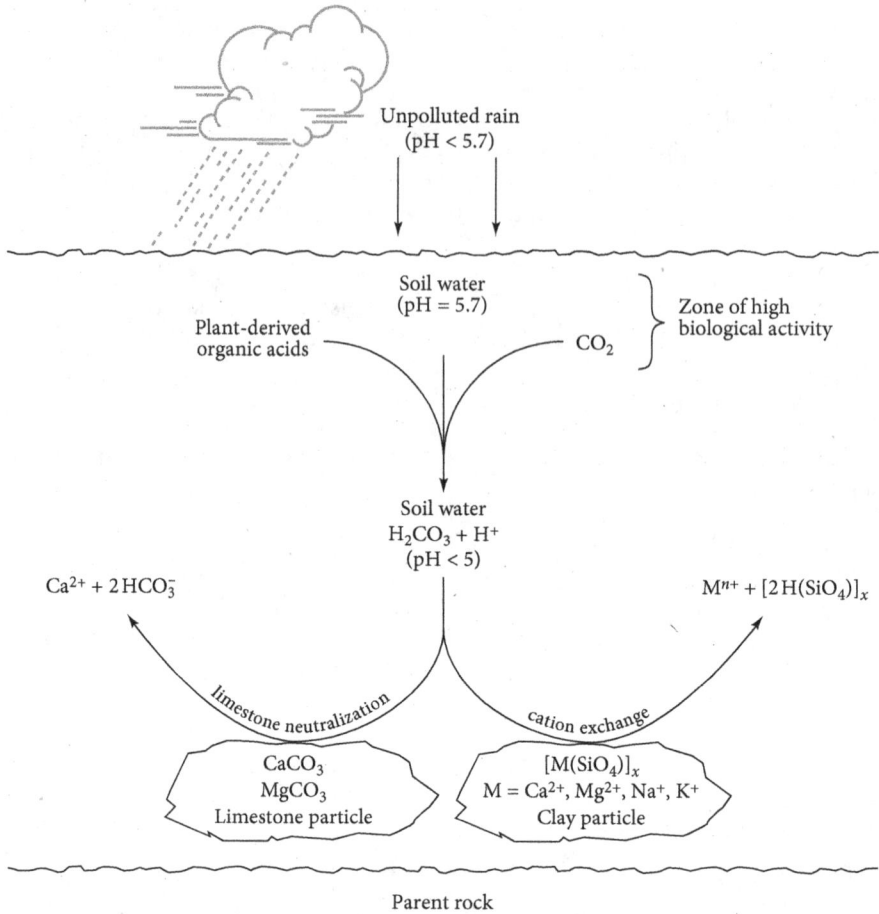

Figure 13.1 Percolation of rainwater through soil and neutralization by limestone and clays. *Source:* Adapted from Spiro et al., *Chemistry of the Environment, 3rd Ed.,* University Science Books, © 2012, all rights reserved.

and provide new ion-exchange sites. In some of the weathering reactions, Al^{3+} is also released from the aluminosilicate framework of clays. Because $Al(OH)_3$ has a very low solubility, it precipitates at pH values above ~4.2 and stays bound to the soil particles. If soil acidification exceeds the CEC, so that nearly all the base cations are replaced by protons (β declines toward zero), then the pH drops to < 4.2 and Al is solubilized:

$$Al(OH)_3 + 3\,H^+ \rightarrow Al^{3+} + 3\,H_2O \qquad (13.14)$$

QUESTION 14

Calculate the solubility of $Al(OH)_3$ in pure water and the pH of the resulting solution.

Q14 ANSWER The dissolution equation is:

$$Al(OH)_3 \rightarrow Al^{3+} + 3\,OH^-$$

$$K_{sp} = [Al^{3+}][OH^-]^3 = s^4 = 3 \times 10^{-34}$$

$$s = 4.2 \times 10^{-9}\,\text{M}$$

$$\text{pH} = 14.0 - \text{pOH} = 14.0 - [-\log(4.2 \times 10^{-9}\,\text{M})] = 14.00 - 8.38 = 5.62$$

13.4.2 *Acid Deposition and Watershed Buffering*

The atmosphere receives substantial inputs of SO_2 and NO_x from both natural and anthropogenic sources. These emissions are cleared from the air within a few days by oxidation reactions and subsequently transferred to the soil, either directly by dry deposition in aerosols or indirectly by wet deposition in rainfall. Such reactions are vital to the health of the biosphere because they cleanse the air of noxious fumes. If SO_2 and NO_x accumulated in the atmosphere, the air would quickly become toxic.

However, the cleansing of the atmosphere transfers sulfuric and nitric acids to the soils. Hence, soils can be described as sinks or reservoirs for atmospheric pollutants. Pollutants emitted to the air flow through the environment, mediated by a series of physical and chemical processes, as illustrated schematically in Figure 13.2. The first step is air transport and then deposition onto the soil (1). Soils, in their capacity as chemical filters, may adsorb, neutralize, or otherwise retain and store the pollutant. When buffering capacities are diminished, the soil may release the pollutant to rivers and lakes (2a) or to the groundwater (2b). Eventually, pollutants are discharged to the oceans through stream (3a) and subsurface (3b) flow and are deposited in the ocean sediment (4), their ultimate repository.

The effectiveness of the soil as a chemical filter of acidic inputs depends on its buffering capacity and the rate of acid deposition. Although the buffering capacity of most soils is sufficient to neutralize naturally occurring acids, over time the capacity can be overwhelmed by high inputs of acid deposition. Figure 13.3 illustrates schematically the course of events as the soil is continuously acidified. If limestone is present, the pH is initially maintained at ~8, through the dissolution of carbonate. When the limestone is dissolved away, then the pH drops to ~5.5 as acid displaces the base cations in the soil. When these are all displaced,

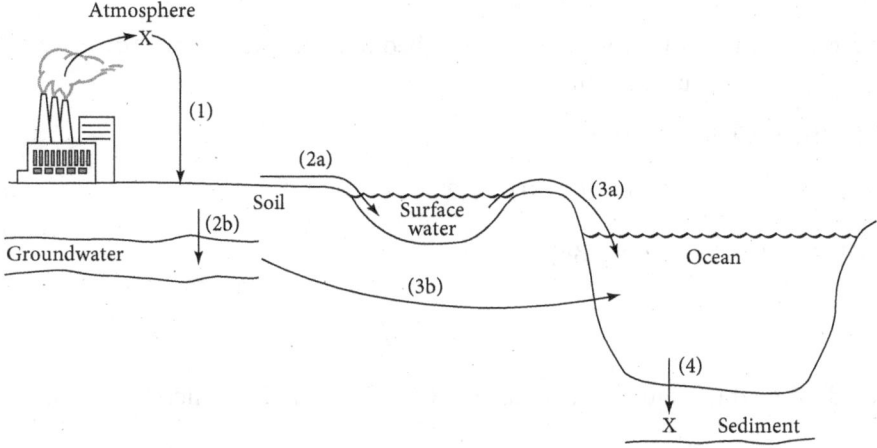

Figure 13.2 Flow of pollutant X from sources to sinks.
Source: Adapted from Spiro et al., *Chemistry of the Environment, 3rd Ed.*, University Science Books, © 2012, all rights reserved.

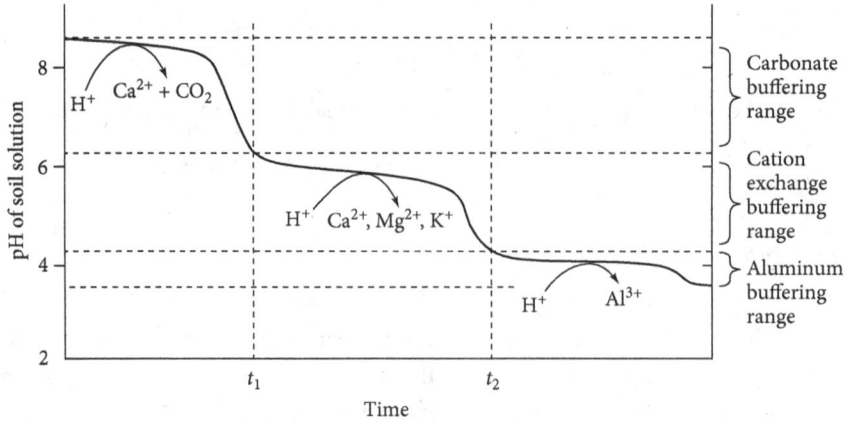

Figure 13.3 Scheme showing the progression of decline in soil solution pH over time in response to atmospheric acid inputs. For a given soil, the time scales over which the soil solution passes from one buffering range to the next depends on the intensity of acid deposition and the concentration of exchangeable cations. Arrows indicate the chemistry by which soils retain H^+ from the soil solution by exchanging it for cations. The period from t_1 to t_2 is the time over which soils lose 90%–95% of their original exchangeable base cations.
Sources: Adapted from Stigliani & Shaw, "Energy Use and Acid Deposition: The View from Europe," *Annual Review of Energy*, Vol. 15, 1990, pp. 201–216; and Spiro et al., *Chemistry of the Environment, 3rd Ed.,* University Science Books, © 2012, all rights reserved.

the pH falls to ~4, where $Al(OH)_3$ is gradually dissolved. Thus, there are three distinct buffer regions, as the soil is titrated by protons. Figure 13.3 can also represent the fate of a watershed that is subject to continuing inputs of acid. The pH can stay fairly constant, at either the carbonate or the silicate buffer pH, for extended periods of time, and then drop rapidly when the buffer capacity is exceeded. It is hard to predict when this will happen. The length of time depends on many factors, including the rate of deposition, the nature of the soil, the size of the watershed, and the flow characteristics of the lake or groundwater.

QUESTION 15

If the concentration of H^+ in a water sample in the watershed described in Figure 13.3 measures 6.3×10^{-6} M, what buffering region is the watershed in?

Q15 ANSWER $pH = -\log[H^+]$ (from Chapter 12)

$$pH = -\log(6.3 \times 10^{-6}\,M) = 5.2$$

This would be in the cation exchange buffering region.

*QUESTION 16

A sample from the watershed described in Figure 13.3 measures 1.5×10^{-8} M H^+. Which buffering region is the watershed in?

One case where the historical record of acidification has been well-documented is the watershed of Big Moose Lake in the Adirondack Mountains of New York. This area has received some of the highest inputs of acid deposition in North America because it is downwind from western Pennsylvania and the Ohio Valley, historically the industrial heartland of the United States. Pollutants carried by the westerly winds are trapped in the mountains and deposited via wet and dry deposition.

The difficulties in recognizing acidification while it is occurring are illustrated in Figure 13.4, which shows the historical trends in Big Moose lake water pH (dashed curve), SO_2 emissions upwind from the lake (solid line), and the extinction of different fish species. The Adirondack watersheds lie on a granitic rock and are without limestone. The pH of the lake remained nearly constant at ~5.6, the silicate buffer value, over the entire period from 1760 to 1950. Then, within the space of 30 yr, from 1950 to 1980, the pH declined more than one whole pH unit to ~4.5. The decline in pH lagged behind the rise in SO_2 emissions by some 70 yr, and the peak years of sulfur emissions preceded the decline in pH by 30 yr. The deposition rate is estimated to have been ~2.5g of sulfur m^{-2} yr^{-1} during the peak period. These quantities, deposited year after year, were large enough to deplete the capacity for base cation exchange in the watershed. Thus, beginning in ~1950, atmospheric acid deposition moved through the buffer-depleted soils of the watershed and percolated into the lake with diminished neutralization. At that point, acid-sensitive fish species (e.g., smallmouth

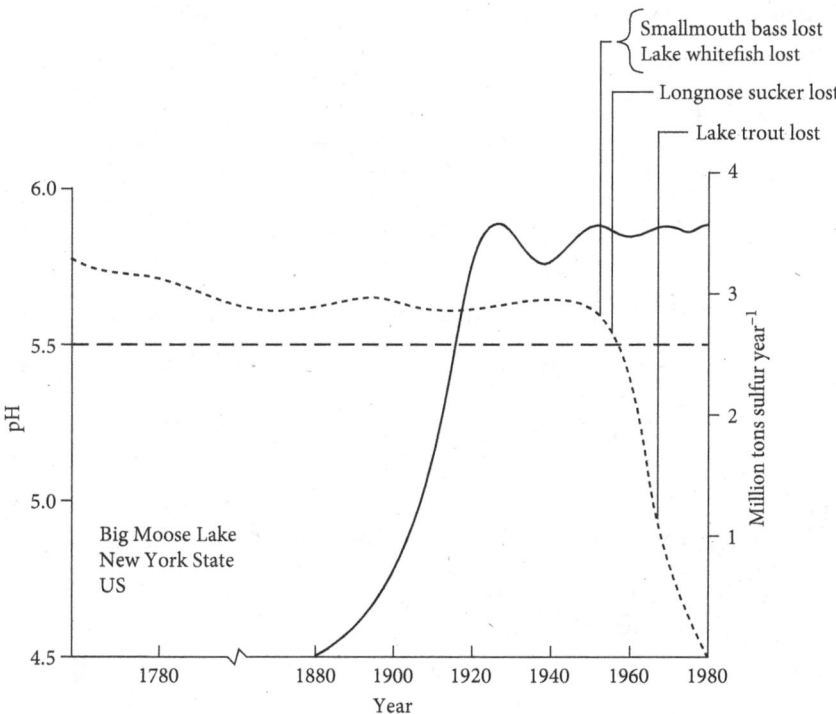

Figure 13.4 Trends in lake water pH (dashed curve); upwind emissions of SO_2 from the U.S. industrial Midwest (solid curve); and fish extinctions for the period from 1760 to 1980.
Sources: Adapted from Stigliani, "Changes in Valued Capacities of Soils and Sediments as Indicators of Nonlinear and Time-Delayed Environmental Effects," *Environment Monitoring Assessment*, Vol. 10, 1988, pp. 245–307, and Spiro et al., *Chemistry of the Environment, 3rd Ed.,* University Science Books, © 2012, all rights reserved.

bass, whitefish, and longnose sucker) began to disappear, followed in the late 1960s by the more acid-resistant lake trout.

From the shape of the historical pH trend, we can see that Big Moose Lake was the subject of an inadvertent titration experiment conducted over four generations of industrial activity. The coal-driven industrialization of the Ohio Valley, upwind from the lake, supplied the acid inputs, mostly as H_2SO_4 formed from SO_2 released during coal combustion (see Chapter 9). The soils of the watershed supplied buffering chemicals. Thus, the watershed's natural buffering capacity delayed recognition of the deleterious effects of coal-burning for about three generations. During this time, there was no direct evidence of how pollution was affecting the pH of lake water or fish mortality. As this example suggests, polluting activities may be far displaced in time from their environmental effects.

QUESTION 17

With reference to Figure 13.4, assume that the average annual deposition of sulfur in the watershed of Big Moose Lake between 1880 and 1920 was 0.80 g S m^{-2}, and from 1921 to 1950 it was 2.5 g S m^{-2}. Calculate the cumulative acid equivalents (equiv) per square meter that were deposited in the watershed over the entire period from 1880 to 1950.

Q17 ANSWER The cumulative deposition of sulfur in Big Moose Lake from 1880 to 1950 was:

$$41 \text{ yr} \times \frac{0.80 \text{ g S}}{m^2 \text{ yr}} = 33 \frac{\text{g S}}{m^2}$$

$$30 \text{ yr} \times \frac{2.5 \text{ g S}}{m^2 \text{ yr}} = 75 \frac{\text{g S}}{m^2}$$

$$\text{Total: } (33 + 75) \frac{\text{g S}}{m^2} = 108 \frac{\text{g S}}{m^2}$$

Moles of S were:

$$\frac{108 \text{ g S}}{m^2} \times \frac{1 \text{ mol S}}{32.07 \text{ g S}} = 3.37 \frac{\text{mol S}}{m^2}$$

1 mol of S produces 2 mol of H^+ (H_2SO_4). Thus, cumulative acid equivalents per m^2

$$3.37 \frac{\text{mol S}}{m^2} \times \frac{2 \text{ eq } H^+}{1 \text{ mol S}} = 6.74 \frac{\text{eq } H^+}{m^2}$$

---

*QUESTION 18

Sulfur deposition caused concern in another watershed in the northeast United States. Between 1880 and 1920, 0.63 g S m^{-2} and from 1921 to 1950, 3.3 g S m^{-2}. Calculate the cumulative acid equivalents per square meter deposited over the watershed during the entire period.

QUESTION 19

Assume that the Big Moose Lake watershed soils buffered acidity via exchange reactions with base cations on clay mineral surfaces. Assume further that base saturation (β) declined from 50% in 1880 to 5% in 1950. Calculate the total cation exchange capacity (CEC_{tot}) and the buffering capacity ($CEC_{tot} \times \beta$) in 1880 (in units of equiv m^{-2}). Also, assume that the silicate buffer rate (br_{Si}), which replenishes the base cations in soil from weathering of silicate, was negligible over this period.

Q19 ANSWER If β declined from 50% to 5% over the period from 1880 to 1950, then 6.74 mol H$^+$ m^{-2} replaced base cations at 45% of the cation exchange sites. Therefore:

$$CEC_{tot} = \frac{100}{45} \times 6.74 \frac{eq}{m^2} = 15.0 \frac{eq}{m^2}$$

Since buffering capacity = $CEC_{tot} \times \beta$

$$\text{buffering capacity (1880)} = 15.0 \frac{eq}{m^2} \times 0.50 = 7.5 \frac{eq}{m^2}$$

*QUESTION 20

Calculate the buffering capacity of the Big Moose Lake watershed soils in 1950.

*QUESTION 21

Similar to Q17, assume that the watershed soils buffered acidity via exchange reactions with base cations on clay mineral surfaces. Assume further that base saturation (β) declined from 50% in 1880 to 5% in 1950. Calculate the CEC_{tot} and the buffering capacity ($CEC_{tot} \times \beta$) in 1880 (in units of equiv m^{-2}). But now assume br_{Si} was equal to 0.02 equiv m^{-2} yr^{-1} over the time scale from 1880 to 1950, instead of being negligible.

13.4.3 Global Acidification

Thus far, we have been discussing acidification on local and regional scales. But acid deposition on a global level from industrial sources is of the same order of magnitude as deposition from natural sources (Table 13.2). Acid deposition is not the whole story, because there are also natural and anthropogenic sources of alkaline chemicals in the atmosphere. These include ammonia (NH_3) and alkaline particles derived from ash, as well as windblown alkaline minerals. These chemicals have been crudely estimated to neutralize between 20% and 50% of the generated acidity. The atmosphere has acted as an acidic medium throughout geologic time, but natural sources of acidity, although of the same order of magnitude as anthropogenic sources, are spread evenly across the globe. Polluting sources are concentrated near industrial and urban centers, with levels of acidity exceeding 50–100 times the natural background. It is the excessive concentration of acidity in particular regions that causes problems for the biosphere.

TABLE 13.2 Estimated Natural and Anthropogenic Sources of Global Atmospheric Acidity

Source	10^{12} mol H$^+$ generated per year
Natural	
Unpolluted rainwater	1.0
Lightning[a]	1.4
Volcanoes[b]	1.3
Biogenic sulfur	4.1
Total natural	**7.8**
Pollution	
Coal combustion/metal smelting[b]	5.8
Combustion processes[a]	1.4
Total from pollution	**7.2**

[a]Refers to acidity generated from NO_x emissions.
[b]Refers to acidity generated from SO_2 emissions.
Sources: Adapted from Schlesinger, *Biogeochemistry: An Analysis of Global Change*, Academic Press, Inc., 1991, and Spiro et al., *Chemistry of the Environment, 3rd Ed.*, University Science Books, © 2012, all rights reserved.

QUESTION 22

Show how the numbers given in Table 13.2 were calculated for natural sources. Make the following assumptions: the pH of unpolluted rainfall is 5.7, and annual global precipitation is 496,000 km^3 yr^{-1}; production of NO$_2$ from lighting is 20×10^{12} g N yr^{-1}; production of SO$_2$ from volcanoes is 20×10^{12} g S yr^{-1}; biogenic production of dimethyl sulfide (DMS) and H$_2$S is 65×10^{12} g S yr^{-1}.

Q22 ANSWER Natural components of global atmospheric acidity:

Unpolluted rainwater: pH = 5.7:

$$[H^+] = 10^{-pH} = 10^{-5.7} \frac{\text{mol H}^+}{\text{L}}$$

annual global precipitation =

$$496{,}000 \frac{\text{km}^3}{\text{yr}} \times \frac{1\text{ L}}{1{,}000\text{ mL}} \times \frac{\text{mL}}{\text{cm}^3} \times \frac{(100{,}000\text{ cm})^3}{\text{km}^3} \times 10^{-5.7} \frac{\text{mol H}^+}{\text{L}} = 1.0 \times 10^{12} \frac{\text{mol H}^+}{\text{yr}}$$

Lightning:

$$20 \times 10^{12} \frac{\text{g N}}{\text{yr}} \times \frac{1\text{ mol N}}{14.0\text{ g N}} \times 1.4 \times 10^{12} \frac{\text{mol N}}{\text{yr}}$$

Since 1 mol N = 1 mol H^+ (from HNO_3), $1.4 \times 10^{12} \frac{\text{mol } H^+}{\text{yr}}$

Volcanoes:
Since 1 mol S = 2 mol H^+ (from H_2SO_4)

$$20 \times 10^{12} \frac{\text{g S}}{\text{yr}} \times \frac{1 \text{ mol S}}{32.07 \text{ g S}} \times \frac{2 \text{ mol } H^+}{1 \text{ mol S}} = 1.2 \times 10^{12} \frac{\text{mol } H^+}{\text{yr}}$$

Biogenic sulfur:

$$65 \times 10^{12} \frac{\text{g S}}{\text{yr}} \times \frac{1 \text{ mol S}}{32.07 \text{ g S}} \times \frac{2 \text{ mol } H^+}{1 \text{ mol S}} = 4.1 \times 10^{12} \frac{\text{mol } H^+}{\text{yr}}$$

Total natural H^+:

$$1.0 \times 10^{12} \frac{\text{mol } H^+}{\text{yr}} + 1.4 \times 10^{12} \frac{\text{mol N}}{\text{yr}} + 1.2 \times 10^{12} \frac{\text{mol } H^+}{\text{yr}} + 4.1 \times 10^{12} \frac{\text{mol } H^+}{\text{yr}} = 7.7 \times 10^{12} \frac{\text{mol } H^+}{\text{yr}}$$

*QUESTION 23

Calculate moles of hydrogen ions per year generated from coal-burning and smelting, assuming this activity generates $\sim 93 \times 10^{12}$ g S yr^{-1}. Assume combustion processes from all activities generate $\sim 20 \times 10^{12}$ g N yr^{-1}.

QUESTION 24

Calculate the acid-neutralizing capacity of the atmosphere due to the generation of ammonia (NH_3). Assume that natural and anthropogenic sources each generate $\sim 60 \times 10^{12}$ g N yr^{-1}. Calculate the total *net* generation of H^+ year^{-1} from both natural and anthropogenic sources.

Q24 ANSWER Natural and human sources of NH_3 each generate a total of 60×10^{12} g N yr^{-1}:

$$60 \times 10^{12} \frac{\text{g N}}{\text{yr}} \times \frac{1 \text{ mol N}}{14.00 \text{ g N}} = 4.3 \times 10^{12} \frac{\text{mol N}}{\text{yr}}$$

Since 1 mol N in NH_3 neutralizes 1 mol H^+, natural and human sources of ammonia *each* neutralize 4.3×10^{12} mol of H^+ yr^{-1}. As calculated in Q22 and *Q23, natural sources generate 7.7×10^{12} and anthropogenic sources generate 7.2×10^{12} mol H^+ yr^{-1}:

$$\text{net natural sources } \frac{\text{mol } H^+}{\text{yr}} = (7.7 \times 10^{12} - 4.3 \times 10^{12}) \frac{\text{mol } H^+}{\text{yr}} = 3.4 \times 10^{12} \frac{\text{mol } H^+}{\text{yr}}$$

$$\text{net human sources } \frac{\text{mol } H^+}{\text{yr}} = (7.2 \times 10^{12} - 4.3 \times 10^{12}) \frac{\text{mol } H^+}{\text{yr}} = 2.9 \times 10^{12} \frac{\text{mol } H^+}{\text{yr}}$$

$$\text{net natural + human } \frac{\text{mol H}^+}{\text{yr}} = (3.4 \times 10^{12} + 2.9 \times 10^{12}) \frac{\text{mol H}^+}{\text{yr}} = 6.3 \times 10^{12} \frac{\text{mol H}^+}{\text{yr}}$$

*QUESTION 25

Calculate the average moles H^+ m^{-2} yr^{-1} from natural sources. (The area of the globe is 510×10^{12} m^2.) Calculate the average moles H^+ m^2 yr^{-1} industrialized regions of the globe. (Assume that the area of industrialized regions is 11.75×10^{12} m^2.)

13.5 Organic and Inorganic Carbon Cycles

Earlier, we discussed the carbon cycle that results from the biological processes of photosynthesis and respiration (see Chapter 7), and the disturbance of this cycle by the burning of fossil fuels. In reality, the cycle of photosynthesis and respiration is embedded in a much more complex biogeochemical cycle, involving sedimentation and burial, as well as the tectonic processes of the earth's crust.

This larger cycle (Figure 13.5) is actually two cycles, linked by the CO_2 molecule. One cycle is organic and involves the photosynthesis–respiration–combustion balance. It also

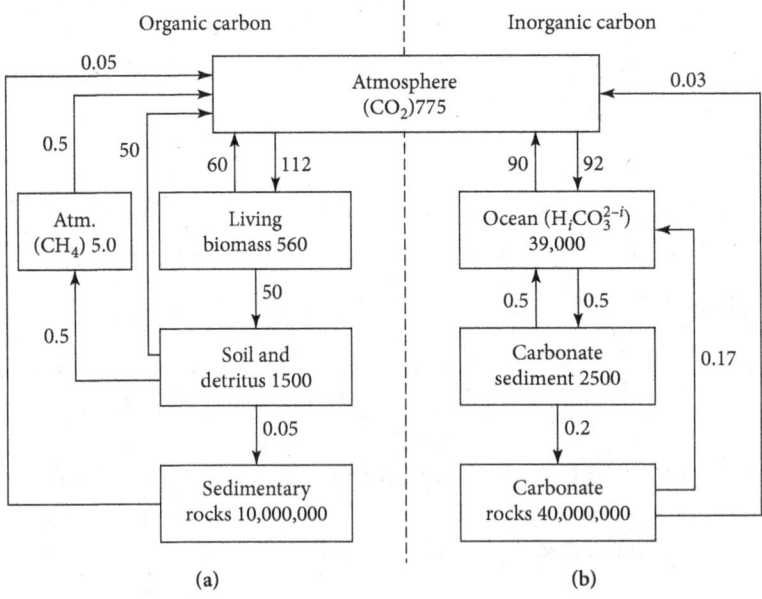

Figure 13.5 Reservoirs (boxes) and annual flows (arrows) of carbon (in units of 10^{15} g) through organic (a) and inorganic (b) cycles. Outputs from the atmosphere to living biomass and the ocean (total of ~204 × 10^{15} g) are roughly 4 × 10^{15} g larger than inputs to the atmosphere from the living biomass, soils and detritus, and the ocean (total of ~200 × 10^{15} g). This finding is attributed to the net accumulation of carbon in these reservoirs owing to anthropogenic emissions of CO_2 from the burning of fossil fuels. *Source:* Adapted from Spiro et al., *Chemistry of the Environment, 3rd Ed.,* University Science Books, © 2012, all rights reserved.

includes the burial of reduced carbon compounds, together with their reoxidation when the rock bearing them is exposed to the atmosphere through geological uplift. The other cycle is inorganic and involves the weathering of silicate rocks, the precipitation of calcium carbonate in the ocean, and finally the conversion of the carbonate to calcium silicates plus CO_2 through tectonic activity. The geological phases of these cycles operate very slowly, on time scales of millions of years, but they ultimately control the level of CO_2 in the atmosphere.

Most of the earth's carbon is trapped in carbonate rock. This reservoir is 40×10^{15} g, four times larger than the reduced carbon reservoir in rocks. The rate of carbonate rock formation from sediment is 0.2×10^{15} g yr^{-1}, but most of this flow is reversed by the redissolution of carbonate in the ocean. Calcium carbonate dissolves in the deep ocean because of increasing solubility at low temperatures and high pressures. The rate at which the carbonate rock is recycled to silicate and CO_2 through tectonic activity is only 0.03×10^{15} g yr^{-1}. Dividing this into the carbonate rock reservoir gives a residence time of 1.3 billion yr. Thus, the inorganic carbon cycle operates for a very long time indeed.

QUESTION 26

Carbonate rocks are the largest sink for carbon on the planet. From Figure 13.5, calculate the residence time of carbonate sediment, the source for the formation of carbonate rocks.

Q26 ANSWER The reservoir of carbonate sediment has $2,500 \times 10^{15}$ g of carbon, but it has two sources of removal (ocean and carbonate rocks)

We use the overall removal of carbon per year:

$$0.5 \times 10^{15} \text{ g} + 0.2 \times 10^{15} \text{ g} = 0.7 \times 10^{15} \text{ g}$$

to calculate the residence time:

$$\text{Residence time} = \frac{2,500 \times 10^{15} \text{ g}}{0.7 \times 10^{15} \text{ g yr}^{-1}} = 3,600 \text{ yr}$$

QUESTION 27

Is the carbonate sediment reservoir changing over time? If so, how long would it take to change by 10%.

Q27 ANSWER Overall addition of carbon per year is 0.5×10^{15} g, so there is a net loss of:

$$0.5 \times 10^{15} \text{ g} - 0.7 \times 10^{15} \text{ g} = -0.2 \times 10^{15} \text{ g yr}^{-1}$$

At that rate, the reservoir would shrink by 10% in:

$$0.10 \times \frac{2,500 \times 10^{15} \text{ g}}{0.2 \times 10^{15} \text{ g}} = 1,200 \text{ yr}$$

The chemistry of the organic carbon cycle was laid out earlier (see Chapter 4). Green plants harness solar photons and store chemical energy by reducing CO_2 to carbohydrates via photosynthesis:

$$CO_2 + H_2O \rightarrow CH_2O + O_2 \qquad (13.15)$$

Energy is released through the reverse reaction, during respiration or combustion. The annual exchange of carbon between the living biomass, the soil and detritus, and the atmosphere is large, 110×10^{15} g (Fig. 13.5), amounting to flows that are ~11% and 3% of their respective reservoirs. (The regular seasonal variations in this flow can be seen in the sawtooth pattern of the atmospheric CO_2 record.) It is important to keep in mind that approximately one-half of the global gross primary productivity occurs from photosynthesis by algae in the oceans, so this contribution of carbon to the atmosphere is through ocean processes in addition to the growth of plants on land.

Also, the CO_2 exchange between the oceans and atmosphere is controlled by two different "pumps." The "biological pump" is driven by photosynthesis and respiration, whereas the "solubility pump" is driven by the higher solubility of CO_2 in cold than in warm waters. The oceans are a sink for CO_2 in cold waters, and they release CO_2 to the atmosphere in warm waters, especially in upwelling areas. The solubility pump accounts for ~40% of the ocean–atmosphere exchange.

About one-half of the return flow of carbon to the atmosphere goes by way of decomposition of soils and marine sediments, a reservoir of reduced carbon that is nearly three times the size of the living biosphere. (A small part of this latter flow occurs by way of methane (CH_4) release to the atmosphere and subsequent oxidation to CO_2.) A tiny fraction of the reduced carbon, 0.05×10^{15} g yr^{-1}, is buried in sedimentary rock. It is eventually oxidized back to CO_2 when the rock is recycled in the crust and exposed to the atmosphere. Although the rate of burial is low, the amount of reduced carbon accumulated in rock over the eons is enormous, 10^{22} g; the turnover time is 200 million yr. It is this part of the cycle that we are accelerating by burning fossil fuels. Even though fossil fuel deposits account for <1% of buried reduced carbon, the rate of oxidation is so much faster than the deposition rate that the CO_2 reservoir size is being significantly increased.

13.6 Conclusions

The chemical interactions between the rocks and minerals in the lithosphere and water on the earth provide critical services to our planet, whether through absorption of acidic inputs from the atmosphere or CEC of soils and minerals. Unfortunately, the excess pollutants being added to the atmosphere in the form of SO_x, NO_x, and CO_2 are changing the earth in ways that exceed the buffering capacity of natural systems. We have observed the acidification of freshwater lakes and even the vast reaches of the world's oceans in the past 200 yr. Progress has been made to decrease acidic inputs from coal burning in some locations, but unfortunately, more needs to be done to address this in other parts of the world. In addition, CO_2 emissions into the atmosphere need to be curtailed globally, as not enough is being done to limit these emissions around the world.

14

OXYGEN AND LIFE

14.1 Life, Redox Reactions, and Energy

Oxygen and life are intertwined in their origins and evolution. Photosynthetic microorganisms gave rise to the current levels of oxygen, which drives atmospheric chemistry and many biochemical processes. Life is powered by redox reactions, in which electrons are transferred from 1 molec (the reductant) to another (the oxidant), often with the release of energy. Oxygen is a powerful oxidant, and the most important biological redox reaction is *respiration*, in which O_2 oxidizes carbohydrates to carbon dioxide and water:

$$(CH_2O) + O_2 \rightarrow CO_2 + H_2O + energy \tag{14.1}$$

Many microorganisms (*aerobes*) and all higher life forms depend on respiration and die without oxygen. However, there are other microorganisms (*anaerobes*) that can utilize other oxidants, even though less energy is produced.

Air-breathing organisms have a plentiful supply of oxygen, but in aquatic systems, higher life forms have a limited supply to depend on, because of the low solubility of O_2 in water (9 mg L^{-1} of pure water at $20°C$, and less at higher temperatures). When oxygen runs out, anaerobes take over.

14.2 Biological Oxygen Demand

The O_2 available for respiration in water depends not only on its solubility and how quickly it can be renewed by wind and water flow, but also on the supply of reduced molecules available for oxidation. This supply is called the biological (or biochemical) oxygen demand (BOD). It is a key indicator of water quality because when the BOD exceeds the oxygen availability, healthy aquatic life becomes impossible. Fish and other water creatures die and anaerobic bacteria take over, usually producing noxious chemicals (H_2S, NH_3, and others).

BOD is defined operationally as the number of milligrams of O_2 taken up by a 1L sample of water on incubation in the dark for five days at $20°C$. Table 14.1 shows some typical BODs for various water discharges.

TABLE 14.1 Typical BODs for Various Processes

Type of Discharge	BOD (mg O_2 L^{-1} Wastewater)
Domestic sewage	165
All manufacturing	200
Chemicals and allied products	314
Paper	372
Food	747
Metals	13

Source: Adapted from Spiro et al., *Chemistry of the Environment, 3rd Ed.,* University Science Books, © 2012, all rights reserved.

QUESTION 1

Calculate the BOD if 25 mg of sugar is dissolved in 2.5L of water, using the empirical formula for carbohydrates (CH_2O). How does this BOD compare with the O_2 solubility at 20°C?

Q1 ANSWER The reaction with oxygen is:

$$(CH_2O) + O_2 \rightarrow CO_2 + H_2O$$

so 1 mol of CH_2O consumes 1 mol of O_2. The molar masses are:

$$\text{Carbohydrates: } 12\frac{g}{mol} + 2\frac{g}{mol} + 16\frac{g}{mol} = 30\frac{g}{mol} \, CH_2O$$

$$\text{Oxygen: } 2 \times 16\frac{g}{mol} = 32\frac{g}{mol} \, O_2$$

$$\text{BOD} = 25 \text{ mg } CH_2O \times \frac{1 \text{ mol } CH_2O}{30 \text{ g}} \times \frac{1 \text{ mol } O_2}{1 \text{ mol } CH_2O} \times \frac{32 \text{ g } O_2}{1 \text{ mol}} \times \frac{1}{2.5 \text{ L}} = 10.7 \text{ mg } L^{-1}$$

This value exceeds the O_2 solubility, 9 mg L^{-1} by 20%.

---

*QUESTION 2

The empirical formula CH_2O is accurate for glucose ($C_6H_{12}O_6$), but not quite accurate for table sugar, sucrose, $C_{12}H_{22}O_{11}$. For 25 mg of sugar, recalculate the BOD with each of the two actual formulas. Explain why CH_2O is often used as the representative formula for carbohydrates in aquatic ecosystem calculations.

---

* Answers to starred questions can be found at the end of the book.

*QUESTION 3

Raw sewage also contains ammonium ions, which oxidize more slowly via nitrification bacteria, so a second BOD is obtained after eight days of incubation (nitrosomonas).

$$2\,NH_4^+ + 3\,O_2 \rightarrow 2\,NO_2^- + 4\,H^+ + 2\,H_2O$$

Calculate this second BOD if 1 L of wastewater contains 10 mg of ammonium ions.

*QUESTION 4

During an oil spill, 1,000 kg of n-propanol (C_3H_8OH) is discharged into a coastal area near Houston containing about 5.0×10^8 L of water. By how much does the BOD (in mg L^{-1}) of this waterbody increase?

*QUESTION 5

In lakes and oceans, most biological activity takes place in the *euphotic* zone, limited by the depth to which light can reach. Here algae, phytoplankton, and cyanobacteria grow via photosynthesis and are consumed by animals and bacteria. Because of gravity, some dead organisms fall below the euphotic zone where bacterial decomposition continues in the deeper layers, consuming dissolved O_2. There is little mixing of water between the warm surface layer and the cold deeper layers (thermal stratification). Suppose a lake with a cross-sectional area of 1 km^2 and a depth of 50 m has a euphotic zone that extends 15 m below the surface. How much biomass (in g of C) can be decomposed by aerobic bacteria in the water column of the lake below the euphotic zone if the average temperature is 20°C, at which the O_2 solubility is 8.9 mg L^{-1}?

14.3 Oxidation Levels and Water

Elements can exist in multiple oxidation states, depending on the number of electrons removed or added to the valence shell of the atoms. Because redox reactions frequently involve electron transfer between complex molecules, it is important to keep track of the atomic oxidation states in order to balance the reaction equations.

The general rules for assigning oxidation numbers (states) are given in Table 14.2.

QUESTION 6

In the nitrification reaction in Q3, what are the oxidation numbers for each element? Is this a redox reaction? How do you balance this reaction?

Q6 ANSWER

$$2\,NH_4^+ + 3\,O_2 \rightarrow 2\,NO_2^- + 4\,H^+ + 2\,H_2O$$

TABLE 14.2 Assigning Oxidation Numbers

1. The oxidation number of an element that is not combined with another element is 0.
2. The overall charge of a molecule is equal to the sum of the oxidation numbers of the atoms in the species.
3. The oxidation number of hydrogen is +1 when combined with nonmetals and –1 when combined with metals.
4. The oxidation numbers of alkali metals and alkaline earth metals is equal to their group number (1 and 2, respectively).
5. The oxidation number for halogens is –1, unless the halogen atom is in combination with oxygen or another halogen higher in the group. The oxidation number of fluorine is always –1.
6. The oxidation number of oxygen is usually –2. Exceptions include when it forms compounds with fluorine (see item 5 above), or is found as peroxides (O_2^{2-}), superoxides (O_2^-), and ozonides (O_3^-).

Source: Adapted from Spiro et al., *Chemistry of the Environment, 3rd Ed.,* University Science Books, © 2012, all rights reserved.

Reactants:

- The oxidation number for O is zero, as according to rule 1 in Table 14.2, the oxidation number of an element that is not combined with another element is zero.

- The oxidation number for H is +1, as rule number 3 states that hydrogen is +1 when combined with a nonmetal.

- The oxidation number of N is –3, as hydrogen contributes +4 charges total, but the overall charge of the molecule is +1.

Products:

- The oxidation number for H is +1, as is evident from the H^+ ion as well as the fact that H is bonded to a nonmetal in H_2O.

- According to rule 6, O in this reaction has an oxidation number of –2.

- The oxidation number for N is +3, since O contributes –4 overall, but the total charge of the molecule is –1.

Because the oxidation numbers change for some elements (N and O), this is a redox reaction. To balance a redox reaction, both mass and charge balance need to be reached on two sides of the reaction. In the above reaction, 12 electrons are lost from two NH_4^+ ions, which are gained by three O_2 molecules, so the charge balance is met. The mass balance is also met for each element.

*QUESTION 7

The following molecules or molecular ions may occur in the environment. List the oxidation states of the elements in each one.

(a)   Acids that occur in atmospheric aerosols, acid rain, and various water bodies:

Nitric acid (HNO_3), phosphoric acid (H_3PO_4), dihydrogenphosphate ($H_2PO_4^-$), hydrogenphosphate (HPO_4^{2-}), formic acid (HCOOH), acetic acid (CH_3COOH), and oxalic acid (HOOCCOOH).

(b)   Species that are excellent oxidizing agents (gain electrons easily):

MnO_4^-, CrO_3, $Cr_2O_7^{2-}$, OsO_4

(c)   Species that are excellent reducing agents (lose electrons easily):

Li, Na, $NaBH_4$, and $LiAlH_4$

(d)   Sodium can be oxidized to different extents by excess oxygen or ozone. Sodium oxide (Na_2O), sodium peroxide (Na_2O_2), and sodium superoxide (NaO_2) are ionic solids at room temperature with a different stability. Sodium ozonide (NaO_3) is extremely unstable; however, if potassium is left undisturbed in air for years it accumulates a covering of superoxide (KO_2) and ozonide (KO_3). Potassium ozonide is a sensitive explosive that has to be handled at low temperatures in an atmosphere consisting of inert gas. Write the oxidation states of oxygen in the various oxide compounds mentioned above.

The general rules for balancing a redox reaction are summarized in Table 14.3.

TABLE 14.3 Balancing Redox Equations

1. Identify the species being oxidized and reduced from the change in their oxidation numbers.
2. Separate out the (unbalanced) oxidation and reduction half-reactions.
3. Balance all of the elements in each half-reaction, except O and H.
4. If the reaction occurs in an acidic solution, balance O by using H_2O and then balance H by using H^+. If in basic solution, balance O with H_2O, then balance H by adding H^+ to the side of each half-reaction that needs an H. Now add the number of OH^- molecules necessary to react away the H^+ atoms to form water, then add the same number of OH^- molecules to the other side of the equation.
5. Balance electrical charges by adding electrons (to the left side for reduction reactions and to the right side for oxidation reactions) until the charges on the two sides of the arrow are equal.
6. Multiply all species in either one or both half-reactions by factors that provide equal numbers of electrons in both half-reactions.
7. Add the two half-reactions together, and make sure that the number of charges and atoms balance each other out.

Source: Adapted from Spiro et al., *Chemistry of the Environment, 3rd Ed.*, University Science Books, © 2012, all rights reserved.

QUESTION 8

Use the information in Table 14.3 to balance the following redox reaction that occurs under acidic conditions:

$$Cr_2O_7^{2-} + HNO_2 \rightarrow Cr^{3+} + NO_3^-$$

Q8 ANSWER First, we assign oxidation numbers to each atom so we can see what is being reduced and what is being oxidized. Use information in Table 14.2 to assign oxidation numbers:

$Cr_2O_7^{2-}$: Oxidation number for O is –2 and Cr is +6

HNO_2: Oxidation number for H is +1, O is –2, and N is +3

Cr^{3+}: Oxidation number is +3

NO_3^-: Oxidation number of O is –2 and N is +5

Looking at the change in oxidation numbers:

Cr is going from +6 → +3, which means that Cr is gaining electrons and is being reduced.

N is going from +3 → +5, which means that N is losing electrons and is being oxidized.

Now we separate the full reaction into half-reactions to deal with the oxidation and reduction parts separately:

$$\text{Red: } Cr_2O_7^{2-} \rightarrow Cr^{3+}$$

Step 1: Balance all atoms except O and H:

$$Cr_2O_7^{2-} \rightarrow 2\,Cr^{3+}$$

Step 2: Reaction occurs in acidic solution, so use H_2O to balance O and then H^+ to balanced H:

$$14\,H^+ + Cr_2O_7^{2-} \rightarrow 2\,Cr^{3+} + 7\,H_2O$$

Step 3: Balance charge by adding electrons:

$$6\,e^- + 14\,H^+ + Cr_2O_7^{2-} \rightarrow 2\,Cr^{3+} + 7\,H_2O$$

Now we work with the oxidation part of the reaction:

$$\text{Ox: } HNO_2 \rightarrow NO_3^-$$

Step 1: Balance all atoms except O and H:

$$HNO_2 \rightarrow NO_3^-$$

Step 2: Reaction occurs in acidic solution, so use H_2O to balance O and then H^+ to balanced H:

$$H_2O + HNO_2 \rightarrow NO_3^- + 3\,H^+$$

Step 3: Balance charge by adding electrons :

$$H_2O + HNO_2 \rightarrow NO_3^- + 3\,H^+ + 2\,e^-$$

Step 4: Need equal number of electrons for both half-reactions. In this case, we need to multiply the oxidation half-reaction by three to equal the six electrons in the reduction reaction:

$$3 H_2O + 3 HNO_2 \rightarrow 3 NO_3^- + 9 H^+ + 6 e^-$$

Now we need to add the two half-reactions together and make sure that the atoms and charge balance each other on the reactants and products side:

$$6 e^- + 14 H^+ + Cr_2O_7^{2-} \rightarrow 2 Cr^{3+} + 7 H_2O$$

$$+ 3 H_2O + 3 HNO_2 \rightarrow 3 NO_3^- + 9 H^+ + 6 e^-$$

$$5 H^+ + Cr_2O_7^{2-} + 3 HNO_2 \rightarrow 2 Cr^{3+} + 4 H_2O + 3 NO_3^-$$

All of the atoms and charges are balanced, so we have correctly balanced this redox equation.

*QUESTION 9

Balance the following reaction that occurs in acidic solution, using the information in Table 14.3:

$$PbO_2 + I_2 \rightarrow Pb^{2+} + IO_3^-$$

QUESTION 10

Using the information in Table 14.3, balance the following redox reaction that occurs in basic solution:

$$O_2 + Cr^{3+} \rightarrow H_2O_2 + Cr_2O_7^{2-}$$

Q10 ANSWER First, we assign oxidation numbers to each atom so we can see what is being reduced and what is being oxidized. Use information in Table 14.2 to assign oxidation numbers:

O_2: Oxidation number is 0

Cr^{3+}: Oxidation number is +3

H_2O_2: Oxidation number for H is +1 and O is −1 (peroxide)

$Cr_2O_7^{2-}$: Oxidation number of O is −2 and Cr is +6

Looking at the change in oxidation numbers:

O is going from $0 \rightarrow -1$, which means that O is gaining electrons and is being reduced.

Cr is going from $+3 \rightarrow +6$, which means that Cr is losing electrons and is being oxidized.

Now we separate the full reaction into half-reactions to deal with the oxidation and reduction parts separately:

$$\text{Red: } O_2 \rightarrow H_2O_2$$

Step 1: Balance all atoms except O and H:

$$O_2 \rightarrow H_2O_2$$

Step 2: Reaction occurs in basic solution, so use H_2O to balance O and then H^+ to balance H. Then, add enough OH^- to react away the H^+ and turn it into the water, then add the equal number of moles of OH^- to the opposite side:

$$2H^+ + O_2 \rightarrow H_2O_2$$

$$2OH^- + 2H^+ + O_2 \rightarrow H_2O_2 + 2OH^-$$

$$2H_2O + O_2 \rightarrow H_2O_2 + 2OH^-$$

Step 3: Balance charge by adding electrons:

$$2e^- + 2H_2O + O_2 \rightarrow H_2O_2 + 2OH^-$$

$$\text{Ox: } Cr^{3+} \rightarrow Cr_2O_7^{2-}$$

Step 1: Balance all atoms except O and H:

$$2Cr^{3+} \rightarrow Cr_2O_7^{2-}$$

Step 2: Reaction occurs in basic solution, so use H_2O to balance O and then H^+ to balance H. Then, add enough OH^- to react away the H^+ and turn it into the water, then add an equal number of moles of OH^- to the opposite side. Then, subtract excess water molecules since they show up on both sides of the equation:

$$7H_2O + 2Cr^{3+} \rightarrow Cr_2O_7^{2-} + 14H^+$$

$$14OH^- + 7H_2O + 2Cr^{3+} \rightarrow Cr_2O_7^{2-} + 14H^+ + 14OH^-$$

$$14OH^- + 7H_2O + 2Cr^{3+} \rightarrow Cr_2O_7^{2-} + 14H_2O$$

$$14OH^- + 2Cr^{3+} \rightarrow Cr_2O_7^{2-} + 7H_2O$$

Step 3: Balance charge by adding electrons:

$$14OH^- + 2Cr^{3+} \rightarrow Cr_2O_7^{2-} + 7H_2O + 6e^-$$

Step 4: Need equal number of electrons for both half-reactions. In this case, we need to multiply the reduction half-reaction by three to equal the six electrons in the oxidation reaction:

$$6e^- + 6H_2O + 3O_2 \rightarrow 3H_2O_2 + 6OH^-$$

Now we need to add the two half-reactions together and make sure that the atoms and charge balance each other on the reactants and products side:

$$6e^- + 6H_2O + 3O_2 \rightarrow 3H_2O_2 + 6OH^-$$

$$+ \ 14OH^- + 2Cr^{3+} \rightarrow Cr_2O_7^{2-} + 7H_2O + 6e^-$$

$$\overline{3O_2 + 8OH^- + 2Cr^{3+} \rightarrow 3H_2O_2 + Cr_2O_7^{2-} + H_2O}$$

All of the atoms and charges are balanced, so we have correctly balanced this redox equation.

*QUESTION 11

Balance the following redox reaction that takes place in basic solution, using the information in Table 14.3:

$$Pb^{2+} + IO_3^- \rightarrow PbO_2 + I_2$$

QUESTION 12

Write a balanced chemical equation for the reduction of NO_2^- to NH_3 by H_2.

Q12 ANSWER First, balance the number of electrons transferred from oxidant to reductant. Since N changes from +3 to –3, six electrons are transferred. H changes from 0 to 1, so six H atoms, or three H_2 molecules, are required to lose the electrons:

$$NO_2^- + 3\,H_2 \rightarrow NH_3$$

Because the reaction is in water, it is permissible to add H_2O or H^+ or OH^- to either side of the reaction, as needed. Seeing that nitrite had two O atoms, we balance these by adding two water molecules to the right-hand side:

$$NO_2^- + 3\,H_2 \rightarrow NH_3 + 2\,H_2O$$

The total H count on the right-hand side is now seven, which we balance by adding one H^+ to the left-hand side. This also balances the charge:

$$NO_2^- + 3\,H_2 + H^+ \rightarrow NH_3 + 2\,H_2O$$

Now we add OH^- to react away the H^+ to form water and add an equal number of OH^- atoms to the products:

$$OH^- + NO_2^- + 3\,H_2 + H^+ \rightarrow NH_3 + 2\,H_2O + OH^-$$

$$NO_2^- + 3\,H_2 + H_2O \rightarrow NH_3 + 2\,H_2O + OH^-$$

Finally, water molecules are on both sides of the reaction, so simplify the equation to get a balanced equation:

$$NO_2^- + 3\,H_2 \rightarrow NH_3 + H_2O + OH^-$$

14.4 Half-reactions, Reduction Potentials, and Redox Free Energy

Redox reactions can be treated as a combination of a reduction and an oxidation half-reaction. Frequently, these half-reactions can be carried out separately in the electrode compartments of an electrochemical cell, whose cell potential is a direct measure of the free energy change associated with the redox reaction. The cell potential is the difference between the two half-reaction potentials. Refer to Chapter 3 for more details on calculations.

14.5 Natural Sequence of Biological Reductions

When water is depleted of oxygen, organisms that depend upon aerobic respiration cannot survive and anaerobic bacteria take over. These bacteria utilize oxidants other than O_2. These alternative oxidants are less powerful than O_2 and cannot produce as much energy. Nevertheless, bacteria are quite capable of surviving on lower energy processes; in doing so, they can fill ecological niches that are not available to aerobic organisms. The oxidizing power of anaerobic environments in the biosphere is mainly controlled by 5 molecules. In decreasing order of energy produced, they are nitrate (NO_3^-), manganese dioxide (MnO_2), ferric hydroxide [$Fe(OH)_3$], sulfate (SO_4^{2-}), and under extreme conditions, carbohydrate (CH_2O) itself.

The oxidizing power of a molecule depends on the specific reaction being carried out and is generally measured as the *reduction potential* associated with the reduction of the oxidant (see Table 3.1 in Chapter 3). The environmental oxidants we are considering are listed in Table 14.4. These potentials, E_h, are different from $E°$ values discussed in Chapter 3 because they apply when the pH is 7.0, which is more realistic for environmental samples than the defined standard condition of 1 M [H^+]. These half-reactions do not give reversible potentials at electrodes, but the metabolic activity of the vast array of microbes in soils and water ensures that electron transfer does occur on a timescale of hours or days. Consequently, all redox-active materials respond to the reduction potential established by the microbial activity.

TABLE 14.4 Thermodynamic Sequence for Reduction of Important Environmental Oxidants at pH 7.0 and 25°C

Reaction	$E_h(V)^\star$
Disappearance of O_2 $O_2 + 4H^+ + 4e^- \leftrightarrow 2H_2O$	0.812
Reduction of NO_3^- to N_2 $NO_3^- + 6H^+ + 5e^- \leftrightarrow \frac{1}{2}N_2 + 3H_2O$	0.747
Reduction of MnO_2 to Mn^{2+} $MnO_2 + 4H^+ + 2e^- \leftrightarrow Mn^{2+} + 2H_2O$	0.526
Reduction of Fe^{3+} to Fe^{2+} $Fe(OH)_3 + 3H^+ + e^- \leftrightarrow Fe^{2+} + 3H_2O$	−0.047
Formation of H_2S $SO_4^{2-} + 10H^+ + 8e^- \leftrightarrow H_2S + 4H_2O$	−0.221
Formation of CH_4 $CO_2 + 8H^+ + 8e^- \leftrightarrow CH_4 + 2H_2O$	−0.244

$^\star E_h$(V) is the $E°$ value recalculated for pH 7

Sources: Adapted from Schlesinger, *Biochemistry: An Analysis of Global Change,* 2nd Ed., 1997, Academic Press, and Spiro et al., *Chemistry of the Environment, 3rd Ed.,* University Science Books, © 2012, all rights reserved.

QUESTION 13

Using Table 14.4, write a balanced equation for the oxidation of methane by O_2 and calculate the free energy released at pH 7.

Q13 ANSWER To write a balanced reaction, we equalize the electrons transferred by multiplying the O_2 reduction half-reaction by two and subtract the CO_2 to methane half-reaction:

$$2(O_2 + 4H^+ + 4e^- \rightarrow 2H_2O) \qquad E_h = 0.812\ V$$

$$^-(CO_2 + 8H^+ + 8e^- \rightarrow CH_4 + 2H_2O) \qquad E_h = -(-0.244\ V)$$

$$2O_2 + 8H^+ + 8e^- \rightarrow 4H_2O \qquad E_h = 0.812\ V$$

$$+\ CH_4 + 2H_2O \rightarrow CO_2 + 8H^+ + 8e^- \qquad E_h = 0.244\ V$$

$$2O_2 + CH_4 \rightarrow CO_2 + 2H_2O \qquad \Delta E_h = 1.056\ V$$

From Chapter 3:

$$\Delta G_h = -nF\Delta E_h$$

$$n = 8\ mol\ e^-$$

$$F = 96{,}500\ coulombs\ (C)\ (mol\ e^-)^{-1}$$

$$1\ V = 1\ J/C$$

$$\Delta G_h = -(8\ mol\ e^-) \times \frac{96{,}500\ C}{mol\ e^-} \times \frac{J}{V\,C} \times 1.056\ V = -815{,}000\ J = -815\ kJ$$

This is the energy released per mole of CH_4 consumed, or per mole of CO_2 produced. The energy released per mole of O_2 consumed is half this value.

*QUESTION 14

Using Table 14.4, write a balanced equation for the oxidation of Fe^{2+} by NO_3^- and calculate the free energy released, at pH 7.

*QUESTION 15

In anaerobic wetlands, high concentrations of organic carbon may serve as natural buffers against nitrate and sulfate, which might otherwise pollute the water. Use reactions from Table 14.4 to explain why.

*QUESTION 16

According to Table 14.4, what reaction would take place if $Fe(OH)_3$ were in contact with natural water that contains H_2S? How much energy would be released?

14.6 Concentration Dependence of the Potential: pH and E_h

What happens to the reduction potential when conditions are not standard? As in all chemical reactions, the driving force for electrochemical processes depends on the concentrations of reactants and products. When concentrations of reactants or products deviate from the standard condition of 1 mol, or 1 atm in the case of gas (note: the concentrations of solids or liquids and of the water solvent are invariant, and are defined as unity), the reaction potential can be calculated with the Nernst equation:

$$E = E° - \frac{RT}{nF}\ln Q \qquad (14.2)$$

where $E°$ is the standard potential, R is the ideal gas constant (8.314 J mol^{-1} K^{-1}), T is the temperature in Kelvin, F is the Faraday constant (96,500 C (mol e$^-$)$^{-1}$), n is the number of moles of electrons being transferred in the reaction, and Q is the reaction quotient—the concentration expression in the form of equilibrium constant. It is convenient to convert ln to log by multiplying by 2.303, and when T is the standard temperature, 25°C (298K), the value of 2.303 RT F^{-1} is 0.0591. The Nernst equation becomes:

$$E = E° - \frac{0.0591 \text{ V}}{n}\log Q \qquad (14.3)$$

QUESTION 17

The standard potential for the reduction of hydrogen ion is zero, by definition (see Table 3.1 in Chapter 3):

$$2\,H^+\,(aq) + 2\,e^- \rightarrow H_2\,(g) \qquad E = 0.000 \text{ V}$$

What is the E_h value (pH 7, but otherwise standard conditions)?

Q17 ANSWER For the half-reaction:

$$Q = \frac{P_{H_2}}{[H^+]^2}$$

$$E = E° - \frac{0.0591 \text{ V}}{2}\log\frac{P_{H_2}}{[H^+]^2} = E° - \frac{0.0591 \text{ V}}{2}[\log P_{H_2} - 2\log[H^+]]$$

Because the pressure of H_2 is under standard conditions, 1 atm, the log $P(H_2)$ = log (1 atm) = 0. Also, the pH = $-$log [H$^+$]. Therefore:

$$E = E° - \frac{0.0591 \text{ V}}{2}[\log (1 \text{ atm}) - 2\log[H^+]] = E° - \left(\frac{0.0591 \text{ V}}{2}[0 - 2\log[H^+]]\right) = E° - 0.0591 \text{ V} \times \text{pH}$$

It can be seen that with every unit pH increase (decreasing acidity), the hydrogen electrode potential decreases by 0.0591 V. At pH 7:

$$E_h = 0.000 \text{ V} - 0.0591 \text{ V} \times 7 = -0.414 \text{ V}$$

QUESTION 18

Calculate E_h when the temperature is lowered to 0°C.

Q18 ANSWER   At 0°C, T = 273 K and RT F^{-1} becomes:

$$\frac{2.303 \, RT}{F} \log Q = \frac{2.303 \times 273 \text{ K} \times 8.314 \dfrac{J}{\text{mol K}}}{96{,}500 \dfrac{C}{\text{mol}} \times 1 \dfrac{J}{V \, C}} \times \log Q = 0.0542 \log Q$$

Using the relationship derived in Q17:

$$E_h = 0.000 \text{ V} - [7 \times (0.0542 \text{ V})] = -0.379 \text{ V}$$

E_h becomes less negative (less reducing) as the temperature is lowered.

*QUESTION 19

The standard potential for the oxygen reduction electrode is 1.23 V (see Table 3.1 in Chapter 3). Confirm the E_h value for this reaction in Table 14.4:

$$O_2 (g) + 4 H^+ (aq) + 4 e^- \rightarrow 2 H_2O (l) \qquad E° = 1.23 \text{ V}$$

Also, calculate the reduction potential in strong base, with [OH$^-$] = 1 M. Is O_2 a weaker oxidant under basic or neutral conditions?

*QUESTION 20

Recalculate the potential in Q19 if the pH is 7, but the oxygen is supplied by the air (which is 21% O_2).

*QUESTION 21

Even though the reduction potentials depend on pH for both hydrogen and oxygen, the standard cell potential for a H_2/O_2 fuel cell is independent of pH. Explain.

QUESTION 22

Bacteria utilizing the oxidants listed in Table 14.4 generally use organic molecules as their fuel and oxidize them to CO_2. The composition of the fuel can be approximated with the carbohydrate formula, CH_2O. Their oxidation can be represented in reverse as:

$$CO_2 + 4 H^+ + 4 e^- \rightarrow CH_2O + H_2O$$

The potential is essentially the same as that for H_2 reduction: $E_h = -0.414$V (see Q17).

This potential is low enough that any of Table 14.4 oxidants can be utilized, but the energy produced diminishes as their reduction potentials decrease. Even CO_2 itself can be an oxidant, if methane is produced, via methanogenic bacteria. How much energy is then available when considering the reaction of carbohydrate to methane and CO_2?

Q22 ANSWER The CO_2/CH_4 half-reaction (Table 14.4) requires eight electrons per CO_2, so we multiply the CO_2/CH_2O half-reaction by 2, and then subtract it from the CO_2/CH_4 half-reaction to obtain the cell reaction:

$$CO_2 + 8H^+ + 8e^- \rightarrow CH_4 + 2H_2O \qquad E_h = -0.244 \text{ V}$$

$$-2(CO_2 + 4H^+ + 4e^- \rightarrow CH_2O + H_2O) \qquad -(E_h = -0.414 \text{ V})$$

$$CO_2 + 8H^+ + 8e^- \rightarrow CH_4 + 2H_2O \qquad E_h = -0.244 \text{ V}$$

$$+ 2CH_2O + 2H_2O \rightarrow 2CO_2 + 8H^+ + 8e^- \qquad E_h = 0.414 \text{ V}$$

$$2CH_2O \rightarrow CH_4 + CO_2 \qquad E_h = 0.170 \text{ V}$$

$$\Delta G_h = -(8 \text{ mol } e^-) \times \frac{96{,}500 \text{ C}}{\text{mol } e^-} \times \frac{J}{VC} \times 0.170 \text{ V} = -131{,}000 \text{ J} = -131 \text{ kJ}$$

The energy available from methane production is only 16% of that available from aerobic oxidation (-815 kJ, see Q13). Nevertheless, methanogenic bacteria flourish when no better oxidant is available than CO_2.

However, the energy calculation should take into account that the concentrations of reactants and products are not the standard concentrations. The Nernst equation provides corrections for non-standard concentrations:

$$\Delta E = \Delta E_h - \frac{0.0591 \text{ V}}{n} \log Q = 0.170 \text{ V} - \frac{0.0591 \text{ V}}{8} (\log[CH_4] + \log[CO_2] - 2\log[CH_2O])$$

The organic matter concentration might be as high as 1 M in lake and ocean sediments, where methanogenic bacteria are generally found, but the concentrations of the products, CH_4 and CO_2, are much less than the standard concentrations (1 atm). Typical values might be 5 mM for CH_4 and 50 mM for CO_2. Using these values:

$$\Delta E = 0.170 \text{ V} - \frac{0.0591 \text{ V}}{8} (\log[5 \times 10^{-3} \text{ M}] + \log[50 \times 10^{-3} \text{ M}] - 2\log[1.0])$$

$$= 0.170 \text{ V} - \frac{0.0591 \text{ V}}{8} (-2.3 - 1.30 - 0) = 0.197 \text{ V}$$

$$\Delta G_h = -(8 \text{ mol } e^-) \times \frac{96{,}500 \text{ C}}{\text{mol } e^-} \times \frac{J}{VC} \times 0.197 \text{ V} = -152{,}000 \text{ J} = -152 \text{ kJ}$$

A slight improvement over the energy for standard conditions.

QUESTION 23

The standard reduction potential is 0.77V for:

$$Fe^{3+} + e^- \rightarrow Fe^{2+}$$

Water seeping from abandoned mines is strongly acidic and frequently contains iron. If a sample of this water contains both Fe^{3+} and Fe^{2+} at concentrations of 8.0 and 0.4 mM, respectively, what is the reduction potential of the solution.

Q23 ANSWER

$$E = E° - \frac{0.0591\text{ V}}{n} \log\frac{[\text{Fe}^{2+}]}{[\text{Fe}^{3+}]} = 0.77\text{ V} - \frac{0.0591\text{ V}}{1} \log\frac{[0.4\text{ mM}]}{[8.0\text{ mM}]} = 0.85\text{ V}$$

QUESTION 24

Since there are no protons in the $\text{Fe}^{3+}/\text{Fe}^{2+}$ half-reaction, it would appear to have the same potential at pH 7 and 0. However, Fe^{3+} is not stable at pH 7, because it precipitates as Fe(OH)_3:

$$\text{Fe}^{3+} + 3\,\text{H}_2\text{O} \rightarrow \text{Fe(OH)}_3 + 3\,\text{H}^+$$

The reduction reaction becomes:

$$\text{Fe(OH)}_3 + 3\,\text{H}^+ + \text{e}^- \rightarrow \text{Fe}^{2+} + 3\,\text{H}_2\text{O}$$

E_h for this reaction is given as -0.047 V in Table 14.4. From this value, use the Nernst equation to calculate the solubility product for Fe(OH)_3. $K_{sp} = [\text{Fe}^{3+}][\text{OH}]^3$

Q24 ANSWER In the presence of Fe(OH)_3:

$$[\text{Fe}^{3+}] = \frac{K_{sp}}{[\text{OH}^-]^3}$$

For the $\text{Fe}^{3+}/\text{Fe}^{2+}$ half-reaction:

$$E = E° - \frac{0.0591\text{ V}}{n} \log\frac{[\text{Fe}^{2+}]}{[\text{Fe}^{3+}]} = E° - \frac{0.0591\text{ V}}{n} \log\frac{[\text{Fe}^{2+}][\text{OH}^-]^3}{[K_{sp}]}$$

$$= 0.77\text{ V} - \frac{0.0591\text{ V}}{1}(\log [\text{Fe}^{2+}] + 3\log [\text{OH}^-] - \log K_{sp})$$

At pH 7, under otherwise standard conditions ($[\text{Fe}^{2+}] = 1$ M), $[\text{OH}^-] = 10^{-7}$ M:

$$E_h = 0.77\text{ V} - \frac{0.0591\text{ V}}{1}(\log [1\text{ M}] + 3\log[10^{-7}] - \log K_{sp}) = -0.047\text{ V}$$

$$-0.047\text{ V} = 0.77\text{ V} - (0.0591\text{ V})(-21 - \log K_{sp})$$

$$-0.82\text{ V} = -(0.0591\text{ V})(-21 - \log K_{sp})$$

$$13.8 = -21 - \log K_{sp}$$

$$-\log K_{sp} = 34.8$$

$$K_{sp} = 1.5 \times 10^{-35}$$

QUESTION 25

If well water contains Fe^{2+}, it oxidizes on exposure to air and precipitates as Fe(OH)_3. How much Fe^{2+} will remain if the sample is left standing in air, the sample comes to equilibrium, and the pH stays at 7?

Q25 ANSWER The cell potential for the reaction:

$$O_2 + 4\,Fe^{2+} + 10\,H_2O \rightarrow 4\,Fe(OH)_3 + 8\,H^+$$

is obtained by multiplying:

$$Fe(OH)_3 + 3\,H^+ + e^- \rightarrow Fe^{2+} + 3\,H_2O \qquad\qquad E_h = -0.047 \text{ V}$$

by four and subtracting it from the oxygen reduction half-reaction to obtain the cell reaction:

$$4\,Fe^{2+} + 12\,H_2O \rightarrow 4\,Fe(OH)_3 + 12\,H^+ + 4\,e^- \qquad E_h = 0.047 \text{ V}$$

$$+ \quad O_2 + 4\,H^+ + 4\,e^- \rightarrow 2\,H_2O \qquad\qquad E_h = 0.812 \text{ V}$$

$$\overline{O_2 + 4\,Fe^{2+} + 10\,H_2O \rightarrow 4\,Fe(OH)_3 + 8\,H^+ \qquad E_h = 0.859 \text{ V}}$$

At equilibrium $Q = K$ (the equilibrium constant) and

$$E_{cell} = \left(\frac{0.0591 \text{ V}}{n}\right) \log K, \, n = 4$$

And because E_h is referenced to pH 7, we replace $[H^+]$ with $[H^+]/[10^{-7}]$ in the expression for K. Since $pH = -\log[H^+]$, this substitution expresses the deviation from pH 7:

$$\log K = \log\left(\frac{\left(\frac{[H^+]}{[10^{-7}]}\right)^8}{[Fe^{2+}]^4 \, pO_2}\right) = (-8 \times pH) - (-7 \times 8) - 4\log[Fe^{2+}] - \log pO_2$$

Using the Nernst equation relationship:

$$\log K = 0.859 \text{ V} \left(\frac{4}{0.0591 \text{ V}}\right) = 58.1$$

In air, $\log pO_2 = \log 0.21 = -0.678$
To obtain $[Fe^{2+}]$, we rearrange to:

$$4\log[Fe^{2+}] = -58.1 + (-8 \times pH) + 56 + 0.678$$

If the pH remains at 7, then:

$$\log[Fe^{2+}] = \frac{-57.4}{4} = -14.4$$

and $[Fe^{2+}] = 4.4 \times 10^{-15}$ M, which is very low indeed. Essentially, no Fe^{2+} is left in solution.

But if the water becomes acidic, as a result of $Fe(OH)_3$ formation, the equilibrium Fe^{2+} concentration increases. If the pH is lowered to 2, then:

$$4\log[Fe^{2+}] = -58.1 - 16 + 56 + 0.678 = -17.4$$

and $[Fe^{2+}] = 4.4 \times 10^{-5}$ M, a concentration that is commonly found in environmental samples.

---

*QUESTION 26

Runoff from an agricultural field contains NO_3^- from excess fertilizer. It enters a pond that has sufficient BOD to use up the dissolved O_2, but contains 10^{-5} M Mn^{2+}. What reaction is expected, based on Table 14.4? If the pond is big enough that $[Mn^{2+}]$ is unaffected by the reaction, and the pH is 7, what is the equilibrium NO_3^- concentration at the runoff entry point? What happens if the pond becomes acidic and the pH decreases to 5.5?

14.7 Water as an Ecological Medium: Limiting Nutrient

Although organic matter contains mostly C, H, and O, other elements are also essential for biological growth. These include N, P, and S; several other minor elements (Si, Cl, and I); and several metallic elements (such as Fe). Growth is limited by the element in the shortest supply, relative to its abundance in biological tissue. In natural bodies of water, the surrounding terrain usually produces an adequate supply of the minor elements, while C, H, and O are available from the water itself, and from the atmosphere. However, N and/or P can be in short supply. Although N_2 makes up 80% of the atmosphere, it is unavailable except when captured by N_2-fixing bacteria, living in symbiotic association with certain plants. P has no atmospheric source; it occurs as phosphate ion, PO_4^{3-}, which, being multiple-charged, is strongly absorbed by particulate matter and sediments.

The quality of water (clarity, low BOD) is compromised by excessive biological activity (*eutrophication*), fed by excessive flows of nutrients. Thus, restricting the limiting nutrient, usually N or P, is important for maintaining water quality. Freshwater bodies are generally adequately supplied with N, through runoff of NO_3^- from the soil. P is then the limiting nutrient, and its control (e.g., in sewage treatment and elimination from laundry detergent) has been shown to improve water quality. In the oceans, however, the nitrate concentration is low, and "dead zones" at river outflows, coastal regions with overabundant organic matter, have been associated with excessive nitrate inputs from agricultural fertilization.

QUESTION 27

In a pond, the algal composition follows the atomic ratio of 106:16:1 for C:N:P. What is the limiting nutrient if the pond contains the following amounts of nutrients: total C = 30 mg L^{-1}, total N = 1.20 mg L^{-1}, and total P = 0.24 mg L^{-1}? If it is known that two-thirds of the phosphorus in the pond originates from the use of a phosphate fertilizer, will banning this fertilizer deter the eutrophication process?

Q27 ANSWER First, we calculate the moles of each nutrient in the pond:

$$C = \frac{30 \text{ mg}}{L} \times \frac{mmol}{12 \text{ mg}} = 2.5 \frac{mmol}{L}$$

$$N = \frac{1.20 \text{ mg}}{L} \times \frac{mmol}{14 \text{ mg}} = 0.086 \frac{mmol}{L}$$

$$P = \frac{0.24 \text{ mg}}{L} \times \frac{\text{mmol}}{31 \text{ mg}} = 0.0077 \frac{\text{mmol}}{L}$$

The three nutrients' atomic ratios can be calculated by dividing each concentration by 0.0077 mmol L^{-1}:

$$C = \frac{2.5 \text{ mmol}}{L} \times \frac{L}{0.0077 \text{ mmol}} = 325$$

$$N = \frac{0.086 \text{ mmol}}{L} \times \frac{L}{0.0077 \text{ mmol}} = 11$$

$$P = \frac{0.0077 \text{ mmol}}{L} \times \frac{L}{0.0077 \text{ mmol}} = 1$$

So, the ratio is: $C:N:P = 325:11:1 = 106:3.59:0.33$

Since algal growth requires the ratio $C:N:P = 106:16:1$, N is the limiting nutrient here by a small amount.

If it is known that two-thirds of the P comes from human source (fertilizing), and this amount of P is banned, then in the pond, the ratio becomes:

$$C:N:P = 106:3.59:0.11$$

At this point, comparing to the desired atomic ratio for algal growth $106:16:1$, we see that P is now the limiting nutrient. Therefore, limiting the source of P (fertilizer use) will deter eutrophication.

*QUESTION 28

In the Chesapeake Bay, excess nitrogen occurs in the spring. The phytoplankton composition follows the atomic ratio of $106:16:1$ for $C:N:P$. A water sample is taken that contains the following amounts of nutrients: total $C = 45$ mg L^{-1}, total $N = 2.50$ mg L^{-1}, and total $P = 0.40$ mg L^{-1}? Do you think this water sample was taken during the spring season?

QUESTION 29

The atmosphere and hydrosphere are interfaced. What takes place in one sphere may affect the composition of another, having ecological and public health consequences. In anaerobic aquatic environments, what gases can be generated at a pH of 7.0? Which of these is abundant in the atmosphere or a greenhouse gas? And is there the possibility to produce any toxic gases?

Q29 ANSWER When oxygen is absent, several oxidants may act to oxidize organic carbon in aquatic systems. From Table 14.4, NO_3^- and SO_4^{2-} may lead to the production of nitrogen (N_2) and hydrogen sulfide (H_2S) gases, respectively. In addition, in the absence of oxidants, partially reduced carbon can be disproportionate to produce methane (CH_4) and carbon dioxide (CO_2).

Of these gases, N_2 is abundant in the atmosphere; CH_4 and CO_2 are greenhouse gases; and H_2S is a toxic gas.

14.8 Conclusions

Aquatic chemistry is dependent on microorganisms and redox reactions, especially those of dissolved O_2. O_2 is the strongest natural oxidant and is required by all higher life forms and by aerobic microorganisms. But the supply of O_2 is limited by its low solubility in water. When the supply runs out, these organisms die and anaerobic bacteria take over. Biological oxygen demand is a measure of the health of a water system, expressing its potential for using up the O_2 supply. Anaerobic bacteria can utilize less powerful oxidants for their metabolism: nitrate, MnO_2, Fe_2O_3, and sulfate, in order of oxidizing power. When none of these are available, CO_2 can be used as an oxidant, producing methane. Reduction potentials can be used to determine which reactions can occur, to calculate concentrations of oxidized and reduced species in the water, using the Nernst equation. The pH of water in the environment is around 7, so it is important to be able to calculate and use reduction potentials under realistic conditions, not under standard conditions, with the $[H^+]$ or $[OH^-]$ at 1 M. Excess nutrients, principally nitrate and phosphate from sewage treatment and agricultural runoff, can feed algal blooms, which then use up the O_2 when they die. Keeping the water healthy requires controlling the limiting nutrient, which can be determined from the water composition.

WATER POLLUTION AND WATER TREATMENT

15.1 Water Use and Water Quality: Point and Nonpoint Sources

Pictures of polluted oceans, rivers, lakes, or streams are prevalent on the Internet and in the news. Polluted water is an ever-growing problem in today's world as populations grow and more waste is created. Water pollution issues are different around the world, depending on the materials emitted by various sources into water bodies. However, there are ways to clean the water, whether it derives from your home, dormitory, apartment, or work.

The quality of surface and groundwater is of concern from two overlapping, but distinct, points of view: (1) human health and welfare and (2) the health of aquatic ecosystems. Both aspects of water quality are enhanced by minimizing the effects of human activities, but the specific issues and control measures depend on factors such as where the waste is generated and how it is distributed.

Water quality is as important an issue as water quantity. Although much of the water supply is returned to water bodies such as rivers and lakes after use, its quality becomes increasingly degraded. These effects are summarized in Table 15.1.

TABLE 15.1 Effects on Water Quality from Water Use

Water Use	Effects on Water Quality
Domestic/commercial	Decreases dissolved oxygen; pollutes water with trace pollutants, organics, heavy metals, and inorganics
Industrial/mining	Decreases dissolved oxygen; pollutes water with toxic heavy metals and organics; causes acid mine drainage
Thermoelectric	Increases water temperature (thermal pollution)
Irrigation/livestock	Causes salinization of surface and groundwaters; decreases dissolved oxygen (near feedlots)

Source: Adapted from Spiro et al., *Chemistry of the Environment, 3rd Ed.*, University Science Books, © 2012, all rights reserved.

QUESTION 1

Look at Table 15.1 and think about the community you live in. What water uses can have effects on the water quality? What are those effects?

Q1 ANSWER From a residential area, such uses could be watering lawns, bathing, washing cars, and so on. These will increase the concentrations of trace pollutants in the waters as well as increase the number of organics, heavy metals, and inorganics. If the city has a lot of industry and mining, then these processes use water and will pollute the water with toxic heavy metals and organics. Acid drainage can occur from mines as well. Thermal pollution results from energy production, as the turbines are cooled with water. Finally, in agricultural areas, there can be pollution from runoff from pesticides that can pollute the water supply and there can be salinization of the surface and groundwater systems. Many of these uses of water also cause an issue with the dissolved oxygen in the water, thus adversely affecting the wildlife.

Thermal pollution from a factory can also affect water quality. This happens when water is released from a factory that is hotter, or colder, than the body of water it is entering. The wide changes in temperature can disrupt delicate ecosystems that are highly temperature-dependent. Even small temperature fluctuations can have drastic changes. Energy changes can be calculated using $q = mc\Delta T$, where q is the energy change, m is the mass, c is the specific heat, and ΔT is the change in temperature.

QUESTION 2

A factory is built next to a lake with a volume of 50. km^3. The factory releases 80. °C water at a rate of 2.0×10^3 L hr^{-1}. Assuming no heat is lost to the surroundings and is only transferred between factory and lake water, what would the effect be on the lake after 1 hr? Assume the initial temperature of the water is 20. °C.

Q2 ANSWER First, convert the size of the lake from km^3 to L.

$$50.\ km^3 \times \frac{(1{,}000\ m)^3}{(1\ km)^3} \times \frac{(100\ cm)^3}{(1\ m)^3} \times \frac{(1\ mL)}{(1\ cm)^3} \times \frac{(1\ L)}{1{,}000\ mL} = 5.0 \times 10^{13}\ L$$

Then, calculate the temperature change using the fact that heat lost is equal to the heat gained.

$$-q_{lost} = q_{gained}$$

Using the relationship $q = mc\Delta T$ (where $\Delta T = T_{final} - T_{initial}$), the following relationship can be developed. The factory water side of the equation on the left is negative, because the heat is given off from the hot water and transferred to the lake water. The final temperature is the same for both the factory and the lake water.

$$-(m_{factory\ water})(c)(T_f - 80.°C) = (m_{lake\ water})(c)(T_f - 20.°C)$$

In this equation, m is the mass of the substance and c is the specific heat capacity. Because both substances are water, the specific heat capacity term cancels out. If we assume a density of 1 g mL^{-1} or 1 kg L^{-1}, we can say the mass is the same as the volume.

$$-(2.0 \times 10^3 \text{ kg}) (T_f - 80. \text{ °C}) = (5.0 \times 10^{13} \text{ kg}) (T_f - 20. \text{ °C})$$

$$-(2.0 \times 10^3 \text{ kg } T_f - 1.6 \times 10^5 \text{ kg °C}) = (5.0 \times 10^{13} \text{ kg } T_f - 1.0 \times 10^{15} \text{ kg °C})$$

Combining like terms and dividing to isolate T_f gives:

$$\frac{1{,}000{,}000{,}000{,}160{,}000 \text{ kg °C}}{50{,}000{,}000{,}002{,}000 \text{ kg}} = \frac{50{,}000{,}000{,}002{,}000 \text{ kg °C } T_f}{50{,}000{,}000{,}002{,}000 \text{ kg}}$$

$$T_f = 20.0000000024\text{°C}$$

This value is not much different from 20°C; however, over time the temperature will continue to increase and thus affect the water quality.

*QUESTION 3

A factory is built next to a lake with a volume of 50. km^3. The factory releases 80. °C water at a rate of 2.0×10^3 L hr^{-1}. Assuming no heat is lost to the surroundings and is only transferred between factory and lake water, what would the effect be on the lake after 10 hr? (Assume no water is drained during this time and the initial temperature of the water is 20. °C.)

*QUESTION 4

A factory is built next to a lake with a volume of 50. km^3. The factory releases 90. °C water at a rate of 5.0×10^3 L hr^{-1}. Assuming all the heat from the factory water is transferred to the lake water, what would the effect be on the lake after 1 hr? Assume the temperature of the lake is 20. °C.

15.2 Regulation of Water Quality

Governments around the world try to regulate water quality in the interests of public health and environmental protection. In the United States, the legal instruments for regulation are two basic laws, the Clean Water Act (CWA) of 1972 and the Safe Drinking Water Act (SDWA) of 1974, both of which have been amended several times. Under these acts, the U.S. Environmental Protection Agency (EPA) is required to set standards that protect human health from the harmful effects of contaminants.

Figure 15.1 shows examples of point and nonpoint sources of water pollution. Point sources are those that have a single source that directly pollutes the water source, whereas nonpoint source solutions are those that do not have a single source and their origins cannot be traced back to one origin.

---

* Answers to starred questions can be found at the end of the book.

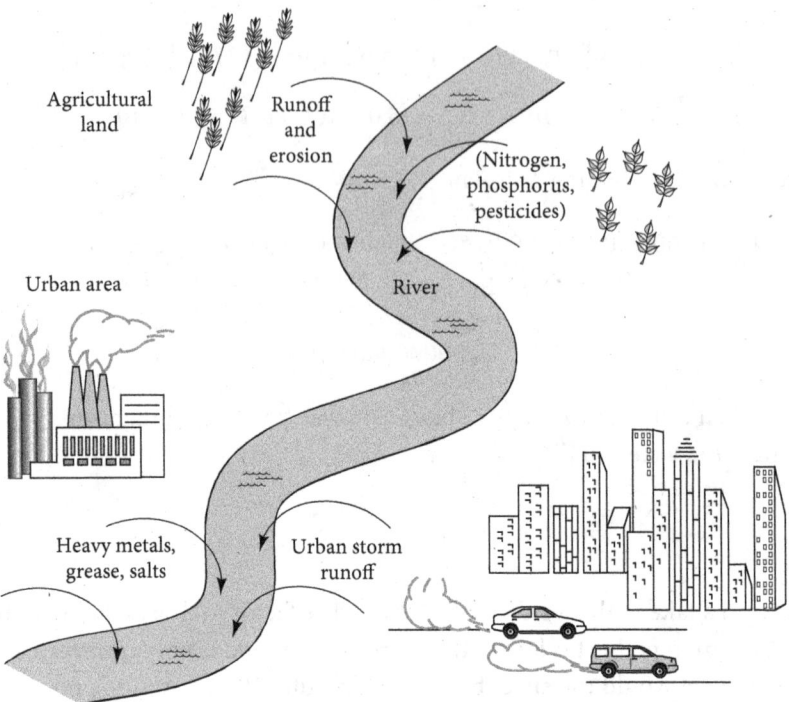

Figure 15.1 Examples of point and nonpoint sources of water pollution. The point source in the image is the factory; all other sources of water pollution are non-point.
Source: Adapted from Spiro et al., *Chemistry of the Environment, 3rd Ed.,* University Science Books, © 2012, all rights reserved.

QUESTION 5

Explain how the regulation of water quality can cause disputes between states and other countries.

Q5 ANSWER Generally, most waterways (rivers, lakes, streams, and so on) do not lie exclusively in one state or country. Even if a lake is located within the boundaries of a particular municipality, the waters that feed it do not necessarily come from within that area. Regulation by one state might limit the number of pollutants, but the other state could be the one contaminating the area from the pollution that moves downstream. It is important that all considerations are made when looking at these regulations and determining all parties who contribute water pollutants.

QUESTION 6

Why are point-source and nonpoint-source pollution important considerations for water quality?

Q6 ANSWER Point-source pollutants have a more exact area of entry than nonpoint-source pollutants. Nonpoint-source pollutants are more difficult to control as they derive from more diffuse locations. Nonpoint sources require more collaboration between entities to control and could thus cause more disputes.

*QUESTION 7

Describe different causes of nonpoint pollution.

15.3 Water and Sewage Treatment

Cities and towns must treat their water supplies for domestic and commercial uses to ensure freedom from disease and to eliminate odors, turbidity (cloudiness of a liquid), and other contaminants. This includes treating both wastewater and sewage.

Figure 15.2 shows a process for the treatment of municipal wastewater. The raw wastewater first goes through primary treatment, which includes screening to remove large solid waste and sedimentation, allowing suspended particles to separate. The solid waste (primary sludge) is treated by chemicals or is dried and then incinerated or disposed of. The liquid portion of the waste is then moved to secondary treatment, where it is treated to remove organics that are dissolved and aerobic digestion is performed. The solid is treated as before and the liquid is then moved to receiving waters.

QUESTION 8

Calculate the percent of biological oxygen demand (BOD) removed through secondary wastewater treatment.

Q8 ANSWER According to the figure, 200 mg L^{-1} BOD starts the process. At the end of the process, 25 mg L^{-1} BOD remains. To calculate the percent removed, use the following relationship:

$$\frac{\text{starting amount} - \text{final amount}}{\text{starting amount}} \times 100\%$$

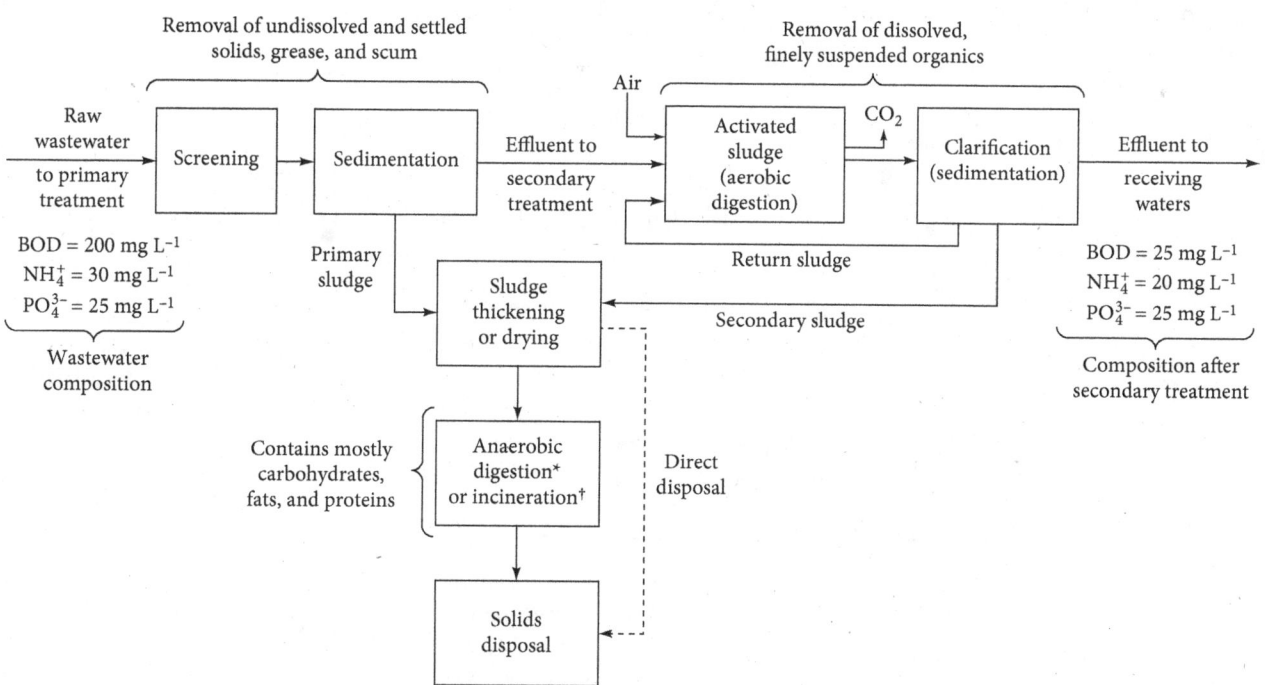

* Typically 50% of the sludge can be digested anaerobically to produce methane gas.
† Dried sludge can be burned as low-quality fuel with a heat value of 13.5 kJ g^{-1}.

Figure 15.2 Primary and secondary treatment of municipal wastewater.
Source: Adapted from Spiro et al., *Chemistry of the Environment, 3rd Ed.*, University Science Books, © 2012, all rights reserved.

$$\text{BOD:} \quad \frac{200\frac{mg}{L} - 25\frac{mg}{L}}{200\frac{mg}{L}} \times 100\% = 87.5\% \text{ removed}$$

*QUESTION 9

Calculate the percent of ammonium (NH_4^+) removed through secondary wastewater treatment.

*QUESTION 10

Calculate the percent of phosphate (PO_4^{3-}) removed through secondary wastewater treatment.

Secondary treatment does not do a great job of removing the ammonium and phosphate from the wastewater. Tertiary treatment must be completed to remove those ions from the water. Figure 15.3 shows the tertiary treatment for BOD, phosphate, and ammonium ions. Lime (calcium hydroxide) is added to the secondary effluent. The phosphate is precipitated and removed. Excess hydroxide is then added to remove the ammonium ions. Carbonate is then added, as well as activated charcoal and chlorine, in order to obtain the final effluent.

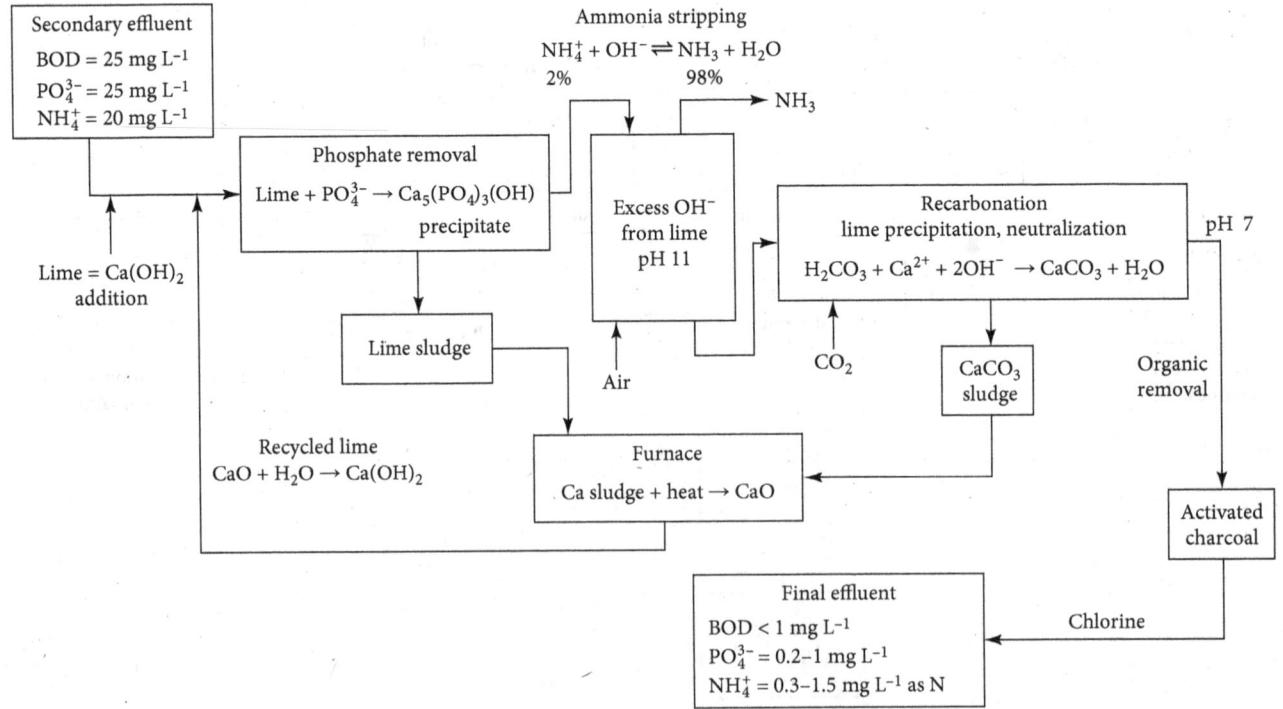

Figure 15.3 Process for removing BOD, phosphate, and ammonium from the secondary effluent. *Source:* Adapted from Spiro et al., *Chemistry of the Environment, 3rd Ed.,* University Science Books, © 2012, all rights reserved.

QUESTION 11

Calculate the efficiency range for phosphate removal from the secondary effluent as illustrated in Figure 15.3.

Q11 ANSWER According to the figure, 25 mg L^{-1} phosphate starts the process and 0.2 – 1 mg L^{-1} phosphate remains. Calculate the efficiency at the two extrema of the range.

Use the efficiency equation:

$$\text{efficiency} = \frac{\text{starting amount} - \text{final amount}}{\text{starting amount}} \times 100\%$$

For 1 mg L^{-1}:

$$\frac{25\frac{mg}{L} - 1\frac{mg}{L}}{25\frac{mg}{L}} \times 100\% = 96\% \text{ efficient at removal}$$

For 0.2 mg L^{-1}:

$$\frac{25\frac{mg}{L} - 0.2\frac{mg}{L}}{25\frac{mg}{L}} \times 100\% = 99.2\% \text{ efficient at removal}$$

Thus, the process is 96%–99.2% efficient.

*QUESTION 12

Calculate the efficiency range for ammonium removal from the secondary effluent as illustrated in Figure 15.3.

QUESTION 13

Calculate the mass of phosphate remaining after 500 L of the secondary effluent has been treated. Assume maximum efficiency.

Q13 ANSWER For the maximum efficiency, we want the amount of phosphate remaining in the effluent to be minimal, which is 0.2 mg L^{-1}. Multiply this by 500 L to calculate the mass of phosphate.

$$\frac{0.2 \text{ mg}}{1 \text{ L}} \times 500 \text{ L} = 100 \text{ mg phosphate}$$

*QUESTION 14

Calculate the mass of ammonium remaining in 750 L of the secondary effluent after treatment, assuming minimum efficiency.

QUESTION 15

If a wastewater plant produces 5.0×10^3 kg of C per day, how many kg of carbon dioxide would be produced? How many moles of CO_2 would be produced and how many liters of CO_2 would be produced at standard temperature and pressure (STP)?

Q15 ANSWER First, we need to calculate the percent of carbon in CO_2.

$$MM_{CO_2} = 12.01 \frac{g}{mol} + 2\left(16.00 \frac{g}{mol}\right) = 44.01 \frac{g}{mol}$$

$$\%C = \frac{12.01\ g}{44.01\ g} \times 100\% = 27.3\%$$

Using the percent of carbon in the final product, we can calculate the total mass of CO_2.

$$(\text{percent carbon}) \times (\text{mass of } CO_2) = \text{mass of carbon}$$

$$(0.273) \times (\text{mass of } CO_2) = 5.0 \times 10^3\ kg$$

$$\text{mass of } CO_2 = \frac{5.0 \times 10^3\ kg}{0.273} = 1.8 \times 10^4\ kg$$

To convert to moles, first convert kilogram to gram by multiplying by 1,000.

$$1.8 \times 10^4\ kg \times \frac{1,000\ g}{1\ kg} = 1.8 \times 10^7\ g$$

Then, divide by the molar mass of CO_2 (44.01 g mol^{-1})

$$\frac{1.8 \times 10^7\ g}{44.01\ g/mol} = 4.2 \times 10^5\ mol\ CO_2$$

Recall that 1 mol of any gas at STP occupies a volume of 22.4 L.
Multiply the number of moles by 22.4 L mol^{-1} to get the total liters of gas.

$$4.2 \times 10^5\ mol \times 22.4 \frac{L}{mol} = 9.3 \times 10^6\ L$$

There is a lot of gas, so we have to consider what occurs with all the organic waste.

*QUESTION 16

If a wastewater plant produces 500. kg of C per hour for a day, how many kilograms of carbon dioxide would be produced? How many moles of CO_2 would be produced and how many liters of CO_2 would be produced at STP?

15.4 Desalinization of Seawater

Although the earth contains a high percentage of water available for use, not all is suitable, especially ocean and seawater. Removing salt from seawater is an obvious method of extending our supply of fresh water, but one that has until recently been prohibitively costly. The problem is the high latent heat of water, which requires a large quantity of energy to distill the water away from the salt. On a global scale, the sun provides this energy. Indeed, the hydrological cycle is a giant solar still that provides us with fresh water, gratis. But if this energy had to be produced artificially, the cost would be prohibitive, even with advanced engineering of distillation columns to maximize heat recovery. Some distillation-based desalinization plants have been operating in the Middle East, where oil has been abundant.

QUESTION 17

How much energy is required to raise the temperature of 2.001 kg of water from 25°C to 100°C where the water can boil? Recall the heat capacity of water is 4.184 J g^{-1} °C^{-1}.

Q17 ANSWER First, convert from kilogram to gram by multiplying by 1,000.

$$2.001 \text{ kg} \times \frac{1,000 \text{ g}}{1 \text{ kg}} = 2,001 \text{ g}$$

Use the relationship $q = mc\Delta T$ to calculate the heat necessary to boil the water. Remember, $\Delta T = T_{final} - T_{initial}$

$$q = (2,001 \text{ g}) \times \left(4.184 \frac{\text{J}}{\text{g °C}}\right) \times (100°C - 25°C) = 6.3 \times 10^5 \text{ J}$$

$q = 6.3 \times 10^5$ J or 630 kJ, which is the energy required to completely boil 2.001 L of water.

As this is not the only process needed to distill water, one can see that the process of distillation is incredibly energy-intensive.

*QUESTION 18

How much energy is required to raise to the temperature of 4.001 kg of water from 35°C to 95°C? Recall the heat capacity of water is 4.184 J g^{-1} °C^{-1}.

However, the prospects for desalinization have greatly improved with the advent of technology for reverse osmosis (RO). Osmotic pressure is the difference in pressure between two solutions separated by a semipermeable membrane that permits the solvent, but not the solute, to flow across. For seawater in contact with freshwater, the osmotic pressure is equivalent to a 240 m head of water. This pressure could be harnessed as a source of usable energy, or, alternatively, equivalent pressure could be applied to force the seawater through a semipermeable membrane, producing salt-free water and leaving the salt behind. This process is RO. Much less energy is required to counter the osmotic pressure than to distill the water.

The osmotic pressure (Π) is a function of the solute concentration in molarity (M) and the absolute temperature (T) in Kelvin. The following equation can be used to calculate the osmotic pressure:

$$\Pi = MRT \tag{15.1}$$

The constant R in this equation is 0.08206 L atm mol^{-1} K^{-1}.

QUESTION 19

Calculate the osmotic pressure of a 0.415 M solution at 75°C.

Q19 ANSWER Use (15.1) to calculation the osmotic pressure. Here, we use the universal gas constant for R, 0.08206 L atm mol^{-1} K^{-1}.

$$\Pi = (0.415\ M)\left(0.08206\ \frac{L\ atm}{mol\ K}\right)(75°C + 273) = (0.415\ M)\left(0.08206\ \frac{L\ atm}{mol\ K}\right)(348\ K) = 11.9\ atm$$

*QUESTION 20

If the osmotic pressure of a solution at 55°C is 6.25 atm, what is the concentration of the solution?

QUESTION 21

In what areas could RO be used?

Q21 ANSWER Communities on large bodies of seawater could benefit a lot from having an RO plant. It is also important that the community have a source of energy to complete the RO.

Figure 15.4 is a diagram of a RO plant in Tampa, FL, USA. The plant is near a power plant and they can take advantage of the warmer water produced from the power plant. It takes less energy to force warmer water through the semipermeable membrane.

QUESTION 22

If 19 million gallons (gal) of 3.0 M salt water are mixed with 1.4 billion gal of freshwater, what is the new concentration of the water? Assume the volumes are additive.

Q22 ANSWER First, add the volumes in order to calculate the final volume.

$$1,400,000,000\ gal + 19,000,000\ gal = 1,419,000,000\ gal$$

In order to complete this problem, we can use the dilution equation:

$$C_1 \times V_1 = C_2 \times V_2$$

$$(\text{concentration 1})(\text{volume 1}) = (\text{concentration 2})(\text{volume 2})$$

$$(3.0\ M)(19,000,000\ gal) = (x)(1,419,000,000\ gal)$$

$$x = 0.040\ M$$

*QUESTION 23

If 22 million gal of saltwater is mixed with 1.6 billion gal of freshwater, the concentration is 0.0608 M. What was the original concentration of the saltwater? Assume the volumes are additive.

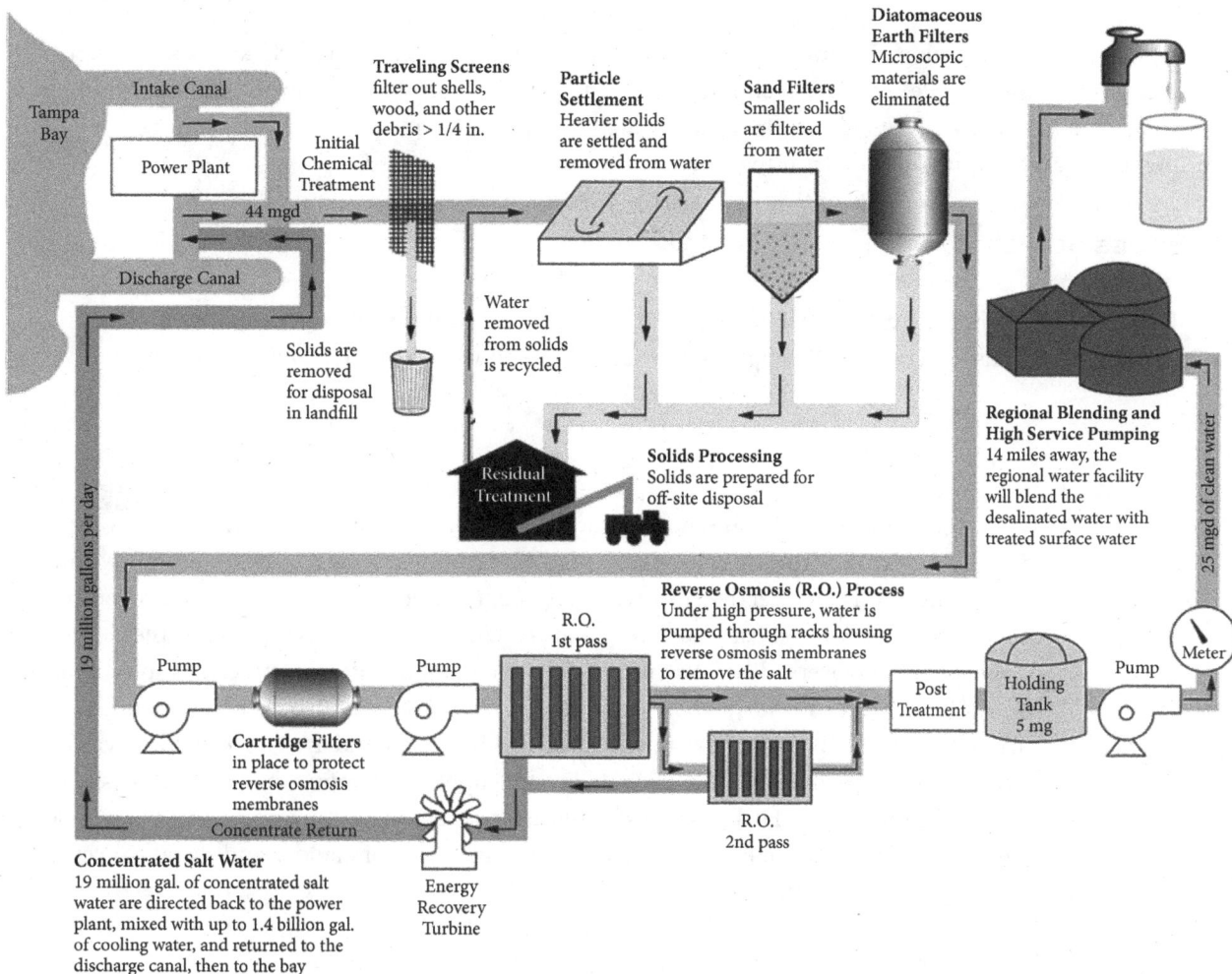

Figure 15.4 RO plant in Tampa, FL, USA.
Source: Adapted from Spiro et al., *Chemistry of the Environment, 3rd Ed.,* University Science Books, © 2012, all rights reserved.

15.5 Health Hazards: Pathogens and Disinfection

The spread of pathogenic microorganisms through the water supply is the most serious pollution hazard to human health. Waterborne pathogens are ubiquitous throughout the world. Even waters that are untouched by humans can be contaminated by animal wastes. Hikers in the wilderness who drink untreated water from seemingly pristine streams often become infected by Giardia microbes. The most serious problems, however, are created by the contamination of drinking water by human wastes. The failure to treat sewage and separate it from drinking water takes an enormous toll in the form of disease around the world. Unsanitary water is one of the most pervasive human problems, especially in developing countries.

QUESTION 24

Why are pathogens a greater problem in developing countries?

Q24 ANSWER Developing countries do not have the monetary resources to help keep clean water; this technology is very expensive. Technology is not advanced to produce readily available drinking water. Also, in many communities, the water available serves many purposes such as bathing, cooking water, areas for waste, and drinking water.

*QUESTION 25

Using well water helps greatly alleviate many of the pathogens. Water is filtered through the earth and the pathogens are removed through that process. Explain why the pathogens are removed.

Besides natural filtration, another way to destroy contaminants in the water is through the use of chlorine. The chlorine smell might be familiar from its use in many swimming pools. Chlorine is an effective and relatively inexpensive disinfectant with a proven record of success. It is used in most water treatment systems around the world. Nevertheless, its use has become controversial because it can introduce organochlorine molecules into drinking water. Hypochlorous acid (HOCl) is not only an oxidant but also a chlorinating agent. In particular, hydroxybenzenes are readily attacked by HOCl and are converted to a variety of chlorinated compounds. Most chlorinated water supplies, therefore, have trace levels of chloroform. Chloroform ($CHCl_3$) is a suspected liver carcinogen in humans, and there is some epidemiological evidence for a modest increase in the risk of bladder and rectal cancer from drinking chlorinated water.

QUESTION 26

Ozone and chlorine dioxide are sometimes used instead of chlorine in the disinfection process. Calculate the percent of chlorine atoms in chlorine (Cl_2), ozone (O_3), and chlorine dioxide (ClO_2). Why might chlorine lead to more chloroform production than the other two compounds?

Q26 ANSWER First determine the molar mass of each compound.

$$\text{Ozone, } O_3: 3 \times 16.00\,\frac{g}{mol} = 48.00\,\frac{g}{mol}$$

$$\text{Chlorine, } Cl_2: 2 \times 35.45\,\frac{g}{mol} = 70.90\,\frac{g}{mol}$$

$$\text{Chlorine dioxide: } ClO_2: \left(1 \times 35.45\,\frac{g}{mol}\right) + \left(2 \times 16.00\,\frac{g}{mol}\right) = 35.45\,\frac{g}{mol} + 32.00\,\frac{g}{mol} = 67.45\,\frac{g}{mol}$$

Next, divide the mass of chlorine in each compound by the total mass of the compound.

$$\text{Ozone: } \frac{0 \text{ g Cl}}{48.00 \text{ g ozone}} \times 100\% = 0\%$$

$$\text{Chlorine: } \frac{70.90 \text{ g Cl}}{70.90 \text{ g chlorine}} \times 100\% = 100\%$$

$$\text{Chlorine dioxide: } \frac{35.45 \text{ g Cl}}{67.45 \text{ g chlorine dioxide}} \times 100\% = 52.56\%$$

Chlorine molecules contain more chlorine atoms by percentage than the other water disinfectants, which could lead to more chloroform formation.

15.6 Organic and Inorganic Contaminants

Surface and groundwater can be contaminated by the migration of chemicals from poorly maintained landfills, industrial waste sites, accidental spills, and leaks in storage tanks, especially those buried in the ground. Regulations for the cleanup of leaky underground oil and gasoline storage tanks have become major preoccupations of home and gas station owners in the United States.

Drinking wells can be contaminated by trace amounts of petroleum fractions, chlorinated solvents, or polychlorinated biphenyls (PCBs). These sparingly soluble organic compounds can escape confinement and migrate through the soil; they often accumulate in underground pools, from which they slowly enter the water table over a long period of time. Leaks from gasoline storage tanks have contaminated many wells with the water-soluble additive methyl tert-butyl ether (MTBE), leading to its elimination from the gasoline supply.

Generally, organic contaminants are measured on the parts per million (ppm) scale. We introduced the idea of ppm [and parts per billion (ppb)] in Chapter 9 related to air pollution in the atmosphere, but this is a useful way to talk about contaminants in water as well. Remember that one ppm would indicate that for every 1 million molecules (molec)/liters (L)/milliliters (mL) of water, 1 molec/L/mL would be a contaminant.

QUESTION 27

If 50. mL of gasoline (octane) were spilled in a pond that contained 5.0×10^4 L of water, what would the contamination amount be in ppm?

Q27 ANSWER First convert 50. mL to L by dividing by 1,000.

$$50. \text{ mL} \times \frac{1 \text{ L}}{1,000 \text{ mL}} = 0.050 \text{ L}$$

Set up a fraction for the volume of gasoline to volume of water and set that equal to volume of gasoline per 1 million L of water

$$\frac{0.050 \text{ L gasoline}}{5.0 \times 10^4 \text{ L water}} = \frac{x \text{ L gasoline}}{1,000,000 \text{ L water}}$$

Solve this equation for x.

$$x = 1.0 \text{ ppm, by volume}$$

*QUESTION 28

If 5.0 L of gasoline (octane) were spilled in a pond that contained 2.0×10^5 L of water, what would the contamination amount be in ppm.

*QUESTION 29

If 25 mL of gasoline (octane) were spilled in a pond that contained 5.0×10^5 L of water, what would the contamination amount be in ppb?

When gasoline spills, the amount can be similar to the above problems, and thus, it is important to monitor the spills to prevent the migration of contaminants to our lakes and streams.

Gasoline is one source of pollution in waterways. Another is the runoff of fertilizers from fields, especially near farmland. Fertilizers can increase the level of nitrate ions (NO_3^-) in the groundwater. The main nitrate health hazard is "blue baby syndrome," a condition of respiratory failure in babies having excessive nitrate in their diet. Some of the nitrate is reduced (adding electrons) by anaerobic bacteria in the stomach to nitrite ion (NO_2^-) (see Chapter 14 for information on oxidation-reduction reactions). The nitrite oxidizes (causes a loss of electrons) the Fe^{2+} ion in hemoglobin to Fe^{3+}, which is unable to bind O_2. The Fe^{3+}-containing hemoglobin is called methemoglobin, and the condition is known as methemoglobinemia. In adults, the methemoglobin is rapidly reduced back to the Fe^{2+} form, but in babies this process is slow. Nitrate-induced methemoglobinemia is now a rare condition in industrialized countries but remains a concern in developing countries.

QUESTION 30

Write the reduction reaction for nitrate to nitrite. Balance in acid if necessary.

Q30 ANSWER $NO_3^- + 2e^- \rightarrow NO_2^-$
The oxidation state of N in nitrate is +5 and in nitrite is +3.

Now the half-reaction must be balanced:

$$NO_3^- + 2e^- \rightarrow NO_2^-$$

Add $2H^+$ to the reactants side to balance the charge:

$$NO_3^- + 2e^- + 2H^+ \rightarrow NO_2^-$$

Then, add one molecule of water to the product's side to balance the oxygen and hydrogen:

$$NO_3^- + 2e^- + 2H^+ \rightarrow NO_2^- + H_2O$$

*QUESTION 31

Write the oxidation reaction for iron (II) to iron (III).

15.7 Low-Cost Water Technology for Developing Countries

More than 1 billion people in the world do not have access to clean drinking water, and almost 2.2 million children die yearly as a result. Standard techniques for purifying water (e.g., chlorination, distillation, boiling, sedimentation, use of high-tech filters) are often too costly for those living in developing countries. In addition, these methods may require energy supplies and maintenance that are not available in rural locales.

Many nonprofit organizations are active in developing and implementing low-tech appropriate technologies. One is the KlarAqua system for household water. It is made of two plastic buckets and three clay filters, making it simple to use and manufacture. The ceramic filters can be manufactured by local potters and are coated with colloidal silver to kill bacteria. Another device is the LifeStraw, a plastic straw with coarse (100 μm polyethylene mesh) and fine (15 μm polyester mesh) filters, to remove particles including microbe clusters. Water is sucked through these filters and then through iodine-impregnated beads to kill microbes, and finally through activated carbon to remove iodine. A LifeStraw costs ~\$3.50 and can process ~700 L of water.

QUESTION 32

If a child needs 1.00 L of water per day for a year, how much would it cost to save the 2.20 million children who die each year from water issues if you used a LifeStraw?

Q32 ANSWER Each child requires 365 L of water per year. First, calculate the cost per child by determining how many LifeStraws each child needs.

$$\frac{365 \text{ L}}{700 \frac{\text{L}}{\text{LifeStraw}}} = 0.521 \text{ LifeStraws}$$

$$0.521 \text{ LifeStraws} \times \frac{\$3.50}{1 \text{ LifeStraw}} = \$1.83 \text{ per child}$$

$1.83 \times 2,200,000$ children $= \$4,015,000$. This is the amount of money to provide clean water for the children who die each year due to waterborne illness.

---

*QUESTION 33

How many LifeStraws would be necessary to provide 2.5 million children with 1,095 L of water each year (3.00L a day)?

*QUESTION 34

How many children could be provided 5.00 L of water each day of a month for a donation of $50.00?

15.8 Trace Pollutants

Many chemicals that do not naturally occur in the environment are found at trace levels (parts per trillion [ppt]or ppb concentrations) in both waterways and drinking water sources. These pollutants include pharmaceutical and personal care products in addition to chemicals used in industry and manufacturing. The pharmaceuticals that humans consume, and their metabolites, are flushed down the toilet and enter the environment. Almost 10,000 different drugs are in use today, and it is unknown which of these are persistent, bioaccumulative, or toxic to wildlife. Pharmaceuticals detected in the environment include ibuprofen, antibiotics, antidepressants, anticonvulsants, tranquilizers, antipsychotics, nitroglycerin, and steroids. Researchers in Japan even found the antiviral drug Tamiflu contaminating rivers downstream of sewage treatment facilities deriving from urinary excretion of the drug. There is concern that birds exposed to the waterborne contaminant might develop and spread drug-resistant strains of seasonal and avian flu.

QUESTION 35

Imagine you live in a city of 250,000 people. Each day, each person on average flushed 10 mg of medicine down the toilet. How much medicine accumulated in the water in that month in the city?

Q35 ANSWER First, determine the total mass in mg per day for the city by multiplying 100 mg by 250,000 people.

$$250,000 \text{ people} \times \frac{10 \text{ mg}}{\text{person}} = 2,500,000 \text{ mg of pharmaceuticals per day}$$

Per month:

$$\frac{2,500,000 \text{ mg}}{\text{day}} \times \frac{30 \text{ days}}{\text{month}} = 7.5 \times 10^7 \text{ mg} = 75,000 \text{ g} = 75 \text{ kg}$$

QUESTION 36

Imagine that the mass flushed down the toilet went to one lake that had 5.0×10^5 L of water. What would the concentration of pharmaceuticals in the water be in ppm?

Q36 ANSWER First, convert the volume of water to mL which then can be changed to grams since the density of water is 1.0 g mL^{-1}

$$5.0 \times 10^5 \text{ L} \times \frac{1,000 \text{ mL}}{1 \text{ L}} \times \frac{1 \text{ g}}{1 \text{ mL}} = 5.0 \times 10^8 \text{ g}$$

Set up a proportion to determine ppm.

$$\frac{75,000 \text{ g medicine}}{500,000,000 \text{ g water}} = \frac{x \text{ g medicine}}{1,000,000 \text{ g water}}$$

$$x = 150 \text{ ppm}$$

*QUESTION 37

If 2,500 g of medicine were flushed down the toilet and ended up in a lake that was 5.0×10^5 L of water, what would the ppm concentration of the medicine be in the lake?

*QUESTION 38

If the measurement of medicine found in lake water was 1.5 ppm, how much medicine had reached the lake? The lake is 2.5×10^4 L of water.

15.9 Perchlorate Pollution

Perchlorate (ClO_4^-) occurs naturally at low levels in the environment but is also an item of commerce. Perchlorate salts are used as oxidizers in fireworks, explosives, flares, and rocket propellants. Perchlorates are highly soluble and can contaminate groundwater around manufacturing facilities and rocket test sites. Perchlorate contamination of drinking water and agricultural wells has been discovered in several California communities near rocket facilities. Elevated perchlorate levels were also found in Colorado River water, below a Nevada ammonium perchlorate manufacturing facility. The water is used for irrigation of Arizona and California fields, raising concern about contamination of produce.

Perchlorate is a potential health hazard for children and fetuses because it interferes with iodine uptake by the thyroid gland and can decrease the production of thyroid hormones.

The EPA is currently evaluating whether to include perchlorate as part of the national primary drinking water standard, while California and Massachusetts have already set drinking water standards to 6 and 2 μg L^{-1}, respectively.

Between 2001 and 2005, the EPA sampled drinking water supplies across the United States and found that across 26 states, 4.1% of systems had a perchlorate level > 4 μg L^{-1}.

QUESTION 39

Would you expect the perchlorate level to be high in the area you live?

Q39 ANSWER Answers will depend on your area but if you have areas where there are lots of fireworks such as amusement parks, explosions, rocket launches, or similar activities, the level of perchlorate is much higher than in other areas.

QUESTION 40

If 50.00 g of potassium perchlorate was released into a waterway with 7.5×10^4 L of water, what would the ppm concentration of perchlorate be?

Q40 ANSWER First, determine the percent composition of perchlorate in potassium perchlorate. The formula for the compound is $KClO_4$.

$$\text{MM } KClO_4 = K\left(1 \times 39.10\,\frac{g}{mol}\right) + Cl\left(1 \times 35.45\,\frac{g}{mol}\right) + O\left(4 \times 16.00\,\frac{g}{mol}\right) = 138.55\,\frac{g}{mol}$$

$$\text{MM } ClO_4^- = Cl\left(1 \times 35.45\,\frac{g}{mol}\right) + O\left(4 \times 16.00\,\frac{g}{mol}\right) = 99.45\,\frac{g}{mol}$$

$$\text{percent perchlorate} = \frac{99.45\,\frac{g}{mol}}{138.55\,\frac{g}{mol}} \times 100\% = 71.78\%$$

Multiply the percent of perchlorate (in decimal form) by the total mass of potassium perchlorate.

$$0.7178 \times 50.00\,g = 35.89\,g$$

Convert 75,000 L to g

$$(7.5 \times 10^4\,L)\left(\frac{1,000\,mL}{mL}\right)\left(\frac{1\,g}{1\,mL}\right) = 7.5 \times 10^7\,g$$

Now complete a ratio to determine the ppm

$$\frac{35.89\,g}{7.5 \times 10^7\,g} = \frac{x\,g}{1,000,000\,g}$$

$$x = 0.478\,ppm$$

*QUESTION 41

Ammonium perchlorate is used in fireworks. How much ammonium perchlorate could be used in an area with a watershed of 1.0×10^5 L and a limit of 4.0 ppm perchlorate?

15.10 Conclusions

Water is available in abundance on planet earth, and pure water is supplied continuously by the hydrological cycle, which is powered by the sun. The bounty of freshwater is distributed unevenly around the globe; even where it is abundant, the resource is often mismanaged through profligate use or through contamination from wastes. Access to unpolluted, pathogen-free water is a critical need for much of the world's population.

The problem of water pollution is, thus, one largely of capacity depletion: acid buffer capacity in soil–water interactions and oxidation capacity in receiving waters (see Chapter 14). Reduction of water pollution thus requires strategies that focus not only on a decrease in pollutant emissions but also on maintenance and replenishment of water resources. In some cases, this may be achieved by redirecting water pollutants for other applications. For example, provided they are free of toxic chemicals, sewage wastes, instead of being discharged to water bodies, can be applied as fertilizer to the land, which has a much higher oxidizing capacity.

PART V

BIOSPHERE

16

NITROGEN AND FOOD PRODUCTION

16.1 Nitrogen Cycle

The nitrogen cycle is critical to food production. The central process in biological energy production is the photosynthetic conversion of CO_2 to reduced carbon compounds, which are then used as fuel by all manner of life (see earlier discussions of the carbon cycle and photosynthesis). However, organisms need more than carbon; as discussed earlier in the context of aquatic ecosystems, they also require nitrogen, phosphorus, and several other elements in smaller amounts. These elements are normally available in soil in sufficient quantities to support an adequate level of natural plant growth.

Possibilities for increasing productivity beyond natural limits, however, are often constrained by the supply of nitrogen available to the plant. Roughly 80% (by mixing ratio) of the atmosphere consists of molecular nitrogen (N_2), which is an extremely stable and unreactive form of nitrogen. In order to participate in biological reactions, nitrogen must be fixed; that is, it must be combined with other elements. The cycling of nitrogen through the environment is illustrated in Figure 16.1 and quantified in Table 16.1. Overall, roughly 40% of fixed nitrogen is from natural sources (lightning, algae, and soil bacteria), whereas 60% is from anthropogenic sources (see Fig. 16.1). Some N_2 is fixed nonbiologically through reaction with O_2 at sufficiently high temperatures in combustion or lightning. The nitrogen oxides formed in the atmosphere are converted to nitric acid (HNO_3) and washed out in rain, thus providing the soil with a supply of nitrate (NO_3^-). Plants can utilize nitrate in the production of protein and other essential organic nitrogen compounds.

QUESTION 1

Calculate the percent contribution of fixed nitrogen by natural sources, based on Figure 16.1 and Table 16.1.

Q1 ANSWER In Figure 16.1, the natural sources of fixed nitrogen include bacteria and algae (37.1%) and lightning (2.8%):

$$37.1\% + 2.8\% = 39.9\% \text{ nitrogen fixation from natural sources}$$

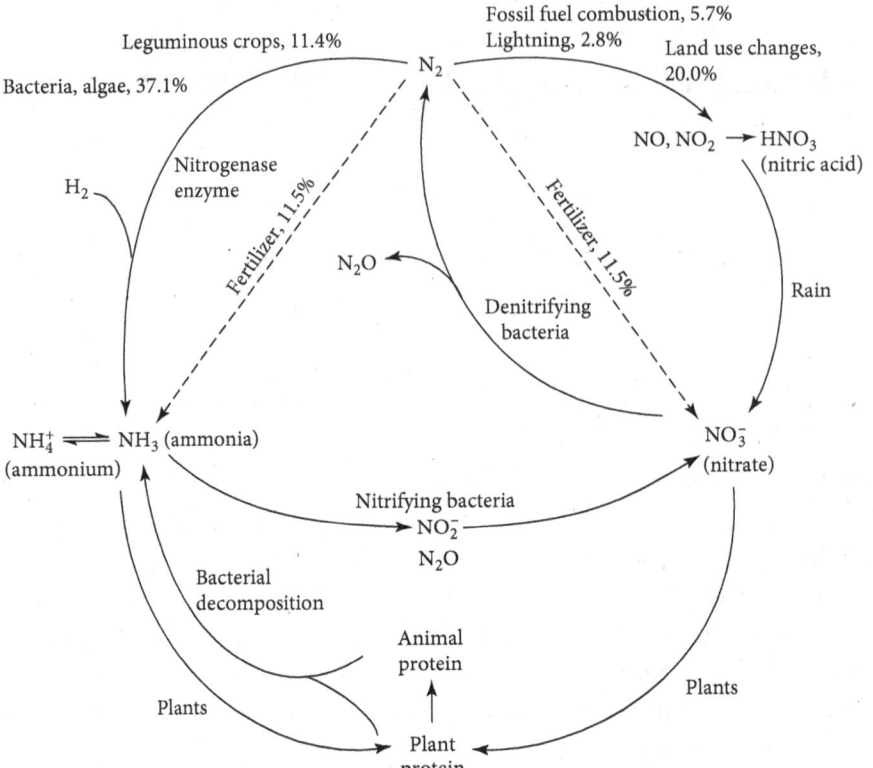

Figure 16.1 The nitrogen cycle showing percentage contributions to nitrogen fixation from natural and human sources. (The total contribution of synthetic fertilizers, assuming an equal share of ammonia and nitrate, is 23%.)
Source: Adapted from Spiro et al., *Chemistry of the Environment, 3rd Ed.,* University Science Books, © 2012, all rights reserved.

From Table 16.1, the natural sources of fixed nitrogen include lightning (10 Tg yr^{-1}) and bacteria and algae (130 Tg yr^{-1}). While the fixed nitrogen totals are 140 Tg natural and 233 Tg anthropogenic.

The total percent fixed nitrogen from natural sources can be calculated by adding the natural sources together and dividing by the total fixed nitrogen:

$$\frac{(10 + 130) \text{ Tg yr}^{-1}}{(233 + 140) \text{ Tg yr}^{-1}} \times 100\% = 37.5\% \text{ fixed nitrogen from natural sources}$$

Data from both the figure and table show that approximately 40% of the fixed nitrogen in the world derives from natural sources.

*QUESTION 2

Calculate the percent contribution of fixed nitrogen by anthropogenic sources, based on Figure 16.1 and Table 16.1.

---

* Answers to starred questions can be found at the end of the book.

TABLE 16.1 Global Sources of Fixed Nitrogen

Anthropogenic	$(Tg\ yr^{-1})^a$
Nitrogen fertilizer	103
Legumes and other plants	40
Fossil fuels	20
Land-use changes	
Biomass burning	40
Wetland draining	10
Land clearing	20
Total anthropogenic	**233**
Natural	$(Tg\ yr^{-1})^a$
Lightning	10
Algae, soil bacteria	130
Total natural	**140**

[a]Teragram = Tg = 10^{12} g

Sources: Data from Vitousek et al., "Human Alteration of the Global Nitrogen Cycle: Causes and Consequences," *Ecological Applications,* Vol. 7, No. 3, 1997, pp. 737–750 and Fertilizer Association, *Assessment of Fertilizer Use by Crop at the Global Level,* 2017, available at https://www.ifastat.org/plant-nutrition; adapted from Spiro et al., *Chemistry of the Environment, 3rd Ed.,* University Science Books, © 2012, all rights reserved.

But the amount of nitrogen available through the pathway of nitrogen oxides and nitrate is insufficient to support the abundant plant life we know. The majority of naturally occurring nitrogen fixation is accomplished by certain bacteria and blue-green algae (cyanobacteria) that are able to reduce N_2 to NH_3. Even though the overall reaction (the Haber process):

$$N_2 + 3\,H_2 \xrightarrow{\text{Fe or Ru Catalyst}} 2\,NH_3 \qquad \Delta G = -94 \text{ kJ mol}^{-1} \tag{16.1}$$

is downhill in energy (exothermic), the triple bond in N_2 is so strong (941 kJ mol^{-1}) that additional energy must be provided by the organisms to overcome the activation barrier.

Plants can use ammonia directly for their nitrogen source; animals gain their nitrogen by eating plants. When plants and animals die, the reduced nitrogen in their tissues is converted to ammonia by bacterial decomposition and added to the ammonia pool. The ammonia can be used as fuel by other bacteria (*Nitrosomonas*), which convert NH_3 to NO_2^- (nitrite), using O_2 as an oxidant. Still other bacteria (*Nitrobacter*) oxidize the nitrite further to NO_3^- (nitrate). The overall process of nitrogen oxidation is called nitrification. Plants can also utilize NO_3^- from this biological process and the nonbiological process discussed earlier as a nitrogen source. Thus, there is continual cycling between oxidized and reduced forms of fixed nitrogen in soils.

If the fixed nitrogen were not returned to the atmosphere, the atmospheric pool of N_2 eventually would be depleted. But the nitrogen cycle is closed by denitrifying bacteria, which use NO_3^- instead of O_2 as the oxidant in metabolic reactions, reducing the nitrate back to N_2. Both nitrification and denitrification processes release some N_2O as a by-product. Nitrous oxide (N_2O) is a greenhouse gas (see Chapter 7) and the main source of stratospheric NO, a principal player in ozone (O_3) destruction chemistry (see Chapter 10).

QUESTION 3

Reductive nitrogen fixation requires the breaking of the triple bond in nitrogen (N≡N) by the reaction:

$$N≡N + 3\,H_2 \rightarrow 2\,NH_3 \tag{16.1}$$

Imagine that the overall reaction occurs by the sequential breaking of each bond in molecular nitrogen by the reactions:

$$N≡N + H_2 \rightarrow HN=NH \tag{16.2a}$$

$$HN=NH + H_2 \rightarrow H_2N–NH_2 \tag{16.2b}$$

$$H_2N – NH_2 + H_2 \rightarrow 2\,NH_3 \tag{16.2c}$$

Reaction (16.1) requires an activation energy of 941 kJ mol^{-1} to break the triple bond. The most energy-demanding step in the process is reaction (16.2a), which requires an activation energy of ~527 kJ mol^{-1}. How much slower would the reaction be if it had to go the route of reaction (16.1) versus reaction (16.2a)? Calculate the ratio of the rate constants (k_{2a}/k_1) at 27°C for reactions (16.1) and (16.2a) given that k is proportional to $e^{-Ea/RT}$, where E_a is the activation energy, $R = 8.314$ J K^{-1} mol^{-1}, and T is measured in Kelvin.

Q3 ANSWER Let E_{a1} be the activation energy for reaction (16.1) equal 941 kJ mol^{-1}. Similarly, let E_{a2a} be the activation energy for reaction (16.2a) equal 527 kJ mol^{-1}. Since the rate constant k is proportional to $e^{-Ea/RT}$, the following relationship holds:

$$\frac{k_{2a}}{k_1} = \frac{e^{\frac{-E_{a2a}}{RT}}}{e^{\frac{-E_{a1}}{RT}}}$$

The rate constants for reactions (16.1) and (16.2a) are k_1 and k_{2a}, respectively. Rearranging the above equation and the temperature given in the problem (27°C), we obtain:

$$\frac{k_{2a}}{k_1} = e^{\frac{(E_{a1} - E_{a2a})}{RT}} = e^{\frac{\left(941\frac{kJ}{mol} - 527\frac{kJ}{mol}\right)}{\left(8.314\frac{J}{mol\,K}\right)(27°C + 273)K}} = e^{\frac{\left(414\frac{kJ}{mol}\right)}{\left(8.314\frac{J}{mol\,K}\right)300\,K}}$$

Thus:

$$\frac{k_{2a}}{k_1} = e^{\frac{\left(414\frac{kJ}{mol}\right)}{\left(2,478\frac{J}{mol}\right)\times\left(\frac{1\,kJ}{1,000\,J}\right)}} = e^{166} = 1.22 \times 10^{72}$$

This shows that the rate constant for (16.2a), breaking the first single bond of N_2, is much, much faster than reductive nitrogen fixation in (16.1).

QUESTION 4

From the value of (k_{2a}/k_1) calculated in Q3, is it surprising that microorganisms can fix nitrogen at ambient temperatures, whereas the Haber process requires temperatures between 400°C and 600°C? Calculate the temperature at which k_1 equals the value of k_{2a} at 27°C.

Q4 ANSWER Reaction (16.2a) is approximately 10^{72} times faster than reaction (16.1) at 27°C, enabling microorganisms to fix nitrogen at ambient temperatures, assuming reaction (16.2a) is the rate-determining step. The temperature T' at which k_1 equals the value of k_{2a} at 27°C can be calculated by looking at the relationships of the ratio of the activation energy to the temperature for each reaction:

$$\frac{E_{a1}}{T'} = \frac{E_{a2a}'}{300 \text{ K}}$$

Solving for :

$$T' = \frac{300 \text{ K} \times E_{a1}}{E_{a2a}} = \frac{300 \text{ K} \times 941 \frac{\text{kJ}}{\text{mol}}}{527 \frac{\text{kJ}}{\text{mol}}} = 536 \text{ K} = 263°\text{C}$$

The temperature of k_1 (263°C) when it is equal to k_{2a} (27°C) is much higher. This calculation is an approximation. Note that the temperature calculated is a little low, considering that the Born–Haber process takes place at between 400°C and 600°C.

*QUESTION 5

Which reaction (16.2a), (16.2b), or (16.2c) is the rate-determining step in the overall mechanism for the conversion of N_2 to NH_3?

16.2 Agriculture: Fertilizer and the Green Revolution

The major natural sources of nitrogen for agriculture are the nitrogenase-containing bacteria, some of which grow in symbiosis with a limited variety of crop plants, most notably those of the legume family. These symbiotic bacteria are contained in the nodules of such legumes as beans, peas, alfalfa, and clover. When the plants die, most of the nitrogen is returned to the soil in fixed form, where it is available for other kinds of plants; a small fraction is returned to the atmosphere via the denitrification reaction. The fertilizing ability of legumes is the reason for crop rotation, an ancient agricultural practice in which legumes are planted alternately with cereals, grains, and other vegetables to maintain the productivity of the nonlegume plants. In the absence of crop rotation, plants that do not fix nitrogen deplete soil's nitrogen stores quickly, unless these stores are replenished by the addition of fertilizer.

One traditional fertilizer is animal manure, but increasingly in recent decades, it has been supplanted by artificial fertilizers produced industrially. The industrial production of nitrogen fertilizer (dashed lines in Fig. 16.1) is accomplished via the Haber process, in which reaction (16.1) is carried out over an iron catalyst. Even with a catalyst, the reaction requires high pressures (100 atm) of the reacting gases, and high temperatures (500°C). (On the contrary, the nitrogenase enzyme operates at ambient pressure and temperature, but energy input in the form of MgATP is required.) The resulting ammonia can be injected directly into crop-bearing soils or, more conveniently, it can be added as ammonium nitrate salt, produced by air oxidation of half the ammonia to HNO_3, which is recombined with the remaining ammonia:

$$NH_3 + 2O_2 \rightarrow HNO_3 + H_2O \tag{16.3}$$

$$NH_3 + HNO_3 \rightarrow NH_4NO_3 \tag{16.4}$$

The Haber process [reaction (16.1)] requires a source of hydrogen gas. Currently, the most economical process for obtaining hydrogen is from methane reformation:

$$CH_4 + 2\,H_2O \rightarrow 4\,H_2 + CO_2 \qquad\qquad (16.5)$$

Thus, fertilizer production is a major consumer of natural gas and contributes to global CO_2 emissions.

The production of nitrogen fertilizers has increased dramatically in the past eight decades and is now almost 2.5 times the estimated nitrogen-fixation rate of leguminous crops (103 versus 40 Tg yr^{-1} of fixed N). The total of these two anthropogenic sources (leguminous crops and synthetic fertilizer) roughly equals the estimated natural microbial nitrogen-fixation rate of 130 Tg yr^{-1}, and the nitrogen oxide contribution from fossil fuel combustion adds another 20 Tg yr^{-1}.

In addition, land-use changes (e.g., biomass burning from slash-and-burn agriculture, land clearing, draining of wetlands) result in significant amounts of additional fixed nitrogen. Land clearing kills trees and other vegetation, which in turn increases the amount of plant protein decomposed into NH_3. Wetlands are a natural sink for nitrates, as they harbor denitrifying bacteria that reduce nitrate to N_2. Removing wetlands in essence reduces the biosphere's capacity to denitrify nitrate. Quantifying the contribution of land-use changes is highly uncertain, but researchers estimate it to be on the order of 70 Tg yr^{-1}. Thus, human activities now appear to dominate the global nitrogen cycle, although this was not the case as recently as 1970. The relationship of these fixation rates to the global fluxes and pools of nitrogen is illustrated in Figure 16.2. Human activity has increased the rates of both nitrogen fixation and denitrification. The sum of the increased fixation on land (350 Tg yr^{-1}) and increased runoff of fixed nitrogen from land to the oceans (40 Tg yr^{-1}), however, does not appear to be entirely balanced by the increase in denitrification on land (160 Tg yr^{-1}) and in the oceans (110 Tg yr^{-1}). This result suggests that fixed nitrogen is now accumulating in the terrestrial pool (see Q7). The flows in the oceans show that inputs of fixed nitrogen are less than the output via denitrification. The uncertainties are too large to place great confidence in this balance, but the result is consistent with the fact that nitrogen is a limiting nutrient, particularly in the deep ocean.

Most of the global fixed nitrogen is in the ocean, where it is divided nearly equally between organic and inorganic forms. The inorganic ions, NH_4^+ and NO_3^-, do not accumulate in soils; because they are soluble, they are quickly taken up by plants and bacteria or are washed out of the soil and transported to the oceans. The terrestrial fixed nitrogen is mostly organic matter.

The application of fertilizer can dramatically improve agricultural yields. The near quadrupling of yields for corn and other grains between 1950 and 2020 made the United States and Canadian Midwest the breadbasket of much of the world. Moreover, crop yields have increased everywhere as a result of fertilizer availability and improved crop strains.

A major issue is whether the productivity gains of world agriculture can be sustained. So far, crop yields can be improved only by fertilization. When sufficient fertilizer has been added to support plant growth at its maximum rate, further additions either have no effect or actually inhibit growth. The curves of diminishing returns for corn in various soils are shown in Figure 16.3. Per capita world grain production leveled off in ~1980, and actually declined slightly during the 1990s.

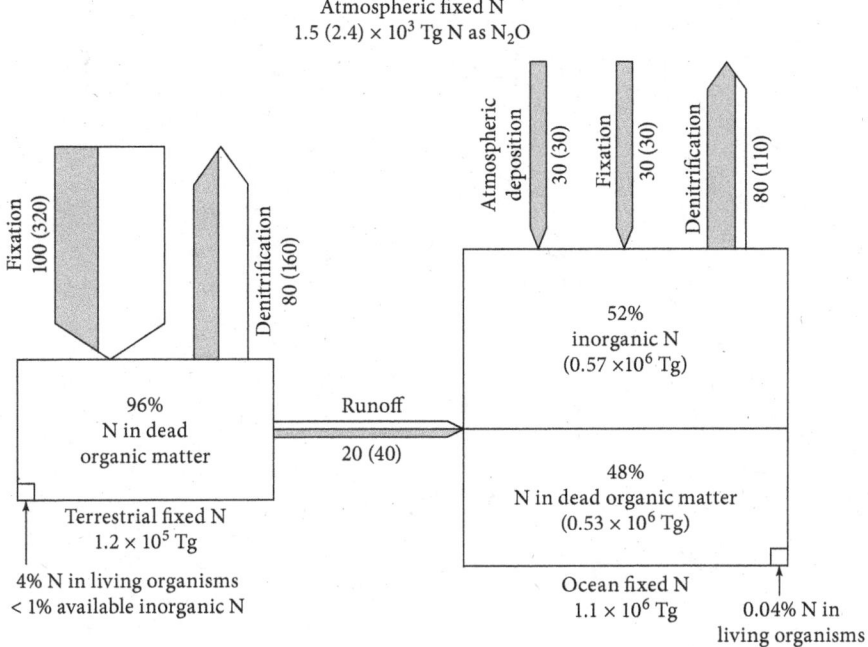

Figure 16.2 Global fluxes of fixed nitrogen (arrows) are in units of teragrams of N per year (Tg N yr⁻¹); pools of nitrogen, in atmosphere, land, and oceans are in units of teragrams of N. Shaded portions of arrows and the accompanying number (not in parentheses) indicate preindustrial flows of fixed nitrogen. The arrows in total, and the accompanying numbers (in parentheses), refer to current nitrogen flows. Atmospheric deposition to the oceans includes fixed nitrogen caused by lightning and net transfers of nitrogen from the land to the oceans via rainfall. *Sources:* Adapted from Kinsig & Socolow, "Human Impacts on the Nitrogen Cycle," *Physics Today*, 1994; Vitousek et al., "Human Alteration of the Global Nitrogen Cycle: Causes and Consequences," *Ecological Applications*, Vol. 7, No. 3, 1997, pp. 737–750; Schlesinger, *Biogeochemistry: An Analysis of Global Change*, Academic Press, 1997; adapted from Spiro et al., *Chemistry of the Environment, 3rd Ed.*, University Science Books, © 2012, all rights reserved.

QUESTION 6

Given the flows of nitrogen shown in Figure 16.2, during preindustrial times was there any net accumulation in the stocks of nitrogen on land or in the oceans?

Q6 ANSWER The balance of fluxes of N on land can be expressed as:

$$\text{(fixation)} - \text{(denitrification)} - \text{(runoff)} = \text{net accumulation}$$

In preindustrial times, this relationship, in units of Tg N yr⁻¹, was:

$$(100 - 80 - 20)\frac{\text{Tg N}}{\text{yr}} = 0\frac{\text{Tg N}}{\text{yr}}$$

Thus, there was no net accumulation on land.
For the oceans, the following balance must hold:

$$\text{(atmospheric deposition)} + \text{(fixation)} + \text{(runoff)} - \text{(denitrification)} = \text{net accumulation}$$

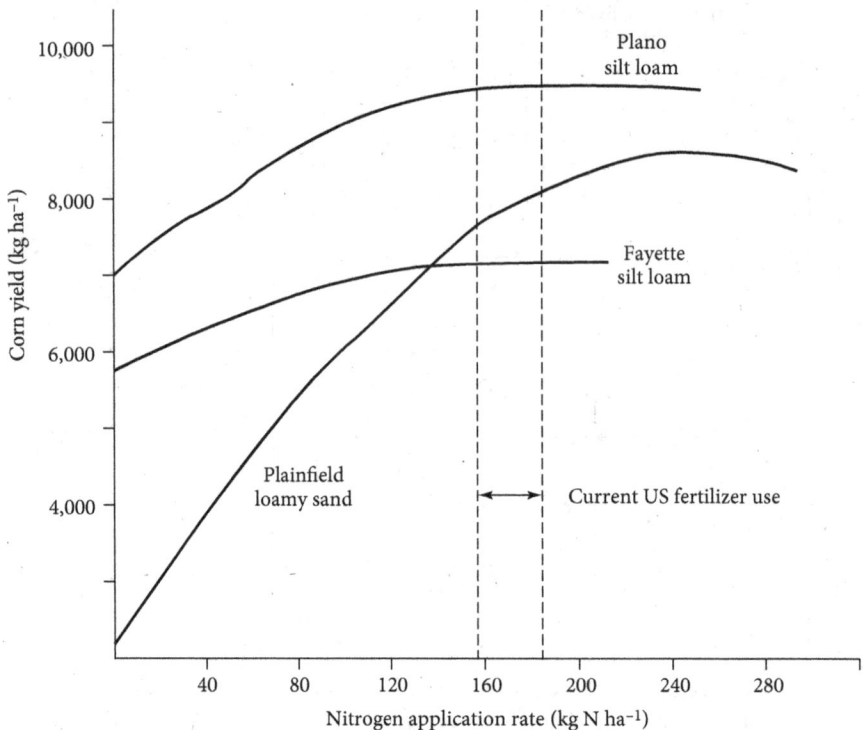

Figure 16.3 Yield response of corn to nitrogen inputs applied to three soils.
Sources: Adapted from Oberle & Keeney, "A Case for Agricultural Systems Research," *Journal of Environmental Quality*, Vol. 20, 1990, pp. 4–7. Current U.S. fertilizer use was updated based on fertilizer use for corn in Iowa for the period 1985–1995 from data presented in Ney et al., "Iowa Greenhouse Gas Action Plan," *University of Iowa Report to the Iowa Department of Natural Resources*, 1996. Adapted from Spiro et al., *Chemistry of the Environment, 3rd Ed.*, University Science Books, © 2012, all rights reserved.

In preindustrial times:

$$(30 + 30 + 20 - 80)\frac{\text{Tg N}}{\text{yr}} = 0\frac{\text{Tg N}}{\text{yr}}$$

Thus, there was no net accumulation in the oceans.

*QUESTION 7

Is there any accumulation under current nitrogen flows? If so, calculate the net stock changes. Describe the environmental effects of increased nitrogen stocks in the terrestrial biosphere from the point of view of water quality of lakes, rivers, groundwaters, and coastal marine areas.

QUESTION 8

Figure 16.3 shows the relationship between corn yield and nitrogen fertilizer application. With respect to corn grown on Plainfield loamy sand, estimate the yields for four cases: (1) no fertilizer applied, (2) 100 kg N applied, (3) 160 kg (current usage) applied, and 4) 200 kg applied. Given that corn is ~1.3% nitrogen by weight, calculate the percent of applied nitrogen contained in the harvested corn for the latter three cases. Are we reaching the point of diminishing returns with respect to nitrogen application? Where does the residual nitrogen go?

Q8 ANSWER According to Figure 16.3, the yield of corn grown on Plainfield loamy sand is:

No fertilizer application: 200 kg ha^{-1}

Fertilizer application = 100 kg N: 5,500 kg ha^{-1}

Fertilizer application = 160 kg N: 7,200 kg ha^{-1}

Fertilizer application = 200 kg N: 8,000 kg ha^{-1}

Given that corn is about 1.30% N by weight, we can calculate the weight of the nitrogen contained in the harvested corn:

No fertilizer application:

$$\text{N in corn} = 0.0130 \times 200 \frac{\text{kg}}{\text{ha}} = 2.6 \frac{\text{kg}}{\text{ha}}$$

Fertilizer application = 100 kg N:

$$\text{N in corn} = 0.0130 \times 5,500 \frac{\text{kg}}{\text{ha}} = 71.5 \frac{\text{kg}}{\text{ha}}$$

Fertilizer application = 160 kg N:

$$\text{N in corn} = 0.0130 \times 7,200 \frac{\text{kg}}{\text{ha}} = 93.6 \frac{\text{kg}}{\text{ha}}$$

Fertilizer application = 200 kg N:

$$\text{N in corn} = 0.0130 \times 8,000 \frac{\text{kg}}{\text{ha}} = 104 \frac{\text{kg}}{\text{ha}}$$

The percentage of N in fertilizer that ends up in the corn can be calculated using the following formula:

$$\frac{\text{N in corn with fertilizer application} - \text{N in corn with no fertilizer application}}{\text{N in fertilizer application}} \times 100$$

Thus,

Fertilizer application = 100 kg N:

$$\%\text{N in corn} = \frac{(71.5 - 2.6)\frac{\text{kg N}}{\text{ha}}}{100\frac{\text{kg N}}{\text{ha}}} \times 100 = 68.9\%$$

Fertilizer application = 160 kg N:

$$\%\text{N in corn} = \frac{(93.6 - 2.6)\frac{\text{kg N}}{\text{ha}}}{160\frac{\text{kg N}}{\text{ha}}} \times 100 = 56.9\%$$

Fertilizer application = 200 kg N:

$$\%\text{N in corn} = \frac{(104 - 2.6)\,\frac{\text{kg N}}{\text{ha}}}{200\,\frac{\text{kg N}}{\text{ha}}} \times 100 = 50.7\%$$

As more fertilizer is added, an increasing fraction of the nitrogen is left over in the field. For 100 kg N ha^{-1} application, the residual nitrogen is:

$$[100\ \text{kg N} \times (1 - 0.689)] = 31.1\ \text{kg/ha}$$

For 160 kg N ha^{-1}, the residual nitrogen is:

$$[160\ \text{kg N} \times (1 - 0.569)] = 69.0\ \text{kg/ha}$$

For 200 kg N ha^{-1}, the residual nitrogen is:

$$[200\ \text{kg N} \times (1 - 0.507)] = 98.6\ \text{kg/ha}$$

Thus, the percent increase in yield between 100 kg N ha^{-1} and 200 kg ha^{-1} is about:

$$\frac{(8{,}000 - 5{,}500)\ \text{kg ha}^{-1}}{5{,}500\ \text{kg ha}^{-1}} \times 100\% = 45.5\%$$

but the percent increase in residual nitrogen is about:

$$\frac{(98.6 - 31.1)\ \text{kg ha}^{-1}}{31.1\ \text{kg ha}^{-1}} \times 100\% = 217\%$$

Thus, the application rate is reaching a point of diminishing returns. The residual nitrogen must be absorbed and processed by the environment. Much of it ends up in groundwater, or in nearby lakes and rivers, where it causes problems associated with eutrophication.

*QUESTION 9

Figure 16.3 shows the relationship between corn yield and nitrogen fertilizer application. With respect to corn grown on "Plano silt loam," estimate the yields for four cases: (1) no fertilizer applied, (2) 100 kg N applied, (3) 170 kg (current usage) applied, and (4) 200 kg applied. Given that corn is ~1.3% nitrogen by weight, calculate the percent of applied nitrogen contained in the harvested corn for the latter three cases. Are we reaching the point of diminishing returns with respect to nitrogen application with the Plano silt loam-like we are with Plainfield loamy sand?

16.3 Nutrition

We now turn our attention to the biochemical pathways through which the food we eat keeps us functioning: the chemistry of biological metabolism.

The major nutritional categories are carbohydrates, fats, and proteins. Carbohydrates are sugar molecules linked together in a long chain; fats are triglycerides of fatty acids, which have long hydrocarbon chains bonded to a glycerol unit; and proteins are composed of strings of amino acids joined by peptide bonds; each amino acid contains a characteristic side chain.

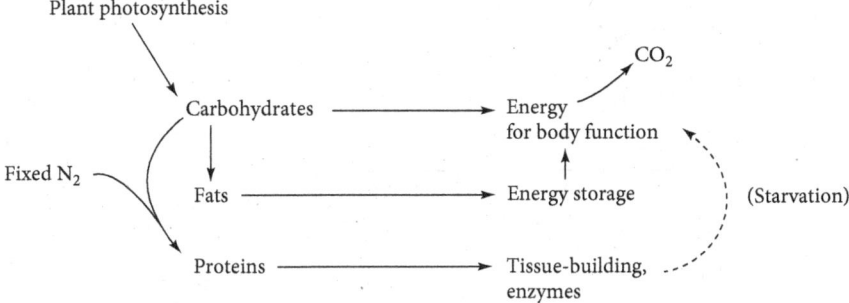

Figure 16.4 Flowchart of biochemical functions of carbohydrates, fats, and proteins.
Source: Adapted from Spiro et al., *Chemistry of the Environment, 3rd Ed.,* University Science Books, © 2012, all rights reserved.

16.3.1 *Energy and Calories*

A simplified flowchart for the main biochemical processes is shown in Figure 16.4. Carbohydrates are the immediate products of photosynthesis and also the immediate sources of biological energy in the process of respiration. Most of the energy that we need to keep our various bodily functions going is obtained from the oxidation of carbohydrates. Fats represent a form of biological energy storage. In comparison with carbohydrates, fats contain less oxygen and more carbon and hydrogen, whose oxidation is the source of our energy. The average energy content of fats is 9 calorie (cal) g^{-1}, compared with 4 cal g^{-1} for carbohydrates. Also, fats are immiscible with water, whereas carbohydrates are hydrophilic. Carbohydrates are usually found in association with a quantity of water that is approximately four times their weight. Consequently, the conversion of carbohydrates to fats represents a concentration of energy in lightweight portable form. A person who weighs 70 kg (154 lb) contains ~16% fat. This value represents enough energy for bodily requirements for 30 days. When we eat more food than we require for our energy needs, the excess calories are stored as fat. However, the biological oxidation of fats is slower than that of carbohydrates. When we need energy, we burn off our carbohydrate supply first and then call upon the fat stores.

16.3.2 *Proteins*

Aside from an energy supply, we need to maintain the biochemical machinery of the body itself, which is the realm of protein chemistry. Proteins make up most of the body's structural tissues and also the enzymes, the biological catalysts that carry out the thousands of reactions that are necessary for the maintenance of life. There are 20 different kinds of amino acids, each with a different chemical side chain (Fig. 16.5). Each kind of protein molecule is made up of a fixed sequence of these amino acids. Many of the amino acids can be synthesized by the body from a variety of starting materials as long as there is an adequate supply of protein nitrogen in the diet. There are, however, eight amino acids (Fig. 16.5) that the body cannot synthesize: valine, leucine, iso-leucine, threonine, lysine, methionine, phenylalanine, and tryptophan.

These essential amino acids, the ones that cannot be synthesized by the body, must be obtained directly from the diet and in amounts that allow us to maintain a proper balance of amino acids overall. For example, the ratio of tryptophan to lysine needs to be 0.6:2.6 = 0.23, because the average human protein contains 0.6 tryptophans and 2.6 lysines per 100 amino acids (Table 16.2). If this ratio is exceeded, the amount of protein that can be made by the body is limited by the lysine; the extra tryptophan is simply burned up. Whichever essential amino acid is present in the lowest amount relative to the frequency with which it is used is the one that limits the total amount of protein produced.

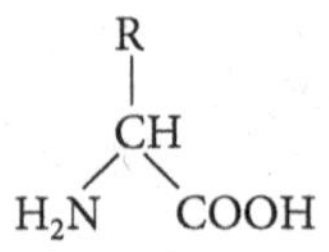

$$
\begin{array}{c}
R \\
| \\
CH \\
\diagup \quad \diagdown \\
H_2N \qquad COOH
\end{array}
$$

General formula

Amino Acid[a]	R (Side Chain)
1. Glycine	—H
2. Alanine	—CH_3
3. Serine	—CH_2OH
4. Aspartic acid	—CH_2COOH
5. Glutamic acid	—CH_2CH_2COOH
6. Asparagine	—CH_2CONH_2
7. Glutamine	—$CH_2CH_2CONH_2$
8. Arginine	—$CH_2CH_2CH_2NHC(NH)NH_2$
9. Cysteine	—CH_2SH
10. Tyrosine	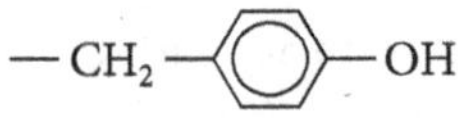 —CH_2—⟨◯⟩—OH
11. Proline	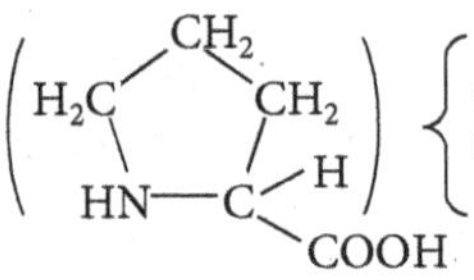 (ring structure) General formula including R group
12. Histidine	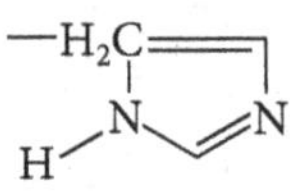 (ring structure)

Essential Amino Acids[b]	
13. Valine	—$CH(CH_3)_2$
14. Leucine	—$CH_2CH(CH_3)_2$
15. Isoleucine	—$CH(CH_3)CH_2CH_3$
16. Threonine	—$CH(OH)CH_3$
17. Lysine	—$CH_2CH_2CH_2CH_2NH_2$
18. Methionine	—CH_2CH_2—S—CH_3
19. Phenylalanine	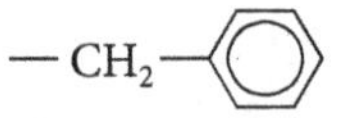 —CH_2—⟨◯⟩
20. Tryptophan	—CH_2—(indole ring)

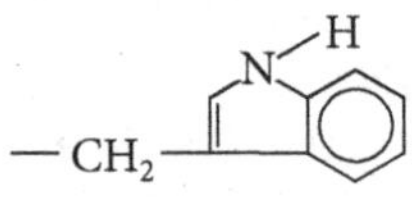

[a]Amino acids 1–12 can be synthesized by the human body if sufficient protein nitrogen is present in the diet.
[b]Amino acids 13–20 cannot be synthesized and must be obtained directly from food.

Figure 16.5 The amino acids and their chemical structures.
Source: Adapted from Spiro et al., *Chemistry of the Environment, 3rd Ed.,* University Science Books, © 2012, all rights reserved.

Different food sources contain different amounts of protein; but they also vary in the amino acid composition of their protein. As might be expected, the amino acid composition of animal protein is fairly close to human protein. For this reason, milk and meat provide a rather close approximation to the amino acid balance we need (Table 16.2). Plant protein, on the other hand, is farther from the human composition. The cereal plants, particularly wheat,

TABLE 16.2 Essential Amino Acid Content in Common Foods

Essential Amino Acid	Out of 100 Amino Acids, Number Present in . . .				
	Human Protein	Cow's Milk	Meat	Beans	Wheat
Tryptophan	0.6	0.5	0.5	0.4	0.6
Phenylalanine	3.1	3.8	2.7	2.8	2.8
Lysine	2.6	3.3	3.4	2.8	1.1
Threonine	2.0	2.0	2.0	1.6	1.3
Methionine	3.5	2.8	3.3	2.2	3.2
Leucine	4.0	4.2	3.3	3.1	3.1
Isoleucine	2.5	2.8	2.3	2.2	1.6
Valine	3.2	3.6	2.7	2.5	2.0

Sources: Data from Deathrage, *Food for Life*, Plenum Press, 1975; President's Science Advisory Committee, *The World Food Problem, Report of the Panel on World Food Supply, Vol. II*, Author, 1967. Adapted from Spiro et al., *Chemistry of the Environment*, 3rd Ed., University Science Books, © 2012, all rights reserved.

are quite deficient in lysine. The lysine frequency in wheat is less than one-half of the human average, which means more than twice as much wheat protein as milk protein is needed to supply human needs. Even though the remaining amino acids are closer to the correct proportions, they cannot be utilized without sufficient lysine. For this reason, wheat protein is often referred to as low-quality protein in comparison with the high-quality protein contained in meat, eggs, and milk.

Different plants show different patterns of deviation from the ideal amino acid balance. While wheat is deficient in lysine, beans have lysine in abundance. However, beans are deficient in methionine, which wheat has in abundance. In this respect, wheat and beans are complementary. If the two are mixed in equal amounts, the amino acid balance is greatly improved. It takes less total vegetable protein for the mixture than for either one alone to provide human requirements. A diet of one-half wheat protein and one-half bean protein provides a protein mixture that is only 10% less efficient than milk protein in providing the right balance. It is no accident that beans and rice or wheat products are traditionally eaten together in many parts of the world. In order to be effective complements, the different proteins must be mixed in the same meal, so that they are digested together. Thus, vegetarians must pay attention to balancing different protein sources.

Although meat is an important component in the diets of relatively prosperous nations, most people in the world cannot afford to buy it, yet it is possible to do without meat altogether and still remain healthy. Moreover, although meat is a high-quality source of protein, it represents a considerable waste of biological energy because animals typically store a rather small fraction of the food they consume in the form of meat. It takes 10 g of plant protein from the feed and hay that cows eat to produce 1 g of beef protein that humans consume. More serious is the large consumption of water associated with meat production, because of the irrigation requirements of animal feed (see Chapter 11). The energy use of producing meat versus plant protein needs to be taken into account as well. Most countries do not have enough plant protein to spare for meat production on a large scale. If there is sufficient grass to support grazing animals, they can make a net contribution to the human food supply. However, in meat-producing countries, it is common to feed animals high-quality plant food (e.g., cereals and soybeans) to help them grow more quickly.

For problems Q10 and Q11 refer to the following table:

Grams of Corn and Bean Protein Required to Give Daily
Requirements of Essential Amino Acids

Essential Amino Acids	Corn	Beans
Tryptophan	70.0	47.9
Lysine	72.2	33.0
Threonine	37.5	44.3
Leucine	27.4	44.8
Isoleucine	49.0	39.3
Valine	38.8	44.9
Phenylalanine	35.0	39.3
Methionine	33.3	56.7

QUESTION 10

Given that corn consists of 7.8% protein and only 60% of it is absorbed by the digestive tract, calculate the number of grams of corn that must be ingested for an adequate protein intake. There are 368 kcal in 100g of corn. How many calories must be consumed daily in order to get adequate protein? Given that the average American consumes 2,000–3,000 kcal daily, can an all-corn diet supply the proper balance of protein and energy?

Q10 ANSWER On an all-corn diet, the human digestive tract must absorb 72.2g of corn protein day^{-1} (see table corresponding to this problem) to satisfy total protein requirements. This would ensure an adequate intake of lysine, which requires the highest corn intake among all amino acids, even though some of the other essential amino acids could be satisfied by ingestion of far less corn. Actual corn consumption also has to allow for the fact that only 60% of the corn is absorbed by the digestive tract.

Thus, the daily requirement for corn protein is:

$$\frac{72.2 \text{ g}}{0.60 \text{ absorbed}} = 120 \text{ g corn protein}$$

Since corn protein makes up only 7.8% by weight:

$$\text{corn consumed} = \frac{120 \text{ g corn protein}}{0.078} = 1{,}538 \text{ g corn}$$

Daily consumption of calories contained in 1,538 g of corn is:

$$\text{kilocalories} = 1{,}538 \text{ g corn} \times \frac{368 \text{ kcal}}{100 \text{ g corn}} = 5{,}660 \text{ kcal}$$

An all-corn diet would be too rich in calories. The excess not needed by the body would be stored as fat. If calorie consumption were reduced, an inadequate supply of protein would result.

*QUESTION 11

Given that beans are 24% protein and 78% of the protein is absorbed by the digestive tract, calculate the number of grams of corn and beans that must be consumed to supply adequate protein for a person whose diet is half corn and half beans. If there are 338 kcal in 100g of beans, calculate the number of calories consumed in this diet (see Q10 for additional data for corn).

16.4 Conclusions

In comparison to preindustrial times, the amount of added nitrogen from anthropogenic sources is impacting the earth's environment. More of the nitrogen is concentrated in the land and being lost from the ocean. Nitrogen inputs are important for growing crops to feed our increasing population, but overfertilization can lead to eutrophication of waterways and other environmental impacts. People need a balance of different types of food to create the proteins that make up our body and provide energy too, and this requires nitrogen inputs to feed our growing population.

17

PEST CONTROL

17.1 Insecticides and Resistance

Farmers struggle constantly against insects that thrive on food crops. These insects consume an estimated 30% of crops worldwide. Farmers employ many insecticides in the struggle. Insecticides also play an important role in combatting insect-borne diseases (like malaria and bubonic plague). Most households keep pesticides for control of fleas, ants, and so on.

However, the insects put up a counter struggle by developing resistance, thereby reducing the insecticides' effectiveness. Through mutations and genetic variations, some of the target insects block an insecticide via one biochemical mechanism or another; perhaps they make an enzyme to break down the insecticide or they make a transport protein that carries the insecticide out of their cells. Whatever the mechanism, the resistant insects survive while the rest of the insects die. Subsequently, the resistant insects reproduce and rebuild the population, which is then impervious to the insecticide. When a new insecticide is introduced, it may be effective for a while, until resistance builds up again.

QUESTION 1

Suppose an insecticide is applied to a crop to eradicate fruit flies. Assume that one out of one million fruit flies possesses an enzyme that breaks down the insecticide into nontoxic metabolic products. Assume further that all the normal fruit flies die off quickly, and the population of the resistant flies increases geometrically (i.e., 1, 2, 4, 8, ...). If a new generation occurs every 23.5 days, in how many days will a new, resistant fruit fly population be restored?

Q1 ANSWER If the population doubles at each generation, n, then the number of generations required to reach a million insects is:

$$2^n = 10^6$$

To solve for n, we can take the logarithm of each side:

$$\log (2^n) = \log (10^6)$$

or:

$$n \log 2 = 6$$

---

so:

$$n = \frac{6}{\log 2} = \frac{6}{0.30} = 20$$

Since a new generation occurs in 23.5 days, it would take:

$$20 \times 23.5 \text{ days} = 470 \text{ days}$$

a little over a year, to produce a million insects. So, the pesticide would be ineffective for the next planting season.

*QUESTION 2

How long would the insecticide remain effective if a new generation occurs in 60 days?

*QUESTION 3

Resistance can be delayed by limiting the pesticide dose so that the insect population is sufficiently reduced to limit crop damage while allowing the nonresistant population to outgrow the resistant insects after the crop is out of danger. Suppose that the population is reduced by 99%. How long would it take the population to recover, if a new generation occurs every 23.5 days? What would the fraction of resistant insects be at that time?

17.2   Persistent Insecticides: Organochlorines

Some insecticides are slow to break down in the environment, thereby retaining their effectiveness for a long time. These are generally low-volatility organic molecules with low solubility in water. Many also have multiple chlorine atoms, making them resistant to attack by microbes, which have few enzymes capable of attacking C–Cl bonds. The first and most famous (and notorious) of these is DDT, p-dichlorodiphenyltrichloroethane (Fig. 17.1).

Introduced by the Allies during World War II to control typhus and malaria outbreaks, DDT has since saved millions of additional lives through disease-vector control. Its discov-

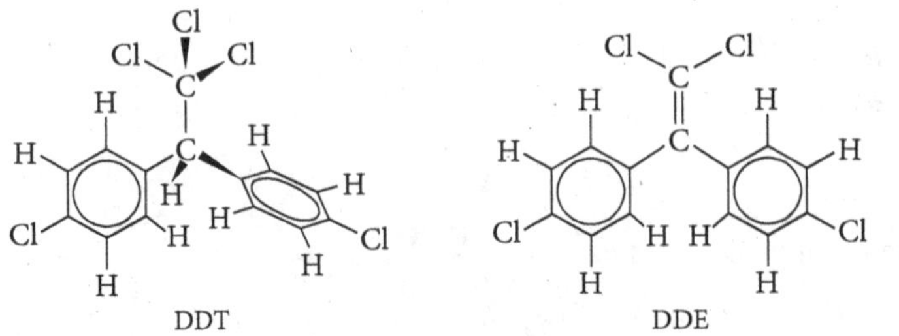

Figure 17.1 Structures of DDT and its dehydrochlorination product, DDE.
Source: Adapted from Spiro et al., *Chemistry of the Environment, 3rd Ed.*, University Science Books, © 2012, all rights reserved.

---

* Answers to starred questions can be found at the end of the book.

erer, the Swiss chemist Paul Muller, won the Nobel Prize for Medicine and Physiology in 1948. After the war, DDT was the first pesticide to come into widespread agricultural use. DDT readily penetrates the waxy outer coating of insects, but its toxicity to animals, including humans, is low, because they absorb much less DDT in their tissues.

However, DDT accumulates in aquatic and terrestrial food chains (see Section 17.4), and can disrupt ecosystems. In addition, DDT and its breakdown product DDE (*p*-dichloro-diphenyltrichloroethylene, Fig. 17.1) cause thinning and breakage of bird eggshells. A dramatic decline in wild bird populations, particularly of bald eagles and the peregrine falcons, led Rachel Carson to publish *Silent Spring* in 1962, launching the modern environmental movement. DDT was banned in the United States and other industrial countries in the 1970s. By then, however, DDT's use had been declining for over a decade because insect resistance had limited its usefulness. Many insects had evolved to produce high levels of an enzyme, *DDTase*, which catalyzes the removal of HCl from DDT, forming DDE.

Other organochlorine compounds were found to be insecticidal and replaced DDT in agriculture, but they too rapidly encountered insect resistance. Most were eventually banned as well, because of their persistence and harmful ecological effects, and because of concerns about potential health effects, especially as possible hormone mimics (see Chapter 18). They are still in widespread use, often illegally, in developing countries.

17.3 Molecular Shape and Biological Activity: DDT and DDE

DDT kills insects by binding to the protein that forms sodium channels in the nerves, holding the channels open. The nerves fire uncontrollably, leading to spasms and death. DDE is ineffective and harmless to insects.

QUESTION 4

DDE looks similar to DDT (Fig. 17.1). Why is it ineffective?

Q4 ANSWER The three-dimensional (3D) shapes of the molecules are actually quite different. In DDE, the ethylene substituents (Cl atoms and chlorophenyl rings) are all coplanar because ethylene has a double bond (the C atoms have sp^2 hybrid orbitals), but the ethane C atoms of DDT are tetrahedral (sp^3 hybrid orbitals) and the substituents are directed above and below the plane of the paper. (Remember that ethylene is a C=C double bond and ethane is a C–C single bond.) To bind properly, the insecticide must fit the 3D shape of the binding site on the protein. The shape of DDE is different enough that it does not bind, or binds too weakly, to have much effect at the concentrations encountered by the insects. The molecular shape is a central aspect of all kinds of biological activity, including the design of drugs for disease therapies.

---

*QUESTION 5

Another DDT breakdown product, DDD, retains the 3D shape, but one of the ethane Cl substituents is replaced by H. Pose a reason it may be ineffective.

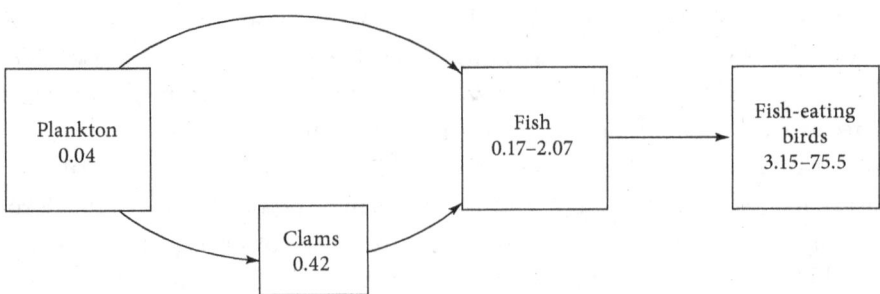

Figure 17.2 Accumulation of DDT in a particular aquatic food chain (in units of ppm, parts per million). *Source:* Adapted from Spiro et al., *Chemistry of the Environment, 3rd Ed.,* University Science Books, © 2012, all rights reserved.

17.4 Bioaccumulation

Being persistent, organochlorine pesticides are taken up by organisms. They build up in the food chain as larger organisms eat the smaller ones. They accumulate in fat tissue and are only slowly broken down and excreted. Figure 17.2 shows how DDT builds up in the aquatic food chain, leading to high levels in fish and birds that eat them.

QUESTION 6

Figure 17.2 shows the definite numbers for plankton, at the bottom of the food chain, and for the clams that eat them. Why are wide ranges given for fish and birds?

Q6 ANSWER There are many kinds of fish, some of which feed on the plankton, some on the clams, some on smaller fish, and some on bigger fish. The biggest fish would generally have the highest levels. Similarly, the birds come in different sizes and eat one another as well as the fish. The birds at the top of the food chain, hawks, falcons, and eagles accumulate the most DDT.

---

*QUESTION 7

What might a terrestrial food chain, analogous to Figure 17.2, look like?

Organochlorine molecules deposit in fat tissues because they have low solubility in water (they are hydrophobic), but dissolve readily in oils. Fats, and also lipids, are esters of fatty acids, which have long nonpolar hydrocarbon chains. They are oily and also have polar groups. It has been found that the solubilities of molecules in the solvent 1-octanol $[CH_3(CH_2)_6CH_2OH]$ are close to their solubilities in fats. The relative solubilities in fats versus water is conveniently approximated by the octanol/water partition coefficient $K_{ow} = [S]_o/[S]_w$, where $[S]_o/[S]_w$ are the concentrations of the molecule S when it is equilibrated between the two liquids. Table 17.1 lists the log K_{ow} values for a number of insecticides and herbicides.

TABLE 17.1 Solubilities of Selected Pesticides

	Solubility in Water (mg L^{-1})	Octanol–Water Partition Coefficient (log K_{ow})
Organochlorine Insecticides		
DDT	0.0028	6.0
Aldrin	0.08	5.8
Chlordane	0.20	5.1
Kepone	3.71	4.9
Mirex	0.05	6.4
Organophosphate Insecticides		
Parathion	19	3.7
Malathion	144	2.7
Carbamate Insecticides		
Carbaryl	73	2.4
Aldicarb	6,017	1.2
Herbicides		
Atrazine	38	2.6
Metolachlor	519	3.0

Sources: Taken from average values reported in Mackay et al., *Physical-Chemical Properties and Environmental Fate Handbook*, Chapman & Hall/CRCnetBASE, 2000. Adapted from Spiro et al., *Chemistry of the Environment, 3rd Ed.*, University Science Books, © 2012, all rights reserved.

QUESTION 8

If a 100 mL sample of river water contains 10 mg of a chemical that is 10 times more soluble in octanol than in water, how much would be extracted by shaking the sample with 10 mL of octanol?

Q8 ANSWER A 10 times greater solubility implies that:

$$K_{ow} = 10 = \frac{[S]_o}{[S]_w}$$

The water concentration, $[S]_w$, is:

$$[S]_w = \frac{10 \text{ mg}}{100 \text{ mL}} = 0.1 \frac{\text{mg}}{\text{mL}}$$

so:

$$[S]_o = 10 \times 0.1 \frac{\text{mg}}{\text{mL}} = 1 \frac{\text{mg}}{\text{mL}}$$

10 mL of octanol would contain:

$$10 \text{ mL} \times 1 \frac{\text{mg}}{\text{mL}} = 10 \text{ mg}$$

10 mg of the chemical.

*QUESTION 9

Calculate the amount of the herbicide atrazine (K_{ow} in Table 17.1) in an agricultural pond (from runoff from the fields) with a million liters of water, if 10 mL of octanol extracts 1.5 mg of atrazine from 100 mL of the pond water.

*QUESTION 10

From Table 17.1, predict which organochlorine pesticide, DDT or Kepone, would accumulate to a greater extent in food chains.

QUESTION 11

Consider the food chain in Figure 17.2. If fats and lipids make up 10% of the plankton mass, and if they are in equilibrium with the surrounding water with respect to the DDT solubility, predict the DDT concentration in the water.

Q11 ANSWER Because the plankton contain 0.04 ppm DDT, the concentration in their fat tissue (10% or 0.1 of their total tissue) would be:

$$\frac{0.04 \text{ ppm}}{0.1} = 0.4 \text{ ppm}$$

Using:

$$\log K_{ow} = 6.0 = \log \frac{[\text{DDT}]_{\text{fat}}}{[\text{DDT}]_{\text{water}}}$$

The concentration of DDT in the water is predicted to be:

$$[\text{DDT}]_{\text{water}} = \frac{[\text{DDT}]_{\text{fat}}}{K_{ow}} = \frac{0.4 \text{ ppm}}{10^6} = 0.4 \times 10^{-6} \text{ ppm}$$

The water solubility is 0.0028 mg L^{-1}, or 2.8×10^{-3} ppm (the same units as mg L^{-1}, since mg = 10^{-3} g and 1L of water weighs 10^3 g, so mg L^{-1} = 10^{-3} g/10^3 g = 10^{-6}, i.e., 1 ppm).

This is much larger than the equilibrium value calculated; all the DDT is expected to leach out into the water, so the leaching rate must be slow. The 0.04 ppm content of DDT in algae means that the algae took in DDT from some other source than water (likely from DDT in the air or adsorbed on dust particles). Assuming the measurement represents a steady state (i.e., the value is not changing from day to day), then the rate of uptake must be the same as the water leaching rate.

QUESTION 12

In view of the water solubility, how can one account for the 10-fold higher DDT concentration in clams than algae?

Q12 ANSWER Clearly, the clams are not in equilibrium with the water either. Again, assuming a steady state, the leach rate must be the same as the rate of uptake from eating algae.

QUESTION 13

Suppose the clams have 10% of fat in their tissues, like plankton, and that they eat their weight in plankton every day. What can you say about the rate at which clams lose DDT? If they were transferred to DDT-free water, how long would it take to have clean clams?

Q13 ANSWER Assuming a steady state, the clams would take in 0.04 ppm DDT each day from the plankton and would have to lose the same amount. Since the clams have 0.42 ppm DDT, on average, the excretion rate is:

$$\frac{0.04 \text{ ppm}}{0.42 \text{ ppm}} \times 100\% = 10\%$$

If they were transferred to clean water, they would lose 10% of their DDT per day.

*QUESTION 14

If plankton in the Figure 17.2 food chain had 0.04 ppm Kepone instead of DDT, what is the expected concentration of Kepone in the water?

QUESTION 15

What would you expect the Kepone concentration in clams to be in the same food chain?

Q15 ANSWER If the leach rate were the same for Kepone as for DDT, the level in clams would also be 0.42 ppm. But because K_{ow} is more than 10 times larger for Kepone than for DDT (Table 17.1), the leach rate is also likely to be greater. If the leach rate were proportional to K_{ow}, the expected clam concentration would be about the same as for the plankton, 0.04, and there would be little buildup of Kepone in the food chain.

As a result of the food chain accumulation of DDT and other persistent organic molecules, fish have been banned from human consumption in a number of lakes. Elsewhere, the rate at which fish can be safely consumed may be limited, in order to avoid reaching dangerous exposure levels. These levels are determined for a number of contaminants by environmental agencies (e.g., the U.S. Environmental Protection Agency) from a variety of toxicological data.

The cleansing effect of stopping DDT exposure can be seen in the steady decline in human tissue levels, illustrated in Figure 17.3 by data on breast milk in Canadian women in the years after DDT was banned.

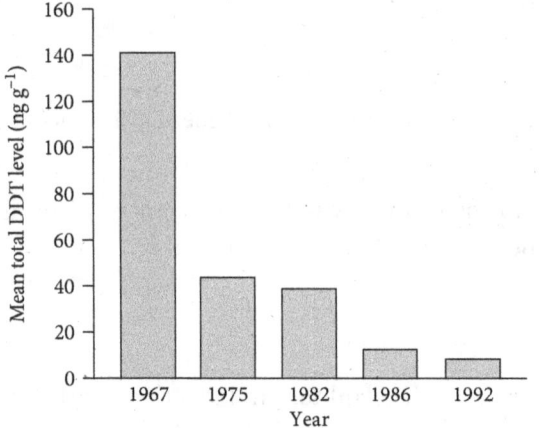

Figure 17.3 DDT levels in the breast milk of Canadian women.
Sources: Adapted from *The State of Canada's Environment*, 1996, Canada—Government of Canada, and Spiro et al., *Chemistry of the Environment, 3rd Ed.*, University Science Books, © 2012, all rights reserved.

17.5 Acetylcholinesterase Inhibitors: Organophosphates and Carbamates

Insecticides have been developed that are less persistent because they are more water-soluble (Table 17.1) and break down rapidly into harmless products, once released into the environment. Because they do not last long, they must be highly potent to be practical. The two classes of widely used less-persistent insecticides, organophosphates and carbamates (Fig. 17.4), are powerful neurotoxins. Indeed, the organophosphate insecticides are in the same chemical family as nerve gas chemical warfare agents.

The organophosphates and carbamates both work by inhibiting the enzyme *cholinesterase*, which breaks down *acetylcholine*, the neurotransmitter responsible for firing motor nerve cells. The cells continue to fire until acetylcholine is gone. If the cholinesterase is inhibited, nerve firing continues uncontrollably, leading to paralysis and death.

Cholinesterase works by binding acetylcholine at its active site (Fig. 17.5), where the OH group of a serine amino acid (one of the enzymes building blocks) is poised to attack the acetyl group, forming an acetyl enzyme *intermediate* and releasing choline. A water molecule then attacks the intermediate, forming acetic acid and releasing the enzyme for another round of acetylcholine destruction.

The insecticides inhibit the enzyme by mimicking acetylcholine. They bind to the active site and induce the serine to carry out a displacement reaction, just as acetylcholine does. Instead of an acetyl group, the enzyme ends up with a bound *phosphoryl* group in the case of an organophosphate insecticide or a bound *carbamyl* group in the case of a carbamate insecticide (Fig. 17.5). These groups are much less susceptible to attack by water than is the acetyl group. In both cases, the atom being attacked, carbon in the case of carbamyl and phosphorus in the case of phosphoryl, is attached to additional electronegative groups that donate electron density and diminish the reactivity of the C or P atom with the incoming water molecule. Consequently, the enzyme is blocked by the carbamyl or phosphoryl group from binding acetylcholine.

The potency of the inhibitor depends on the rate of the first reaction, in which the enzyme's active site serine is captured. The better the *leaving group* (X for the organophosphates, OR for the carbamates, Fig. 17.5), the faster the reaction. For example, fluoride is an excellent leaving group because the fluoride ion is quite stable in water (as opposed to methoxide, which is a strong base and is harder to displace). Organophosphate molecules with fluoride substituents are extremely powerful cholinesterase inhibitors, and the nerve gases are molecules of this class (an example is sarin, methylisopropoxy fluorophosphate).

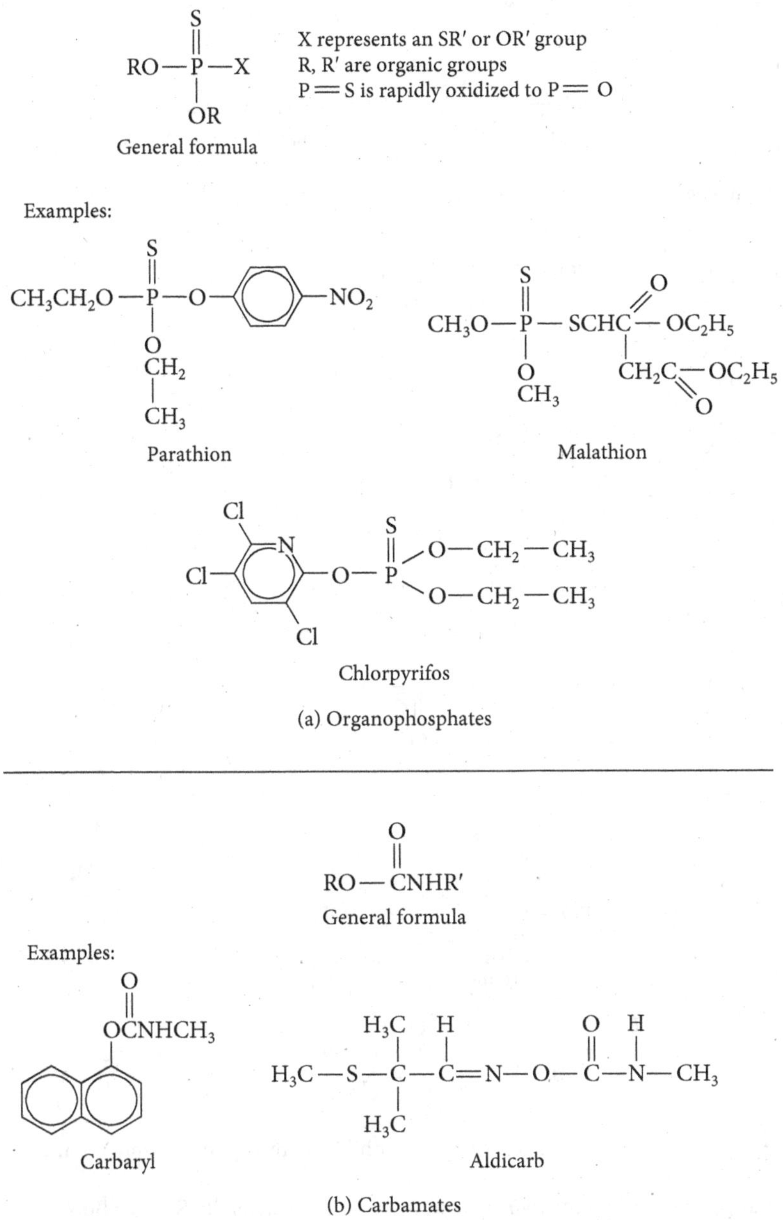

$$RO-\underset{\underset{OR}{|}}{\overset{\overset{S}{||}}{P}}-X$$

General formula

X represents an SR′ or OR′ group
R, R′ are organic groups
P = S is rapidly oxidized to P = O

Examples:

$$CH_3CH_2O-\underset{\underset{CH_3}{\underset{|}{CH_2}}}{\overset{\overset{S}{||}}{P}}-O-\bigcirc-NO_2$$

Parathion

$$CH_3O-\underset{\underset{CH_3}{\underset{|}{O}}}{\overset{\overset{S}{||}}{P}}-SCHC-OC_2H_5$$
$$CH_2C-OC_2H_5$$

Malathion

Chlorpyrifos

(a) Organophosphates

$$RO-\overset{\overset{O}{||}}{C}NHR'$$

General formula

Examples:

$$\overset{\overset{O}{||}}{OCNHCH_3}$$

Carbaryl

$$H_3C-S-\underset{\underset{H_3C}{\underset{|}{C}}}{\overset{\overset{CH_3}{\overset{|}{}}}{C}}-\overset{\overset{H}{|}}{C}=N-O-\overset{\overset{O}{||}}{C}-\overset{\overset{H}{|}}{N}-CH_3$$

Aldicarb

(b) Carbamates

Figure 17.4 Chemical structures of (a) organophosphate and (b) carbamate insecticides.
Source: Adapted from Spiro et al., *Chemistry of the Environment, 3rd Ed.,* University Science Books,
© 2012, all rights reserved.

In order to use organophosphates as insecticides without massively poisoning people and other animals, the reactivity of the organophosphate must be toned down. One way is to make sure the P substituents are poor leaving groups, like the RO⁻ groups in parathion and chlorpyrifos (Fig. 17.4), which require protonation (there are few protons in water at neutral pH), unlike F⁻. Another is to replace the P=O group with a P=S group (producing a *phosphorothioate*). The S atom significantly deactivates the P atom toward attack, slowing the rate of reaction with cholinesterase. However, once inside the insect, the S atom is rapidly removed by oxidative enzymes, and the molecule is converted back to a more potent organophosphate neurotoxin. Animals also have oxidative enzymes but at levels much lower than insects. Thus, the

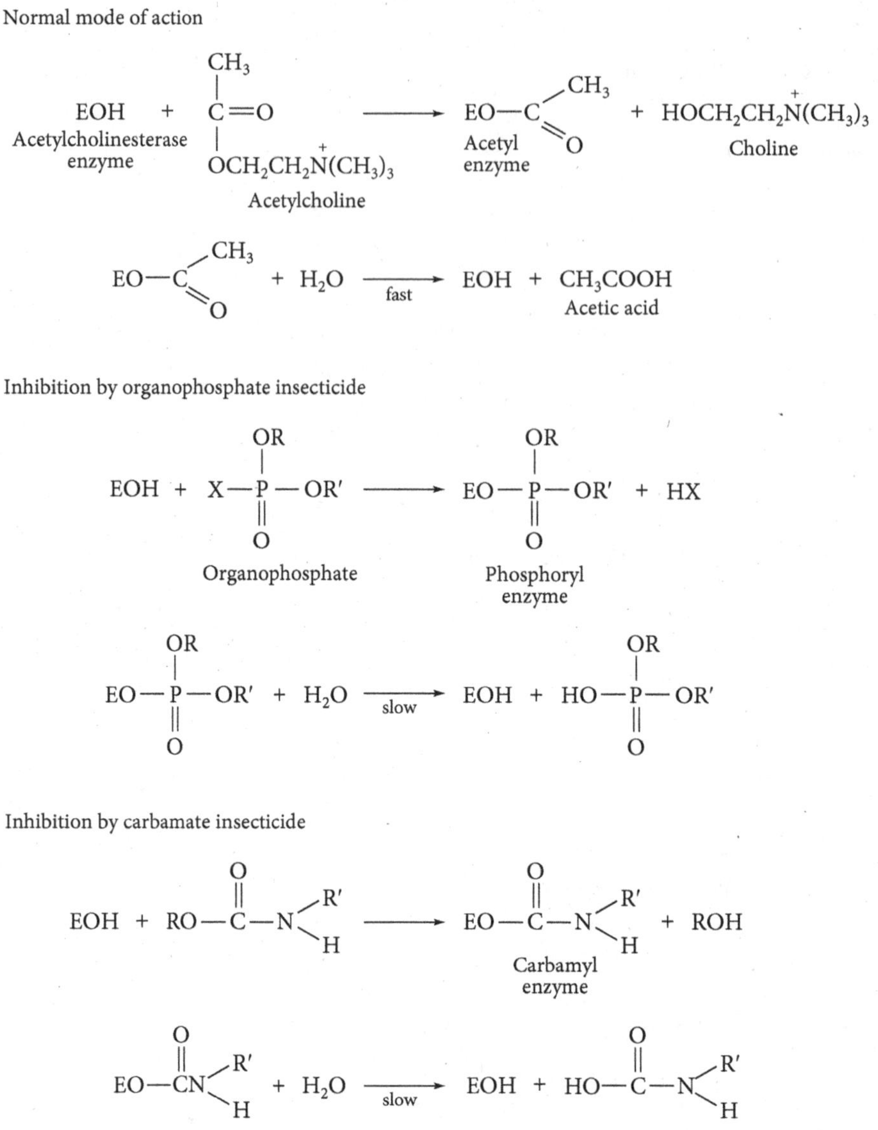

Normal mode of action

Inhibition by organophosphate insecticide

Inhibition by carbamate insecticide

Figure 17.5 Equations for the mechanism of cholinesterase and its inhibition by organophosphates and carbamates.
Source: Adapted from Spiro et al., *Chemistry of the Environment, 3rd Edition,* University Science Books, © 2012, all rights reserved.

phosphorothioate turns the insect's biochemistry on itself while decreasing toxicity toward other species. All the common organophosphate insecticides have P=S bonds (Fig. 17.4).

QUESTION 16

Why does changing O to S deactivate the organophosphate group?

Q16 ANSWER S, being less electronegative than O, draws less electron density away from the P atom, where the lone pair electrons of the serine OH group have to attack. The electrostatic interaction between the P and the incoming O is less favorable.

*QUESTION 17

Consider a para-substituted phenyl diethylphosphate molecule:

$$X\text{-phenyl-OP(OC}_2\text{H}_5)_2$$
$$\overset{\displaystyle O}{\overset{\displaystyle \|}{}}$$

Predict the trend in neurotoxicity as the X substituent becomes more electron withdrawing.

Although the molecular design has reduced their toxicity, the organophosphate and carbamate insecticides are generally much more toxic than organochlorine insecticides, and their use is being curtailed because of health concerns. There is currently considerable controversy about the widely used organophosphate chlorpyrifos (Fig. 17.4). The U.S. EPA banned it for household use because of evidence of neurotoxicity in children and it is being phased out of production by the parent company that produces the chemical since it was banned for agricultural use in the state of California.

Farmers do have alternatives to broad-scale insecticide spraying, although they may require more labor and careful management. They can rotate crops, interweave different crops to help hold insects at bay, and employ biological controls, including releasing insect predators and using traps baited with insect hormones. These are some of the methods used on organic farms, and increasingly on conventional farms as well.

17.6 Herbicides

Weeds also limit crop yields by competing for nutrients and crowding out the crop plants. The intensive labor required to weed crops has been increasingly displaced by the use of chemical herbicides. In addition, herbicides have enabled "no-till" agriculture, which minimizes erosion by reducing disturbance of the soil. Crops can be planted without plowing by injecting seeds directly into the ground after weeds have been controlled by herbicides. Herbicides that have been widely used are shown in Figure 17.6.

The chlorophenoxy compounds 2,4-D and 2,4,5-T have been used, respectively, on lawns and for clearing bush. However, 2,4,5-T has been banned because of contamination during the manufacture with toxic dioxin molecules. Pentachlorophenol is used as a wood preservative. Paraquat has been used extensively in marijuana eradication campaigns and to combat louse infestations, but is quite toxic, and has poisoned people, especially in developing countries. Until recently, the most widely used crop herbicide has been atrazine, a triazine molecule. Its toxicity is low, but there have been concerns about apparent correlations of high well-water concentrations in agricultural areas with cancer and birth defects. In addition, atrazine has been implicated as a possible factor in the decline of amphibian populations. In some areas, atrazine was replaced by metolachlor, which degrades more rapidly in the field.

QUESTION 18

Atrazine is a persistent molecule, breaking down slowly in the environment. Looking at its structure would you also expect it to bioaccumulate?

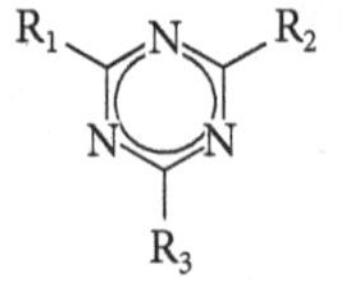

General formula for triazines

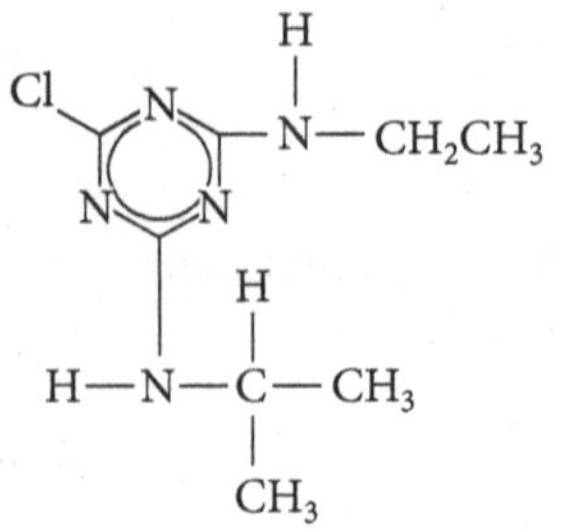

Atrazine

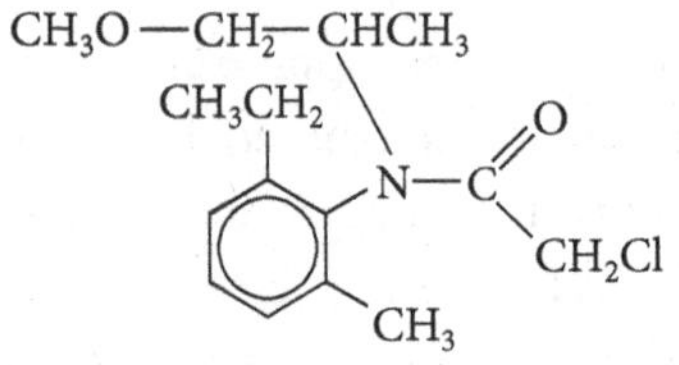

Metolachlor

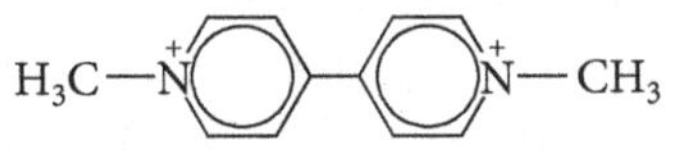

Paraquat

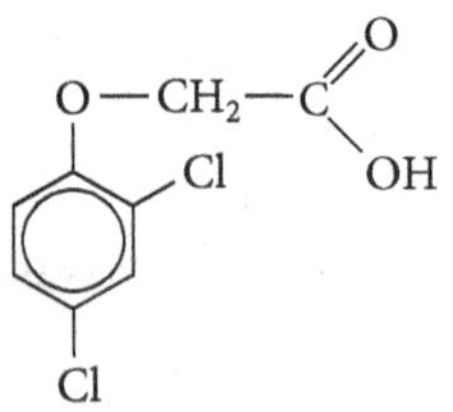

2,4-D
(2,4-Dichlorophenoxyacetic acid)

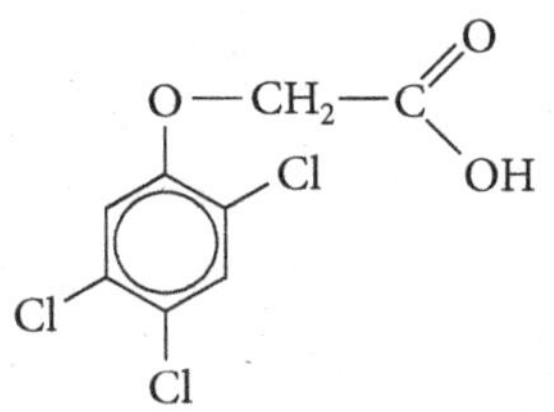

2,4,5-T
(2,4,5-Trichlorophenoxyacetic acid)

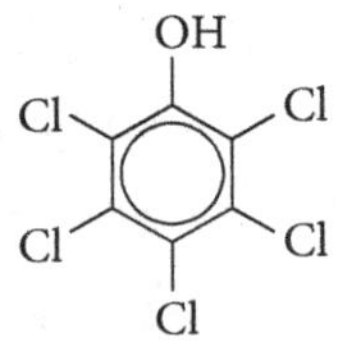

PCP
(Pentachlorophenol)

Figure 17.6 Structures of commonly used herbicides.
Source: Adapted from Spiro et al., *Chemistry of the Environment, 3rd Ed.,* University Science Books, © 2012, all rights reserved.

Q18 ANSWER Atrazine has many N atoms, whose lone pairs attract water, making it quite water-soluble (Table 17.1). Consequently, it is excreted rapidly and is not stored in fat tissues, so it does not bioaccumulate.

In recent years, however, atrazine has been largely displaced by glyphosate (trade name Roundup), a simple amino acid derivative:

$$[N\text{-(phosphonomethyl)-glycine, } {}^{-}HO_3PCH_2N^{+}H_2CH_2COOH]$$

It is water-soluble and is readily metabolized by soil microorganisms. It has low toxicity, but there is some evidence pointing to a correlation with non-Hodgkin lymphoma. It has been classed as a probable carcinogen by the International Agency for Research on Cancer, but a large agricultural health study supported by the U.S. National Cancer Institute found no significant association between glyphosate exposure and cancer. The issue remains unsettled.

In plants, glyphosate binds and inhibits an enzyme that is critical for the biosynthesis of aromatic amino acids, essential constituents of protein. All plants share this biosynthetic pathway, but animals do not. Consequently, glyphosate is a very broad-spectrum herbicide.

It is often used to clear brush and fight invasive plants, and also to prepare fields for no-till agriculture. Having both positive and negative charges, glyphosate adsorbs strongly to the surface of soil particles, which have complementary charges, and is not taken up by plants that are subsequently planted.

Glyphosate cannot be used to kill weeds among crops, since the crop plants would also be killed. However, glyphosate-resistant crops have been developed through genetic engineering, allowing glyphosate to keep them clear of weeds.

17.7 Genetically Modified Organisms

The enzyme inhibited by glyphosate is 5-enolpyruvylshikimate-3-phosphate (EPSP) synthase. The EPSP synthase catalyzes the joining of two metabolites, shikimate-3-phosphate (S3P) and phosphoenolpyruvate (PEP), to make EPSP, with the release of phosphate ion (P_i) (see Fig. 17.7). The EPSP product is subsequently converted to the aromatic amino acids required by other enzymes.

The EPSP protein provides a binding site that brings S3P and PEP into a reactive orientation, shown in Figure 17.8. Glyphosate, which resembles PEP, with the same negatively charged ends, can displace PEP from the binding site. However, it cannot react with S3P, since a $-NHCH_2-$ fragment intervenes at the reactive phosphate-C bond of PEP. The reaction is blocked and the enzyme remains inactive while glyphosate is bound.

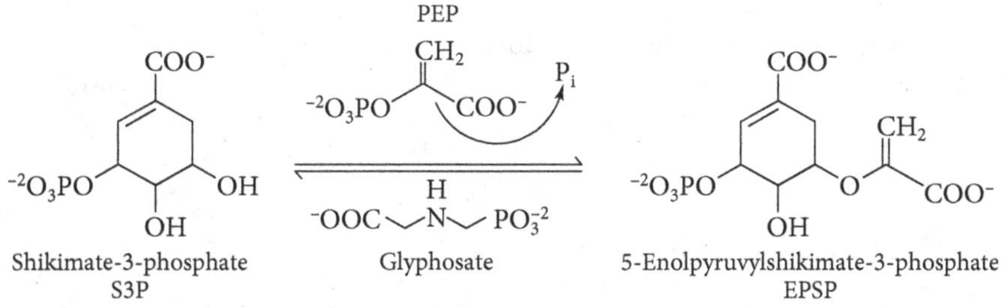

Figure 17.7 Catalytic transfer of the enolpyruvyl moiety from PEP to S3P, forming the products EPSP and phosphate (P_i). Glyphosate inhibits EPSP synthase.
Sources: Adapted from Schönbrunn et al., "Interaction of the Herbicide Glyphosate with its Target Enzyme 5-Enolpyruvylshikimate-3-Phosphate Synthase in Atomic Detail," *Proceedings of the National Academy of Sciences,* Vol. 98, 2001, pp. 1376–1380 and Spiro et al., *Chemistry of the Environment, 3rd Ed.,* University Science Books, © 2012, all rights reserved.

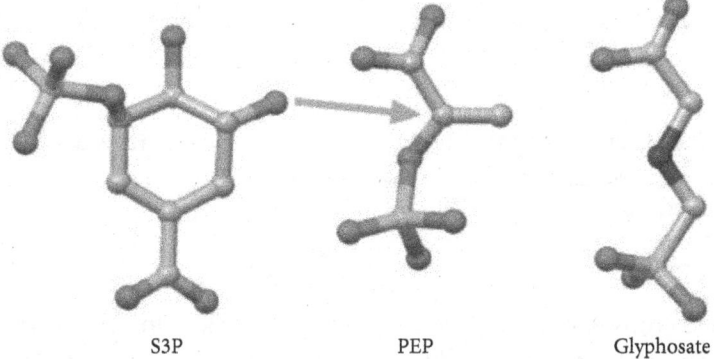

Figure 17.8 3D view of the reaction at the EPSP synthase catalytic site and of glyphosate (which is unreactive) bound in a PEP-like orientation.
Source: Adapted from Goodsell, *PDB-101 Molecule of the Month: EPSP Synthase and Weedkillers,* 2018, PDB: Protein Data Base, available at https://doi.org/10.2210/rcsb_pdb/mom_2018_2.

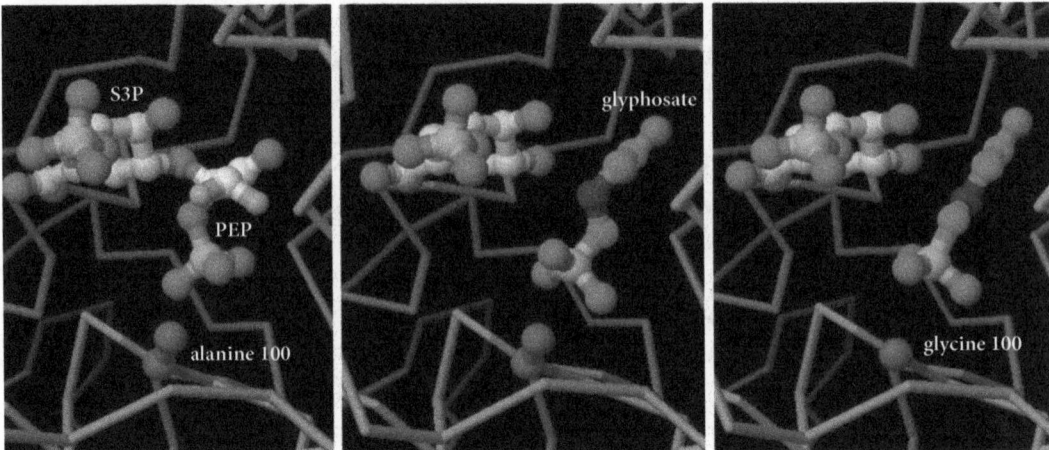

Figure 17.9 The active site of EPSP synthase from three crystal structures showing how S3P and PEP are bound in a reactive configuration (left panel), and how glyphosate is bound to the wild type enzyme (right panel) or to the engineered enzyme, with alanine replacing glycine at position 100 of the amino acid chain of the enzyme. The chain is folded into the 3D structure shown by the green tubing. Glycine has only an H atom at its central carbon atom (green sphere, right panel), while alanine has a methyl group at this position (second green sphere, central and left panels).
Source: Adapted from Goodsell, *PDB-101 Molecule of the Month: EPSP Synthase and Weedkillers*, 2018, PDB: Protein Data Base, available at https://doi.org/10.2210/rcsb_pdb/mom_2018_2.

EPSP synthase can be genetically engineered to reduce its affinity for glyphosate, but not for PEP. One way of doing this is illustrated by the enzyme crystal structures shown in Figure 17.9.

QUESTION 19

From the structures in Figure 17.9, explain how genetic engineering can reduce the affinity for glyphosate, but not for PEP.

Q19 ANSWER Glyphosate can bind in its fully extended form to the naturally occurring enzyme (right panel), but when alanine replaces glycine at position 100 through alteration of the plant's DNA, the extra methyl group gets in the way. Glyphosate is forced to distort in order to bind and its binding affinity is decreased. However, the methyl group does not interfere with the smaller PEP, so the modified enzyme remains fully functional.

*QUESTION 20

Continued use of glyphosate on engineered crops results in the development of glyphosate-resistant weeds, which then nullify the effectiveness of glyphosate. Suggest a mechanism for these weeds to resist glyphosate.

Several genetically modified crops have been developed that are resistant to other herbicides. In addition, genetic engineering has been used to develop plants that are insect resistant. In this case, the gene encoding for a natural insecticide, called Bt, produced by the soil bacterium *Bacillus thuringiensis*, is inserted into the plant DNA. Because the plants produce

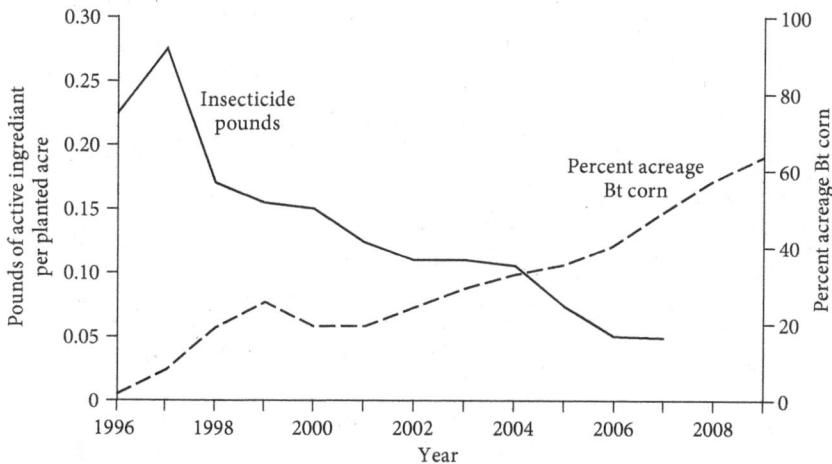

Figure 17.10 Pounds of insecticide used per acre planted and the percent acres of Bt corn over time, in the United States.
Sources: Adapted from National Research Council Committee on the Impact of Biotechnology on Farm-Level Economics and Sustainability, *Impact of Genetically Engineered Crops on Farm Sustainability in the United States*, 2010, National Academy of Sciences Press, and Spiro et al., *Chemistry of the Environment*, 3rd Ed., University Science Books, © 2012, all rights reserved.

their own insecticide, the need for insecticide spraying is reduced. Indeed, the use of insecticides on U.S. cornfields has diminished greatly since Bt corn was introduced (Fig. 17.10).

Of course, genetic engineering comes with costs. Resistant weeds and insects inevitably evolve, rendering the strategy ineffective. The remedy often adopted is to introduce a different herbicide-resistant or insecticide-resistant crop, and the cycle repeats. In the case of herbicide-resistant crops, the sprayed herbicide can drift on the wind and damage nearby nonresistant crops. And, of course, genetic alteration is unpopular with many consumers, at least partly because of its association with industrial agriculture. Nevertheless, genetically modified crops are popular with farmers, especially in developing countries with warm climates, where insects and weeds flourish.

17.8 Conclusions

Farmers constantly struggle against insects that invade their crops and weeds that choke them. The discovery of DDT, an insect neurotoxin, gave them a powerful weapon against pests, and farmers sprayed it widely on crops. But DDT, and its breakdown product DDE, had serious ecosystem effects, because they are fat-soluble and buildup in the food chain. Raptors, at the top of the food chain, were particularly affected, since their eggshells thinned and broke. Their numbers plummeted, leading to DDT being banned for agriculture. DDT and other organochlorine pesticides were replaced by acetylcholinesterase inhibitors, which are likewise neurotoxins, but are more water-soluble and do not persist in the environment. However, they are more toxic to humans and are hazardous, especially to agricultural workers. A variety of herbicides have also been widely used to control weeds. Glyphosate is particularly toxic to plants but came into widespread use with the development of genetically modified crops that resist glyphosate. Crops resistant to other herbicides have also been developed through genetic engineering. But resistance inevitably develops, to either pesticides or herbicides, through mutations and evolutionary pressure. Farmers are turning to more sustainable practices, requiring more careful and systematic management.

18

TOXICITY OF CHEMICALS

18.1 Acute and Chronic Toxicity

It is useful to distinguish between two types of effects: (1) an acute effect, in which there is a rapid and serious response to a high but short-lived dose of a chemical, and (2) a chronic effect, in which the dose is relatively low but prolonged, and a time lag occurs between the initial exposure and the full manifestation of the effect. Acute poisons interfere with essential physiological processes, leading to a variety of symptoms of distress and, if the interference is sufficiently severe, to death. Chronic exposure to chemicals can have more subtle effects, possibly setting in motion a chain of biochemical events that lead to disease states, including cancer.

Acute toxicity is relatively easy to gauge. At high enough levels, the effects of toxins on bodily function are obvious and fairly consistent across individuals and species. These levels vary enormously for different chemicals. Everything is toxic at some level, and the difference between toxicity and no observable effect is a matter of dose. The most widely used index of acute toxicity is LD_{50}, the lethal dose for 50% of a population. This number is obtained by graphing the number of deaths among a group of experimental animals, usually rats, at various levels of exposure to the chemical, and interpolating the resulting dose-response curve to the dose at which one-half of the animals die (Fig. 18.1). The dose is generally expressed as the weight of the chemical per kilogram of body weight, on the assumption that toxicity scales inversely with the size of the animal. Table 18.1 lists LD_{50} values for several substances, showing nine orders of magnitude variation between the most toxic (botulin toxin, the agent responsible for botulism) and the least toxic (sugar). Among insecticides, we see that DDT is lethal at a dose that is ~30 times higher than parathion, but 12 times lower than malathion, when measured in rats or mice. An example of chemical toxicity rating is given in Table 18.2.

QUESTION 1

Caffeine is thought to be safe under normal use. However, in large amounts, it is poisonous to animals (moderately toxic to rats according to Tables 18.1 and 18.2) and humans. Assuming that humans and rats are equally sensitive to the toxic effects of caffeine, how many cups of coffee drunk consecutively would be lethal to 50% of a group of 150-lb humans? Assume one cup of strong coffee contains ~140 mg of caffeine. Use the data presented in Table 18.1 for the LD_{50} value of caffeine for rats. Is it likely that a person could die from drinking too much coffee?

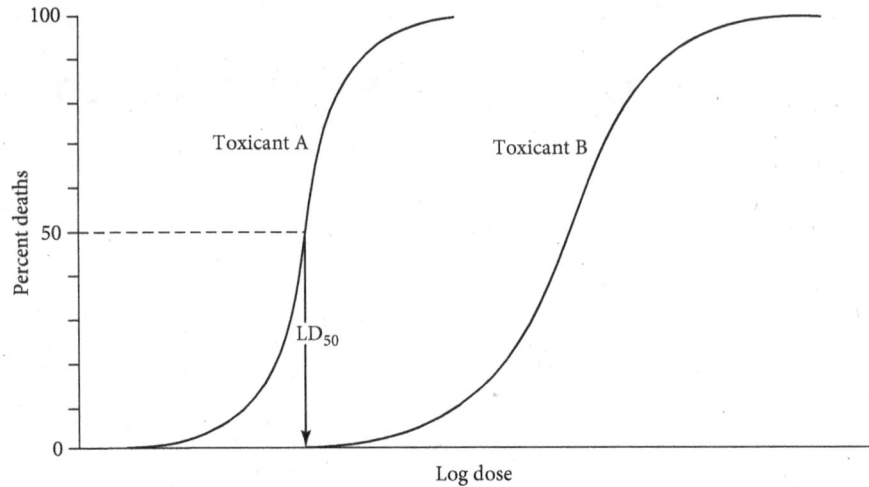

Figure 18.1 Illustration of dose-response curve in which the response is the death of the organism; the cumulative percentage of deaths of organisms is plotted on the *y*-axis.
Sources: Adapted from Manahan, *Environmental Chemistry, 8th Ed.,* 2005, Lewis Publishers, and Spiro et al., *Chemistry of the Environment, 3rd Ed.,* University Science Books, © 2012, all rights reserved.

TABLE 18.1 The LD_{50} Values of Selected Chemicals

Chemical	LD_{50} (mg kg^{-1})[a]
Sugar	29,700
Ethanol	14,000
Vinegar	3310
Sodium chloride	3000
Atrazine	1870
Malathion (insecticide)	1200
Aspirin	1000
Caffeine	130
DDT (insecticide)	100
Arsenic	48
Parathion (insecticide)	3.6
Strychnine	2
Nicotine	1
Aflatoxin-B	0.009
Dioxin (TCDD)	0.001
Botulin toxin	0.00001

[a]For rats or mice

Sources: Data from Buell & Gerard, *Chemistry in Environmental Perspective,* 1994, Prentice Hall; adapted from Spiro et al., *Chemistry of the Environment, 3rd Ed.,* University Science Books, © 2012, all rights reserved.

TABLE 18.2 Rating of Chemical Toxicity

Toxicity Rating	Commonly Used Term	LD$_{50}$ Single Oral Dose for Rats (g/kg)	LD$_{50}$ Skin for Rabbits (g/kg)	Probably Lethal Dose for Humans
1	Extremely toxic	≤0.001	≤0.005	Taste (1 grain)
2	Highly toxic	0.001–0.05	0.005–0.043	1 teaspoon (4 ml)
3	Moderately toxic	0.05–0.5	0.044–0.340	1 oz (30 g)
4	Slightly toxic	0.5–5.0	0.35–2.81	1 pint (250 g)
5	Practically non-toxic	5.0–15.0	2.82–22.6	1 quart (500 g)

Source: Adapted from Mumford & Carson, *Hazardous Chemicals Handbook, 2nd Ed.,* 2002, Butterworth-Heinemann.

Q1 ANSWER Caffeine has an LD$_{50}$ value in rats of 130 mg kg^{-1}. Assuming that humans are equally sensitive, a lethal dose to a 150-lb human can be calculated. That weight in kg is:

$$150 \text{ lb} \times \frac{1 \text{ kg}}{2.205 \text{ lb}} = 68.0 \text{ kg}$$

The amount of caffeine that would be toxic to 50% of the population is:

$$68.0 \text{ kg} \times 130 \frac{\text{mg}}{\text{kg}} = 8.84 \times 10^3 \text{ mg}$$

Assuming one cup of coffee contains 140 mg of caffeine, the amount of coffee drank consecutively that would be lethal to 50% of the population would be:

$$8.84 \times 10^3 \text{ mg} \times \frac{\text{cup}}{140 \text{ mg}} = 63.2 \text{ cups}$$

Thus, it is not likely that a human could die from drinking too much coffee.

*QUESTION 2

Although consuming a moderate amount of sugar is thought to be safe, large amounts can be poisonous to humans. Assume that rats and humans are equally sensitive to the toxic effects of sugar, consuming many regular-sized Hershey's candy bars would be lethal to 50% of a group of 175-lb humans? One regular-sized Hersey's chocolate bar has ~25.0 g of sugar. Use the data presented in Table 18.1 for the LD$_{50}$ value of sugar for rats. Is it likely that a person could die from eating too many Hershey's chocolate bars?

---

* Answers to starred questions can be found at the end of the book.

*QUESTION 3

Aflatoxin B, a potent toxin, is produced by a common mold, especially prevalent in hot climates, which often contaminates food crops. In Kenya, maize (corn) flour is a staple that is often contaminated with aflatoxin. The average daily consumption of maize flour is 400g per person. In one study, the aflatoxin contamination level was found to be 130 parts per billion (ppb). At that level, how much aflatoxin would the average Kenyan, weighing 65 kg, consume per day? How does this compare with the LD_{50} value in Table 18.1?

QUESTION 4

Consider the following LD_{50} value of a metal contained in two types of grains that animals eat: 2 mg kg^{-1} and 0.40 g kg^{-1}. Using the toxicity rating in Table 18.2, what toxicity rating is each value for rats (via eating) and for rabbits (via skin contact), respectively? What is the implication of this comparison?

Q4 ANSWER Using Table 18.2, a 2 mg kg^{-1} (0.002 g kg^{-1}) LD_{50} would be considered highly toxic for rats via oral uptake of this metal, but extremely toxic for rabbits via skin contact, respectively. A 0.40 g kg^{-1} LD_{50} would be considered moderately toxic for rats via oral intake, but slightly toxic for rabbits via skin contact, respectively. This implies that the toxicity rating is a relative concept, which depends on the intake pathway and type of organism that intakes the toxin. Also, what is less toxic for one biological species may be much more toxic for another.

Chronic effects are much more difficult to evaluate, especially at the low exposure levels that are likely to be encountered in the environment. In an experimental setting, the lower the dose, the fewer the animals that show any particular effect. To obtain statistically significant results, a study might have to include a prohibitively large number of animals if they are tested at low doses. The only available recourse is to evaluate the effects of a series of high doses, and then to extrapolate the dose–response curve to the expected incidence at low doses. But extrapolation to levels that people are exposed to may have to extend over several orders of magnitude, and there is no assurance that the actual dose–response function is linear in the extrapolated range. The biochemical mechanisms that control effects may be different at high and low doses. The controversy over this issue is especially heated in the context of animal testing for cancer.

The other approach to evaluating health risks is epidemiology, which includes the study of human exposure to chemicals in the workplace or in the environment and the effect on the health of a population. Epidemiology, in principle, can provide data that is most directly relevant to risk estimation. However, the variables in epidemiological studies are difficult to control; despite sophisticated statistical analysis, it is hard to ascertain that a particular effect is not influenced by some other factors, such as smoking or poor diet, rather than by the chemical under study. Not surprisingly, results are more reliable when the risk is larger than when it is small. For example, the smoking-related risk of lung cancer is easy to demonstrate statistically, because lung cancer incidence is 10–20 times higher in smokers than in nonsmokers.

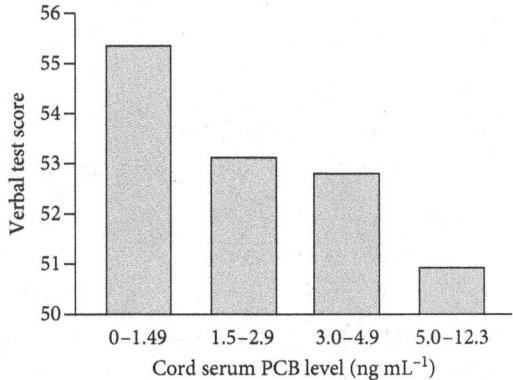

Figure 18.2 Test outcomes (McCarthy verbal test scores) of the 1990 Lake Michigan case study of four-year-old children; the children's scores are graphed versus the PCB concentrations in the umbilical cord serum at birth.
Sources: Adapted from Jacobsen et al., "Effects of in Utero Exposure to Polychlorinated Biphenyls and Related Contaminants on Cognitive Functioning in Young Children," *The Journal of Pediatrics*, Vol. 116, 1990, pp. 38–44 and Spiro et al., *Chemistry of the Environment, 3rd Ed.,* University Science Books, © 2012, all rights reserved.

Both experimental and epidemiological approaches are important for examining a special class of toxic effects that is of concern: prenatal effects on the fetus. Screening for such effects with experimental animals is now routine. In addition to obvious birth defects, the possibility of developmental deficits resulting from prenatal exposure to high doses of chemicals is of concern. The occurrence of fetal alcohol syndrome is a flagrant example of a high-dose effect (one drink contains 15g of alcohol), but there is also concern about more subtle effects from environmental exposures to chemicals at lower doses. For example, a study of families living on the Lake Michigan shore who regularly ate fish caught in the lake found that low verbal test scores of four-year-olds were associated with maternal exposure to polychlorinated biphenyls (PCBs) (Fig. 18.2).

QUESTION 5

Describe how epidemiology is used to study chronic diseases (e.g., cancer). What is the limitation of this approach?

Q5 ANSWER One area of epidemiology is the statistical study of human exposure to chemicals in the workplace or in the environment, and the effect on the health of a population. Epidemiology, in principle, can provide data that are most directly relevant to risk estimation. The problem is that the variables in epidemiological studies are difficult to control. Despite sophisticated statistical analysis, it may be hard to ascertain that a particular effect is not influenced by some other confounding factor, such as smoking or poor diet, rather than by the chemical under study alone.

*QUESTION 6

How can animal studies be used to study chronic diseases? What limitations do animal studies have for studying these diseases?

QUESTION 7

Look closely at the data plotted in Figure 18.2. Although there is definitely a correlation between higher PCB exposure and lower verbal test scores, can scientists definitively say that one caused the other?

Q7 ANSWER No. The human body is complex, and although the higher the concentration of PCBs in the umbilical cord blood, the lower the verbal test score, there are other potential environmental exposures, such as heavy metals, that could also cause or co-cause lower verbal test scores. The data definitely demonstrate a correlation, but not necessarily causation.

18.2 Cancer

Of all the possible effects of chemicals in the environment, none is more feared than cancer, and none has generated more controversy. The public's fear of cancer has driven regulatory agencies to set very low tolerances for many chemicals in various environmental settings, from food to drinking water to toxic waste sites. These standards continue to stir debate; they are claimed to be too lenient by many environmental activists and too strict by manufacturers and others who might have to pay for required cleanups. Because of the uncertainties associated with the available data, as discussed in Section 18.1, it is very hard to establish the truth of the matter. In the absence of hard evidence, there is a great deal of room for subjective factors that influence our perceptions of risk.

18.2.1 *Mechanisms*

Cancers occur when cells divide uncontrollably, eventually consuming vital tissues. The normal mechanisms that limit cell growth and division are disrupted, which can happen in many different ways. The common thread is that mutations occur in the cell's deoxyribonucleic acid (DNA) at positions that specify the synthesis of key regulatory proteins. It has been shown that several such mutations are required to transform a normal cell into a cancerous one. This requirement explains why there is a long latency period, often 20 yr or more, between exposure to a cancer-causing substance and the actual occurrence of cancer. Because of the probabilistic nature of mutations, the risk of cancer increases with age. Although children and young adults can and do develop cancers, most cancers are primarily diseases of old age. One of the causes of increasing cancer incidence is simply the increase in life expectancy during the last century.

Cancer-inducing chemicals are called carcinogens. Carcinogens can operate in two ways: they can be mutagens, inducing mutations by attacking the DNA bases, or they can be promoters, which increase the cancer probability indirectly. For example, promoters can act by increasing the rate of cell division. The more often cells divide, the greater the probability that cancerous mutations will accumulate. Thus, alcohol is a promoter of liver cancer because its consumption in excessive amounts causes cell proliferation in the liver, which is the organ that handles alcohol metabolism.

The body has a variety of ways of ridding itself of foreign chemicals (xenobiotics). One of the most important is the hydroxylation of lipophilic organic compounds. For example, when benzanthracene is hydroxylated (Fig. 18.3), not only does a hydroxy group increase water solubility

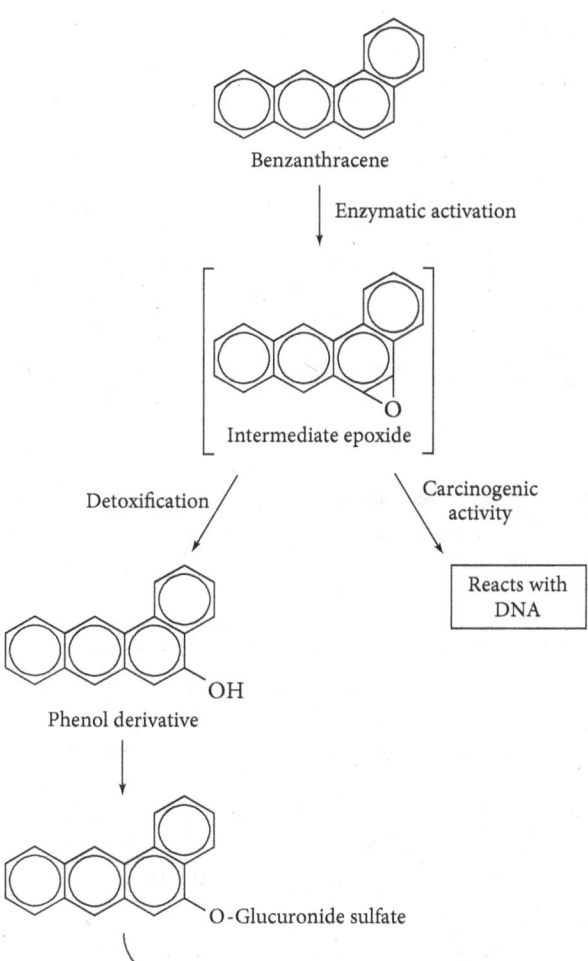

Benzanthracene

↓ Enzymatic activation

Intermediate epoxide

Detoxification /   Carcinogenic activity \

Reacts with DNA

OH
Phenol derivative

O-Glucuronide sulfate

→ Excreted

Figure 18.3 Activation of benzanthracene, a PAH.
Sources: Adapted from Heidelberger "Chemical Carcinogenesis," 1975, *Annual Review of Biochemistry,* Vol. 44, pp. 79–121, and Spiro et al., *Chemistry of the Environment, 3rd Ed.,* University Science Books, © 2012, all rights reserved.

but it also serves as a point of attachment for other hydrophilic groups (e.g., glucuronide sulfate), which increases the water solubility further and promotes excretion by the kidneys. Hydroxylation is accomplished by inserting one of the oxygen atoms of O_2 into a C–H bond. This reaction is tricky because the highly reactive oxygen atom must be generated exactly where it is needed; otherwise, it will attack any molecule in its vicinity, adding to the supply of free radicals.

The reaction is carried out by a class of enzymes, cytochrome P450, which contains a heme group to bind the O_2 (just as hemoglobin does) and an adjacent binding site for the xenobiotic molecule. However, the immediate product is sometimes not the hydroxylated molecule, but rather an epoxide precursor (Fig. 18.3), which is a potent electrophile. Since this precursor is generated inside the cell, it has a chance of diffusing into the nucleus and reacting with the DNA before it rearranges to the hydroxylated product. This finding is the reason why polycyclic aromatic hydrocarbon (PAH) compounds like benzanthracene are carcinogenic. Another possibility is that the hydroxylated product can itself be a precursor to a reactive agent. For example, hydroxylation of dimethylnitrosamine (Fig. 18.4), another carcinogen, releases formaldehyde (CH_2O), leaving an unstable intermediate that is a source of methyl carbocation (CH_3^+), a powerful electrophile, which can react readily with DNA if generated nearby.

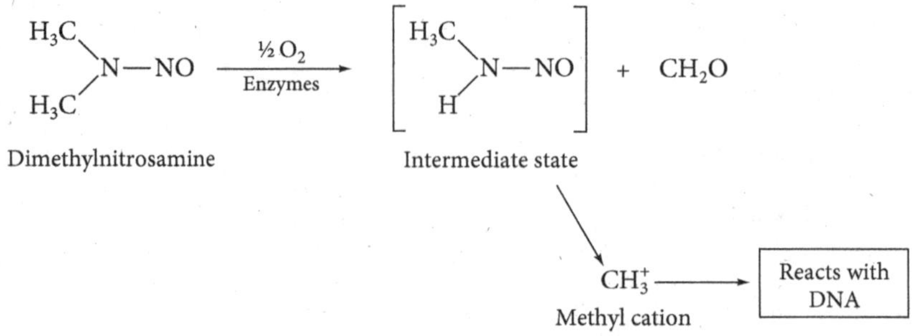

Figure 18.4 Activation of dimethylnitrosamine in the body.
Source: Adapted from Spiro et al., *Chemistry of the Environment, 3rd Ed.,* University Science Books, © 2012, all rights reserved.

PAHs and nitrosamines are anthropogenic carcinogens, but there are many natural ones as well. Aflatoxins, which are complex products of a mold that infests peanuts, corn, and other crops, are powerful carcinogens (see Q3).

QUESTION 8

Describe how molecular biology mechanisms (e.g., Figs. 18.3 and 18.4) are used to study chronic diseases (e.g., cancer). What is the limitation of this approach?

Q8 ANSWER Detailed chemical mechanisms of molecular biology (such as the activation of benzanthracene and dimethylnitrosamine in Figs. 18.3 and 18.4, respectively) can be used to elucidate the effects of carcinogens at the molecular level. Great strides have been made in probing the mechanisms of action by various classes of chemicals, but it is not yet possible to translate this understanding into a quantitative estimate of physiological effects.

*QUESTION 9

Why are lipophilic xenobiotics generally more harmful and pervasive than hydrophilic xenobiotics in the environment?

*QUESTION 10

Ample evidence indicates that some cancers are caused by exposure to carcinogenic chemicals. Using benzanthracene and dimethylnitrosamine as examples (Figs. 18.3 and 18.4), explain how the body actively participates in its own destruction.

18.2.2 *Cancer Incidence and Testing*

Despite extensive epidemiological studies, it is not easy to elucidate the contribution of environmental chemicals to cancer incidence, for the reasons mentioned above (see Section 18.2.1). However, some cancer causes are firmly established by epidemiological data. The most striking evidence is the historical data on lung cancer and smoking (Fig. 18.5). A many-fold rise in U.S. lung cancer mortality tracked the increase in cigarette smoking, with a lag of sev-

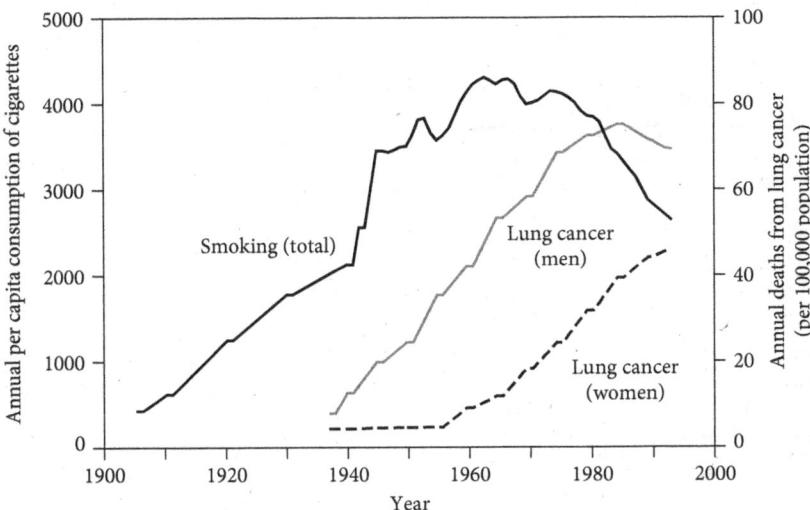

Figure 18.5 Cigarette smoking and lung cancer in the United States; death rates are averages for all ages. *Sources:* Adapted from United States Department of Agriculture, *Tobacco Outlook*, Economic Research Service of USDA, 2007; National Cancer Institute Surveillance Epidemiology and End Results, *Cancer Statistics Fast Stats*, National Cancer Institute, 2009 (available at http://seer.cancer.gov/faststats/); Garfinkel & Silverberg, "Lung Cancer and Smoking Trends in the United States over the Past 25 Years," in Davis & Hoel (Eds.), *Trends in Cancer Mortality in Industrial Countries*, Vol. 41, No. 3, pp. 137–145, 1990, The New York Academy of Sciences; adapted from Spiro et al., *Chemistry of the Environment, 3rd Ed.*, University Science Books, © 2012, all rights reserved.

eral decades, and this happened in different historical periods for men and women. Lung cancer deaths have been decreasing since the 1980s, tracking the decline in smoking. However, smoking still accounts for 80% and 90% of cancer deaths in women and men, respectively.[1]

Diet also plays a role in cancer, as suggested by data showing marked changes in the pattern of cancer incidence when people migrate from one part of the world to another (Fig. 18.6). The rates and types of cancers contracted by migrating ethnic groups change when their diets are modified. It is thought that high levels of salt or smoked fish in the Japanese diet may account for excess stomach cancers, while high fat in the U.S. diet might be responsible for a higher rate of colon cancer. However, the actual contributions of dietary components to cancer incidence (or to protection from cancer) have been hard to quantify.

Data on occupational exposure have firmly implicated several industrial chemicals. For example, vinyl chloride causes liver cancer, benzene causes leukemia, and asbestos causes mesothelioma, a cancer of the lining of the lung. However, exposure of people at large to these chemicals is far lower than in an occupational setting, and the hazard at these lower levels can only be guessed by extrapolation.

Alternatively, carcinogenic risk can be estimated from test data. Because many carcinogens are mutagens, carcinogens can be screened by using the Ames bacterial test. The suspected carcinogen is administered to mutant bacteria that are unable to grow in the absence of the amino acid histidine in the culture medium. Certain additional mutations will produce a revertant organism, capable of growing again in the histidine-deficient medium. The stronger the mutagen, the greater the number of revertant organisms produced. Thus, the number

---

[1] American Lung Association, *Lung Cancer Fact Sheet*, available at https://www.lung.org/lung-health -diseases/lung-disease-lookup/lung-cancer/resource-library/lung-cancer-fact-sheet.

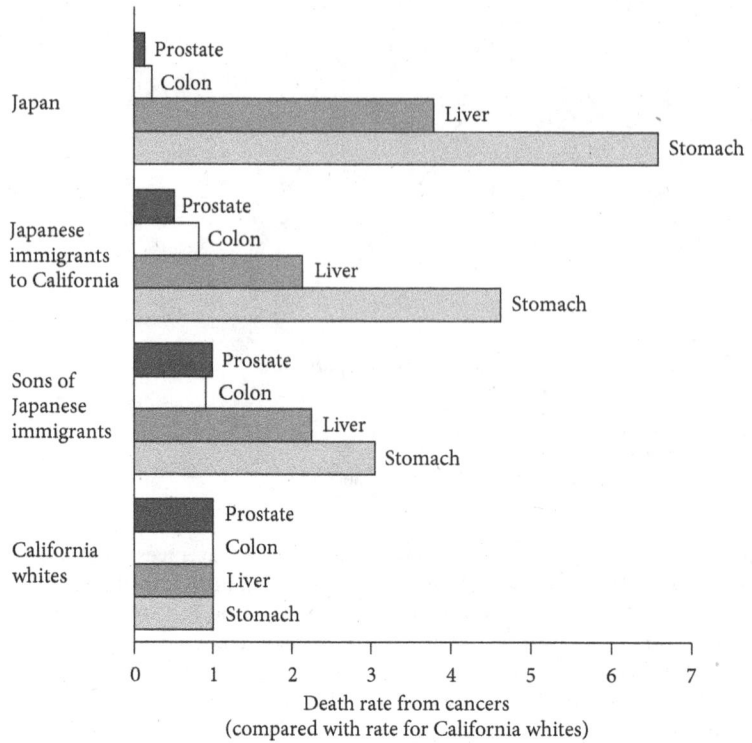

Figure 18.6 Change in death rates from various cancers with migration from Japan to the United States. *Sources:* Adapted from Wynder & Hirayama , "Comparative Epidemiology of Cancers of the United States and Japan," *Preventive Medicine*, Vol. 6, No. 4, 1977, pp. 567–594 and Spiro et al., *Chemistry of the Environment, 3rd Ed.,* University Science Books, © 2012, all rights reserved.

of revertant colonies is a measure of the mutation rate, which can be determined at various concentrations of the test substance. (The test can also be used to monitor complex mixtures for mutagenic activity in order to separate and identify the active ingredient.) Because some chemical compounds are not mutagens until they are metabolically activated, in order to assay carcinogenicity, the Ames test requires adding a rat-liver extract, which contains the cytochrome P450 enzymes responsible for oxidative activation of the carcinogen by the hydroxylation mechanisms described in the preceding section. Because bacteria are very different from people, the test cannot reliably distinguish all human carcinogens or evaluate their potency. It is, however, an inexpensive and widely used screening method.

The main source of carcinogenicity data used in risk assessment has been animal tests, usually involving rats and mice. Cancers are counted over the lifetime of the animal at various chronic doses, and the results are extrapolated to typical exposure levels to obtain an estimate of the cancer risk. Because of the need to obtain statistically significant results on a limited number of animals, most of the data are at the maximum tolerated dose (MTD), defined as the dose expected to produce a 10% decrement in weight gain, compared to control animals.

There are arguments about how to extrapolate from high to low doses. In the absence of actual data (usually unavailable for the reasons discussed above), a linear model is used, involving a straightforward proportionality of dose and effect.

Despite their deficiencies, animal tests can serve as a rough guide for a possible comparison of carcinogenic hazards from different substances. Ames and Gold (Table 18.3) have proposed an index for this purpose, human exposure dose/rodent potency (HERP). It is calculated by estimating the lifetime exposure for an average person and dividing by the

TABLE 18.3 Comparison of Possible Cancer Hazards from Human Exposure to Rodent Carcinogens

HERP %[a]	Risk Agent
0.0003	Tap water, 1 L day^{-1} (chloroform, 17 μg; average US intake, 1987–1992)
0.0003	Carbaryl (carbamate insecticide, see Section 17.4), 2.6 μg day^{-1} (1990 US average)
0.008	Aflatoxin, 18 ng day^{-1} (US average, 1984–1989)
0.002	DDT, 14 μg day^{-1} (US average before 1972 ban)
0.03	Orange juice, 140 g day^{-1} (*d*-limonene, 4.3 mg)
0.1	Coffee, 13.3 g day^{-1} (caffeic acid, 24 mg)
0.4	Conventional home air, 14 h day^{-1} (formaldehyde, 600 μg)
0.5	Wine, 28 g day^{-1} (ethanol)
2.1	Beer, 250 g day^{-1} (ethanol)
6.8	Butadiene, 66 mg day^{-1} (rubber industry workers, 1978–1986)
14	Phenobarbital, 60 mg day^{-1} (one sleeping pill)

[a]Human exposure dose/rodent potency= HERP. Cancer hazard based on a typical person's average daily exposure to the substances over a lifetime. Calculated as a percentage of the TD$_{50}$ for rats or mice (whichever is more potent), corrected for body weight.

Sources: Adapted from Gold et al., "Natural and Synthetic Chemicals in the Diet: A Critical Analysis of Possible Health Hazards," *Issues in Evironmental Science & Technology,* Vol. 15, pp. 95–128, and Spiro et al., *Chemistry of the Environment, 3rd Ed.,* University Science Books, © 2012, all rights reserved.

rodent TD$_{50}$, the dose for inducing cancer in 50% of the animals. Although HERP assumes the applicability of linear extrapolation from high-dose animal test data, it can nevertheless give a broad perspective on the relative magnitude of different hazards. The results suggest, for example, that exposure to pesticide residues or chlorinated tap water is much less hazardous than exposure to such common items in our diet as wine, beer, or coffee. When natural chemicals found in food are tested in rodents, half of them are found to induce cancers, the same fraction as for synthetic chemicals. This high rate suggests that the high doses themselves may increase the cancer rate in rodents and that the HERP values may exaggerate the cancer hazards from chemicals.

QUESTION 11

Describe how epidemiological data linked smoking to lung cancer (see Fig. 18.5).

Q11 ANSWER Large data sets were used to track people and their health. All kinds of data were taken, including lifestyle, what people ate, health indicators, and so on. By using sophisticated statistical analyses, a connection was made between smoking and lung cancer incidence. Figure 18.5 demonstrates this well: as the incidence of smoking went down, the number of men with lung cancer also went down. The number of women with lung cancer was still increasing because the number of women smoking was still increasing.

*QUESTION 12

How was epidemiological data used to link food choices to different types of cancer (see Fig. 18.6)?

QUESTION 13

Describe how the Ames bacterial test is used to study chronic diseases (e.g., cancer). What is the limitation of this approach?

Q13 ANSWER One tool for studying carcinogens is the Ames bacterial test. Chemicals are studied for mutagenicity (many carcinogens are also mutagens). The test chemical is administered to mutant bacteria that are unable to grow unless another mutagenic chemical reverses the mutation, allowing it to grow again. The stronger the mutagen, the greater the number of revertant organisms produced. Because bacteria are very different from people, the test cannot distinguish reliably all human carcinogens. However, it is an inexpensive and useful screening method.

QUESTION 14

How does the HERP (Table 18.3) help us understand the toxicity of substances to humans?

Q14 ANSWER HERP is calculated by the average human exposure of a substance over a lifetime divided by the concentration of the substance that causes cancer in 50% of rodents. Using this method, one can compare the relative toxicity of different substances, some that might be surprising. For example, conventional home air has a HERP of 0.4% (due to formaldehyde) in comparison to coffee, which is only 0.1%. HERP helps people in assessing the riskiness of ingesting different chemicals.

*QUESTION 15

The HERP value for aflatoxin is low, 0.008, despite it being a potent carcinogen. The reason is that the average exposure in the United States is only 18 ng (18×10^{-9} g) day^{-1}. Using the data in Q3, what would be the HERP value in Kenya?

*QUESTION 16

What are the deficiencies relating to HERP in Table 18.3?

18.3 Hormonal Effects

Increasing concern has focused on the biochemical role of environmental chemicals that mimic hormone functions. Hormones are messenger molecules, excreted by various glands that circulate in the bloodstream and powerfully influence the biochemistry of specific tissues. Hormone activity is initiated by binding to receptor proteins in the target cells.

There are two kinds of hormones, water- and lipid-soluble, with entirely different mechanisms of action. Water-soluble hormones (e.g., insulin) are peptides and proteins. They bind to receptor proteins embedded in the target cell membrane, analogous to the neurotransmitter receptors. This binding induces the activation of enzymes inside the cell, which catalyze the synthesis of interior messenger molecules; in turn, these second messengers bind and activate proteins that control metabolic processes.

The lipid-soluble hormones are steroids, derivatives of cholesterol. They diffuse through cell membranes and are picked up at the inside surface by specific receptor proteins that are dissolved in the interior fluid (cytosol) of the target cell. Hormone binding changes the shape of the receptor protein and enables it, after diffusion to the nucleus, to turn specific genes on or off (Fig. 18.7). Thus, the steroid hormones act by inducing or inhibiting the synthesis of enzymes and regulatory proteins.

Chemicals from outside the body (xenobiotics) can also bind to hormone receptors if they have the proper shape and distribution of electrical charges. This problem is unlikely for peptide hormones because water-soluble xenobiotics are quickly excreted. But lipophilic xenobiotics, which are stored in the fat tissue, might bind to steroid hormone receptors. If the resemblance to the hormone is close enough, the xenobiotic can turn on the same biochemical machinery, but if the resemblance is only partial, then binding may not activate the receptor. In that case, the xenobiotic blocks the hormone and depresses its activity; it is an antihormone. Either way, there

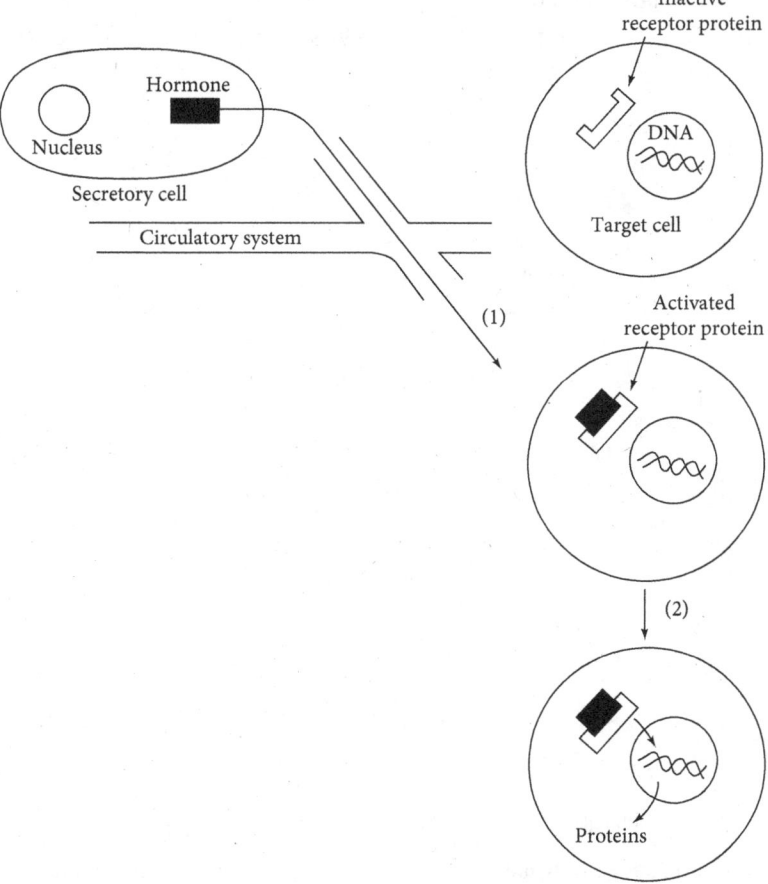

Figure 18.7 Mechanism of steroid hormone function. (1) Activation of the receptor protein. (2) Induction of biochemical reactions associated with gene activation on DNA; synthesis of proteins by activated genes. *Source:* Adapted from Spiro et al., *Chemistry of the Environment, 3rd Ed.,* University Science Books, © 2012, all rights reserved.

is potential for upsetting the biochemical balance controlled by the hormone. A mechanism of this kind is probably responsible for DDT's disruption of calcium deposition in birds' eggs.

The sex hormones are in the steroid class; estrogens and androgens induce and maintain the female and male sexual systems. Estrogenic xenobiotics have provoked more concern because of the association of estrogen with breast cancer in women. Estrogen binding to receptors in the breast stimulates the proliferation of breast cells; as we have seen in the preceding section, cell proliferation promotes mutagenesis and cancer. A link between breast cancer and estrogen has been established in laboratory animals, and there is a statistical association, albeit equivocal, between estrogen therapy and breast cancer incidence. The incidence of breast cancer has been rising, and many industrial chemicals in the environment have been found in laboratory assays to be estrogenic (Fig. 18.8).

BPA (Fig. 18.8) has become a major health concern because, although weakly estrogenic (it binds to the estrogen receptor 10,000 times less tightly than estradiol), laboratory studies have nevertheless indicated a variety of developmental effects on rodents from BPA at levels similar to those found in humans. There is particular concern about children, whose BPA levels are higher than adults.

Human exposure to BPA reflects its very wide use in polycarbonate plastic and epoxy resin. Polycarbonate is a tough clear plastic, used particularly in beverage bottles. BPA epoxy has many uses, but concerns about human exposure center on its use in the lining of food cans. BPA units are integral parts of both polymers, but traces can be released by hydrolysis of the ester

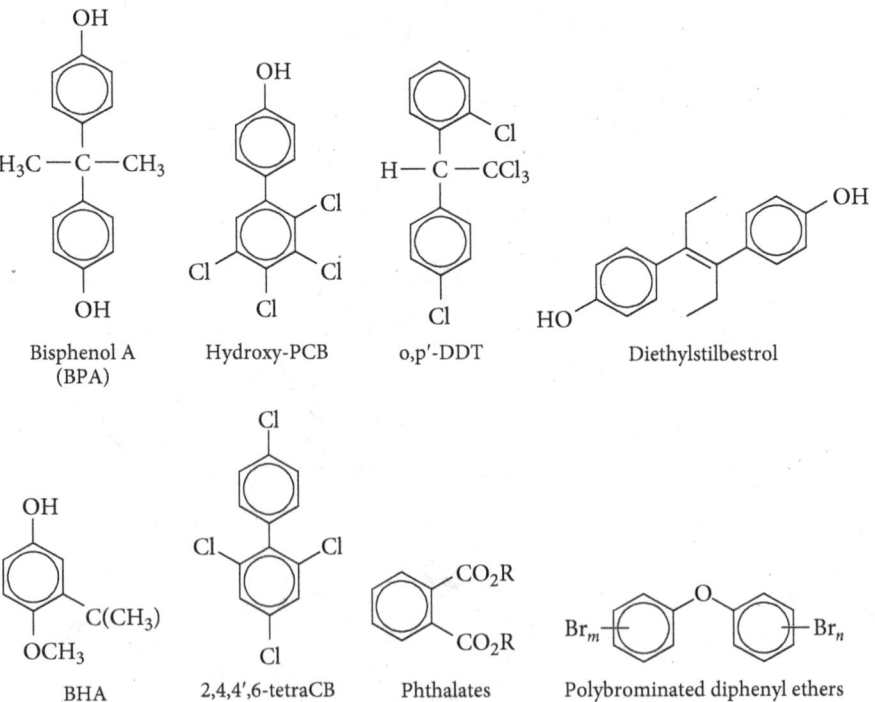

Figure 18.8 Structures of dialkylphthalates, bisphenol A (BPA), polybrominated diphenyl ethers, and other synthetic estrogens. The R groups on the phthalate molecule denote a limited number of alkyl groups that render the molecule estrogenically active. (CB = chlorobiphenyl.)
Sources: Adapted from Committee on Hormonally Active Agents in the Environment, Board on Environmental Studies and Toxicology, National Research Council, *Hormonally Active Agents in the Environment,* 1999, National Academy Press; and Spiro et al., *Chemistry of the Environment, 3rd Ed.,* University Science Books, © 2012, all rights reserved.

(polycarbonate) or ether (epoxy) bonds. These reactions are accelerated at high temperatures or in the presence of acid or base. (Consequently, polycarbonate vessels should not be put in a microwave, or used to store acidic fruit juices.)

A largely ignored compound, BPAF, is identical to BPA, except the hydrogen atoms on the methyl groups are replaced with fluorines. Recent studies have found that BPAF may be even more potent than BPA in altering the effects of steroid hormones (e.g., estrogens) in the body. BPAF is currently used in many plastics, electronic devices, and optical fibers.

QUESTION 17

Explain what a hormone is and describe its mechanism of action. What properties of certain environmental chemicals cause them to disrupt hormonal balance?

Q17 ANSWER Hormones are messenger molecules excreted by various glands. They circulate in the bloodstream and strongly influence the biochemistry of specific tissues. Hormone activity is initiated by its binding to receptor proteins in the target cells. Hormones are either water-soluble or lipid-soluble. Peptides and proteins are water-soluble hormones. They bind to receptor proteins embedded in the target cell membrane, which in turn induces the activation of secondary messengers inside the cell. Lipid-soluble hormones are steroids that diffuse across cell membranes and bind directly with receptor proteins inside the cell. Hormone binding changes the shape of the receptor protein and enables it to turn on specific genes after diffusion to the nucleus.

---

*QUESTION 18

Give two examples of deleterious health effects attributable to hormonal dysfunction caused by lipophilic xenobiotics.

QUESTION 19

Why might the chemical structure of BPAF be more potent than BPA in altering the effects of steroid hormones?

Q19 ANSWER The hydrogens on the methyl groups of BPA are replaced with fluorine atoms in BPAF. The more electronegative F atoms could cause BPAF to bind more strongly to receptor proteins inside the cell.

QUESTION 20

Plasticizers are important functional additives in plastic manufacturing. Around 80% of final polyvinyl chloride (PVC) products contain plasticizers. It is estimated that PVC can contain up to 50% by weight of plasticizers and 90% of the plasticizers are made from phthalates. In 2015, the weight of PVC wastes generated was 15 million metric tons across the world. Estimate how many tons of phthalates were contained in this waste.

Q20 ANSWER

$$(15 \times 10^6 \text{ tons}) \times 0.80 \times 0.50 \times 0.90 = 5.4 \times 10^6 \text{ metric tons} = 5.4 \text{ million metric tons}$$

18.4 Persistent Organic Pollutants: Dioxins, PCBs, and Perfluorooctanesulfonic Acid

There are many organic compounds in the environment, some of which are toxic. We have already dealt with air pollutants and insecticides and herbicides. There are numerous organic waste products from industrial operations or from the use and disposal of manufactured products that may contaminate air, land, and water, from effluents, from leakage of waste dumps, or from accidental spills and fires. A worldwide treaty (the Stockholm Convention) was negotiated in 1995 to phase out persistent organic pollutants (POPs). The treaty went into force in 2004. The 12 chemicals initially covered by the treaty are dioxins and furans, PCBs, and nine organochlorine pesticides (aldrin, chlordane, eldrin, dieldrin, heptachlor, hexachlorobenzene, mirex, toxaphene, and DDT, although there is an exemption for developing countries using DDT for malaria control). In 2009, 16 chemicals were added to the treaty, including perfluorooctanesulfonic acid (PFOS), brominated and chlorinated flame retardants (pentabromodiphenyl ether, octabromodiphenyl ether, hexabromodiphenyl ether, and penta-chlorobenzene), pesticides (lindane and chlordecone), and by-products of the manufacturing of lindane (α- and β-hexachlorocyclohexane).

Having dealt with several of these chemicals in the above sections on pesticides and hormones, we turn our attention to dioxins and furans, PCBs, and PFOS-related compounds.

18.4.1 *Dioxins and Furans*

The term dioxin is shorthand for a family of polychlorinated dibenzodioxins (Fig. 18.9), sometimes abbreviated as PCDDs. The polychlorinated dibenzofurans (PCDFs) have a similar structure. These chemicals are not made intentionally, but are formed as contaminants in several large-scale processes, including 1) combustion, 2) paper pulp bleaching with chlorine, and 3) manufacture of certain chlorophenol chemicals. It was this last process that brought dioxin its initial notoriety as a contaminant of the herbicide 2,4,5-T, a component of Agent Orange, used extensively as a defoliant in the Vietnam War. The herbicide was made by reacting chloro-acetic acid with 2,4,5-trichlorophenol (2,4,5-T), which was itself formed by reacting 1,2,4,5-tetrachlorobenzene with sodium hydroxide [reaction sequence (1) in Fig. 18.9]. During this prior reaction, which was carried out at a high temperature, some of the trichloro-phenoxide condensed with itself [reaction (2) in Fig. 18.9] to form 2,3,7,8-tetrachlorodibenzo-dioxin (TCDD); Agent Orange contained ~10 parts per million (ppm) of this material. It was subsequently shown that control of the reaction temperature and of the trichlorophenoxide concentration could keep the TCDD contamination to 0.1 ppm, but it was too late to save the herbicide, which was banned in the United States in 1972.

2,3,7,8-tetrachlorodibenzodioxin turned out to be extraordinarily toxic to laboratory animals, producing birth defects, cancer, skin disorders, liver damage, suppression of the immune system, and death from undefined causes. The LD_{50} (see Table 18.1) in male guinea pigs was only 0.6 $\mu g\ kg^{-1}$. In laboratory animals, low doses have been found to be teratogenic and to lead to developmental abnormalities.

As evidence on toxicity has accumulated, however, the risk to humans from dioxin has become less clear. The variation in toxicity among chemical species turned out to be large, with LD_{50} values that were orders of magnitude higher in other animals than in guinea pigs. In humans, exposure to high levels of PCDDs causes chloracne, a painful skin inflammation, but these levels have only been encountered in accidental industrial exposures. The EPA carried out a dioxin assessment, which concluded that dioxin is likely to increase cancer incidence in humans; epidemiological data on industrial workers indicate an association of

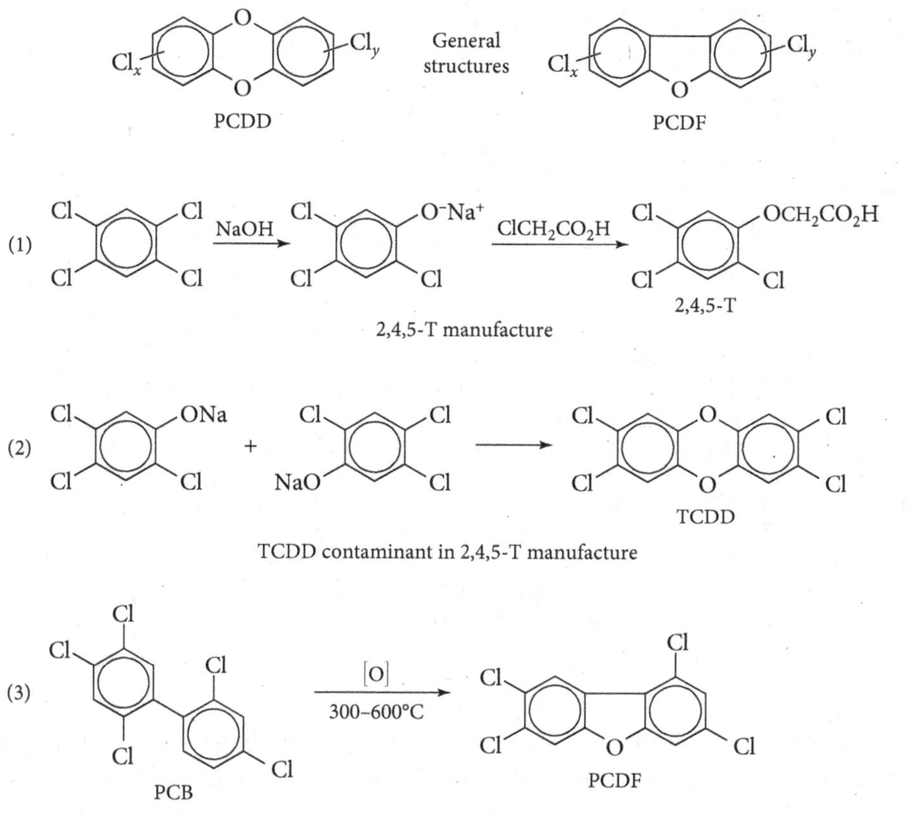

Figure 18.9 Polychlorinated dibenzodioxins, PCDFs, and PCBs; chemical structures and reactions. *Source:* Adapted from Spiro et al., *Chemistry of the Environment, 3rd Ed.,* University Science Books, © 2012, all rights reserved.

cancer incidence with increasing exposure levels. The World Health Organization (WHO) has classified TCDD as a known human carcinogen. However, there is a continuing controversy about the magnitude of the risk.

Rapid strides have been made in understanding TCDD's complex biochemistry. The molecule binds strongly to a receptor protein that is present in all animal species. This receptor, called Ah (for aryl hydrocarbon), is activated by a number of planar aromatic molecules (its natural substrate is still uncertain); the binding of TCDD is particularly strong. Like a hormone receptor (Fig. 18.7), the Ah receptor interacts in complex ways with the cell's DNA. One effect is the induction of a cytochrome P450 enzyme (a variant labeled 1A1), which is responsible for hydroxylating a number of xenobiotics, including PAHs (but not TCDD itself, since its chlorine atoms deactivate the ring toward oxidation). There are additional effects on a variety of biochemical pathways, which are currently under study. It remains uncertain, however, whether all of TCDD's toxic effects originate in its binding to the Ah receptor.

The main source of dioxin in the environment is combustion. When material containing chlorine is combusted, dioxin is produced in traces; because the volume of material combusted annually is huge, these traces add up to a substantial aggregate environmental load. Combustion produces a wide range of PCDD congeners (molecules with the same structure, but with varying numbers and positions of chlorine substituents), as well as PCDFs (Fig. 18.9). In the context of combustion products, "dioxin" means the aggregate of PCDDs and PCDFs,

TABLE 18.4 International Toxicity Equivalency Factors for PCDDs and PCDFs

Congener	PCDD Series	PCDF Series
2,3,7,8-	1 (defined)	0.1
1,2,3,7,8-	0.5	0.05
2,3,4,7,8-		0.5
1,2,3,4,7,8-	0.1[a]	0.1[b]
1,2,3,4,6,7,8-	0.01	0.01[c]
Octachloro	0.001	0.001

[a]Same value for 1,2,3,6,7,8- and 1,2,3,7,8,9-congeners.
[b]Same value for 1,2,3,6,7,8-, 1,2,3,7,8,9-, and 2,3,4,6,7,8-congeners.
[c]Same value for 1,2,3,4,7,8,9-congener.

Sources: Adapted from Bunce, *Environmental Chemistry, 2nd Ed.*, 1994, Wuerz Publishing, Ltd.; and Spiro et al., *Chemistry of the Environment, 3rd Ed.*, University Science Books, © 2012, all rights reserved.

also abbreviated to PCDD/Fs. Both classes of molecules are toxic, but the toxicity varies among the congeners. Toxicity is assumed to be roughly proportional to the strength of binding to the Ah receptor. The most toxic dioxin is TCDD; toxicity decreases progressively when Cl atoms are removed from the 2, 3, 7, and 8 positions, or when they are added to the remaining positions on the rings. These alterations reduce the "fit" of the molecule to the binding site on the Ah receptor. A similar toxicity pattern is observed for the PCDF congeners, but the toxicity is about an order of magnitude lower for the PCDFs than for the PCDDs. In order to gauge the effects of exposure to these chemicals, a scale of international toxicity equivalence factors (I-TEFs) has been established based on toxicity relative to TCDD, which is assigned a value of 1 (Table 18.4). This factor is 0.1 for 2,3,7,8-PCDF, for example, and 0.001 for the octachloro congeners of either series. With these factors, we can convert the distribution of both classes of molecules (PCDD/Fs) into a single toxicity equivalent quantity (TEQ), expressed in grams of TCDD equivalents. For example, 1.0 g each of TCDD and 2,3,7,8-PCDF would have a TEQ value of 1.1 g.

QUESTION 21

How was TCDD involved in the decision to ban 2,4,5-trichlorophenol?

Q21 ANSWER TCDD is a contaminant in 2,4,5-trichlorophenol manufacture, formed as a by-product at high temperatures. Although TCDD can be minimized by lowering the reaction temperature, the dioxin contaminant had gained 2,4,5-trichlorophenol such notoriety, especially from its massive use as the defoliant Agent Orange during the Vietnam War, that its use was eventually banned.

QUESTION 22

Why do the geometric dimensions of TCDD make it so toxic? Why is toxicity diminished in other congeners where chlorines are removed from the 2, 3, 7, 8 positions, or when chlorines are added to the remaining positions on the rings?

Q22 ANSWER The planar geometry of TCDD appears to be responsible for its strong binding to Ah, a receptor protein present in all animal species. Because Ah interacts in complex ways with the cell's DNA, the disruption of its mechanisms of action by TCDD may explain TCDD's toxic effects.

Relative to TCDD, the toxicity of congeners declines markedly when chlorine atoms are removed from the 2, 3, 7, and 8 positions, or when they are added to the remaining positions on the rings. These alterations reduce the fit of the molecule to the binding site on the Ah receptor, thus lessening interference with Ah's biochemical function.

*QUESTION 23

Why do you suppose that the toxicity of TCDF is about one-tenth that of TCDD?

*QUESTION 24

If a solid sample is found to be contaminated with 1 ppm of both 2,3,7,8- and 1,2,3,7,8-PCDD, what is the TEQ concentration?

*QUESTION 25

Describe why animal testing is not always the best way to learn about human health effects from chemical exposure based on the data in Table 18.3 and what you learned in the text.

The total U.S. PCDD/F emission rate from combustion is estimated to be ~5 kg yr^{-1} TEQ. But this total is spread out over an enormous area, and the atmospheric concentrations are very low. Exposure to breathing dioxin-laden air is minimal, even if one lives next to an incinerator. As with other hydrophobic materials, exposure to dioxins is determined by bioaccumulation mechanisms. The main exposure route (95%) for humans is dietary: meat, dairy products, and fish (Table 18.5). The dioxins deposit on hay and feed crops consumed by cows, which concentrate the dioxins in their fat tissues. Likewise, fish concentrate dioxins from algae, which absorb dioxins from atmospheric fallout, and also from local pollution sources (e.g., pulp bleaching plants or sewage and wastes). As a result, we all have detectable concentrations of dioxins in our fat tissue, although the most prevalent one is the octachloro congener, which is not very toxic. The average daily dose of PCDD/Fs is estimated to be roughly 0.1 ng (nanogram = 10^{-9} g) TEQ day^{-1} in the United States. This dose is not far from the levels at which biochemical effects can be detected in laboratory animals. If the dioxin deposition rate is declining, as the sedimentary record indicates, then the average exposure should also decline. A large decline in human tissue levels has indeed been observed for samples taken in several countries.

TABLE 18.5 Average Content of TCDD in the American Food Supply

Food	TCDD Concentration (pg g^{-1})[a]	Average TCDD Intake (pg per person per day)[a]
Ocean fish	500	8.6
Meat	35	6.6
Cheese	16	0.31
Milk	1.8	0.20
Coffee	0.1	0.04
Ice cream	5.5	0.04
Cream	7.2	0.01
Sour cream	10	0.01
Cottage cheese	2.1	0.01
Orange juice	0.2	0.01
Total		15.9

[a]1 picogram (pg) = 10^{-12} g

Sources: Data from Henry et al., "Exposures and Risks of Dioxin in the U.S. Food Supply," 1992, *Chemosphere,* Vol. 25, pp. 235–238; adapted from Spiro et al., *Chemistry of the Environment, 3rd Ed.,* University Science Books, © 2012, all rights reserved.

QUESTION 26

Propose a reason why sour cream has a TCDD concentration of 10 pg g^{-1}, whereas milk only has 1.8 pg g^{-1} using the data in Table 18.5.

Q26 ANSWER TCDD tends to concentrate in fat, and sour cream has more overall fat than milk.

*QUESTION 27

Using the data from Table 18.5, what kind of diet should you eat if you are concerned about TCDD exposure?

18.4.2 Polychlorinated Biphenyls

As the name implies, PCBs are made by chlorinating the aromatic compound biphenyl (see Fig. 18.9 for the molecular structure of a specific PCB). A complex mixture results in variable numbers of chlorine atoms substituted at various positions of the rings; a total of 209 congeners is possible. The PCBs were manufactured in the United States from 1929 to 1977, with a peak production of ~100,000 tons a year in 1970. They were used mainly as the coolant in power transformers and capacitors because they are excellent insulators, are chemically stable, and have low flammability and vapor pressure. In later years, they were also used as heat-transfer fluids in other machinery and as plasticizers for PVC and other polymers; they found additional uses in carbonless copy paper, as de-inking agents for recycled newsprint, and as weatherproofing agents. As a result of industrial discharges and the disposal of all these products, PCBs were spread widely in the environment.

Because PCBs are chemically stable, they persist in the environment, and because they are lipophilic, they are subject to bioaccumulation, as are DDT and dioxins. The PCB concentrations at the top of the food chain are significant in many localities. For example, herring gull eggs on the shores of Lake Ontario contained >160 ppm of PCBs in 1974. Since then, however, the level has declined by a factor of 16, reflecting the termination of PCBs in all open uses, those in which disposal cannot be controlled. Production was drastically curtailed in 1972 and halted in 1977. The PCB-containing transformers continue in service, but their disposal is regulated and spent PCBs are stored or incinerated.

In laboratory studies, PCBs are less toxic than PCDDs and PCDFs, but they probably operate by the same mechanism, binding to the Ah receptor. The most toxic PCBs are those that have no Cl atoms in the ortho positions of the ring and can, therefore, adopt a coplanar configuration of the rings, as in PCDDs and PCDFs. Coplanarity is inhibited in ortho-substituted biphenyls by the steric interaction of the substituent with the ortho H atoms on the other ring. If substituents occupy three or four of the ortho positions, they bump into each other, and the rings are necessarily twisted away from each other. The PCBs with this substitution pattern are the least toxic. Even if PCBs are less toxic to humans and other animals than PCDDs and PCDFs, they are much more abundant in the environment. The studies like the one illustrated in Figure 18.2, which indicates a connection between PCB exposure *in utero* and subsequent learning deficits, are cause for concern.

QUESTION 28

Name the uses of PCBs before they were banned in the late 1970s.

Q28 ANSWER PCBs were used mainly as coolants in power transformers and capacitors because they are excellent insulators, chemically stable, and have low flammability and vapor pressure. They were also used as heat-transfer fluids in other machinery, and as plasticizers for PVC and other polymers. Additional uses were in carbonless copy paper, as de-inking agents for recycled newsprint, and as weatherproofing agents.

*QUESTION 29

With reference to Figure 18.2, how would you surmise that PCBs ended up in umbilical cord serum of pregnant women living near Lake Michigan in 1986?

18.4.3 *Perfluorooctanesulfonic Acid (PFOS) and Perfluorooctanoic Acid (PFOA)*

PFOS and PFOA are fully fluorinated, organic compounds and are the two perfluorinated chemicals (PFCs) that have been produced in the largest amounts in the United States. Of PFCs, they have also received the most attention in recent years due to their persistence, bioaccumulation, and toxicity. They are commonly used as simple salts (such as potassium or

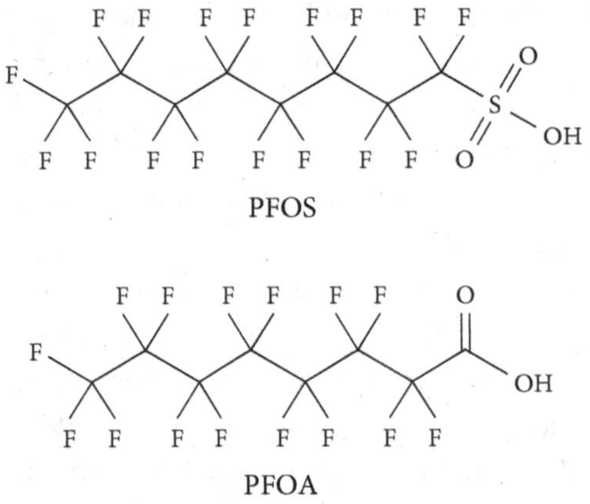

Figure 18.10 Chemical structures of PFOS ($C_8F_{17}SO_3H$) and PFOA ($C_8F_{15}O_2H$). *Source:* Adapted from Spiro et al., *Chemistry of the Environment, 3rd Ed.,* University Science Books, © 2012, all rights reserved.

ammonium) of perfluoroalkyl sulfonate and perfluoroalkyl carboxylate, respectively, and as surface-active agents in a variety of applications such as firefighting foams, coating additives, and cleaning products[2].

PFOS and PFOA are manmade compounds and have been used for decades (Fig. 18.10); concerns about their toxicity began in the early 2000s. In 2000, the 3M Company decided to phase out the production of PFOS products to be replaced by perfluorobutane sulfonate (PFBS) and shorter chain PFCs. In 2009, PFOS was listed as a new persistent organic pollutant (POP) by the Conference of Parties, Stockholm Convention. The United States Environmental Protection Agency (EPA) has classified PFOA as a "likely carcinogen." In the United States, PFOS and PFOA were phased out of production in 2002 and 2015, respectively, but are still present in some imported products.

Both PFOS and PFOA are stable end-products resulting from the degradation of precursors (such as perfluoroalkyl sulfonamides) through a variety of physical and biotic transformation pathways. Because of their structures and bonding, PFCs, including PFOS and PFOA, are chemically and biologically stable in the environment and resist typical environmental degradation processes, including atmospheric photooxidation, direct photolysis, and hydrolysis. As a result, these chemicals are extremely persistent in the environment, detected in air, water, and soil around fluorochemical facilities.

Studies have found PFOS and PFOA in the blood samples of the general human population and wildlife, indicating that their exposure is widespread. In the early 2000s, PFOA and PFOS were routinely found in blood samples in the U.S. population in the parts per billion range, although those concentrations have decreased in recent years. Potential pathways of exposure include ingestion of food and water (including drinking water), use of commercial products, or inhalation from long-range air transport of PFC-containing aerosols. Toxicology studies show that PFOS and PFOA are readily adsorbed after oral exposure; accumulate primarily in the serum, kidney, and liver; and have half-life time in humans ranging from two to nine years. This half-life leads to continued exposure and increased body burdens that could result in adverse health outcomes. Rodent studies have raised concerns about potential developmental, reproductive, and other systemic effects of PFOS and PFOA.

---

[2] U.S. Environmental Protection Agency, *Emerging Contaminants and Federal Facility Contaminants of Concern,* available at https://www.epa.gov/fedfac/emerging-contaminants-and-federal-facility -contaminants-concern

QUESTION 30

Using chemical bonding and reference data, how would you explain PFOS and PFOA are extremely persistent in the environment, thus posing potential adverse effects on the environment and human health?

Q30 ANSWER The C–F bond is one of the strongest single bonds in organic molecules, with a bond dissociation energy (BDE) at around 485 kJ mol^{-1}. The very high electronegativity of the F atom makes the C–F attraction (partially ionic, via the $C^{\delta+}$—$F^{\delta-}$ dipole) strong and the C–F bond length short. In comparison, the DEs for C–Cl, C–Br, and C–I bonds are 327, 285, and 213 kJ mol^{-1}, respectively. This very high BDE makes it difficult for PFOS and PFOA to be broken down under common environmental conditions such as photolysis, hydrolysis, or biodegradation.

*QUESTION 31

What specific properties make PFOS and PFOA persistent in water and soil? And do you expect they exist in the gas phase or other medium in the atmosphere?

18.5 Toxic Metals

The biosphere has evolved in close association with all the elements of the periodic table, and indeed, organisms harnessed the chemistry of many metal ions for essential biochemical functions at the early stages of evolution. As a result, these elements are required for viability, although in small doses. At elevated doses, however, they can be harmful.

But the main environmental concern is with heavy metals that are nonessential but have been widely dispersed through human activity. Especially concerning are mercury, cadmium, arsenic, and lead.

18.5.1 *Mercury*

The environmental toxicity of mercury is associated almost entirely with eating fish; this source accounts for some 94% of human exposure. When mercury is present in sediments, sulfate-reducing bacteria generate methylmercury and release it into the waters above, where it is absorbed by fish from the water passed across their gills or from their food supply. The CH_3Hg^+ ion forms CH_3HgCl in the saline milieu of biological fluids and this neutral complex passes through biological membranes, distributing itself throughout the tissues of the fish. In the tissues, the chloride is displaced by protein and peptide sulfhydryl groups. Because of mercury's high affinity for sulfur ligands, the methylmercury is eliminated only slowly. Therefore, it is subject to bioaccumulation when little fish are eaten by bigger fish. The phenomenon is the same as for DDT and other lipophiles, but the mechanism is different because Hg accumulates in protein-laden tissue rather than in fat.

Biomethylation of mercury occurs in all sediments, and fish everywhere have some mercury. But the levels are greatly elevated in bodies of water whose sediments are contaminated by mercury from waste effluents. The worst case of environmental mercury poisoning occurred in the 1950s in the fishing village of Minamata, Japan. A plastics plant used Hg^{2+} as a catalyst and discharged mercury-laden residues into the bay, where the fish accumulated methylmercury to levels approaching 100 ppm. Thousands of people were poisoned by the contaminated fish, and hundreds died from it. Those affected suffered numbness in the limbs, blurring and even loss of vision, and loss of hearing and muscle coordination, all symptoms of brain dysfunction resulting from the ability of methylmercury to cross the blood–brain barrier. Likewise, methylmercury can pass from mother to fetus, and a number of Minamata infants suffered intellectual disability and motor disturbance before the cause of the poisoning was identified. Based on this incident and others, the recommended limit for mercury in fish for human consumption has been set at 0.5 ppm. Fortunately, cessation of waste mercury discharge lowers the levels of mercury in the local fish.

Much of the world's discharge of mercury into the aqueous environment stems from chlor-alkali plants. These plants manufacture Cl_2 and NaOH, large-volume commodities that are mainstays of the chemical industry. A mercury pool electrode was used in this process, but these installations have largely been phased out and replaced by an alternative process using cation-exchange membranes.

Even though local discharges have caused the most serious mercury contamination, it has been discovered that fish have elevated mercury levels even in lakes that are quite remote from any local source. Thus, mercury is transported over long distances, a consequence of the fact that there are two volatile forms, metallic mercury, Hg^0, and dimethylmercury, $(CH_3)_2Hg$ (Fig. 18.11). Both are formed in the same milieu as methylmercury. Bacteria have a detoxification system that rids their environment of methylmercury by converting it to Hg^0, which is volatilized. The $(CH_3)_2Hg$ is produced in the same biomethylation process as CH_3Hg^+. Long-range transport means that mercury discharges can have regional, and even global consequences.

Small amounts of metallic mercury are used in batteries, switches, lamps (including energy-efficient compact fluorescent lamps), and other electrical equipment; improper use and disposal (e.g., batteries in municipal incinerators) add to the global load of Hg vapor. Mercury emissions have decreased 95% in the United States, thanks largely to its elimination from batteries, and to emission controls on municipal and medical waste incinerators. Similar reductions have been achieved in other developed countries.

But adding to the global load is the use of mercury to extract gold or silver from ores, a practice used for centuries in Central and South America and applied on a large scale today in the gold fields of Brazil. This process releases massive amounts of mercury into the environment because the extracted gold is recovered by heating the amalgam to drive off the mercury. Large quantities of amalgam washings have contaminated parts of the Amazon River sediment with mercury. This practice is estimated to account for 2% of global atmospheric mercury emissions.

Mercury is also present in coal and is released to the air when the coal is burned. The concentrations are low, but because so much coal is burned in power plants, the total amount is significant. The mercury can be largely recovered from the flue gas using filters, and the EPA established a permissible level for Hg emission in 2016. But in 2020, under a new administration, the EPA revised its cost–benefit analysis, which opens the standard to legal challenges. However, according to a study published in *Science* magazine, the new analysis omits the co-benefit of reducing particulate matter (caught in the same filters as the Hg), which by itself exceeds the compliance costs. The political struggle over costs and benefits extends to many other environmental regulations.

Inorganic mercury is not particularly toxic when ingested because neither the metal nor the ions (Hg_2^{2+} and Hg^{2+} and their complexes) penetrate the intestinal wall effectively. How-

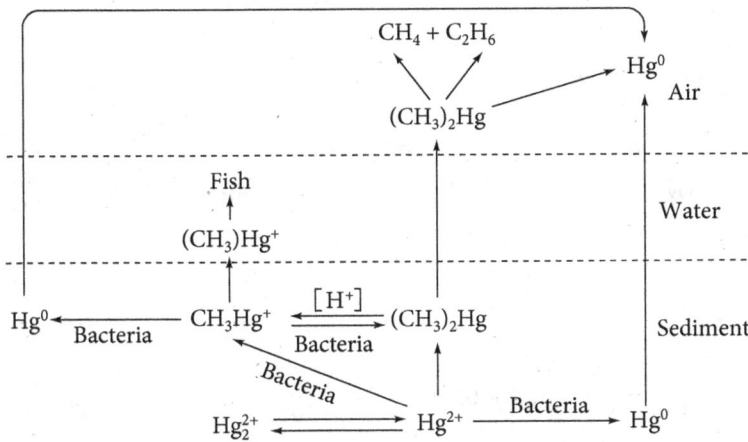

Figure 18.11 The biogeochemical cycle of bacterial methylation and demethylation of mercury in sediments. *Sources:* Adapted from National Research Council, *An Assessment of Mercury in the Environment*, 1978, National Academy Press; and Spiro et al., *Chemistry of the Environment, 3rd Ed.*, University Science Books, © 2012, all rights reserved.

ever, Hg^0 is highly toxic when inhaled; in atomic form, it is able to pass through the lung membranes into the bloodstream and across the blood–brain barrier. In the brain, it can presumably be oxidized and bound to protein sulfhydryl groups because it produces the same neurological effects as methylmercury. For this reason, individuals should avoid handling elemental mercury, and all spills should be cleaned up (with sulfur, which ties up the mercury atoms). The Brazilian gold miners suffer serious health problems from elemental mercury released during the amalgam operations, as silver and gold miners have for centuries.

QUESTION 32

In the human body, the half-life of methylmercury is 70.00 days and for Hg^{2+} it is 6.00 days. What is the maximum body accumulation of each type of mercury at a constant ingestion rate of 2.00 mg Hg day^{-1}? *Hint:* The maximum concentration in the body can be calculated from the following equations:

$$\frac{C_{max}}{C_o} = e^{-\lambda}\left[\frac{1}{1 - e^{-\lambda}}\right]$$

$$\lambda = \frac{0.693}{t_{1/2}}$$

where C_{max} is the maximum concentration, C_0 is the initial concentration per time step, and $t_{1/2}$ is the half-life.

Q32 ANSWER Maximum body accumulation (C_{max}) can be calculated from:

$$\frac{C_{max}}{C_o} = e^{-\lambda}\left[\frac{1}{1 - e^{-\lambda}}\right]$$

where $C_o = 2$ mg Hg, $t_{1/2}$ is the half-life, and $\lambda = \dfrac{0.693}{t_{1/2}}$

For methylmercury, $\lambda = \dfrac{0.693}{70.00 \text{ days}} = 0.009900 \text{ day}^{-1}$:

$$C_{max} = (2.00 \text{ mg}) \times e^{-0.009900} \left(\dfrac{1}{1 - e^{-0.009900}} \right) = (2.00 \text{ mg}) \times 0.9901 \times \left(\dfrac{1}{1 - 0.9901} \right) = 201 \text{ mg}$$

For Hg^{2+}, $\lambda = \dfrac{0.693}{6.00 \text{ days}} = 0.116 \text{ day}^{-1}$:

$$C_{max} = (2.00 \text{ mg}) \times e^{-0.116} \left(\dfrac{1}{1 - e^{-0.116}} \right) = (2.00 \text{ mg}) \times 0.891 \times \left(\dfrac{1}{1 - 0.891} \right) = 16.3 \text{ mg}$$

*QUESTION 33

Why is methylmercury so much more toxic than inorganic mercury? Why is there such a marked difference in half-lives?

QUESTION 34

Relate how Hg accumulation in the food chain led to the Minamata disaster. Why is the conversion to methylmercury the key step in mercury toxicity?

Q34 ANSWER In the 1950s, a PVC plant used Hg^{2+} as a catalyst and discharged inorganic mercury residues into Minimata Bay. The mercury settled into the sediments, and, under the prevalent anaerobic conditions, was transformed into methylmercury (CH_3Hg^+) by methanogenic bacteria. The methylmercury was released to the waters above and was absorbed by fish from water passed across their gills, or from bioaccumulation in their food supply. Levels of mercury in the fish approached 100 ppm. In the biological fluids of the fish, methylmercury was transformed into CH_3HgCl, a form that is able to pass across biological membranes, distributing itself throughout protein-laden tissues of the fish. Local villagers, who consumed large quantities of the fish, were subjected to high levels of mercury poisoning.

Conversion to methylmercury is the key step in mercury toxicity because other forms of mercury (Hg^0, Hg_2^{2+}, and Hg^{2+}) are not particularly toxic when ingested. These latter inorganic forms cannot penetrate the intestinal wall effectively (although Hg^0 is highly toxic when inhaled). On the contrary, methylmercury has the ability to cross the blood–brain barrier and cause serious symptoms of brain dysfunction. Moreover, it can pass through the placenta from mother to fetus, causing mental retardation and motor disturbance.

*QUESTION 35

In addition to the Minimata disaster, what are other ways that people are exposed to mercury today?

18.5.2 *Cadmium*

Cadmium is found in the same column of the periodic table as mercury and zinc, but its chemical properties are much closer to zinc than to mercury. This resemblance to zinc accounts for cadmium's distribution, as well as its particular hazards. Cadmium is always found in association with zinc in the earth's crust, and it is obtained as a by-product of zinc mining and extraction; there are no separate cadmium mines. Moreover, cadmium is always present as a contaminant in zinc products. Indeed, one of the pervasive sources of cadmium in the urban environment is zinc-treated (galvanized) steel. The weathering of galvanized steel surfaces produces zinc- and cadmium-laden street dust; although the concentration is low, the total amount of cadmium is substantial. It has been pointed out that proposals to ban cadmium in products (mostly batteries, electroplate, pigments, and plastics stabilizers) in order to reduce environmental exposure might have the opposite effect, because of lowered economic incentive to recover cadmium from zinc-mine residues and from refined zinc itself.

Mimicry of zinc is probably why cadmium is actively taken up by many plants, because zinc is an essential nutrient. Most of our cadmium intake is from vegetables and grains in our diet. However, smokers get an extra dose because of the cadmium concentration in tobacco leaves; heavy smokers have twice as much cadmium in their blood, on average, as nonsmokers.

There is concern that cadmium buildup in agricultural soils may eventually produce dangerous levels in food. Cadmium inputs to soils are mainly from airborne deposition (wet plus dry) and from commercial phosphate fertilizers, which contain cadmium as a natural constituent of phosphate ore. The cadmium burden could be further increased by the use of fertilizer from sewage sludge, a sludge-disposal measure that is increasingly advocated. Sewage is often contaminated by cadmium and other metals; however, there is some evidence that the cadmium is firmly bound in the sludge and might not be released to growing plants.

Soil conditions were certainly a factor in the only known case of widespread environmental cadmium poisoning, which occurred in the Jinzu valley of Japan. Irrigation water drawn from a river that was contaminated by a zinc mining and smelting complex led to high levels of cadmium in the rice. Hundreds of people in the area, particularly older women who had borne many children, developed a painful degenerative bone disease called itai-itai (ouch–ouch), apparently because Cd^{2+} interfered with Ca^{2+} deposition. Their bones became porous and subject to collapse. Sufferers were estimated to have had a cadmium intake of ~60 μg day^{-1}, several times the normal intake.

Although the 70 ppm soil Cd level in Jinzu was elevated, it has been still higher, 300 ppm, in Shipham, England, a Zn mining locale from the seventeenth to nineteenth centuries. Yet, health inventories in Shipham showed only slight effects attributable to Cd. The Shipham soils have high pH, 7.5, and also a high content of calcium carbonate and of hydrous oxides of iron and manganese, which are good absorbers of Cd^{2+}. On the contrary, the Jinzu soils had low pH (5.1) and a low content of hydrous oxides. Thus, the Cd was far more available for plant uptake in Jinzu than in Shipham.

Chronic exposure to cadmium has been linked to heart and lung disease, including lung cancer at high levels, to immune system suppression, and to liver and kidney disease. The Cd^{2+} sequestering protein metallothionein provides protection until its capacity is exceeded. Since metallothionein is concentrated in the kidney, this organ is damaged first by excessive Cd. The downside of metallothionein protection is that Cd is stored in the body and accumulates with age, so that damage from long-term exposure becomes irreversible.

QUESTION 36

Describe how people living in the Jinzu valley of Japan were exposed to excess Cd concentrations from soil sources. Where did the Cd derive from in the first place?

Q36 ANSWER Irrigation water spread on rice fields was contaminated with Cd from a Zn mining and smelting operation. Cd was taken up by the rice and eaten by the people in the valley. Cd was not adsorbed by the soil, which had a relatively low pH; had the soil pH been higher, $Fe(OH)_3$ would have formed that would have adsorbed the excess Cd (Chapter 13).

*QUESTION 37

Describe how acidification acts to increase the risks to the environment and human health posed by Cd.

18.5.3 *Arsenic*

The poisonous properties of arsenic (As) compounds have been known, and exploited, since antiquity, but it is the turn of the twenty-first century that has witnessed inadvertent arsenic poisoning on a mass scale. In Bangladesh and the neighboring Indian province of West Bengal, some 70 million people are at risk of poisoning due to high arsenic levels in the groundwater. In West Bengal alone, up to 200,000 people have been diagnosed with arsenicosis. There are many more cases in Bangladesh, where some 4.5 million people may have been drinking arsenic-laced water for many years.

Ironically, the problem grew out of a United Nations-sponsored effort, starting in the late 1960s, to provide clean drinking water by sinking tube wells into the shallow aquifer that underlies the region. The effort did in fact improve health greatly by reducing the incidence of water-borne diseases. However, the high arsenic content of the reservoir went unrecognized for years. As many as one-half of the 4 million tube wells draw water that exceeds the Bangladesh arsenic standard of 50 ppb. (The U.S. standard was reduced from 50 to 10 ppb in 2001, the guideline recommended by the WHO.) In the more contaminated areas, arsenic levels routinely exceed 500 ppb.

Arsenic in drinking water is a slow poison. The first symptoms are keratoses, discoloring the skin. Later, these develop into cancers, and the liver and kidneys also deteriorate. This process can take up to 20 yr. In its early stages, arsenicosis is reversible, if arsenic consumption is discontinued, but once cancer begins to develop, there is no effective treatment. Poor nutrition renders the Bangladesh villagers more vulnerable than they otherwise might be.

Why is the aquifer laden with arsenic? The answer is not entirely clear, but the likeliest theory is that it is associated with iron-laden sediments from the rivers that drain into the region (the Ganges, Brahmaputra, and Meghna rivers). Arsenic is found naturally in association with sulfide minerals. Iron sulfide (pyrite) is the most widespread source of arsenic, and its weathering leads to the release of arsenate, along with sulfate and ferric hydroxide. The sulfate is washed out to sea, but the more highly charged arsenate adsorbs on the ferric hydroxide and is deposited in river sediments. Typically, these sediments contain 2–6 ppb of arsenic, and the alluvial sediments of Bangladesh and West Bengal, with concentrations of 2–20 ppm, are markedly higher. However, these sediments are also rich in organic matter, and the high BOD leads to reducing conditions, resulting in reduction and solubilization of the iron (Chapter 14), with the release of the adsorbed arsenate.

The Bangladesh and West Bengal sediments are not unique in their tendency to release arsenic into the groundwater. This problem is found in many parts of the world, including regions in Chile, Taiwan, Mexico, and the southwest region of the United States. Indeed, an epidemiological study in Taiwan produced a clear link between arsenic levels in well water and skin cancer incidence. However, only in Bangladesh and West Bengal has the arsenic exposure been so widespread and so long-lasting.

The outlook for the victims of arsenicosis is not encouraging. The villagers are desperately poor and have no access to alternative sources of clean water. There is concern that even labeling the tainted wells could backfire if their users revert to polluted surface waters for their needs. New wells could be dug into a deeper uncontaminated aquifer. In addition, there are several schemes to treat the current well water to remove the arsenic, mostly involving adsorption on $Fe(OH)_3$ particles. However, because there are millions of contaminated wells, the cost of any solution is enormous, and there are numerous administrative and political barriers. The World Bank is coordinating a mitigation plan and its funding, but the massive effort could take 10 yr or more.

QUESTION 38

Why are people in Bangladesh and some parts of India exposed to high levels of arsenic?

Q38 ANSWER In the late 1960s, the United Nations developed a program to provide clean drinking water to the region by sinking tube wells into a low-lying aquifer. Although this reduced the incidence of water-borne diseases, they did not realize that this aquifer had high concentrations of naturally occurring arsenic. Water samples from over half of the wells drilled have arsenic concentrations >50 ppb (50 ppb is the Bangladesh standard) and many of the most contaminated wells have concentrations >500 ppb.

---

*QUESTION 39

How does As get mobilized in aquifers?

---

*QUESTION 40

What can be done to remediate the high arsenic problem in Bangladesh and some parts of India?

18.5.4 Lead

Of all the toxic chemicals in the environment, lead is the most pervasive; it poisons many thousands of people yearly, especially children in urban areas. Once absorbed in the body, lead enters the bloodstream and moves from there to soft tissues. In time, it is deposited in bones, because Pb^{2+} and Ca^{2+} have comparable ionic radii. The lead content of bones increases with age; when bone matter dissolves, as can happen in illness or old age, the lead is remobilized into the bloodstream and can produce added toxic effects.

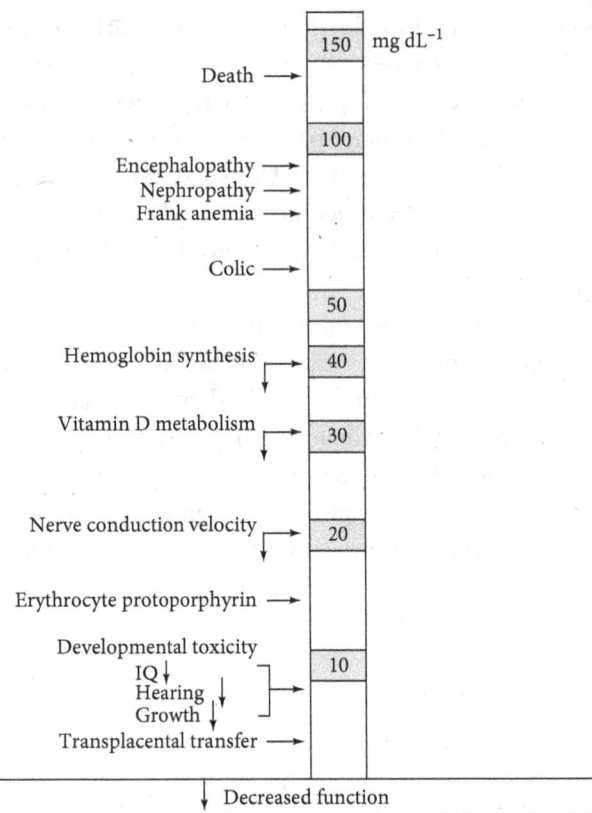

Figure 18.12 Lowest observed effect levels of inorganic lead in children (μg L^{-1}).
Sources: Adapted from Agency for Toxic Substances Disease Registry, *The Nature and Extent of Lead Poisoning in Children in the United States: A Report to Congress,* 1988, ATSDR; and Spiro et al., *Chemistry of the Environment, 3rd Ed.,* University Science Books, © 2012, all rights reserved.

Lead is a neurotoxin. The biochemical mechanism for lead's effects on nerve cells is uncertain, but a diminution of nerve conduction velocity can be detected at relatively low blood lead levels (Fig. 18.12), while higher levels are associated with nerve degeneration. Even at quite low levels, possibly as low as 5 μg dL^{-1}, lead exposure in several epidemiological studies was found to be associated with impairments in growth, hearing, and mental development of children. Between 2018 and 2020[3], 50.5% of all U.S. children had detectable limits of lead in their blood, and 1.9% of children has been lead levels >5 μg dL^{-1}. There is also great concern about exposure in utero because lead can cross the placenta and interfere with fetal development.

Lead poisoning can be treated by intravenous injection of chelating agents (Fig. 18.13). The chelators compete with protein-binding sites and the resulting Pb^{2+}-chelate complexes are excreted by the kidneys. Because the chelating agents can also bind metal ions other than Pb^{2+}, they are administered as the Ca^{2+} complex to avoid stripping Ca and other more weakly binding metals from the body. The more strongly binding Pb^{2+} displaces the Ca^{2+} and is removed selectively.

There are two major exposure routes for lead: drinking water and ingestion of dust. Lead can contaminate water either from lead-based solder used in pipe and fitting connections or from the pipes themselves, which in older houses and water systems are made of lead. In contact with oxygen-bearing water, metallic lead can be oxidized and solubilized:

$$2\,Pb + O_2 + 4\,H^+ \rightarrow 2\,Pb^{2+} + 2\,H_2O \tag{18.1}$$

---

[3] Hauptmann, Niles, Gudin et al., "Individual- and Community-Level Factors Associated with Detectable and Elevated Lead Levels in US Children: Results from a National Clinical Laboratory," 2021, *JAMA Pediatrics,* Vol. 175, No. 12, pp. 1252–1260, https://doi.org/10.1001/jamapediatrics.2021.3518

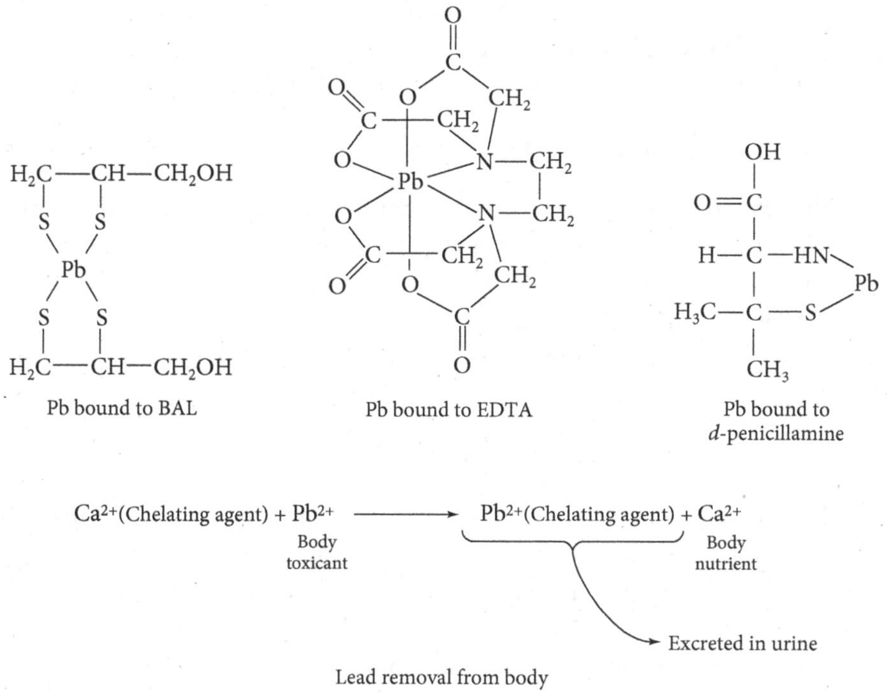

Figure 18.13 Chelating agents for removing lead from the body.
Note: BAL = British anti-Lewisite; EDTA = ethylenediaminetetraacetic acid.
Source: Adapted from Spiro et al., *Chemistry of the Environment, 3rd Ed.,* University Science Books, © 2012, all rights reserved.

Because this reaction consumes two protons per lead ion, the dissolution rate is strongly pH-dependent. The lead hazard is greatest where the water is soft, that is, where there has been little neutralization of the rainwater's acidity. Hard water, on the other hand, generally has a higher pH; in addition, it has a higher carbonate concentration (because of neutralization by $CaCO_3$). Carbonate precipitates Pb^{2+} as the sparingly soluble $PbCO_3$, which inhibits the dissolution of the underlying metal in the pipes and fittings. Some water supply districts, especially those with soft water and old lead pipes, now add phosphate to the drinking water to form a similarly protective coating of lead phosphate. Failing to add phosphate to the water system is what caused Flint, Michigan's serious lead contamination in their drinking water in 2014, after the city switched its water source to the relatively acidic Flint river.

The choice of disinfectant also affects lead pipe solubilization. Because of concerns about chlorinated organic molecules, some water suppliers in the United States have switched from chlorine to chloramine as a disinfectant. Chlorine in water produces hypochlorous acid:

$$Cl_2 + H_2O \rightarrow HOCl + H^+ + Cl^- \tag{18.2}$$

which is both a disinfectant and a chlorinating agent. When ammonia is added, chlorine reacts with it to form chloramine:

$$Cl_2 + NH_3 \rightarrow NH_2Cl + H^+ + Cl^- \tag{18.3}$$

which retains disinfective properties, but is a weak chlorinating agent. However, NH_2Cl is also a weaker oxidant than HOCl. While HOCl is able to oxidize Pb^{2+} to insoluble PbO_2 (the anode of Pb acid batteries), which forms a protective coating on lead pipes, NH_2Cl is unable to do so. Also, the mixture of chlorine and ammonia promotes the dissolution of Cu from pipes, by forming NH_3 complexes with the oxidized Cu^{2+}. The Cu^{2+} can in turn promote the dissolution of Pb (the reduction potential of Cu^{2+} is higher than that of Pb^{2+}):

$$Cu^{2+} + Pb \rightarrow Cu + Pb^{2+} \tag{18.4}$$

Consequently, at the junction of Cu and Pb pipes, chloramine can set up an electrochemical reaction that releases Pb^{2+}. This chemistry is believed to account for a serious episode of lead contamination after the Washington, DC, water company switched to chloramine in 2000, leading to elevated blood lead levels in many of the city's children.

Lead can leach into food and drink stored in pottery if, as has been common, its glaze contains lead oxide:

$$PbO + 2H^+ \rightarrow Pb^{2+} + H_2O \tag{18.5}$$

It is particularly hazardous to drink fruit juices (because of their acidity) or hot drinks (because the rate of dissolution increases with temperature) from such vessels. Although lead-free glazes are now the rule among pottery manufacturers, lead-glazed pottery remains a significant source of dietary lead.

Lead is abundant in dust and is deposited on food crops, or on food as it is being processed. Food and direct ingestion of dust account for most of the average lead intake, estimated to be ~50 μg day^{-1} in the United States. In urban areas or along roadways, the lead levels in dust generally exceed 100 ppm, while in remote rural areas the levels are 10–20 ppm. Ingestion of just 0.1 g of urban dust can provide a 10 μg or higher dose of Pb. It does not take much dust, therefore, whether ingested directly or mixed in with food (the two contributions are estimated to be comparable), to account for the average daily intake. Children are particularly at risk because they play in the dust, they absorb a higher fraction of the lead they take in, and they are exposed to more lead per unit of body mass.

In addition, paint is an important source of lead in dust. Lead salts are brightly colored and have been widely used as pigments and as paint bases. Lead chromate, $PbCrO_4$, has provided the yellow coloring for striping on roads and for school buses, while the "red lead" oxide (Pb_3O_4) is the base for the corrosion-resistant paints on bridges and other metal structures. The hydroxycarbonate $Pb_3(OH)_2(CO_3)_2$ is "white lead," which was widely used as the base of indoor paints, but has now been replaced by titanium dioxide (TiO_2). Nevertheless, older buildings, particularly in the United States, where lead was banned from indoor paint only in 1978 (much of Europe banned it in 1927), still have leaded paint on the walls. Dust and paint chips from the walls are the main source of indoor Pb exposure. Leaded paints have also been widely used on building exteriors, and weathering raises the lead levels in the dust outside the building. Although less dangerous than lead inside the house, the lead outside is still a matter of concern because children often play in the dust, and ingest it, or track it inside. Nearly 22% of children living in inner-city homes built before 1946 have elevated blood lead levels.

Leaded gasoline is another major contributor to lead in dust. Adding tetraethyllead or tetramethyllead to gasoline improves its octane rating by scavenging radicals and inhibiting preignition (see Chapter 8). Most of the lead is emitted in small particles of PbX_2 (X = Cl or Br), formed by reaction with ethylene dichloride or dibromide, which are added to the gasoline to prevent the buildup of lead deposits inside the engine. Most of the particles settle out not far from where they are generated, contaminating the dust near roadways and in urban areas. Lead levels in urban dust correlate strongly with traffic congestion.

Lead additives were phased out of use in the United States starting in the early 1970s, when catalytic converters were introduced for pollution control, because lead particles in the exhaust gases deactivate the catalytic surfaces. Most countries have followed suit, although leaded gas is still available in a few of them.

Although the original motivation for removing Pb from gasoline was to protect catalytic converters, an additional clear benefit was the reduction in human exposure to lead. Everywhere that lead has been phased out, there has been a steady drop in blood lead levels of the populace. Other sources of lead, particularly leaded solder in food cans, have also been curtailed in the same period. Nonetheless, the strongest correlations of blood lead concentrations are with gasoline lead levels. Whereas average blood lead levels of 15 μg dL^{-1} (dL = deciliter; since 1 dL of blood weighs ~100g, 15 μg dL^{-1} is ~0.15 ppm) were common before 1980, they are tending toward 3 μg dL^{-1} in regions that have eliminated lead in gasoline.[4]

QUESTION 41

Describe how acidification acts to increase the risks to the environment and human health posed by Pb. Consider two cases: (1) the dissolution of lead into water supplies serviced by lead pipes and (2) beverages served in lead-glazed cups.

Q41 ANSWER Lead has been used extensively in water pipes and as lead-based glazes in cups and glasses. The dissolution of lead into water supplies serviced by lead pipes, or in beverages served in lead-glazed cups, is strongly pH-dependent. In contact with O_2-bearing water, metallic lead can be oxidized and solubilized:

$$2\,Pb + O_2 + 4\,H^+ \rightarrow 2\,Pb^{2+} + 2\,H_2O$$

This reaction is facilitated in acidic solution. Thus, the lead hazard is greatest in drinking water supplies that are "soft," that is, where there is little neutralization of rainwater's acidity.

Also, lead in lead-glazed cups can leach into beverages via the following reaction:

$$PbO + 2\,H^+ \rightarrow Pb^{2+} + H_2O$$

This reaction is also highly pH-dependent. Thus, the risk of ingesting lead is greatly increased when the beverage is acidic (e.g., fruit juices).

---

*QUESTION 42

Describe the main routes of exposure of Pb.

---

[4] Thomas et al., "Effects of Reducing Lead in Gasoline: An Analysis of the International Experience," *Environmental Science & Technology*, Vol. 33, 1999, pp. 3942–3948.

QUESTION 43

Why are children most susceptible to lead poisoning? How can Pb poisoning be treated?

Q43 ANSWER One reason why children are most susceptible to lead poisoning is that they play in the dust. Ingestion of just 0.1g of urban dust can provide a 10 μg or higher dose of lead. In old houses with peeling lead-based paint, paint chips may be picked up and eaten by infants. A good-sized paint chip can provide an infant with a toxic lead dose. Another reason why children are particularly at risk is that they absorb more of the ingested lead than do adults (30%–40% for children, as opposed to only 7%–15% for adults). The third reason is that they are exposed to more lead per unit of body mass.

Lead poisoning can be treated by chelation therapy. In this method, lead is cleared from the body by intravenous injection of chelating agents (see Fig. 18.13). The chelators bind strongly to lead, and the resulting Pb^{2+} complexes are excreted by the kidneys.

*QUESTION 44

Explain why some water suppliers switched from adding chlorine to adding chloramine to the water, and how this can contaminate the water with lead.

*QUESTION 45

If lead was banned from house paint in 1970 and phased out of gasoline in the 1970s, why are some children still suffering from lead exposure?

18.6 Conclusions

The toxicity of chemicals is an important area of study to understand how different chemicals affect our health. Chronic exposure over long periods of time can cause varying health effects depending on the dose and frequency of exposure, ranging from cancer to hormonal effects. Different tests have been developed to understand the potential toxicity of individual chemicals on humans. They are not always accurate; for example, animal models are not the same biochemically as people and much larger doses of the chemicals must be administered to observe an impact. However, they are being refined as our understanding of biochemical pathways advances. Along with epidemiological studies, these tests provide the basis for evaluating risks and setting regulations. Many toxic chemicals have been introduced to the environment through a variety of pathways. Many are legacies of past industrial processes and have since been discontinued (e.g., PCBs and mercury cells for chlorine production), while others emerge as new chemicals so we must understand how a range of doses of toxic chemicals impact our health, including, but not limited to persistent organic pollutants and toxic metals.

Appendix A

ORGANIC STRUCTURES

Structures like the ones in Fig. 5.4 and 5.5 are how we represent the organic molecules of nature. This appendix offers a brief review of the principles of molecular structure among organic chemicals, to aid the reader in interpreting the structural diagrams found throughout the book.

A.1 Hydrocarbons: Alkanes

Hydrocarbons are organic compounds that contain only carbon and hydrogen. Carbon has four valence electrons (the *valence* electrons are the outer shell of electrons that are available for sharing with other atoms), which it can share with up to four other atoms. These shared electrons are the chemical bonds. They are the glue holding a molecule together. The simplest hydrocarbon is CH_4, *methane*, which is the main constituent of natural gas.

$$
\begin{array}{c}
H \\
| \\
H-C-H \\
| \\
H
\end{array}
$$

Carbon can also share electrons with other carbon atoms. If we replace one of the H atoms in methane with a C atom, and add three more H atoms, the result is *ethane*, C_2H_6, whose structure can be represented by drawing lines between the atoms sharing electrons:

$$
\begin{array}{cc}
H & H \\
| & | \\
H-C-C-H \\
| & | \\
H & H
\end{array}
$$

Replacing an H atom in ethane with another C atom and three more H atoms produces *propane*, C_3H_8:

$$
\begin{array}{ccc}
H & H & H \\
| & | & | \\
H-C-C-C-H \\
| & | & | \\
H & H & H
\end{array}
$$

The next members of this series of molecules are *butane*, C_4H_{10}; *pentane*, C_5H_{12}; *hexane*, C_6H_{14}; and so on (Table A.1). Each time a C atom is added, so are two H atoms, and the general formula for this series, called the *alkanes*, is C_nH_{2n+2}. The n- before the names of the alkanes in Table A.1 designates that the molecules are a straight chain of carbon atoms. Examples of branched alkanes are included in section A.2.

417

TABLE A.1 Selected Alkanes

Number of Carbon Atoms	Chemical Formula of Alkane	Name of Alkane
1	CH_4	methane
2	H_3CCH_3	ethane
3	$H_3CCH_2CH_3$	propane
4	$H_3C(CH_2)_2CH_3$	n-butane
5	$H_3C(CH_2)_3CH_3$	n-pentane
6	$H_3C(CH_2)_4CH_3$	n-hexane
7	$H_3C(CH_2)_5CH_3$	n-heptane
8	$H_3C(CH_2)_6CH_3$	n-octane
9	$H_3C(CH_2)_7CH_3$	n-nonane
10	$H_3C(CH_2)_8CH_3$	n-decane

A.2 Isomers

However, when we come to butane, there are two ways of arranging the bonds:

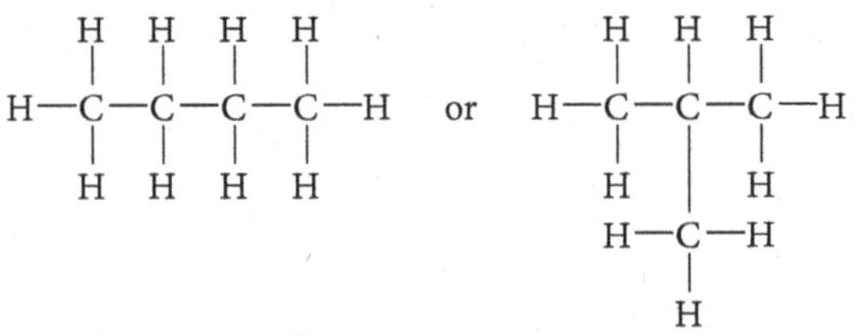

The fourth C atom can be added to the end of the chain of three C atoms, or it can be added to the middle. The two molecules have the same chemical formula, but because the bonding arrangement is different, they have different physical properties (e.g. different melting and boiling points), and different chemical properties (different reactivity). We call them *isomers*. The first isomer is called *n-butane* (n- for "normal"; all straight chain alkanes carry this prefix) and the second *iso-butane*.

As the number of C atoms increase, the number of possible isomers goes up rapidly. To distinguish among these isomers, a systematic naming scheme has been developed. The name is based on the longest chain of C atoms in the molecule, and the position of substitution of a shorter chain is numbered from the end of the longer chain. For example, the systematic name for iso-butane is 2-methylpropane, because the longest chain has three C atoms, and the second C atom of this chain bears a methyl group. The suffix –*yl* indicates a fragment in which a bond to a H atom is replaced by a bond to the rest of the molecule, thus

$$H_3C-H \qquad\qquad H_3C-$$
methane   methyl

(It is convenient to lump the H atoms together with the C atoms to which they are attached, making the C–H bonds implicit.)

Another example of a *branched* chain hydrocarbon is 2,2,4-trimethylpentane.

$$\overset{\displaystyle CH_3 \qquad\quad CH_3}{\underset{\displaystyle CH_3}{H_3\overset{1}{C}-\overset{2}{C}-\overset{3}{C}H_2-\overset{4}{C}H-\overset{5}{C}H_3}}$$

This molecule, also called "iso-octane," is an isomer of n-octane.

The n-alkanes are *straight-chain* molecules, while their isomers are *branched-chain* molecules. Organisms manufacture straight-chain hydrocarbons as the main constituents of *fats* and *lipids*. Fats are energy storage molecules, while lipids are the building blocks of biological membranes. Consequently, petroleum contains a substantial proportion of straight-chain hydrocarbons. But it also contains many branched chain hydrocarbons because the high pressure and temperature conditions of geological deposits induce rearrangement of the C–C bonds. This process is deliberately continued in the chemical reactors of oil refineries, because hydrocarbons with branched chains have higher octane ratings than those with straight chains (see Chapter 5).

A.3 Rings

In the process of attaching C atoms to one another, it is possible to form rings, as well as chains. We can have *cyclo-propane, cyclo-butane, cyclo-pentane, etc.*

cyclo-propane   cyclo-butane   cyclo-pentane

Rings can also be joined together to form *polycyclic* molecules. An example is the biological marker molecule, bacteriohopanetetrol (Figure 5.4), which has three six-membered rings and one five-membered ring, as well as a five-carbon chain. This complicated molecule is made by bacteria, but animals too make polycyclic molecules, such as cholesterol:

QUESTION 1

How many isomers are there of pentane? Draw and name them.

Q1 ANSWER One isomer is obviously the straight chain n-pentane:

$$CH_3-CH_2-CH_2-CH_2-CH_3$$

Another isomer is obtained by putting a methyl group in the middle of a chain of four C atoms:

$$\underset{\displaystyle CH_3}{H_3C-CH-CH_2-CH_3}$$

This is 2-methylbutane. (Note that it does not matter which of the two middle C atoms is attached to the methyl groups, because you get the same molecule by turning the structure around.) Finally, two methyl groups can both be attached to the middle C of a three C chain.

$$CH_3-\underset{\underset{CH_3}{|}}{\overset{\overset{CH_3}{|}}{C}}-CH_3$$

This is 2,2-dimethylpropane. Another name, equally descriptive, would be tetramethylmethane, recognizing that this molecule can be derived from CH_4 by replacing all four H atoms with methyl groups.

There are no other ways of arranging molecules with the formula C_5H_{12}. Pentane has three isomers.

A.4 Unsaturated Hydrocarbons

Carbon can share its four valence electrons two or three at a time, forming double or triple bonds (symbolized by double and triple lines, = and ≡). The simplest hydrocarbon with a double bond is *ethene* (commonly called *ethylene*), $CH_2=CH_2$, while the simplest one with a triple bond is *ethyne* (commonly called *acetylene*), $CH\equiv CH$. The H atoms in these simplest *alkenes* and *alkynes* can be replaced by C atoms to form more complex molecules. For example, substituting an H atom in ethylene and acetylene with a methyl group produces *propene* (often called *propylene*), $CH_2=CHCH_3$, and *propyne*, $CH\equiv CCH_3$, respectively. If there are two or more double bonds or triple bonds, we have a poly-ene, or a poly-yne.

Molecules having C=C or C≡C bonds are said to be *unsaturated*, because the C atoms do not have the full complement of four possible bonding partners. We are all familiar with unsaturated and polyunsaturated cooking oils, because they are less likely than saturated oils to contribute to heart disease. Unsaturated oils contain fat molecules having a double bond in their hydrocarbon chains, while polyunsaturated oils have two or more double bonds in a chain.

A.5 Molecular Shape

Unsaturated molecules differ in shape from saturated molecules because the bond angles differ. Pairs of electrons in the valence shell try to keep as far apart as possible. Thus, if a C atom has four bonding partners, the bonds extend toward the four corners of a tetrahedron (bond angles of about 109°). But if there is a double bond, the two electron pairs in the bond are forced into the same region of space, and together they remain as far away as possible from the other two electron pairs in the single bonds. The three bonds, one double and two single, then point toward the corners of an equilateral triangle (bond angles of about 120°). The result is that the two-carbon unit, with its associated single bonds, is planar. Thus, ethylene can be represented as:

$$\underset{H}{\overset{H}{>}}C=C\underset{H}{\overset{H}{<}}$$

Likewise, the electrons in a triple bond are as far away as possible from the pair in the remaining single bonds, giving a bond angle of 180°. A triple bond is co-linear with the associated single bonds; acetylene can be represented as:

$$H—C\equiv C—H$$

Another important aspect of geometry is the ease of rotation about a bond. Rotation is easy about single bonds; the parts of the molecule connected by a single bond can adopt various interconvertible orientations. But rotation about a double bond is difficult; it does not occur except at very high temperatures. Consequently, there are two separate orientations for the parts of the molecule connected by a double bond, for example:

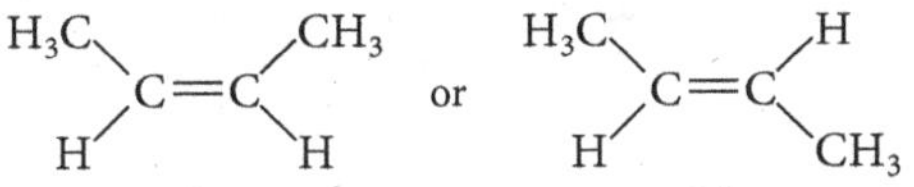

cis-2-butene   trans-2-butene

These two orientations are called *cis* and *trans* (methyl groups on the same or opposite side of the double bond). Because the double bond does not rotate, these are *geometric* isomers, which have different physical and chemical properties because their shapes differ.

A.6 Carbon Framework Representations

To reduce clutter in organic structural diagrams, one can simply draw the carbon framework as a series of connected lines. For example, n-octane and iso-octane can be represented as:

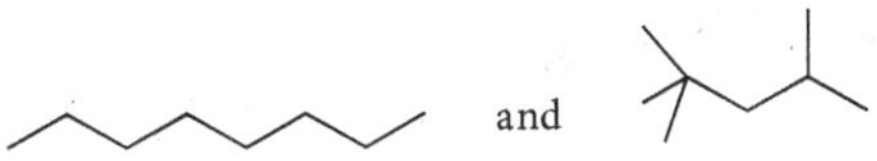

n-octane   iso-octane

The H and C atoms are implicit in these representations. One understands that the lines connect C atoms, which have as many C-H bonds as are needed to complete their complement of four bonds. The kink between lines represents the fact that saturated C atoms have tetrahedral bond angles.

When the carbon framework contains double or triple bonds, this is indicated by drawing double or triple lines at the appropriate place.

cis-2-butene   2-butyne

(Note that the lines are co-linear for 2-butyne, because of the 180° bond angles.)

QUESTION 2

Name the following compounds from the framework representations:

(a)

(b)

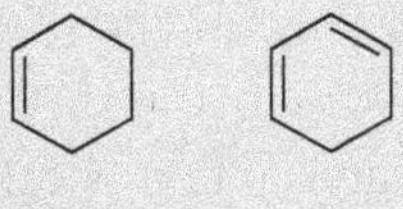

Q2 ANSWER

(a)   cis-3-hexene   2-methyl-trans-3-pentene

(Note that these two molecules have the same number of C and H atoms; they are isomers.)

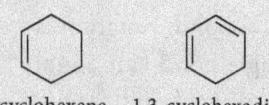

(b)   cyclohexene   1,3-cyclohexadiene

(The bonds are always *cis* in a ring structure, so this designation is not needed.)

A.7 Aromatic Compounds

When the six-membered ring has three double bonds, a different kind of molecule, *benzene*, is formed. The double bonds can be drawn in two alternate, but equivalent ways:

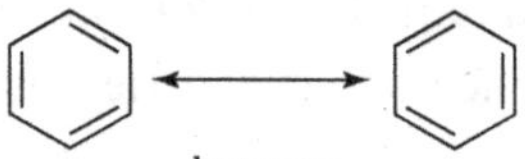

benzene

Normally double bonds are shorter than single bonds, but in benzene, all six carbon-carbon bonds are found to be the same length. The electrons end up adopting both bonding arrangements at the same time, a phenomenon termed *resonance*. We often represent this situation by inscribing a circle in the ring.

Because of resonance, the benzene molecule is more stable and less reactive than we would expect for a molecule with three double bonds.

The hydrogen atoms on benzene can be replaced by carbon atoms, giving rise to a large class of molecules, called (for historical reasons) *aromatic* molecules. For example, methyl-benzene, also called toluene, is a common solvent:

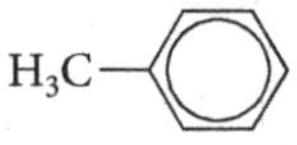

H₃C—⬡

toluene

The lignin molecule in Figure 5.5 is a very complex aromatic molecule.

When two or more of the H atoms on benzene are substituted, the possibility of isomers again arises. For example, there are three isomers of xylene, the molecule obtained by substituting two methyl groups on benzene:

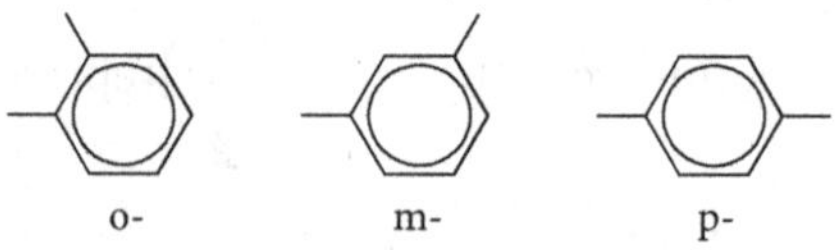

o-   m-   p-

These are designated *ortho-, meta- and para-*, shortened to *o-, m-, and p-*. (The systematic names, however, are 1,2-, 1,3- and 1,4-dimethylbenzene.)

When a benzene molecule is attached to a larger organic molecule, it is called a *phenyl* group. For example, the insecticide DDT is dichlorodiphenyltrichloroethane (see Chapter 17):

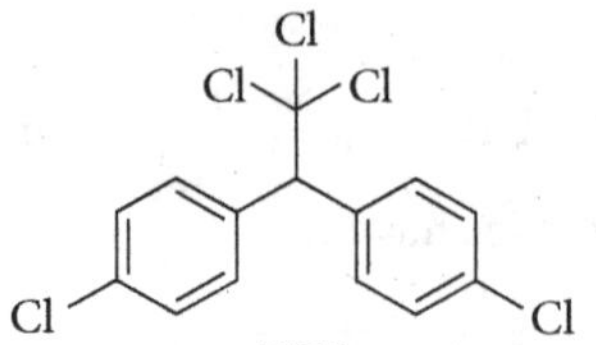

DDT

Just as in the case of the saturated rings in polycyclic alkanes, it is possible to fuse benzene rings to form *polycyclic aromatic hydrocarbons* (PAHs), for example:

naphthalene   anthracene

These PAHs are side products of incomplete combustion. They are of concern as potent carcinogens (see Chapter 9). The ultimate PAH is graphite, whose structure is a continuous sheet of benzene rings (Figure 5.5).

A.8 Hetero-atoms

Organic compounds are based on carbon and hydrogen, but often contain atoms of other elements, collectively called hetero-atoms. The elements most commonly encountered are the halogens, oxygen, nitrogen, sulfur and phosphorus. These hetero-atoms are frequently sites of chemical reactivity, and the chemical groups containing them are called functional groups (Table A.2).

Halogen atoms form only one bond, and can, in principle, be substituted for any of the H atoms in a hydrocarbon. The general symbol for organohalides is RX, where X stands for F, Cl, Br or I, and R represents the rest of the organic molecule. Organohalides are generally less reactive than the parent hydrocarbon. This is one reason why organochlorine pesticides, such as DDT, are of environmental concern. Because of their low reactivity, they last for a long time and can build up in the body tissues of animals, including humans (see Chapter 17).

Oxygen atoms form two covalent bonds, as in H_2O. If an O atom forms a bond with a C atom, as well as with an H atom, we have an *alcohol*, symbolized as ROH. Alcohols are named by adding *–ol* to the end of the name of the organic molecule. The drinkable alcohol is ethyl alcohol, or ethanol, $CH_3–CH_2–OH$. The complex molecule in Figure 5.4 is called bacterio-hopanetetrol because of the four alcohol groups it contains.

If the O atom is bound to two C atoms, we have an *ether*, ROR'. We distinguish R and R', because the two organic fragments which are connected to the O atom need not be the same. An example is

Methyl-t-butyl ether

abbreviated MTBE. (*t-* stands for *tertiary*; *t*-butyl is a way of designating a four-carbon group, in which all the bonds from the connecting C atom are to other C atoms.) MTBE was used as a gasoline additive intended to cut pollution, but gained notoriety because of groundwater contamination from gasoline leaks (see Chapter 15). Notice that the lignin structure (Figure 5.5) has many alcohol and ether groups.

The O atom can also form a double bond with a C atom, and the resulting C=O group is called a *carbonyl*. Carbonyl compounds are distinguished by what is attached to the C atom. *Ketones* have two other C atoms, while *aldehydes* have one C and one H atom. Common examples are:

acetone   acetaldehyde

TABLE A.2 Selected Organic Functional Groups

Name of Class	Functional Group	General Framework of Class
Alkane	None	R—H
Alkene	$-\overset{\vert}{C}=\overset{\vert}{C}-$	$\underset{R''}{\overset{R}{\diagdown}}C=C\underset{R'''}{\overset{R'}{\diagup}}$
Alkyne	$-C\equiv C-$	$R-\!\!\equiv\!\!-R'$
Halide	$-X$ (X = F, Cl, Br, I)	R—X
Alcohol	$-\overset{\vert}{\underset{\vert}{C}}-O-H$	R—OH
Ether	$-\overset{\vert}{\underset{\vert}{C}}-O-\overset{\vert}{\underset{\vert}{C}}-$	$\underset{O}{\overset{R\diagdown\diagup R'}{}}$
Aldehyde	$-\overset{\overset{O}{\|}}{C}-H$	$\underset{R\qquad H}{\overset{O}{\|}}$
Ketone	$-\overset{\overset{O}{\|}}{C}-$	$\underset{R\qquad R'}{\overset{O}{\|}}$
Amine	$-\overset{\vert}{\underset{\vert}{C}}-\overset{\vert}{N}-$	$R-N\underset{R''}{\overset{R'}{\diagup}}$
Imine	$-\overset{\vert}{C}=N-$	$\underset{R}{\overset{R'}{\diagup}}C=N-R''$
Nitrile	$-C\equiv N$	$R-C\equiv N$
Carboxylic acid	$-\overset{\overset{O}{\|}}{C}-O-H$	$R-C\underset{OH}{\overset{\diagup O}{}}$
Ester	$-\overset{\overset{O}{\|}}{C}-O-\overset{\vert}{C}-$	$R-C\underset{O-R'}{\overset{\diagup O}{}}$
Amide	$-\overset{\overset{O}{\|}}{C}-\overset{\vert}{N}-$	$R-C\underset{N}{\overset{\diagup O}{}}\underset{R''}{\overset{R'}{}}$

*For some classes, R, R', and R'' can be H.

Additional functional groups are *carboxylic acids* and *esters*:

$$R-C\underset{OH}{\overset{\diagup O}{}} \qquad R-C\underset{OR'}{\overset{\diagup O}{}}$$

organic acid   ester

The chemistry of the carbonyl group is altered considerably by having an OH or OR group attached to the C atom. *Polyesters* are an important class of polymer molecules in which hydrocarbons are linked by repeating ester groups into long chains. They are widely used in fabrics and in plastic containers.

Finally, two oxygen atoms can share a single bond, while having another single bond each to either H or C. These are *peroxides*, with the general formula R–O–O–R'. If one of the R groups is H, we have a *hydroperoxide*, and the simplest member of the family, H–O–O–H is hydrogen peroxide. The O–O bond is weak, and peroxides are a reactive class of molecules. They play an important role in the chemistry of air pollution (see Chapter 9).

Because sulfur is in the same group of the periodic table as oxygen, the same valence rules apply, and similar functional groups are found in organic sulfur as in organic oxygen compounds. However, sulfur, being a larger atom is able to have more than eight electrons in its valence shell when combining with highly electron-seeking (*electronegative*) elements such as oxygen or fluorine. Oxygen atoms can readily add to organic sulfur compounds, producing *sulfones*, R_2SO; *sulfoxides*, R_2SO_2; or *sulfonic acids*, RSO_3H. Although much less abundant than oxygen in biological tissues, sulfur is a significant constituent of coal and petroleum, and is an important contributor to air pollution when these fossil fuels are burned (see Chapter 9).

Compounds having three bonds to an N atom are *amines*. In addition to *ammonia*, NH_3, they include *primary* amines, RNH_2; *secondary* amines, R,R'NH; and *tertiary* amines R,R',R''N. Compounds with double bonded N are *imines*, R,R'C=NR'' (the Rs can also be H), while compounds with triple bonded N are *nitriles*, RC≡N.

An important class of compounds are the *amides*, in which an N atom is connected to the C atom of a carbonyl group:

$$R-C \underset{\underset{R'}{N}}{\overset{O}{\parallel}} R$$

Polyamide polymers are important as fibers and plastics, while protein molecules, which have *amino acids* linked head to tail through amide linkages,

$$H_2N \underset{\underset{R}{CH}}{\diagdown} \overset{O}{\underset{OH}{C}}$$

are nature's own polyamides.

Phosphorus is in the same group of the periodic table as nitrogen, and the phosphorus analogs of ammines are *phosphines*. However, phosphorus forms much stronger bonds with oxygen than with carbon or hydrogen. In nature, P atoms are always surrounded by O atoms. The most common form is the phosphate ion, PO_4^{3-}, which can condense into polyphosphate ions, such as *tripolyphosphate* (see Chapter 14), by forming P-O-P links. Many important biological molecules are phosphate esters, $ROPO_3^{2-}$; the nucleic acids (see Chapter 18) are polymers held together by *phosphodiester* links, $RO-PO_2-OR'$.

Appendix B

MATHEMATICAL FUNDAMENTALS

B.1 Units and Conversions

When considering environmental issues, the overriding question is "how much"? Whether it is exposure to toxic substances, emission of pollutants, or the amount of energy used, a lot or a little makes the difference between a problem and a non-problem. Often the number need not be precise; knowing something to within a factor of ten (called an order of magnitude) may sometimes be enough, because amounts can vary by many factors of ten.

B.1.a *Exponents*

In order to avoid having to write out many zeros, we can express numbers in *exponential notation*, $n \times 10^y$, where y is the exponent, that is, the number of times n is multiplied by ten. If there is a minus sign in front of it, then y is the number of times n is multiplied by 10^{-1}, i.e. by 0.1. For example, 5.1×10^3 means 5,100, while 5.1×10^{-3} means 0.0051.

In addition, we can use prefixes in front of the units in order to modify the amounts. The commonly used prefixes are kilo- (k), mega- (M), giga- (G), terra- (T), peta- (P), and exa- (E) for $y = 3, 6, 9, 12, 15$ and 18; and centi- (c), milli- (m), micro- (μ), nano- (n), pico- (p) and femto- (f) for $-y = 2, 3, 6, 9, 12$ and 15. The letters in parentheses are accepted abbreviations for the prefixes. Thus 5.1×10^3 meters (abbreviated m) is the same as 5.1 kilometers (km), and 5.1×10^{-3} meters is the same as 0.51 centimeters (cm) or 5.1 millimeters (mm).

B.1.b *Metric Units*

To gauge how much something is, we must specify the units of measurement. For scientific purposes, we use the metric system. Length is specified in meters (m) and kilometers (km); area in hectares (ha) (one hectare is 10,000 square meters); volume in liters (L) (one liter is 1,000 cubic centimeters); and mass in grams (g), kilograms (kg), and metric tons (t) (one metric ton is 1,000 kilograms). This is also common usage around the world, except in the United States, where English units continue to be used: inches (in), feet (ft), and miles (mi) for length; acres (1 acre is 43,560 square feet (ft^2) for area; pints, quarts (qt), and gallons (gal) for volume; ounces (oz), pounds (lb), and tons (t) (sometimes called a short ton is 2,000 pounds (lb)) for weight. We use the metric system in this book, but the following conversion factors may sometimes be useful.

> 1 inch = 2.540 cm
> 1 foot = 0.305 m
> 1 mile = 1.609 km
> 1 acre = 0.405 ha

> 1 quart = 0.946 L
> 1 gallon = 3.785 L
> 1 pound = 0.454 kg
> 1 short ton = 0.9072 t

Temperature is measured in degrees centigrade (°C) in the metric system and degrees Fahrenheit (°F) in the English system. At sea level, water boils at 100°C, but 212°F, and freezes at 0°C, but 32°F. Consequently

$$(°F) = \frac{9}{5}(°C) + 32 \qquad (1)$$

In scientific calculations, we need the absolute temperature (T), which is expressed in Kelvin (K) units. Absolute zero is –273 °C, so

$$K = °C + 273 \qquad (2)$$

Time is measured everywhere in seconds (sec), minutes (min), hours (h), days (d) and years (y).

B.1.c *Energy Units*

Energy comes in many forms and has historically been measured in many different units. The unit that has been chosen as the standard is the joule (J). The joule was originally defined as a unit of work energy. One joule is the work done by a force that accelerates a 1 gram mass at 1 cm/sec^2 for a distance of 1 m. A kilojoule (kJ) is 1,000 J, and an exajoule (EJ) is 10^{18} J.

All forms of energy can be given in joules, but many still prefer to express heat energy in the historically used calorie units, because of their intuitive appeal. One calorie (cal) is the heat energy needed to raise the temperature of 1 g of water by 1 °C (from 14.5 to 15.5 °C). The conversion between the two units is:

$$1 \text{ cal} = 4.184 \text{ J}$$

In common usage, calories are most often associated with measuring the energy value of food. Unfortunately, the nutritionists Calorie (distinguished by the capital letter), is not the same as a calorie, but 1000 time larger (actually a kilocalorie). So if your daily intake is 2000 Cal, you are actually consuming 2×10^6 cal.

Another energy unit commonly used in the U.S. is the British Thermal Unit (BTU), defined as the quantity of heat required to raise the temperature of one pound of water one degree Fahrenheit. A quad (Q) is one quadrillion BTUs (10^{15} BTUs). The conversion factors between joules and BTUs are:

> 1 BTU = 1,055 J = 1.055 kJ
> 1 Q = 1.055 EJ

Power is also related to energy units. Power is the rate at which energy is delivered. The standard unit is the watt (W).

$$1 \text{ W} = 1\frac{\text{J}}{\text{sec}}$$

Conversely, energy is power multiplied by the time over which the energy is delivered.

Thus, electricity is commonly metered out in kilowatt-hours (kWh):

$$1 \text{ kWh} = 1{,}000 \frac{J}{\text{sec}} \times 1 \text{ hr} \times 3{,}600 \frac{\text{sec}}{\text{hr}} = 3.6 \times 10^6 \text{ J}$$

The energy of light waves is proportional to their frequency, which is usually expressed as the wave number ($v = 1/\lambda$) in cm^{-1}. For example, blue light with a 500-nm wavelength has a wave number, or "energy," of $(500 \times 10^{-7} \text{ cm})^{-1} = 20{,}000 \text{ cm}^{-1}$. It is possible to relate the wave number to the equivalent quantity of heat, or chemical energy per mole of photons; thus

$$1 \frac{kJ}{\text{mol}} = 83.6 \text{ cm}^{-1}$$

The electron volt (eV) is a unit commonly used by physicists in describing radiation and elementary particles. It is the amount of energy acquired by any charged particles that carries unit electric charge when it falls through a potential difference of 1 volt. It can be related to the equivalent wave number of electromagnetic radiation:

$$1 \text{ eV} = 8{,}064.9 \text{ cm}^{-1}$$

Appendix C

ANSWERS TO STARRED PROBLEMS

Chapter 1

1.2. DeSimone applied Principle 8—use safer solvents.

1.3. Principles 4—renewable resources and 10—products that degrade after use.

1.4. Principle 8—safer solvents.

1.5. Principles 5—use catalysts and 8—safer reaction conditions.

1.6. Principles 3—design less hazardous chemical syntheses, 5—use catalysts, and 7—maximize atom economy.

1.7. Principles 2—safer products and 3— less hazardous syntheses.

1.8. Principles 1—prevent waste and 2—safer products.

1.9. Principle 2—safer products.

1.10. Principles 1—reduce waste and 7—maximize atom economy.

1.11. Principles 1—reduce waste and 2—safer products.

1.12. Principle 10—design products to degrade and not accumulate in the environment.

1.13. Principle 2—safer products.

1.14. Principles 2—safer products, 4—renewable resources, 8—safer reaction conditions, and 10—design products to degrade and not accumulate in the environment.

1.18. 3.3 yr

1.19. 6.7 yr

1.21. 31.5 TJ

1.24. pH = 11.7

1.27. 7.77×10^8 people

1.29. 97.7 yr

1.30. 100 sec

Chapter 2

2.2. 0.0113%

2.3. 74%

2.6. 28.04%

2.8. Looking at Figure 2.2, the largest growth in coal energy consumption appears to have been in the late 1890s, in petroleum in the 1950s, in natural gas also in the 1950s and then the 2000s, and in nuclear in the 1960s.

2.10. 25 yr

2.15. 71.3 yr

2.17. 116 yr

2.21. 7.5 bbl of oil per year

2.23. 0.68% yr^{-1}

2.25. 150 yr

Chapter 3

3.2. 73%

3.3. 26%

3.7. 80.4%

3.9. 811%

3.10. 1,760%

3.13. 94.5%

3.16. 0.72 V

3.18. Extrapolating the ohmic line to zero current, gives a voltage of ~0.9 V, about 0.1 V less than the initial potential. This drop can be attributed to the activation loss. It would be lowered by finding better catalysts to speed up the electrode reactions.

3.19. The linear portion bends downward after ~1,800 mA cm^{-2}. Increasing access of the gases to the electrodes, e.g., by using more porous electrodes, would improve mass transfer.

3.20. 33%

3.22. 1.10 V

3.23. The vanadium oxidation state is +3 and 0, respectively.

3.24. 1.22 V and the cell reaction is $VO_2^{2+} + V^{2+} + 2H^+ \rightarrow VO^{2+} + V^{3+} + H_2O(l)$

3.26. Collecting, shipping, and processing the ore must require much more energy for Al.

3.28. $1.7 \times 10^3 \dfrac{kWh}{ton}$ (paper cup)

In principle, 70% more energy could be recovered by incineration. For polystyrene, the recoverable energy would be:

$3.3 \times 10^3 \dfrac{kWh}{ton}$ (polysterene cup)

In principle, 11 times as much energy could be recovered.

Chapter 4

4.2. 1,920%

4.4. 14.2%

4.6. $12,220

4.9. 68%

4.11. 0.719 eV; therefore, a wavelength of 1,725 nm does not have enough energy to excite an electron into the conduction band for a Si semiconductor.

4.13. 0.3 kWh less energy

4.16. 37.6%

4.19. 6.3×10^8 W (or 0.63 GW)

4.21. 7.9%

4.23. 570%

4.26. 1,700%

Chapter 5

5.3. 0.74 ppm increase per year

5.5. 677 kJ per mol CO_2, 46.1 kJ produced per g C_3H_8

5.6. −1,257 kJ; 419 kJ of energy given off per mole of O_2; 1,257 kJ given off per mole of ethanol; 27.3 kJ given off per gram of ethanol; 1.59 mol CO_2. Thus, ethanol is about 63% as efficient as gasoline.

5.9. 45.0 kJ g^{-1}; 44.3 kJ g^{-1}. We would expect gasoline (with the shortest chain length) to have higher energy density than kerosene and lubrication oil (increasing hydrocarbon chain length).

5.11. $CO + H_2O \rightarrow CO_2 + H_2$ (Water–gas shift reaction) $CH_4 + 2H_2O \rightarrow CO_2 + 4H_2$ (Net formation of H_2 from CH_4)

5.14. 0.0686 g CO_2 kJ^{-1}; The greenest fuel from a carbon emissions standpoint has the *lowest* mass of CO_2 released per kJ of energy created. Therefore, H_2 is more green than diesel, according to this "green factor." Diesel and gasoline are very similar in comparison.

5.16.

5.18.

5.21. According to the map shown in Figure 5.7, mostly bituminous coal is found in Kentucky. This is the most abundant type of coal found in the United States and contains the highest concentration of sulfur (70.2% has greater than 1% sulfur in content). According to Table 5.3, the heating value (kJ g^{-1}) is highest for the bituminous coal in Kentucky, which makes it favorable for use as fuel at a power plant. The high sulfur content could lead to excess sulfur dioxide emissions, leading to acid rain, but this could be mitigated by removing the sulfur with scrubbers after the coal is burned. Transporting coal from other low-sulfur coal sources in the western United States would use more energy than cleaning the sulfur out of the power plant emissions.

5.23. 2.4×10^7 railcars of carbamate!

Chapter 6

6.2. C-12 is most abundant.

6.4. 15.999 amu

6.6. The biosphere is depleted of C-13; there is more C-12 in the biosphere by 26 parts per thousand.

6.9. (i) $^{235}_{92}U + ^1_0n \rightarrow ^{140}_{56}Ba + 3\,^1_0n + ^{93}_{36}Kr$

(ii) $^{235}_{92}U + ^1_0n \rightarrow ^{144}_{55}Cs + ^{90}_{37}Rb + 2\,^1_0n$

6.11. $2.3 \times 10^{14} \dfrac{g}{cm^3}$

6.13. (i) This isotope will emit a β particle:
$^{24}_{11}Na \rightarrow ^{24}_{12}Mg + ^{\ 0}_{-1}\beta$
(ii) This isotope will capture an electron:
$^7_4Be + ^{\ 0}_{-1}e \rightarrow ^7_3Li$

6.15. $1.41 \times 10^{-12} \dfrac{J}{nucleon}$

6.17. 11.7 yr
6.19. 3,853 yr
6.21. Only hydrogen-3 has a suitable half-life.
6.23. 5.5×10^{-5} mCi
6.25. $209 \dfrac{g}{mol}$; this would be polonium (Po).
6.27. Joe was exposed to more radiation at 3 rad.
6.30. 0.81 J
6.32. -2.74×10^{-11} J
6.34. 65.4%
6.36. Uranium loses electrons.
6.38. 1,218 barriers.
6.40. (i) 1.47×10^{22} atoms decayed **(ii)** 4.24×10^9 J
6.42. 79.5% of C-137 would remain after 10 yr
6.44. -2.82×10^{-12} J
6.46. With a short half-life, the fluorine isotope will not be present in the body for very long. This is good because it will not be continuously emitting positrons. On the other hand, the imaging must occur quite quickly as well to have enough radioactivity for detection. Half of the fluorine isotope will be gone after 2 hr of ingestion.
6.48. Only 6.2% would be left after 1 day.
6.49. 8.3×10^{-4} nm

Chapter 7

7.3. 0.16%
7.5. 2.8×10^{12} light bulbs
7.6. 7.4×10^{14} light bulbs
7.8. Between 25% and 35%
7.9. Between 0% difference between the figure and the average and 10% loss
7.11. 2,000 L less
7.13. UV radiation is less than 400 nm and IR is greater than 700 nm.

7.15. Visible spectrum
7.17. 2.9×10^2 K
7.19. 255 K
7.21. 0.307
7.22. $390 \dfrac{W}{m^2}$
7.24. 0.93%
7.26. 4 vibrations
7.29. 6.2 cm
7.30. 0.38
7.32. 2.6×10^4 L
7.33. 2.8×10^{-4} km^2
7.34. Absorbed by the surface (land and ocean), absorbed by clouds and the atmosphere, reflected by clouds and the atmosphere, and reflected by the surface.
7.36. 70% for absorption.
7.38. 24.7×10^8 TWh
7.40. 5.9% decrease
7.42. $CH_4 = 0.0001892\%$; $N_2O = 0.0000334\%$
7.43. $CO_2 = 48.9\%$; $CH_4 = 170.\%$; $N_2O = 24\%$
7.45. 507 ppm
7.46. $0.00550 \dfrac{ppb}{yr}$
7.47. $CO_2 = 33.9\%$; $CH_4 = 133\%$; $N_2O = 18.5\%$
7.49. 788 GT CO_2
7.51. 50,000,000 yr
7.53. 3.95 yr
7.55. The albedo is between 1.25 and 4.0 times greater in the desert areas than in the forest.
7.57. 0.135 μm
7.58. 0.108 μm
7.59. 1.004%

Chapter 8

8.5. $\ddot{O}=\dot{N}-\ddot{O}:$ NO_2 is more reactive.
$\ddot{O}=N-\ddot{O}-H$
       $:\!\ddot{O}\!:$

8.9. No, this reaction is not spontaneous in the forward direction. $K_{eq} = 4.0 \times 10^{-31}$.
Reactants are favored.
8.10. 1.04×10^{-29}
8.15. 1,118 nm; IR light.
8.21. 0.16 ppb

8.25. Oxidants are secondary pollutants; they are produced in the atmosphere from the compounds directly emitted. Ozone is an example of a secondary pollutant.

8.26. Seven ozone molecules

8.28. $C_2H_5OH + 3O_2 \rightarrow 3H_2O + 2CO_2$

8.30. The substituted ring of 1,2,3-trimethylcyclohexane (b) will have a higher octane number than a straight-chained alkane, but the branched nature of isopropylbenzene (a) will make it more resistant to radical formation. Isopropylbenzene will have a higher octane number.

8.33. Ethane (two carbons) will have the highest vapor pressure, followed by pentene (three carbons), butene (four carbons), and finally heptene (seven carbons).

8.35. Hexane is a six-carbon alkane and although isoprene is smaller (five carbons), it has a higher PA because it has two double bonds.

8.38. The purpose of the lead was to boost the gasoline's octane number and prevent preignition "knocking," so when lead was removed, additives with high octane numbers had to be added. Aromatics and olefins have higher octane numbers than alkenes.

8.39. Ethanol and MTBE have high octane numbers. MTBE has a low vapor pressure; ethanol's is higher, but still lower, on average, than those of the olefins. Both ethanol and MTBE have low PAs.

Chapter 9

9.2. 160 mm Hg

9.3. 0.32 mol O_2

9.6. $5.8 \times 10^{11} \dfrac{\text{molec naphthalene}}{\text{cm}^3 \text{ air}}$

9.7. 409 ppm

9.9. 9.3 yr

9.10. (i) 63 days   **(ii)** $4.9 \times 10^5 \dfrac{\text{molec}}{\text{cm}^3 \text{ sec}}$

(iii) 5.0 ppm

9.12. 3.82×10^6 tons $CaSO_3$

9.14. 4.7 days

9.16. (i) 3 hr

(ii) OH· = 30%; $NO_3\cdot$ = 70%; O_3 = 0.3%

9.18. Beijing = 357 mg; Toulouse = 50 mg; Hilo = 34 mg

9.19. (i) $742 \dfrac{\mu g}{m^3}$

(ii) This value is much larger than the actual concentration of particulate in the atmosphere. The calculated value assumes that all of the particles emitted into the air will stay there for the entire year and accumulate. In reality, the atmospheric lifetime of PM is quite short (about 1 wk), so a correct value needs to account for the deposition and transport. In addition, because of the short lifetime and variety of emission "hotspots," one average value to represent the entire United States does not make a lot of sense; it assumes uniform sources and losses across the entire country.

9.22. $2.49 \times 10^{13} \dfrac{\text{L}}{\text{mol sec}}$

9.25. Carbon is being oxidized.

9.27. 0.038 g

9.29. (i) 11.2 g CO **(ii)** 0.0201 g

Chapter 10

10.2. 3.8%

10.4. 13.9 psi

10.6. 2.28×10^{13} mol ozone

10.8. 5.1×10^{18} molec cm^{-3}

10.9. 2.7×10^{19} molec cm^{-3}

10.11. 0 km = 1.92×10^{19} molec cm^{-3}; $[N_2]$(10 km) = 5.5×10^{18} molec cm^{-3}; $[N_2]$(20 km) = 1.6×10^{18} molec cm^{-3}; $[N_2]$(30 km) = 4.5×10^{17} molec cm^{-3}; $[N_2]$(40 km) = 1.3×10^{17} molec cm^{-3}; $[N_2]$(50 km) = 3.7×10^{16} molec cm^{-3}

10.13. 4.1 Gt C

10.15. 50% of 320-nm light is transmitted through this ozone column.

10.18. 1% increase in transmission of 310-nm light. 6.1% increase in transmission of 295-nm light. 19% increase in transmission of 285-nm light.

10.21. 9,700 DU; there is not enough O_3 in the atmosphere to cut out this much light.

10.23. $8.6 \times 10^{12} \dfrac{\text{molec}}{\text{cm}^3 \text{ hr}}$ of O formed

10.26.

Height (km)	[M] (# cm⁻³)	[O₂] (# cm⁻³)	[O₃]/[O₂]	[O₃] (# cm⁻³)
0	2.46×10^{19}	5.17×10^{18}	2.94×10^{-9}	1.52×10^{10}
5	1.32×10^{19}	1.48×10^{18}	3.42×10^{-9}	5.06×10^{9}
10	7.05×10^{18}	4.24×10^{17}	6.47×10^{-9}	2.75×10^{9}
15	3.77×10^{18}	1.21×10^{17}	5.00×10^{-6}	6.07×10^{11}
20	2.02×10^{18}	3.48×10^{16}	2.31×10^{-5}	8.05×10^{11}
25	1.08×10^{18}	9.97×10^{15}	7.22×10^{-5}	7.20×10^{11}
30	5.79×10^{17}	2.86×10^{15}	1.00×10^{-4}	2.87×10^{11}
35	3.10×10^{17}	8.19×10^{14}	8.84×10^{-5}	7.24×10^{10}
40	1.66×10^{17}	2.35×10^{14}	5.99×10^{-5}	1.40×10^{10}
45	8.87×10^{16}	6.72×10^{13}	5.06×10^{-5}	3.39×10^{9}
50	4.75×10^{16}	1.93×10^{13}	2.50×10^{-5}	4.80×10^{8}

10.29. 427 nm

10.31. $Br + O_3 \rightarrow BrO + O_2$ (Rate = $k[Br][O_3]$); $BrO + O \rightarrow Br + O_2$ (Rate = $k[BrO][O]$); Overall: $O + O_3 \rightarrow 2O_2$; The intermediate in this cycle is BrO. One reaction that could remove Cl from the stratosphere is reaction with methane: $Br + CH_4 \rightarrow HBr + CH_3$

10.34. 4.5×10^{-4}

10.36. Rate = $k[ClO]^2[M]$ = 46 molec/cm³ sec. The key feature is that this reaction is second order in ClO, so it will depend very sensitively on [ClO] and thus will kick in very rapidly as soon as a (photolytic) source of ClO appears. This is responsible for the very rapid O_3 destruction observed immediately after sunlight first illuminates the South Pole stratosphere and triggers the photochemical source of ClO, as reservoir molecules, such as $ClONO_2$ and ClOOCl, release ClO radicals back to the newly illuminated polar stratosphere.

10.38. $HOCl + h\nu \rightarrow Cl + HO$; $Cl_2 + h\nu \rightarrow 2Cl$

10.42. The HFCs have zero ODP because C–F bonds are too strong to be broken photolytically, so they do not release F radicals that could catalytically destroy O_3. However, they have larger GWPs. This is apparently in part due to their sometimes very long lifetimes (e.g., HFC-23), but not exclusively—even though HFC-245fa has a shorter lifetime than HCFC-141b, it has a larger GWP. This is likely due to its more effective absorption of IR light relative to the HCFC.

Chapter 11

11.2. There are approximately 5,100 m³ of water available to each person.

11.5. 1.5×10^6 days

11.7. 1.3%

11.9. 1.98%

11.10. 0.59%

11.12. 1,243 m³ more in the United States

11.15. $10 \frac{km^3}{yr}$

11.16. $4.4 \frac{km^3}{yr}$

11.18. Industrial = 24%; Domestic = 12%; Agricultural= 64%

11.20. 1,200% more water for wheat

11.22. The ratio of the mass of rice to the mass of water is 0.00030. Most of the water is lost through evapotranspiration.

11.23. 3.8 times

11.24. 150% more water

11.26. 29.32%

11.27. 2.38%

11.29. Groundwater: 38%; Surface water: 62%

11.30. Domestic: 7.3%; Industrial/mining: 5.5%; Livestock/aquaculture: 2.8%; Irrigation: 33.0%; Thermoelectric: 51.4%

Chapter 12

12.2. What is unique about water is its extended 3D network of H bonds formed among water molecules (see Section 12.1 above, and the bottom structure in the Fig. 12.4) that greatly enhances the intermolecular forces.

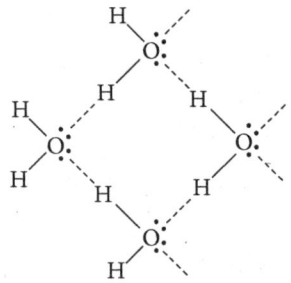

Figure 12.4 Linear H bonding in HF versus 3D network H bonding (bond angles are not the actual angles) in H_2O. *Source:* Adapted from Spiro et al., *Chemistry of the Environment, 3rd Ed.*, University Science Books, © 2012, all rights reserved.

12.4. At 10°C: 14.5; At 25°C: 14.0; At 30°C: 13.8. As the temperature shifts, the pH scale also shifts slightly as seen from the calculations here, due to the change in K_w.

12.6. 4.42

12.8. 180

12.10. 4.22%; A larger percentage of acetic acid is actually dissociated into ions in the more diluted (0.0100M) solution than the more concentrated (0.100M) solution. As the initial acid becomes more dilute ([HA] decreases), the percent dissociation actually increases.

12.11. (a) pH = 11.13 (b) pH = 10.21 (c) pH = 9.25 (d) pH = 8.30 (e) pH = 5.28 (f) pH = 2.32 (g) pH = 1.78

(h)

<!-- graph: pH (y-axis, 0.00 to 12.00) versus Volume HCl added (mL) (x-axis, 0.00 to 80.00) -->

12.13. 485 ppb

12.15. pH = 5.51; while this rainwater (with 800 ppm CO_2 in the air) will be slightly more acidic (pH = 5.51) than if the CO_2 were more stringently controlled (e.g., 538 ppm as in Q14, where rainwater pH = 5.60), its harmful impact on the ecosystem and society will likely be less than the other acid sources [see part (c) in Q9], or the climate warming itself. As the IPCC report suggests, an atmospheric concentration of 800 ppm CO_2 would likely lead to at least a couple of degrees Celsius of warming in the earth's atmosphere by 2100 from the greenhouse effect (see Chapter 14), causing significant changes and potential harm to a variety of the earth systems.

12.17. pH = 10.3

12.18. pH = 11.51

12.20. For $H_2PO_4^-$: pH = 4.6; For HPO_4^{2-}: pH = 9.8

12.22. $[H_2CO_3] = 1.31 \times 10^{-5}$ M; $[HCO_3^-] = 7.09 \times 10^{-4}$ M; $[CO_3^{2-}] = 4.19 \times 10^{-6}$ M

Chapter 13

13.2. 8.0×10^{-4} M

13.4. 9.9×10^{-11} M

13.6. $S = 1.5 \times 10^{-4}$ M; pH = 10.2

13.9. 2.5×10^{-6} M; Removing CO_3^{2-} from the $AgCO_3$ equilibrium will force more products to form, therefore dissolving more of the $AgCO_3$ solid.

13.13. $68.0 \dfrac{cmol}{kg\ soil}$

13.16. 7.8

13.18. $7.80 \dfrac{eq\ H^+}{m^2}$

13.20. $0.75 \dfrac{eq}{m^2}$

13.21. $CEC_{tot} = 11.8 \dfrac{eq}{m^2}$;

buffering capacity (1880) = $5.9 \dfrac{eq}{m^2}$

13.23. $7.2 \times 10^{12} \dfrac{mol\ H^+}{yr}$

13.25. $0.25 \dfrac{mol\ H^+}{m^2\ yr}$

13.28. 5.1 yr

Chapter 14

14.2. Glucose = 10.7 mg/L; Sucrose = 11.2 mg/L; Due to the rather constant mole ratio of C:H:O (about 1:2:1) in the sugar structure, CH_2O can be used as the representative formula for partially reduced carbon (especially carbohydrate) in aquatic ecosystem calculations.

14.3. 26.7 mg O_2 per 1 L wastewater

14.4. BOD increased by 4.8 mg L^{-1}.

14.5. 1.2×10^8 g C

14.7. (a) HNO_3: H (+1), N (+5), and O (−2); H_3PO_4, $H_2PO_4^-$, HPO_4^{2-}: H (+1), P (+5), and O (−2); CH_2O_2: H (+1), C (+2), and O (−2); $C_2H_4O_2$: H (+1), C (0), and O (−2); $C_2H_2O_4$: H (+1), C (+3), and O (−2)
(b) MnO_4^-: Mn (+7) and O (−2); CrO_3: Cr (+6) and O (−2); $Cr_2O_7^{2-}$: Cr (+6) and O (−2); OsO_4: Os (+8) and O (−2)
(c) Li: (0); Na: (0); $NaBH_4$: Na (+1), B (+3), and H (−1); $LiAlH_4$: Li (+1), Al (+3), and H (−1)
(d) Sodium oxide, formula is Na_2O, oxidation state for oxygen is −2. Sodium peroxide, formula is Na_2O_2, oxidation state for oxygen is −1. Sodium superoxide, formula is NaO_2, oxidation state for oxygen is $-\frac{1}{2}$. Potassium ozonide, formula is KO_3, oxidation state for oxygen is $-\frac{1}{3}$.

14.9. $5Pb^{2+} + 4H_2O + 2IO_3^- \rightarrow 8H^+ + 5PbO_2 + I_2$

14.11. $2IO_3^- + 8OH^- + 5Pb^{2+} \rightarrow I_2 + 5PbO_2 + 4H_2O$

14.14. $NO_3^- + 6H^+ + 5e^- \rightarrow \frac{1}{2}N_2 + 3H_2O$; −383 kJ

14.15. Anaerobic bacteria can utilize the organic carbon (represented by CH_2O) to reduce nitrate or sulfate to N_2 or H_2S, which can be volatilized:
$4NO_3^- + 5CH_2O + 4H^+ = 2N_2 + 5CO_2 + 7H_2O$;
$SO_4^{2-} + 2CH_2O + 2H^+ = H_2S + 2H_2O + 2CO_2$

14.16. $8(Fe(OH)_3 + 3H^+ + e^- \rightarrow Fe^{2+} + 3H_2O)$; −134 kJ

14.19. 0.816 V; The results are very similar to those in Table 14.4 (0.812V), essentially no difference since $E°$ was given to only two decimal points. At [OH−] = 1 M; 0.40V; Thus, O_2 is a weaker oxidant in basic rather than in neutral solution.

14.20. 0.806 V

14.21. The cell reaction, $2H_2 + O_2 = 2H_2O$, contains no H^+. The proton dependence of the two half-reactions cancels when they are subtracted to produce the overall reaction.

14.26. $2NO_3^- + 5Mn^{2+} + 4H_2O \rightarrow 5MnO_2 + N_2 + 8H^+$; $[NO_3^-] = 5.56 \times 10^{-7}$ M; at pH 5.5, $[NO_3^-] = 0.556$ M

14.28. Since algal growth requires the ratio C:N:P = 106:16:1, N is the limiting nutrient here and the water sample likely was not taken during the spring.

Chapter 15

15.3. T_f = 20.0000000240°C

15.4. 20.000000007°C

15.7. Agriculture and transportation are large contributors. Industry also plays a major role and any process that produces emissions to the atmosphere that can be returned through rainwater.

15.9. 33.3% removed

15.10. 0% removed

15.12. 92.5%–98.5% efficient

15.14. 1.1×10^3 mg ammonium

15.16. 4.40×10^4 kg; 1.00×10^6 mol; 2.24×10^7 L

15.18. 1.0×10^6 J

15.20. 0.232 atm

15.23. 4.48 M

15.25. As water permeates through the soil, it is traveling through a large filter. The pores in the rock and the soil allow the water to pass through, however, large particles and pathogens cannot permeate through the rocks, and thus, the water is filtered.

15.28. 25 ppm

15.29. 50. ppb

15.31. $Fe^{2+} \rightarrow Fe^{3+} + 1e^-$

15.33. 3.9×10^6 LifeStraws

15.34. 67 children

15.37. 5.0 ppm

15.38. 37.5 g of medicine

15.41. 470 g ammonium perchlorate

Chapter 16

16.2. 62.5% fixed nitrogen

16.5. The activation energy for step 2a of the reaction mechanism is 527 kJ mol^{-1}. It is mentioned that this is the most energy-demanding step of the mechanism as well, which means that it is also the rate-determining step of the mechanism as the other two steps have lower activation energies.

16.7. land $= 120 \dfrac{\text{Tg N}}{\text{yr}}$; oceans $= -10 \dfrac{\text{Tg N}}{\text{yr}}$;

Runoff of excess nitrogen from agricultural land has caused eutrophication of lakes, rivers, estuaries, and coastal waters in all parts of the world which practice intensive agriculture that depends on high use of synthetic nitrogen fertilizer.

16.9. No fertilizer application: 7,000 kg ha^{-1};
Fertilizer application = 100 kg N: 8,500 kg ha^{-1};
Fertilizer application = 160 kg N: 9,200 kg ha^{-1};
Fertilizer application = 200 kg N: 9,250 kg ha^{-1};
Fertilizer application = 100 kg N: 20.0%;
Fertilizer application = 160 kg N: 18.1%;
Fertilizer application = 200 kg N: 14.5%;
Thus, the application rate is reaching a point of diminishing returns.

16.11. 438 g of the corn/bean mixture is needed; = 1,546 kcal

Chapter 17

17.2. 3 yr

17.3. 155 days; 0.01% of the population

17.5. Cl is more electronegative than H and has more electron density. Perhaps the protein has a positively charged group near the position occupied by the DDT Cl atom, contributing to the binding strength.

17.7. It might, for example, start with decaying plants, which feed worms, which feed robins, which feed eagles. Or it might start with grass, which feeds grasshoppers, which feed snakes, which again feed eagles. In each case, an initially low level will build up to high levels in the eagles.

17.9. 58 kg of atrazine

17.10. DDT, with the higher K_{OW} value would partition to a greater extent into fat tissues at each stage of the chain, resulting in a higher level at the top.

17.14. 5×10^{-6} ppm

17.17. As electron withdrawal by X increases, the leaving group, X-phenoxide, becomes a weaker base and easier to displace by the serine OH. Toxicity should increase.

17.20. The DNA alteration that substitutes alanine for glycine 100 could occur by a random mutation in the weeds. The mutated weeds would then survive the application of glyphosate and replace the weeds that are killed. As described for insecticides in

Section 17.1, the evolution of resistance can replace the entire population of susceptible weeds in a few planting seasons.

Chapter 18

18.2. 94.3 chocolate bars. It is not likely that a human could die from eating too many chocolate bars.

18.3. $8.0 \times 10^{-4} \dfrac{\text{mg aflatoxin}}{\text{kg body weight}}$;

This dose is about 9% of the LD_{50} for aflatoxin.

18.6. Animal studies (most often involving mice or rats) have comprised the main source of data on carcinogenic chemicals. Cancers are counted over the lifetime of the animal (a couple of years) at various doses. Because of the need to obtain statistically significant results on a limited number of animals over a short time period, the doses administered are generally much higher than typical levels to which the human population is exposed. It is, thus, necessary to extrapolate the dose–response curve based on actual mouse/rat data to much lower doses, representative of human exposure. There is no unique scientifically proven method of extrapolation. The curve to low doses could be linear, or supra-, or sublinear. Alternatively, there may be a so-called *threshold* response, where the dose of the test chemical must be higher than a given threshold level before initiating a carcinogenic process. Thus, the response to low doses is fraught with great uncertainties. Moreover, the response of mice/rats to chemicals may differ significantly from the human response.

18.9. Lipophilic xenobiotics are generally more harmful and pervasive than hydrophilic xenobiotics, because the former tend to accumulate and store in fat tissue of exposed animals, while hydrophilic chemicals are water-soluble and can be excreted through the kidneys. Thus, the residence times of lipophilic chemicals in the body are generally much longer than those of hydrophilic chemicals. Moreover, a lipophilic chemical can diffuse through cell membranes, and, if its geometry or chemistry is similar to that of a lipophilic hormone, and can interfere with hormone function.

18.10. The body's method of ridding itself of a lipophilic xenobiotic is to convert it to a hydrophilic chemical. One of the most important means for achieving this is via hydroxylation. This is shown in the case of benzanthracene in Figure 18.3. The hydroxylation reaction, however, produces an epoxide intermediate which is a potent electrophile. Since it is generated inside the cell, it has a chance of diffusing into the nucleus and reacting with DNA before it rearranges to the hydroxylated product. A similar mechanism is at work for hydroxylation of dimethylnitrosamine (Fig. 18.4). During the reaction, an unstable intermediate is generated that is a source of methyl cation (CH_3^+), which is a powerful electrophile reacting readily with DNA. The above hydroxylation reactions are catalyzed by cytochrome P450, a liver enzyme.

18.12. Large data sets were collected to understand people's lifestyles and how it connects to their health. Through sophisticated statistical analyses, it was learned that high levels of salt or smoked fish in the Japanese diet caused higher rates of stomach cancer, but higher amounts of fat in the American diet led to high incidences of colon cancer.

18.15. The HERP value in Kenya would be 23, higher than any of the values in Table 18.2, accounting for the great concern about aflatoxin contamination in Kenya.

18.16. First, HERP assumes a linear extrapolation from high-dose animal test data. Lower doses of chemicals do not always exhibit the same linear relationship as high doses. High doses of chemicals in general might cause the cancer rate in rodents, so HERP values might overamplify the cancer risk from chemicals.

18.18. It may bind to a hormone receptor. If the resemblance to the hormone is similar enough, the xenobiotic may turn on the same biochemical mechanisms. If the resemblance is only partial, it may bind to the receptor, but may not activate it. In this case, the xenobiotic blocks the hormone and depresses its activity. In either case, the xenobiotic can upset the biochemical balance controlled by the hormone.

18.23. Although both TCDD and TCDF are planar molecules, the fit of TCDF to the Ah receptor site may be looser than that of TCDD, because the benzene rings in TCDF are drawn closer together through the loss of one of the bridging O atoms (Fig. 18.9).

18.24. The net TEQ would be 1.5 ppm

18.25. Overall, animals exposed to TCDD exhibit different health impacts than humans, and at many different doses.

18.27. A vegetarian or vegan diet would lead to lower TCDD exposures.

18.29. Ingestion of PCB-contaminated fish caught in Lake Michigan

18.31. Both PFOS and PFOA have very low volatility because of their partial ionic nature. Therefore, it is difficult for them to evaporate and they will persist in water and soil. When released directly into the atmosphere, rather than in the pure gas phase, they are expected to adsorb onto aerosol particles, suspend for some duration, and settle to the ground through wet or dry deposition.

18.33. Hg^{2+} is water-soluble, and is excreted quickly from the body by the kidney. On the contrary, methylmercury (CH_3Hg^+) is lipid-soluble so that it can cross cell membranes and accumulate in biological tissues, including sulfur-containing protein (because of mercury's strong affinity to sulfur). Methylmercury is so much more toxic than Hg^{2+} in part because it has a longer residence time in the body. Also importantly, it has the ability to cross the blood–brain barrier and cause serious symptoms of brain dysfunction. Moreover, it can pass through the placenta from mother to fetus, causing intellectual disability and motor disturbance in the child.

18.35. All fish have some level of mercury that humans can ingest. Fish located in contaminated areas often have higher concentrations of mercury that can impact human health. Other potential exposures can occur from the recycling or burning of batteries containing mercury and from the extraction of gold from ore using metallic mercury.

18.37. Acidification aggravates risks posed to the environment and human health by toxic heavy metals, such as cadmium, lead, and mercury. In the case of cadmium, its mobility in soils and sediments is greatly influenced by pH. Thus, when soils highly contaminated by cadmium are acidified (either by acidic deposition or changes in farming practices), the metal is mobilized and may leach to groundwaters, or be taken up by vegetation.

18.39. Arsenic is associated with pyrite, FeS, in the surrounding rock, and is oxidized and released as arsenate during FeS weathering to Fe^{3+} and sulfate. The precipitates as $Fe(OH)_3$ to which arsenate is strongly adsorbed. But the organic matter in the water can reduce $Fe(OH)_3$ to soluble Fe^{2+}, releasing the arsonate into the water supply.

18.40. In order to remediate the problem, deeper wells could be dug or the water from the shallow wells could be treated to remove the arsenic. The problem is that there are millions of these wells, so any potential solution is very costly.

18.42. One route of exposure to lead is through ingestion of drinking water. Probably more significant, however, is the ingestion of lead-contaminated dust. In the past, the major source of lead in dust was the combustion of leaded gasoline.

18.44. Probably more significant, however, is the ingestion of lead-contaminated dust. The mixture of ammonia and chlorine promotes the dissolution of copper from pipes, but NH_3 complexes with oxidized Cu^{2+}. The Cu^{2+} promotes the dissolution of lead because the reduction potential of Cu^{2+} is higher than that of Pb^{2+}: $Cu^{2+} + Pb \rightarrow Cu + Pb^{2+}$. Basically, the chloramine in the water sets up an electrochemical cell at the junction of Cu and Pb pipes that releases Pb^{2+} into the water stream.

18.45. Lead accumulation in dust from these sources has not disappeared. Children playing in the dust are still ingesting it, particularly in poor, heavily trafficked areas, with old housing stock. Also, lead leaching from old plumbing can be a hazard, particularly when the water supply is poorly managed, as was the case in Flint and Washington DC.

PHOTO CREDITS

All photos ©Paul Souders at https://worldfoto.com except where listed below.

Part I

Frontispiece, Polar Bear in Hudson Bay, Canada

Chapter 1: Frontispiece, Polar Bear Tracks on Sea Ice, Svalbard, Norway

Part II

Frontispiece, Wind Turbines in the Oiz Eolic Park, ©Mikel Martinez De Osaba Fotografía/
 Shutterstock.com

Chapter 2: Frontispiece, New York Avenue and U.S. Capitol Building

Chapter 3: Frontispiece, Brooklyn Bridge and Manhattan Skyline

Chapter 4: Frontispiece, Wind Turbines and Cactus at Aruba

Chapter 5: Frontispiece, Oil Refinery, Curacao

Chapter 6: Frontispiece, Nuclear Power Plant, Three Mile Island

Part III

Frontispiece, Katmai National Park, Alaska

Chapter 7: Frontispiece, Polar Bear on Melting Sea Ice, Svalbard

Chapter 8: Frontispiece, Skyline Drive, Shenandoah National Park, Virginia

Chapter 9: Frontispiece, Oil Refinery at Curacao

Chapter 10: Frontispiece, Moon and Livingstone Island, Antarctica

Part IV

Frontispiece, Castle Geyser in Yellowstone National Park, ©www.Bennymarty.com

Chapter 11: Frontispiece, Brown Bear and Salmon, Katmai National Park, Alaska

Chapter 12: Frontispiece, Lemon acid reaction – litmus paper red, ©Milanb/Shutterstock.com

Chapter 13: Frontispiece, Horseshoe Bend on Colorado River

Chapter 14: Frontispiece, Humpback Whale, Alaska

Chapter 15: Frontispiece, Sludge Pollution Pouring into the Baltic Sea and Seagulls, ©Jon Shore/
 Shutterstock.com

Part V

Frontispiece, King Penguins, South Georgia Island

Chapter 16: Frontispiece, Photo by Jonathan Larson (@jrlars) on Unsplash

Chapter 17: Frontispiece, A crop duster applies chemicals to a field of vegetation, ©Grindstone Media
 Group/Shutterstock.com

Chapter 18: Frontispiece, Several barrels of toxic waste at the dump, ©Oliver Sved

Note: **Boldface** pages located in figures; *italicized* pages located in tables.

sea creature's shells and, 103, 291
storage of, 97–99
Carbon monoxide
catalytic converters and, 216–217
concentration in atmosphere, *150*
emission control and, 196, 199
hydrogen production, 90
hydroxyl radicals, 188
indoor air pollution, 217–219
polluting effect of, 203–206, **203**
Carbon tax, 26
Carbonate ion, 103, 207, 275, 280, 282–283, 285, 287–292, 295, 298, 305, 334, 409, 413
Carbonate rock, **82**, 291–292, 297, 298, **304**, 305
Carbonic acid, 275–276, 279, 283, 291
Carbonyl, 187, 191, 423–425
Carboxylic acid, *424*, 424
Carcinogens,
atrazine, 378
benzene, 195–196, 209
chloroform, 340
comparison of risk, 390–394, *393*
formaldehyde, 208, 389
mechanism, 388–390, **389**
nitrosamines, 390
PAHs, 209, 389–390, 423
PFOS, 404
PERC, 7–8
TCDD, 8–9, 398–399
Caribbean, **33**
Carpet, 10–11, 217–218
Cars. *See* Automobiles
Carson, Rachel, 369
Catalysts
in destruction of ozone, 236–237, 242–243
enzymes as, 361
for fuel cells, 50–51
green chemistry principles, 8–10
Haber process, 353, 355
mercury as, 405–406, 408
new diesel technology, 197
for petroleum refining, 92–93
platinum as, 195
for steam reformation, 90
rhodium as, 195
Catalytic converter, 195, 197, 216–217, 415
Cathode, 11, 46, **46**, 48, *49*, 51–53, **52**, **53**
Cation exchange capacity (CEC), 294–296, **296**, 301, 306
Cations, 265, 285, 293–295, *295*–297, 298–299, 301, 406
Cell division, 388–389
Cell proliferation, 388–396
Cellulose, 57, 82
Celsius (°C), 185
Centi (c) prefix, 14, 427
Centimeter (cm), 14, 201–202, 223–224, 427
Cereal, **34**, 355, 362, 364
Cerium (Ce), *49*
Cesium (Cs), *113*, 119, 129, 131, 285

CFCs (chlorofluorocarbons)
green chemistry, 7
as greenhouse gases, 153, 159, *160*,
ozone destruction and, 236, 239
stratospheric ozone and, 221
substitutes for, 244–245, *245*
Chain reactions
bombardment with neutrons, **121**, 121–122
destruction of ozone, 238–239, 242, 244
free radicals, 173, 192, 194
lead additives, 195
nuclear fission, 123, **124**, 127–128, 131
nuclear reactor safety, 129–131
Chapman mechanism, 230–233, 240
Charcoal, 112, 334
Charcoal filtration, *see* Activated charcoal
Cheese, *402*
Chelate, 412
Chelating agents, 9, 412, **413**, 416
Chelation therapy, 416
Chemical contamination. *See* Herbicides; Pesticides
Chemical properties, 10, 409, 418, 421
Chernobyl, 129–131
Chesapeake Bay, 326
Chicken, **258**, 259
Children, 129–130, 195, 211, 343–344, 345, 377, 388, 396, 411–416
Chile, **29**, 411
China
agricultural area, **33**
air pollution in, 211
energy use, *31*, 31–32
energy trends, **29**
nuclear power, 132, **133**
population growth, 17–19, *17*, **18**, *19*
water consumption in, **253**
Chlor-alkali plants, 406
Chloracne, 398
Chloramine, 413–414, 416
Chlordane, *371*, 398
Chlordecone, 398
Chloride, 287, 294–295, *384*, 405
Chlorine (Cl)
abundance, **286**
air emissions, *56*
as disinfectant, 413–414, 416
green chemistry, 8–9
isotopes, 101
organochlorines, *56*, 340, 368–369, 370, *371*, 372, 377, 398–399, 423
ozone destruction and, 221, 229, 236–242, 242–244, **242**
PCDDs and PCDFs, 399–401
in sewage treatment process, 334, **334**, 340–341
Chlorine dioxide, 9, *56*, 340–341
Chloroacetic acid, 398
Chlorofluorocarbons. *See* CFCs (chlorofluorocarbons)
Chlorofluoromethanes, **154**
Chloroform, *190*, 340–341, *393*

Chlorophenol chemicals, 398
Chlorophenoxy compounds, 377
Chlorophenyl rings, 369
Chlorophyll, 70
Chlorpyrifos, **375**, 375, 377
Chocolate, **253**, 385
Cholesterol, 395, 419
Cholinesterase, 374–376, **376**
Chromium (Cr), 10, *49*, *286*, 313–316
Chronic toxicity, 129, 383–388, 390, 394, 409, 416
Cis orientation, 421–422
Cigarette smoke, *118*, 217
Cigarette smoking rates, **391**
Citric acid, 97
Clams, **370**, 370, 373
Clays, 82, 293–295, **296**, 301, 343
Clean Air Act Amendments (1990), 96
Clean Water Act (CWA), 331
Climate
albedo, 166–170, **167**
climate modeling, 158–160
energy flows, 141–145, **142**
international agreements, 171
radiation balance, 145–150
warming, 97, 279
zones, 62, 170
See also Greenhouse effect; Greenhouse gases (GHG)
Climate modeling, 158–160
Cloud particles, 129–131, 166–170
Clouds, 156–157, **157**, 166–169, **167**, **221**, **242**, 242–243
Coal
availability of, **35**
bituminous coal, 59, **84**, 95–96, **95**, *96*,
black-lung disease, 97
burning, 41, 65, 96, 123, 303, 307, 406
combustion energy of, *85*, 87–90
composition of common US, 95–97, **95**, *96*
creation of, 82, **83**
dust, 97, 119, 124
emissions, 97, 206
energy consumption and, **27**, **28**, 29
extraction cost consideration,
fuel energy, 83–85
global atmospheric acidity, 300, *302*, 303
mercury from, 406
production of, 36
power plant efficiency, 40–42
as source of sulfur dioxide, 206, 275, 278, 425
structural units for formation of, **84**
by sulfur content, **95**
Coastal areas, 97, 170–171, 259, 311, 325, 358
Coatings, 8–10, *56*, 62, 216, 343, 369, 404, 413
Cobalt (Co), 115, 137
Coffee, 383, 385, *393*, 393–394, *402*
Cogeneration, 45–46
Colic, **412**